TECHNIQUES IN
PROTEIN CHEMISTRY VII

Edited by

Daniel R. Marshak

Osiris Therapeutics
Baltimore, Maryland

ACADEMIC PRESS
San Diego   New York   Boston   London   Sydney   Tokyo   Toronto

Academic Press Rapid Manuscript Reproduction

Academic Press, Inc.
A Division of Harcourt Brace & Company
525 B Street, Suite 1900, San Diego, California 92101-4495

United Kingdom Edition published by
Academic Press Limited
24-28 Oval Road, London NW1 7DX

Library of Congress Card Catalog Number: 94-230592

International Standard Book Number: 0-12-473555-X (case)
International Standard Book Number: 0-12-473556-8 (comb)

Printed and bound in the United Kingdom
Transferred to Digital Printing, 2011

Contents

Section I
Interactions of Protein with Proteins

Section II
Interactions of Proteins with Ligands

Section III
Behavior of Proteins at Surfaces

Section IV
Modifications to Proteins *in Vivo*

Section V
Manipulation of Sulfur in Proteins

Section VI
Methods Used in Primary Structural Analysis

Section VII
Three Dimensional Protein Structure

Section VIII
Folding and Stability of Proteins

Section IX
Methods and Uses for Synthetic Proteins

Foreword

I am especially grateful to Dan Marshak for editing the seventh volume of *Techniques in Protein Chemistry*. These volumes were conceived as "bench-top" references that would be of ongoing value to practicing protein scientists. Dan has continued this tradition in a truly outstanding manner.

As explained in the Preface, the approach this year has been somewhat different. Rather than the articles being organized by methodology, they have been arranged by concepts. The section on Interactions of Protein with Proteins, for example, includes approaches based on chromatography, mass spectrometry, light scattering, calorimetry, and the use of peptide libraries. It is hoped that this format will serve to alert the reader to alternative approaches that may be available to address a given biological or biochemical problem.

The Ninth Symposium of the Protein Society, held in Boston July 8–12, 1995, was attended by over 1900 participants, the largest meeting the Society has organized to date. We hope the articles presented, which reflect a limited subset of the presentations at that meeting, will reflect the ever increasing diversity, impact, and excitement of protein science.

Dan Marshak has graciously agreed to continue as editor, and will be soliciting contributions from the next annual symposium to be held in San Jose August 3–7, 1996.

Brian W. Matthews
President
The Protein Society

Preface

Techniques in Protein Chemistry VII is the latest in a highly successful chronicle of the most recent developments in proteins. The contributions were selected from presentations at the Ninth Symposium of the Protein Society held in Boston, Massachusetts in July 1995.

Unlike previous volumes in this series, this year's edition is organized not by technique, but by subject area. The goal of this reorganization was to make it easier for the reader who is a practitioner of protein science to find papers that might be of help in the laboratory. For example, contributions on protein–protein interactions have been grouped together even though they use highly diverse methods. The other sections of the book are also organized by areas of chemical or biological similarity, rather than by technique. If you have a problem concerning a sulfur-containing amino acid residue in a protein, reviewing the papers in Section V (Manipulation of Sulfur in Proteins) may be of help. Searching for individual articles in separate sections on mass spectrometry, chemical modifications, protein sequencing, crystallography, etc., can become frustrating. Although no organization scheme is perfect for every reader, this year's book is an attempt to accommodate the broad readership.

Much of the credit for the review of the manuscripts is due the associate editors: Phil Andrews, Gerry Carlson, Steve Carr, Xiaodong Cheng, Lowell Ericsson, Greg Grant, Sheenah Mische, Nick Pace, Len Spicer, and Ken Williams. I do thank them for their contributions. Without their insights into the fields of protein science, this volume would not have its high quality and extreme breadth of coverage. I would like to acknowledge the invaluable assistance of the former editors, particularly John Crabb and Ruth Angeletti for their guidance and encouragement. Finally, I thank my secretary Debra Rizzieri for her assistance.

Proteins are the molecules that create the characteristics of a cell and, ultimately, the phenotype of the organism. As a discipline, protein science contributes an essential component of our understanding of nature. Elucidation of a pattern or mechanism in biology often reveals a beautiful part of our world. It is the genius of the creators of new techniques that ultimately permits these findings. This volume is a tribute to those who labor to invent novel methods and lead scientific discovery in proteins.

Daniel R. Marshak

Acknowledgments

The Protein Society acknowledges with thanks the following organizations which through their support of the Society's program goals contributed in a meaningful way to the ninth annual symposium and thus to this volume.

Applied Biosystems

Aviv Associates, Inc.

Beckman Instruments

BioMolecular Technologies, Inc.

Biosym Technologies, Inc.

Bristol-Myers Squibb

Brookhaven National Laboratory

Dionex Corporation

Finnigan MAT

Fisons Instruments

Hewlett-Packard Company

IntelliGenetics, Inc.

Jasco, Inc.

Merck Sharp & Dohme Research Labs

Michrom BioResources, Inc.

Molecular Simulations, Inc.

Peptides International, Inc.

PerSeptive Biosystems

Pharmacia Biotech

Pickering Laboratories, Inc.

Rainin Instrument Company, Inc.

Shimadzu Scientific Instruments, Inc.

Supelco, Inc.

The Nest Group

Vestec Corporation

Vydac

Waters Corporation

Wyatt Technology

TECHNIQUES IN
PROTEIN CHEMISTRY VII

SECTION I

Interactions of Proteins with Proteins

Identification and Characterization of Protein Ligands to the WW Domain by Western Ligand Blotting

Henry I. Chen and Marius Sudol

Laboratory of Molecular Oncology, The Rockefeller University,
New York, NY

I. Background

When H. Towbin first described in 1979 a novel technique of protein analysis in the landmark article entitled, "Electrophoretic transfer of protein from polyacrylamide gels to nitrocellulose sheets," he astutely noted that the method of "Western blotting" would not be limited to the use of antibodies for detection (1). Indeed, extensive data have indicated that by using a variation of Towbin's original technique, one can detect interactions between a protein and its ligand, whether it is DNA, protein, or lipid in nature. For instance the above technique, otherwise known as "Western ligand blotting" or "Far Western blotting," has been used to probe the interactions between epidermal growth factor (EGF) and its transmembrane receptor (2,3), α-bungarotoxin and the acetylcholine receptor (4), LDL and its receptor (5), tumor necrosis factor and its receptor (6), and too many others to be listed here.

More recently, the field of signal transduction has recognized the importance and prevalence of defined modular protein binding domains such as the Src homology domains SH2 and SH3. These modular domains were originally defined in non-receptor-type protein tyrosine kinases including those of the Src family and were subsequently shown to exert an effect on the kinase activity and on other components of the signalling pathway (7,8,9). In general, they mediate the formation of highly specific protein complexes that prove to be crucial in transducing signals from the cell surface, through the cytoplasm, and ultimately into the nucleus, where event-specific gene transcription occurs. Investigation of the aforementioned domains in mediating protein-protein interaction seems aptly suited to the use of Western ligand blotting. With this method, one can obtain *in vitro* evidence of specific binding between two proteins, which would aid in advancing the existing knowledge of signal transduction mechanisms and functions.

Since the discovery of the SH3 domain and the subsequent cloning of the putative ligands to the c-Abl SH3 domain, namely 3BP-1 and 3BP-2, numerous laboratories have attempted to elucidate the nature and function of the domain (10,11). The SH3 domain was originally identified as a 50 amino acid long region of homology common to the Crk and Src oncogene products and PLC-γ (8,9). Similar to SH2 domains, SH3 domains may in fact mediate the targeting of proteins to their substrates (12-15). At present, the best characterized paradigm for the function of SH3 domains is the transduction of a signal from the EGF receptor to the Ras-MAP kinase pathway via the interaction between the Grb2 SH3 and Sos (16-18). Furthermore, the SH3 domain is believed to be involved in the subcellular localization of proteins. It has been observed that SH3-containing proteins such as α-spectrin, nonmuscle myosin IB, and yeast ABP-1 localize to the cytoskeleton or plasma membrane.

Ligands to SH3 domains in general contain proline-rich sequences with a consensus sequence of Pro-X-X-Pro (19-22). Various groups have reported on the structural requirements for such interactions between the SH3 domain and proline-rich ligands. Although much remains to be uncovered regarding the binding specificity of SH3 domains, considerable work has been done recently to define the cognate binding partners of the SH3 domains in certain proteins such as Crk and Nck. Crk is classified as an adaptor molecule because it possesses an SH2 domain, one or two SH3 domains (depending on the splice variation of the transcript) but no catalytic domain. It has been suggested that c-Crk functions in the cell mainly as a molecular adhesive for bringing together diverse proteins into a complex by binding more than one protein simultaneously via its modular binding domains. Among the known high affinity binders of Crk SH3 are c-Abl (12,23), a novel guanosine exchange factor C3G (24-26), a Sos family member (25), and an unknown 185 kDa protein (25). Nck, a member of the Crk family of adaptor molecules, has been shown to interact via its second SH3 domain *in vivo* with a serine/threonine kinase known as NAK (27). Much of the work done to identify Crk and Abl SH3-binding proteins has been accomplished using Western ligand blotting, exemplifying the versatility and power of the technique (28).

Recent evidence have pointed to the existence of additional distinct protein-binding modules. In 1994, two groups independently reported the identification in the Shc protein of a phosphotyrosine binding domain that is distinct from SH2, which they called the phosphotyrosine interacting domain (PID) or the phosphotyrosine binding domain (PTB) (29,30). They subsequently showed that binding specificity of this new domain also differed from SH2 in that the three residues N-terminal to the phosphotyrosyl residue were important in binding, in contrast to the SH2 dependence on the three C-terminal side residues. The most recent consensus for binding is Asn-X-X-pTyr, which incidentally fails to bind to the Shc SH2 domain, suggesting a different mechanism for binding (31). Clearly, however, phosphotyrosine residues seem to be a common agent in the numerous cellular pathways in which protein-protein interaction is an integral event. Evolution appears to have provided parallel but dissimilar mechanisms for the utilization of this unique docking site by creating variations on a known theme.

The PH domain is another newly characterized structurally conserved protein domain present in a variety of signalling molecules. This domain was originally discovered in the major protein kinase C substrate in platelets, pleckstrin (32,33). Despite the extensive work done in the field, discovery of the cognate ligand to the PH domain has remained elusive. The solution structures of the PH domains in pleckstrin, β-spectrin, and dynamin have been solved (34-36). The N-terminal region of the PH domain consists of a hydrophobic patch flanked by seven antiparallel β sheets, whereas the C-terminal portion contains an amphipathic α helix, both suggesting that the domain may bind to hydrophobic or lipid moieties. Indeed, recent *in vitro* evidence suggests that the C-terminal region of the PH domain including sequences just distal to it may interact with the βγ-subunit of G proteins (37,38). In addition, several researchers have shown that the PH domains of pleckstrin and PLC-γ bind phosphatidylinositol (4,5)-bisphosphate (PtdIns(4,5)P$_2$) (39,40). It has been shown that a functional AH domain (homologous to the PH domain) of the proto-oncogene, c-Akt, was necessary for the PI3 kinase-dependent activation of Akt kinase activity, possibly by binding to phosphatidylinositol 3-phosphate via the AH domain (41).

Following the trend of elucidating new protein-binding modular domains, we reported recently the cloning and characterization of protein ligands for the WW domain of YAP (Yes kinase-associated protein (42,43)), namely WBP-1 and WBP-2 (WW domain-binding protein) (44). The WW domain (named after the two highly conserved tryptophans) appears to be a novel modular protein binding domain with possible subtle similarities to SH3 domains (45). Preliminary NMR spectra of the WW domain of YAP indicates a distinct globular structure consisting of three antiparallel beta-strands and an alpha helix (M. Macias and H. Oschkinat, personal communication). In figures 1 and 2, respectively, we depict the sequence alignments of WW domains from several proteins and the modular nature of the WW domain in representative proteins.

Protein/Species	Sequences of the WW Domains
Yap/Human	V P L P A G W E M A K T S S . G Q R Y F L N H I D Q T T T W Q D P R K A M L S
Yap/Chick	V P L P P G W E M A K T P S . G Q R Y F L N H I D Q T T T W Q D P R K A M L S
Yap/Mouse-1	V P L P A G W E M A K T S S . G Q R Y F L N H N D Q T T T W Q D P R K A M L S
Yap/Mouse-2	G P L P D G W E Q A M T Q D . G E V Y Y I N H K N K T T S W L D P R L D P R F
Nedd4/Human-1	S P L P P G W E E R Q D I L . G R T Y Y V N H E S R R T Q W K R P T P Q D N L
Nedd4/Human-2	S G L P P G W E E K Q D E R . G R S Y Y V D H N S R T T T W T K P T V Q A T V
Nedd4/Human-3	G F L P K G W E V R H A P N . G R P F F I D H N T K T T T W E D P R L K I P A
Nedd4/Human-4	G P L P P G W E E R T H T D . G R I F Y I N H N I K R T Q W E D P R L E N V A
Rsp5/Yeast-1	G R L P P G W E R R T D N F . G R T Y Y V D H N T R T T T W K R P T L D Q T E
Rsp5/Yeast-2	G E L P S G W E Q R F T P E . G R A Y F V D H N T R T T T W V D P R R Q Q Y I
Rsp5/Yeast-3	G P L P S G W E M R L T N T . A R V Y F V D H N T K T T T W D D P R L P S S L
Dmd/Human	T S V Q G P W E R A I S P N . K V P Y Y I N H E T Q T T C W D H P K M T E L Y
Dmd/Ray	T S V Q G P W E R A I S P N . K V P Y Y I N H Q T Q T T C W D H P K M T E L Y
Utro/Human	T S V Q L P W Q R S I S H N . K V P Y Y I N H Q T Q T T C W D H P K S T E L F
FE65/Rat	S D L P A G W M R V Q D T S . G T Y Y W H I P . T G T T Q W E P P G R A S P S
Ess1/Yeast	T G L P T P W T A R Y S K S K K R E Y F F N P E T R H S Q W E P P E G T N P D
ORF1/Human	G D N N S K W V K H W V K G . G Y Y Y Y H N L E T Q E G G W D E P P N F V Q N
38D4/Caeel	R D L L N G W F E Y E T D V . G R T F F F N K E T G K S Q W I P P R F I R T P
Consensus Line:	L P t G W E t t t G t Y Y h N H T t T T t W t P t t

Figure 1. Sequence alignment of various WW domains. YAP, Yes-kinase associated protein. DMD, dystrophin. Utro, utrophin. t, turn-like or polar residue. h, hydrophobic residue. Figure modified from Bork and Sudol (46).

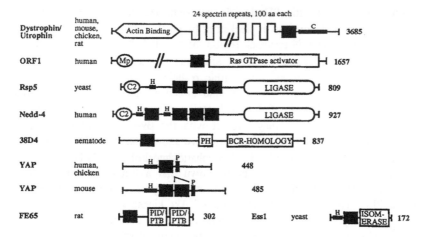

Figure 2. Modular structure of WW domain-containing proteins. Mp, homology to fly muscle protein mp20. C2, a region also found in some forms of protein kinase C, synaptotagmins, and *C. elegans* Unc-13 protein. Ligase, ubiquitin-ligase activity catalytic domain. PH, pleckstrin homology domain. BCR-homology, breakpoint cluster region homology. PID/PTB, phosphotyrosine interaction domain/phosphotyrosine binding. C/H, regions rich in cysteine/histidine. P, proline-rich region in YAP shown to interact to the SH3 domains of Yes, Src, and Nck. Isomerase, homology to proline isomerase activity. Figure modified from Bork and Sudol (46).

As seen in figure 2, the WW domain appears in a variety of structural and signalling proteins, and similar to Src homology domains sometimes present in multiple copies. The coexistence of the WW domain with catalytic regions such as the ubiquitin-ligase domain in Rsp5 and Nedd-4 suggests that the WW domain may possibly aid in bringing substrate molecules into the proximity of the enzymatic machinery. Furthermore, the WW domain may be present together with other modular ligand binding domains such as the PID/PTB and PH domains, potentially relegating such molecules as 38D4 and FE65 to the category of adaptor proteins.

II. Experimental Approaches and Results

A. *Construction of glutathione S-transferase fusion proteins*

First, in order to produce for our studies a large amount of the WW domain of YAP that was also conducive to a rapid purification process, we chose to express the WW domain of human YAP as a fusion protein to glutathione S-transferase (GST). We amplified the gene region encoding for the WW domain of human YAP with the polymerase chain reaction (PCR) using primers engineered with either 5' *Bam* HI or *Eco* RI sites. The purified amplified fragment was then subcloned in between the same restriction sites of the pGEX-2TK vector (Pharmacia), *E. coli* cells transformed with this construct were grown to OD_{600} of 1.0 and subsequently induced with isopropyl β-D-thiogalactoside (1 mM final concentration) for 3 hours. Following harvesting and lysis by sonication in phosphate buffered saline with 1% Triton X-100, the GST fusion protein containing WW-YAP was isolated with glutathione-agarose beads and eluted with 10 mM reduced glutathione. Typically, a yield of approximately 6-8 mg fusion protein per liter of culture could be obtained with this procedure.

B. Labelling of GST fusion proteins

In order to optimize the signal-to-noise ratio in our following binding assays, we required a method to label the fusion protein probes with a high specific activity. The pGEX-2TK vector offered the unique characteristic of a protein kinase A phosphorylation site located between the GST and recombinant regions of the fusion protein. For the phosphorylation reaction we used bovine heart catalytic subunit of cAMP-dependent protein kinase A (PKA) in a large excess of previously published protocols (47). We reconstituted lyophilized PKA (Sigma) to a final concentration of 1500 U/ml in 40 mM dithiothreitol (DTT), which could then be aliquotted and stably stored at -70° C. To improve the specificity of the labelling reaction, we performed the reaction in a minimum volume. By adsorbing 100 μg of fusion protein to 50 μl of glutatione-agarose beads, the protein could be concentrated into a relatively small volume. Phosphorylation by PKA therefore occurred directly on the beads in the presence of 20 μCi of [^{32}P]-γATP, 120 U of protein kinase A catalytic subunit, 10 μl of 1x HMK buffer (20 mM Tris pH 7.5, 100 mM NaCl, 12 mM $MgCl_2$), and 1 mM DTT, totalling a final volume of approximately 150 μl. The reaction was terminated after 40 minutes at 4° C by the addition of 1 ml of stop solution (10 mM sodium phosphate pH 8.0, 10 mM sodium pyrophosphate, 10 mM EDTA, 1 mg/ml bovine serum albumin). After washing the beads with cold PBS, the labelled proteins were eluted by 10 mM reduced glutathione in 50 mM Tris-HCl (pH 8.0) on a flow-thru column. Depending on the subsequent experimental plan, the proteins could then be dialyzed against PBS to remove free glutathione.

C. Coprecipitation of WW-YAP and putative ligands

To gain evidence that a potential ligand to the domain actually existed, we incubated for 16 hours at 4° C either glutathione-agarose adsorbed GST or GST-WW-YAP (100 μg for each precipitation reaction) with lysates (200 μg total protein) from various rat organs to purify for proteins that bound specifically to either GST or WW-YAP.

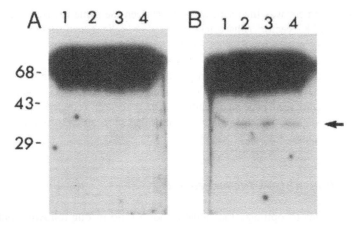

Figure 3. Coprecipitation of putative WW-YAP ligands from cell culture lysates. Lysates of HeLa (lane 1), A431 (lane 2), 3Y1 (lane 3), and v-Src expressing 3Y1 (lane 4) cells were incubated with either GST (*A*) or GST-WW-YAP (*B*), and the complexes were subjected to SDS-PAGE and Western transfer. The blots were then probed with either labelled GST or GST-WW-YAP, respectively.

The complexes were boiled for 2 minutes in protein running buffer with β-mercaptoethanol to dissociate the complexes, resolved by SDS-PAGE, and then Western transferred onto nitrocellulose paper. When the blots were subjected to Western ligand blotting with labelled forms of GST or GST-WW-YAP as a probe (6 μg labelled protein per blot, with total activity of 200 cpm), specific bands of 34 and 38 kDa were present in the blots in which GST-WW-YAP was used for precipitation and probing. These bands were markedly absent from the GST precipitation. When the experiment was repeated using cell culture lysates, a 38 kDa protein specifically binding to WW-YAP was observed (indicated by arrow, figure 3B). The denaturing and reducing conditions of SDS-PAGE and Western transfer did not reduce the interaction between the WW domain and the above two putative ligand below detectable levels.

D. Cloning of WW-YAP ligands

These positive results with ligand blotting encouraged us to search for the ligand of the WW domain using the classical plaque-lift library screening protocol on a 17 day mouse embryo cDNA expression library (Novagen). By inducing bacterial plaques lifted unto nitrocellulose filters with IPTG (isopropyl β-thiogalactoside) and subsequently probing the filters with labelled GST-WW-YAP, we were able to isolate two different partial clones that bound with high specificity to the WW domain of YAP. They failed to bind to GST or to GST-WW-dystrophin. The sequences of the clones, which we have named WBP-1 and WBP-2, showed no significant homology to any known proteins. However, the two clones shared a short region of homology with the common sequence Pro-Pro-Pro-Pro-Tyr, which we have named the PY motif.

PY motif constructs	Binding Affinity
GST- G T P P P P Y T V G	++++
GST- G T A P P P Y T V G	++
GST- G T P A P P Y T V G	0
GST- G T P P A P Y T V G	0
GST- G T P P P A Y T V G	++
GST- G T P P P P A T V G	0

Figure 4. Binding affinity of wildtype and mutant PY fusion proteins with GST-WW-YAP.

E. Binding assays with PY motif

In light of this development, we demonstrated that WBP-1 binds with high specificity to the WW domain through the PY motif. By expressing the PY motif (including five of the flanking amino acids) as a GST fusion protein, we showed through Western ligand blotting that this new motif binds specifically to GST-WW-YAP but not GST or GST-WW-dystrophin. We constructed PY motif mutants using 55-base pair oligomers coding for the decapeptide, NH_2-GTPPPPYTVG-COOH, which were engineered with the appropriate pGEX-2TK subcloning sites and missense mutations in which each residue in the underlined region was alternately replaced with alanine.

Through this strategy of site-directed mutagenesis, we have demonstrated that the second and third prolines, along with the tyrosine residue of the PY motif are essential for binding to the WW-YAP domain in similar assays. Figure 4 above summarizes this data by comparing in relative terms the wildtype PY motif to the various PY mutants with respect to the strength of interaction with the WW domain of YAP. In addition, the PY motif of WBP-1 does not bind to the SH3 domains of arbitrarily chosen proteins, including Yes, Abl, and Grb2. Therefore, the PY motif appears to differ from the PXXP consensus of SH3-binding domains and may in fact represent a novel protein-binding module. Studies are underway to identify additional WW ligands in order to establish a sequence consensus for binding.

In retrospect, the existence of a SH3-like domain that binds polyproline stretches is not completely without precedence. Profilin is an actin/PIP_2-binding protein that has been implicated in cytoskeletal rearrangement. Similar to the SH3 and WW domains, profilin also contains interspecies-conserved aromatic amino residues, including two conserved tryptophan residues. Profilin shows high binding affinity to polyproline chains with at least six consecutive proline residues. Interestingly, many cytoskeletal proteins contain polyproline patches. Spectral fluorescence studies have shown that the two tryptophans are present in a hydrophobic environment partially exposed to the solvent (48). In addition, the spectral emission from these tryptophans changes upon binding to polyprolines, suggesting that the residues are directly or indirectly involved in binding. Substitutions of either of these tryptophans with phenylalanine reduces binding significantly. The similarities between the SH3, WW, and profilin domains suggest that conserved aromatic amino acids may be a common component in domains that bind polyproline motifs. It is likely that there may be additional SH3- and WW-like domains yet to be discovered.

III. CONCLUDING REMARKS

The functional significance of the WW-PY interaction remains uncertain at this point. Studies are underway to dissect the biological role of this new type of protein-protein interaction. As we discussed previously in our most recent paper (44), the phosphorylation state of the tyrosine residue in the PY motif remains a mystery. Since we expressed the fusion proteins used in our experiments in bacteria, an organism that does not phosphorylate proteins on tyrosine residues, we can infer that interaction between the WW domain and the PY motif relies on a non-phosphorylated tyrosine in the latter. A logical extension of this observation is the hypothesis that tyrosine phosphorylation of the PY motif may be involved in inhibiting the WW-PY interaction. The presence of a large negatively charged phosphate moiety on the tyrosine residue may disrupt the formation of hydrogen bonding between the PY motif and the WW domain, either through steric hinderance or charge repulsion. This combination of SH2 and SH3 domain characteristics may represent a means of modulating the extent of protein-protein binding and the propagation of a signal *in vivo*. If this mechanism can be demonstrated *in vivo*, it will represent the premier example of a way to regulate directly the "stickiness" of a modular protein-binding domain.

The technique of Western ligand blotting can provide invaluable data on the role of modular protein-binding domains such as the WW domain *in vitro*. However, its range of usage is limited to binding and competition assays on nitrocellulose paper. In addition, the extent of protein-protein interaction may be severely limited by the denaturing conditions of SDS-PAGE. Indeed, in Western ligand blotting the binding sites need to be dissociated from the endogenous ligand in order for the ligand probe to bind sufficiently for detection. This therefore requires the use of SDS and/or heating prior to electrophoresis. Unfortunately, some proteins fail to renature adequately upon transfer, thus often abolishing any capability for binding. On the other hand, Western ligand blotting is a simple and inexpensive method to analyze different types of protein-protein interactions.

At the present time, however, we focus our work on *in vivo* experiments to delve into the biological function of the WW domain. The widespread presence of the WW domain in a variety of signalling and cytoskeletal proteins suggests that it may be involved in signalling pathways. These signalling pathways are most likely yet to be discovered since none of the proteins in the known pathways contain WW domains. This represents a very exciting time for our lab as we embark on as of yet uncharted regions of cellular signal transduction.

Acknowledgments

We extend our thanks to Dr. Hidesaburo Hanafusa for his support and for valuable comments on this manuscript. We also thank Drs. Peer Bork, Andrea Musacchio, Hartmut Oschkinat, Matti Saraste, and Sahng-June Kwak for helpful discussions. We acknowledge Aaron Einbond and Dan Stieglitz for help in adapting the first two figures to this manuscript. Our work was supported by a grant CA01605 from the National Cancer Institute, by the Council for Tobacco Research, USA, Inc., grant #3035, by the Human Frontier Science Program grant to M.S., and by National Institute of Health grant #5T32GM07739-16 to H.I.C.

References

1. Towbin, H., Staehelin. T., and Gordon, J. (1979) *Proc. Natl. Acad. Sci., USA* **76**, 4350.
2. Fernandez-Pol, J.A. (1982) *FEBS Lett.* **143**, 86.
3. Lin, P.H., Selinfreund, R., and Wharton, W. (1987) *Anal. Biochem.* **167**, 128.
4. Gershoni, J.M. and Palade, G.E. (1983) *Anal. Biochem.* **131**, 1.
5. Daniel, T.O., Schneider, W.J., Goldstein, J.L., and Brown, M.S. (1983) *J. Biol. Chem.* **258**, 305.
6. Smith, R.A., and Baglioni, C. (1989) *J. Biol. Chem.* **264**, 14646.
7. Koch, C.A., Anderson, D., Moran, M.F., Ellis, C., and Pawson, T. (1991) *Science* **252**, 668-674.
8. Mayer, B., Hamaguchi, M., and Hanafusa, H. (1988) *Nature* **332**, 272-275.
9. Stahl, M.L., Ferenz, C.R., Kelleher, K.L., Kriz, R.W., and Knopf, J.L. (1988) *Nature* **332**, 269-272.
10. Cicchetti, P., Mayer, B.J., Thiel, G., and Baltimore, D. (1992) *Science* **257**, 803-806.
11. Ren, R., Mayer, B.J.. Cicchetti, P., and Baltimore, D. (1993) *Science* **259**, 1157-1161.
12. Feller, S.M., Knudsen, B., and Hanafusa, H. (1994) *EMBO J* **13**, 2341-2351.
13. Flynn, D.C., Leu. T.H., Reynolds, A.B., and Parsons, J.T. (1993) *Mol. Cell. Biol.* **13**, 7892-7900.
14. Fumagalli, S., Totty, N.F., Hsuan, J.J., and Courtneidge, S.A. (1994) *Nature* **368**, 871-874.
15. Taylor, S.J. and Shalloway, D. (1994) *Nature* **368**, 867-871.
16. Li, N., Batzer, A., Daly, R., Yajnik, V., Skolnik, E., Chardin, P., Bar-Sagi, D., Margolis, B., and Schlessinger, J. (1993) *Nature* **363**, 85-88.
17. Rozakis-Adcock, M., Fernley, R., Wade, J., Pawson, T., and Bowtell, D. (1993) *Nature* **363**, 83-85.
18. Bar-Sagi, D., Rotin. D., Batzer, A., Mandiyan, V., and Schlessinger, J. (1993) *Cell* **74**, 83-91.
19. Lim, W.A. and Richards, F.M. (1994) *Struct. Biol.* **1**, 221-225.
20. Yu, H., Chen, J.K.. Feng, S., Dalgarno, D.C., Brauer, A.W., and Schreiber, S.L. (1994) *Cell* **76**, 933-945.
21. Knudsen, B.S., Zheng, J., Feller, S.M., Mayer, J.P., Burrell, S.K., Cowburn, D., and Hanafusa, H. (1995) *EMBO J.* **14**, 2191-2198.
22. Wu, X., Knudsen. B., Feller, S.M., Zheng, J., Sali, A., Cowburn, D., Hanafusa, H., and Kuriyan, J. (1995) *Structure* **3**, 215-226.
23. Ren, R., Ye, Z-S, and Baltimore, D. (1994) *Genes & Dev.* **8**, 783-795.
24. Feller, S.M., Knudsen, B., Hanafusa, H. (1995) *Oncogene* **10**, 1465-1473.
25. Knudsen, B.S., Feller, S.M., and Hanafusa, H. (1994) *J. Biol. Chem.* **269**, 32781-32787.

26. Tanaka, S., Morishita, R., Hashimoto, Y., Hattori, S., Nakamura, S., Shibuya, M., Matuoka, K., Takenawa, T., Kurata, T., Nagashima, K., and Matsuda, M. (1994) *Proc. Natl. Acad. Sci. USA* **91**, 3443-3447.
27. Chou, M.M. and Hanafusa, H. (1995) *J. Biol. Chem.* **270**, 7359-7364.
28. Feller, S.M., Knudsen, B., Wong, T.W., Hanafusa, H. (1995) *Methods in Enzymology*, in press.
29. Blaike, P., Immanuel, D., Wu, J., Li, N., Yajnik, V., and Margolis, B. (1994) *J. Biol. Chem.* **269**, 32031-32034.
30. Kavanaugh, W.M. and Williams, L.T. (1994) *Science* **266**, 1862-1865.
31. Kavanaugh, W.M., Turck, C.W., and Williams, L.T. (1995) *Science* **268**, 1177-1179.
32. Haslam, R., Kolde, H.B., and Hemmings, B.A. (1993) *Nature* **363**, 309-310.
33. Mayer, B. and Baltimore, D. (1993) *Trends Cell Biol.* **3**, 8-13.
34. Macias, M.J., Musacchio, A., Ponstingl, H., Nilges, M., Saraste, M., and Oschkinat, H. (1994) *Nature* **368**, 675-677.
35. Yoon, H.S., Hajduk, P.J., Petros, A.M., Olejniczak, E.T., Meadows, R.P., and Fesil, S.W. (1994) *Nature* **369**, 672-675.
36. Ferguson, K.M., Lemmon, M.A., Schlessinger, J., and Sigler, P.B. (1994) *Cell* **79**, 199-209.
37. Inglese, J., Freedman, N.J., Koch, W.J., and Lefkowitz, R.J. (1993) *J. Biol. Chem.* **268**, 23735-23738.
38. Touhara, K., Inglese, J., Pitcher, J.A., Shaw, G., and Lefkowitz, R.J. (1994) *J. Biol. Chem.* **269**, 10217-10220.
39. Cifuentes, M.E., Delaney, T., and Rebecchi, M.J. (1994) *J. Biol. Chem.* **269**, 1945-1948.
40. Harlan, J.E., Hajduk, P.J., Yoon, H.S., and Fesik, S.W. (1994) *Nature* **371**, 168-170.
41. Franke, T.F., Yang, S., Chan, T.O., Datta, K., Kazlauskas, A., Morrison, D.K., Kaplan, D.R., and Tsichlis, P.N. (1995) *Cell* **81**, 727-736.
42. Sudol, M. (1994) *Oncogene* **9**, 2145-2152.
43. Sudol, M., Bork, P., Einbond, A., Kumar, K., Druck, T., Negrini, M., Huebner, K., and Lehman, D. (1995) *J. Biol. Chem.* **270**, 14733-14741.
44. Chen, H.I. and Sudol, M. (1995) *Proc. Natl. Acad. Sci. USA* **92**, 7819-2823.
45. Sudol, M., Chen, H.I., Bougeret, C., Einbond, A., and Bork, P. (1995) *FEBS Lett.* **369**, 67-71.
46. Bork, P. and Sudol, M. (1994) *Trends in Biol. Sci.* **19**, 531-533.
47. Kaelin, W.G. Jr., Krek, W., Sellers, W.R., DeCaprio, J.A., Ajchenbaum, F., Fuchs, C.S., Chittenden, T., Yue, L., Farnham, P.J., Blanar, M.A., Livingston, D.M., and Flemington, E.K. (1992) *Cell* **70**, 351-364.
48. Metzler, W.J., Bell, A.J., Ernst, E., Lavoie, T.B., and Mueller, L. (1994) *J. Biol. Chem.* **269**, 4620-4625.

Electrospray Ionization with High Performance Fourier Transform Ion Cyclotron Resonance Mass Spectrometry for the Study of Noncovalent Biomolecular Complexes

Xueheng Cheng, Steven A. Hofstadler, James E. Bruce,
Amy C. Harms, Ruidan Chen, and Richard D. Smith
Chemical Sciences Department
Pacific Northwest Laboratory
Richland, WA 99352

Thomas C. Terwilliger and Paul N. Goudreau
Genomics and Structural Biology Group
Los Alamos National Laboratory
Los Alamos, NM 87545

I. Introduction

The application of mass spectrometry (MS) in the biological sciences has seen tremendous growth in recent years owing to the effectiveness and wide-spread availability of "gentle" ionization sources that promote the transfer of intact biomolecules into the gas phase. While a number of ionization techniques including fast atom bombardment, liquid secondary ion mass spectrometry, ^{252}Cf desorption, thermospray, matrix assisted laser desorption ionization (MALDI), and electrospray ionization (ESI) have been shown to be effective for various classes of biomolecules [1], recent results (and growing acceptance by the bioanalytical community) suggest a preference towards biomolecular analysis by ESI. The ESI process generates highly charged micro droplets which, upon desolvation, yield (cationic or anionic) intact molecular species. Essentially any compound which can exist as an ion in solution is amenable to ionization by electrospray. ESI-MS has been applied to a wide range of biologically relevant compounds including proteins and peptides, glycoproteins, oligonucleotides, glycolipids, and a number of metabolic intermediates [2-3]. Of specific interest to this work is the ability of the ESI process to promote noncovalently bound complexes into the gas phase and to preserve specific noncovalent biomolecular associations for subsequent mass spectrometric analysis. Several types of noncovalent associations originating in solution, including enzyme-substrate, receptor-ligand, host-guest, protein-nucleic acid, protein quaternary structure, enzyme complexes, and DNA duplex-

TECHNIQUES IN PROTEIN CHEMISTRY VII
Copyright © 1996 by Academic Press, Inc.

13

drug complexes, have been demonstrated to survive transfer into the gas phase by ESI [4-8].

A number of widely utilized ESI-MS approaches have been demonstrated including quadrupole, sector, time-of-flight, quadrupole ion trap, and Fourier transform ion cyclotron resonance (FTICR) mass spectrometry [1]. The versatility of ESI combined with the unique attributes of FTICR detection demonstrated below, make ESI-FTICR a near-ideal platform for the characterization of biomolecules [9]. ESI is well suited for combination with FTICR for the study of large molecules as the m/z range of ions produced by ESI overlaps significantly with the m/z range in which FTICR provides unparalleled mass accuracy and mass resolving power. For example, researchers from several groups have recently acquired mass spectra of small proteins (5 kDa to 20 kDa) which demonstrate resolution in excess of 10^6 with sub ppm mass measurement error [10-11]. Additionally, the ESI-FTICR combination enjoys an expanded mass range compared to other mass spectrometric techniques, and has been used to study species of >> 5 MDa [12,13]. Of particular significance has been the introduction of ion dissociation techniques that facilitate tandem mass spectrometry, or MS/MS^n, in which fragment ions of interest are retained and further dissociated providing increased information on sequence or structural modifications. This formidable combination of extended mass range, mass measurement accuracy, resolving power, and MS/MS^n is exclusive to FTICR mass spectrometry and provides a foundation for studies which would be intractable utilizing conventional mass spectrometric approaches.

In this work we demonstrate the feasibility of mass spectrometric analysis of structurally specific noncovalent biomolecular complexes and highlight a few of the advantages of ESI-FTICR. A few examples from the authors' laboratory will be used to guide the reader through much of the discussion. For the sake of brevity, the reader is referred to the pertinent literature for instrumental and methodological details.

II. RESULTS AND DISCUSSION

Many biologically active complexes are held together by a combination of relatively weak electrostatic, H-bonding and hydrophobic forces; minimizing (or eliminating) their disruption during the ionization process is paramount to the success of any mass spectrometric based analysis. In order to obtain meaningful results it is essential that any perturbations due to solvent denaturation or dissociation in the ESI source are minimized (or eliminated). Equally important is the elimination (or reduction) of contributions due to *nonspecific* solution associations (e.g. random aggregation) which are often observed at high analyte concentrations and that can arise due to the ESI process. (Note that this is different than distinguishing non-specific associations that exist for a given solution.) Contributions arising during ESI can generally be avoided by working at relatively low concentrations, typically in the low μM regime, and by (apparently) proper interface design. Below we highlight some of our findings on solvent conditions and electrospray source parameters which are of preeminent importance for the study of weak noncovalent associations by ESI-MS.

For conventional ESI-MS experiments in which mass measurements of monomeric species are sought, solution conditions employed are generally acidic (for detection of proteins as positive ions), where denaturation and more extensive gas-phase charging allows detection at modest m/z (typically < 2500). However, the preponderance of physiologically relevant noncovalent associations of are unstable under such solution conditions. With many noncovalent interactions of biological interest, it is desirable to maintain aqueous solution conditions near neutral pH. The use of "volatile" buffers, such as ammonium acetate, are generally preferred for this purpose, as the use of nonvolatile salts can interfere with the electrospray process and lead to excessive alkali metal adduction which decreases sensitivity and hinders precise molecular weight measurements. While solution pH, temperature and ionic strength are key factors in buffer selection, the significance of volatile buffering agents cannot be ignored. All of the noncovalent complexes shown in this work are derived from buffer solutions based on 10-20 mM NH_4OAc.

There are several important instrumental variables in the interface region which can be manipulated to either preserve noncovalent complexes or cause their dissociation into individual components. Generally, adjustment of the instrumental variables in the electrospray interface region involves a compromise between providing adequate heating/activation for ion desolvation while minimizing complex dissociation. For example, the application of large voltage differences in the high pressure region of the ESI source is widely used to provide sufficient activation to break covalent peptide bonds to obtain sequence/structure information [14]. Consequently, the interface conditions used to preserve specific noncovalent complexes are inherently more gentle than those conventionally employed for ESI-MS analysis. Similarly, proteins and peptides can be dissociated by intense heating in the desolvation capillary (i.e. thermal induced dissociation) [15]; it follows that relatively low capillary temperatures and low source voltage differences are generally employed to observe noncovalent associations, although this is not true for associations due largely to electrostatic forces that can be quite strong in the gas phase.

In this work we will demonstrate the analysis of noncovalent complexes using a model system based on the gene V protein of the bacteriophage f1. Gene V protein (MW_{ave} = 9689.3) is a member of a class of proteins involved in DNA replication which bind to single-stranded nucleic acids with high affinity and little sequence specificity; the ssDNA intermediate is coated during DNA replication thus forming an ordered superhelical protein-DNA complex. It is known from both gel shift assays and NMR studies that gene V exists as a homodimer under physiological conditions and binds to ssDNA with a stoichiometry of 1 gene V dimer for every 8 bases [16].

A. *Observation of Multimeric Proteins*

As described above, acidic solutions (often containing ≥ 50% methanol or other denaturant) are routinely employed as an electrospray solvent owing to the favorable yield of highly charged protein ions. Presumably, a larger fraction of basic residues are available for protonation in the unfolded protein resulting in very efficient generation of highly charged molecular ions. For example, the mass spectrum in Figure 1a was taken from a 50 μM solution of gene V protein in 100 mM HOAc (pH = 3.4) and is typical of denaturing solution conditions. A number of relatively high charge states are evident and the resulting charge state "envelope" is at a relatively low m/z (i.e. < 2000). While the measured mass of 9689.6 agrees well with the predicted mass of the gene V monomer, there is no evidence of any tertiary structure under these solvent/interface conditions and no peaks corresponding to the homodimer are observed. Alternatively, the spectrum in Figure 1b was acquired from a 50 μM solution of gene V protein in 10 mM NH_4OAc at pH 8.6 in which minimal capillary heating was employed and with only relatively small voltage differences in the high pressure ESI interface region. The key feature of this spectrum is that species corresponding only to the homodimer are observed; i.e. there are no peaks corresponding to monomer, trimer, tetramer, etc. The high resolution mass measurement capabilities of the FTICR can easily distinguish charge states based on the ability to resolve the mass difference of ^{13}C isotopes. Thus, if any species corresponding to $(M+4H)^{4+}$ were present it would be easily discernible from the $(2M+8H)^{8+}$, even though these species have the same nominal m/z, and no other charge states were present. The markedly different spectral features in Figures 1a and b are representative of the spectral features of many noncovalent complexes. First, Figure 1b is dominated by a single charge state (the 8+ species) with minor contributions from the 9+ and 7+ species, while Figure 1a contains an envelope of charge states spanning from 12+ to 6+. Generally, noncovalent complexes are observed occupying only a few charge states and the overall degree of charging is significantly attenuated, presumably due to preservation of some higher order structure and a concomitant lower number of exposed protonation sites. Lower signal-to-noise ratios for the noncovalent complex electrosprayed from a neutral pH buffer solution are not uncommon which can also be attributed to the preservation of some higher order structure and the near neutral pH environment.

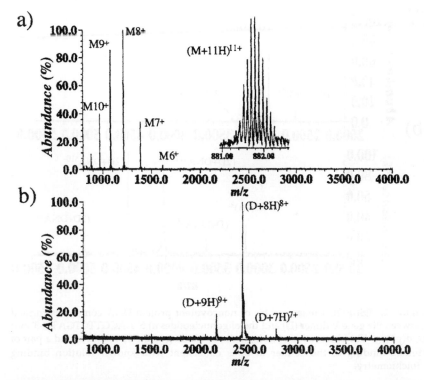

Figure 1. ESI-FTICR mass spectra of 50 μM gene V protein acquired from a) 100 mM HOAC buffer solution and b) 10 mM NH$_4$OAC buffer solution. The acidic solution in a) yields only species corresponding to the multiply charged monomer (M), while the NH$_4$OAC buffer yields exclusively species corresponding to the gene V dimer (D).

B. Observation and Stoichiometry of Noncovalent Protein-DNA Complexes

While observation of the gene V homodimer in the gas phase is encouraging, further studies are needed to determine whether the buffer conditions employed for ESI maintain biological activity and whether the stoichiometry of the active form is consistent with the known solution stoichiometry. In order to address these questions, a 50 μM solution of the gene V protein was mixed with a 25 μM solution of the 12-mer deoxyribonucleotide 5'-AACGTTCTGATC-3', in 10 mM NH$_4$OAc. This solution was electrosprayed using the identical desolvation and interface conditions utilized in Figure 1b. As shown in Figure 2a, binding is observed in which the gene V dimer (D) associates exclusively to a single oligonucleotide. There is no evidence of multiple dimers binding to a single oligonucleotide, consistent with the 8 base/dimer stoichiometry previously established [16]. Additionally, when a 2:1 excess of 12-mer is added, the stoichiometry does not change, instead, free oligonucleotide is observed in addition to the 1:1 protein-DNA complex (not shown).

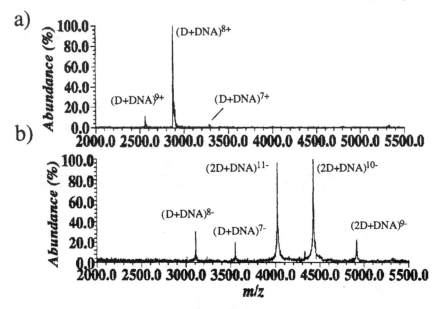

Figure 2. ESI-FTICR mass spectra of noncovalent protein-DNA complexes formed between the gene V dimer (D) and the oligonucleotides a) 5'-AACGTTCTGATC-3' and b) d(pT)$_{18}$. The observation of a single dimer binding to the 12-mer in a) and a pair of dimers binding to the 18-mer in b) is consistent with known solution binding stoichiometry.

   To further explore known solution stoichiometry with observed gas phase stoichiometry, this experiment was repeated with a 40 µM solution of the 18-mer oligonucleotide d(pT)$_{18}$ mixed with a 80 µM solution of gene V protein. The known stoichiometry would predict that the 18-mer would have a sufficiently long binding domain to accommodate 2 gene-V dimers. Desolvation and interface conditions equivalent to those employed in Figures 1b and 2a were employed except the ESI-FTICR was operated in the negative ion mode which, in this instance, yielded improved signal-to-noise compared to operation in the positive ion mode. As shown in Figure 2b, this solution yields primarily complexes between the 18-mer and 2 gene V dimers (2D), even with an excess of DNA present relative to the 4:1 stoichiometry (i.e. 80 µM gene V (M) = 40 µM dimer (D) = 20 µM tetramer (2D)). The relatively narrow charge state distribution and low charge states of the complex serve to make detection of these complexes difficult with most commercially available mass spectrometers. While the ability of the FTICR to make mass measurements over an extended m/z range is an enabling feature for many studies of noncovalent complexes, the true potential of the FTICR detection scheme is evident in the tandem mass spectrometric studies described below.

C. Tandem Mass Spectrometric Measurements of Noncovalent Complexes

Recent progress in affinity based isolation/purification methods often yield biomolecular complexes containing one or more unknown gene product which possess a specifically targeted biological activity. The ability to identify and characterize those species which bind to the target antigen can provide significant leads towards the discovery of novel therapeutic compounds and a basis for broader understanding of complex physiological processes such as the excision and repair of damaged DNA. In order to evaluate the application of tandem mass spectrometric techniques towards the study of noncovalent complexes, the 8+ charge state of the 12-mer/dimer complex (Figure 2a) was isolated in the FTICR trapped ion cell using single frequency quadrupolar axialization [17,18] and subsequently dissociated by sustained off-resonance irradiation collision induced dissociation (SORI-CID) [19]. The resulting MS/MS spectrum in Figure 3a is dominated by the 5+ and 4+ charge states of the gene V monomer species. While in some cases an accurate molecular weight determination of the binding components is adequate for identification of a species which participates in the complex, additional information such as a partial amino acid sequence can be invaluable information to assist in protein identification. A single step of SORI-CID will most likely not provide complete amino acid sequence for peptides > 2kDa but determination of the sequence of a contiguous run of 5 to 10 amino acids is often adequate to recognize a unique DNA motif in the genome of a living organism; specific sequence information can then be used to design specific amplification primers (i.e. DNA probes) for the polymerase chain reaction (PCR). The $(M+5H)^{5+}$ monomer derived from dissociation of the complex was isolated and further dissociated by SORI-CID to produce the MS/MS/MS spectrum shown Figure 3b. The inset shows that even under tandem mass spectrometric conditions, resolution adequate to resolve the ^{13}C isotopic envelope is achieved yielding unambiguous charges state (and thus mass) assignments. (For example, the spacing of the isotope peaks in the inset is 0.5 m/z units; since we know that these peaks correspond to 1 Da differences in mass (due to isotopic constituents; i.e. ^{13}C-^{12}C), this species must be a 2+ charge state.) By comparing the mass differences of the CID fragments with the known masses of the amino acids, partial amino acid sequences are derived [20]. As is typical of SORI-CID, "y" and "b" type ions that arise due to readily assignable peptide bond dissociation, and subsequent H_2O loss, dominate the spectrum during the SORI process; a few of the many assignable fragment peaks are labeled in Figure 3b. FTICR is particularly well suited for such measurements owing to the high mass resolving power, sensitivity, and the ability to perform MS/MS^n (where n $\geq$ 3).

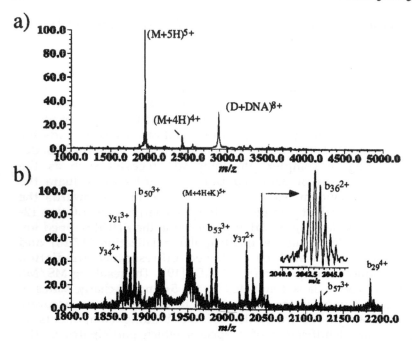

Figure 3. SORI-CID dissociation of the complex from Figure 2a yields a high abundance of the gene V monomer as shown in a). Isolation of the dissociated (M+5H)5+ species with subsequent SORI dissociation yields the MS/MS/MS spectrum in b). Resolution adequate to tresolve the 13C isotope envelope for all product ions yields unambiguous charge state assignment for each CID fragment ion.

III. Conclusions

In this work the gene V protein from bacteriophage f1 was used to demonstrate the potential applications of high performance ESI-FTICR mass spectrometric methods for the study of noncovalently bound biomolecular complexes. The high resolution capabilities of the FTICR detection scheme, as well as the extended m/z range and tandem mass spectrometric capabilities make the technique well suited for the analysis and characterization of noncovalent complexes. A particular strength of this approach is the ability to obtain structural information on species that form such structurally specific complexes; the applications of mass spectrometry are most evident for the identification of post translational modifications. Recent work has also demonstrated the potential of ESI-MS for obtaining information on relative binding strengths [21]. We are presently pursuing the use of these techniques, and others, to rapidly screen combinatorial mixtures in order to identify compounds from large libraries which bind to targeted biomolecules with high affinity [22].

Acknowledgments

The authors wish to thank Drs. Dave Gale and Brenda Schwartz for helpful discussions and the U.S. Department of Energy, and Laboratory Directed Research and Development of Pacific Northwest Laboratory for support of this research. Pacific Northwest Laboratory is operated by Battelle Memorial Institute for the U.S. Department of Energy, through Contract No. DE-AC06-76RLO 1830.

References

1. Burlingame, A. L., Boyd, R. K. and Gaskell, S. J., *Anal. Chem.*, **1994**, *66*, R634-R683
2. Smith, R. D., Loo, J. A., Loo, R. R. O., Busman, M. and Udseth, H. R., *Mass Spectrom. Rev.*, **1991**, *10*, 359-451
3. Fenn, J. B., Mann, M., Meng, C. K., Wong, S. F. and Whitehouse, C. M., *Mass Spectrom. Rev.*, **1990**, *9*, 37-70
4. Ganem, B., Li, Y. T. and Henion, J. D., *J. Amer. Chem. Soc.*, **1991**, *113*, 7818-7819
5. Bruce, J. E., Vanorden, S. L., Anderson, G. A., Hofstadler, S. A., Sherman, M. G., Rockwood, A. L. and Smith, R. D., *Journal of Mass Spectrometry*, **1995**, *30*, 124-133
6. Doktycz, M. J., Habibigoudarzi, S. and Mcluckey, S. A., *Anal. Chem.*, **1994**, *66*, 3416-3422
7. Gale, D. C., Goodlett, D. R., Lightwahl, K. J. and Smith, R. D., *J. Amer. Chem. Soc.*, **1994**, *116*, 6027-6028
8. Smith, R. D., Lightwahl, K. J., Winger, B. E. and Loo, J. A., *Org. Mass Spectrom.*, **1992**, *27*, 811-821
9. Marshall, A. G. and Grosshans, P. B., *Anal. Chem.*, **1991**, *63*, A215-A229
10. Winger, B. E., Hofstadler, S. A., Bruce, J. E., Udseth, H. R. and Smith, R. D., *J. Amer. Soc. Mass Spectrom.*, **1993**, *4*, 566-577
11. Beu, S. C., Senko, M. W., Quinn, J. P. and Mclafferty, F. W., *J. Amer. Soc. Mass Spectrom.*, **1993**, *4*, 190-192
12. Chen, R. D., Wu, A. Y., Mitchell, D. W., Hofstadler, S. A., Rockwood, A. L. and Smith, R. D., *Anal. Chem.*, **1994**, *66*, 3964-3969
13. Smith, R. D., Cheng, X., Bruce, J. E., Hofstadler, S. A. and Anderson, G. A., *Nature*, **1994**, *369*, 137-139
14. Loo, J. A., Edmonds, C. G., Udseth, H. R. and Smith, R. D., *Anal. Chim. Acta*, **1990**, *241*, 167-173
15. Rockwood, A. L., Busman, M., Udseth, H. R. and Smith, R. D., *Rapid Comm. Mass Spectrom.*, **1991**, *5*, 582-585
16. Skinner, M.M., Zhang, H., Leschnitzer, D.H., Guan, Y., Mellamy, H., Sweet, R.M., Gray, C.W., Konings, R.N.H., Wang, A.H.J., Terwilliger, T.C., *Proc. Natl. Acad. Sci. USA*, **1994**, *91*, 2071-2075
17. Hasse, H. U., Becker, S., Dietrich, G., Klisch, N., Kluge, H. J., Lindinger, M., Lutzenkirchen, K., Schweikhard, L. and Ziegler, J., *Int. J. Mass Spectrom. Ion Proc.*, **1994**, *132*, 181-191
18. Schweikhard, L., Guan, S. H. and Marshall, A. G., *Int. J. Mass Spectrom. Ion Proc.*, **1992**, *120*, 71-83
19. Gauthier, J. W., Trautman, T. R. and Jacobson, D. B., *Anal. Chim. Acta*, **1991**, *246*, 211-225

20. Barinaga, C. J., Edmonds, C. G., Udesth, H. R., Smith, R. D. and , *Rapid Commun. Mass Spectrom.*, **1989**, 3, 160-164
21. Schwartz, B. L., Gale, D. C., Smith, R. D., Chilkoti, A. and S., S. P., *J. Mass Spectrom.*, in press
22. Cheng, X., Chen, R., Bruce, J. E., Schwartz, B. L., Anderson, G. A., Hofstadler, S. A., Gale, D. C., Smith, R. D., Gao, J., Sigal, G. B., Mammen, M. and Whitesides, G. M., *J. Amer. Chem. Soc*, in press

A Light Scattering/Size Exclusion Chromatography Method for Studying the Stoichiometry of a Protein-Protein Complex

Jie Wen, Tsutomu Arakawa, Jane Talvenheimo, Andrew A. Welcher,
Tom Horan, Yoshiko Kita, Julia Tseng, Margery Nicolson, and John S. Philo

Amgen Inc., Thousand Oaks, CA 91320

I. Introduction

Interactions of a protein with itself or other macromolecules play important roles in diverse biological processes. These interactions are generally characterized by their affinity and stoichiometry. Traditionally, size exclusion chromatography (SEC) has been used as a simple way to estimate the molecular weight of a complex and therefore determine the stoichiometry. However, one problem when conventional SEC is used for this purpose is that the elution position depends not only on the molecular weight of the complex, but also on its shape. A second problem is that the elution position will change if the complex has any tendency to interact with the column matrix. In addition, when a complex contains carbohydrates or glycosylated proteins, as is the case for many extracellular proteins and polyethylene glycol protein conjugates, SEC may not be able to determine the stoichiometry. This happens because some carbohydrates are so extended that they dominate the hydrodynamic size. Therefore, during the formation of a complex, the binding of an additional protein may not affect the overall hydrodynamic size, or conversely, it may change the carbohydrates' conformation which unduly influences the elution position. This phenomenon has been observed in this laboratory (Jie Wen, unpublished results). In contrast, the molecular weight from a light scattering measurement is independent of the elution position of a protein or complex, and reflects only the polypeptide if the extinction coefficient of the polypeptide alone is used in the analysis (see below). These characteristics make the combination of light scattering with SEC an easy, accurate, and reliable technique. In this paper we will focus on the determination of the stoichiometry of complexes containing at least one glycosylated protein. The reason is that the stoichiometry of these complexes is relatively difficult to determine by other methods and the calculation requires a self-consistent method, as discussed in the *Materials and Methods*.

TECHNIQUES IN PROTEIN CHEMISTRY VII

One important group of protein interactions are those leading to oligomerization of cell surface receptors, which is believed to be the key initiator of signal transduction in many cytokine and growth factor receptors (1,2). We will use two applications to illustrate the procedures of using this technique to study such interactions. First, we will show that two neurotrophins (3,4), brain-derived neurotrophic factor (BDNF) and neurotrophin-3 (NT-3), cause dimerization of the extracellular domains of their receptors, TrkB and TrkC. For "orphan" receptors with unknown ligands, monoclonal antibodies (mAbs) directed against the receptor are often used as "artificial ligands" (5,6) that presumably activate by dimerizing the receptor. In the second example, we will show that mAbs against sHer2, which is the extracellular domain of erbB2/Her2 tyrosine kinase receptor expressed on breast cancer cells, do indeed dimerize their antigen.

II. Materials and Methods

A. *Theoretical Background*

The on-line laser light scattering/SEC system uses three detectors in series after an SEC column (7): a laser light scattering detector, a uv absorbance detector, and a refractive index detector. The light scattering signal *(LS)* is given by

$$(LS) = K_{LS}cM(dn\,/\,dc)^2 \tag{1}$$

where K_{LS} is an instrument calibration constant, *c* is the concentration of a protein or complex in mg/ml, *M* is the molecular weight, and *(dn/dc)* is the refractive increment. Similarly, the refractive index signal, *(RI)*, is given by

$$(RI) = K_{RI}c(dn\,/\,dc) \tag{2}$$

where K_{RI} is again an instrument calibration constant. The *(dn/dc)* of polypeptides is constant and nearly independent of their composition. Hence, for proteins or complexes that contain no carbohydrate, we can determine *M* from the ratio of the two detectors, *(LS)* and *(RI)*.

$$M = K'(LS)\,/\,(RI) \tag{3}$$

where $K' = K_{RI}\,/\,[K_{LS}(dn\,/\,dc)]$. This so called "two-detector method" is only valid for determining the molecular weight or stoichiometry of a protein or a complex without carbohydrate (7).

For a protein or complex containing carbohydrates, the *(dn/dc)* is no longer known or constant because a carbohydrate usually has a different *(dn/dc)* from a protein, and the carbohydrate content is normally unknown. Therefore, we also need to use the signal from an absorbance detector, *(UV)*,

$$(UV) = K_{UV}c\varepsilon \tag{4}$$

where K_{UV} is an instrument calibration constant and ε is the extinction coefficient for 1 mg/ml of protein or complex at 1 cm pathlength. By simply combining equations 1, 2, and 4, we get:

$$M = \frac{K_{RI}^2}{K_{LS}K_{UV}} \frac{(LS)(UV)}{\varepsilon\,(RI)^2} \tag{5}$$

where M and ε are the molecular weight and extinction coefficient of an entire protein or complex, including carbohydrate. This equation is the basis for the "three-detector method". However, in most cases the ε is unknown, especially for a complex, and therefore equation 5 is not commonly used in light scattering analysis. What may be obtained relatively easily is the polypeptide extinction coefficient, ε_p, which can be obtained either from experimental data or estimated with reasonable accuracy from the amino acid composition (8). If we use ε_p and select a wavelength where the carbohydrate does not absorb, it is possible to eliminate all the contributions from the carbohydrates. To demonstrate this, we re-express *(LS)*, *(RI)*, and *(UV)* signals based on the polypeptide concentration, c_p, and the mass and *(dn/dc)* of the polypeptide and carbohydrate components. From equation 1 we obtain:

$$(LS) = K_{LS}c_p\left(\frac{M_p + M_c}{M_p}\right)(M_p + M_c)\left(\frac{M_p(dn\,/\,dc)_p + M_c(dn\,/\,dc)_c}{M_p + M_c}\right)^2$$

$$(LS) = K_{LS}c_p\frac{\left(M_p(dn\,/\,dc)_p + M_c(dn\,/\,dc)_c\right)^2}{M_p} \tag{6}$$

where subscripts p and c stand for the polypeptide and carbohydrate components. Similarly, equation 2 gives:

$$(RI) = K_{RI}c_p\left(\frac{M_p + M_c}{M_p}\right)\left(\frac{M_p(dn\,/\,dc)_p + M_c(dn\,/\,dc)_c}{M_p + M_c}\right)$$

$$(RI) = \frac{K_{RI}c_p\left(M_p(dn\,/\,dc)_p + M_c(dn\,/\,dc)_c\right)}{M_p} \tag{7}$$

From equation 4 we derive:

$$(UV) = K_{UV}c_p\left(\frac{M_p + M_c}{M_p}\right)\left(\frac{\varepsilon_p M_p + \varepsilon_c M_c}{M_p + M_c}\right)$$

but since we have chosen a wavelength where $\varepsilon_c = 0$

$$(UV) = K_{UV}c_p\varepsilon_p \tag{8}$$

Combining equations (6) through (8) and solving for M_p gives:

$$M_p = \frac{K_{RI}^2}{K_{LS}K_{UV}} \frac{(LS)(UV)}{\varepsilon_p(RI)^2} \tag{9}$$

As shown in equation 9, all contributions of the carbohydrate are canceled algebraically, and we can measure the polypeptide molecular weight directly as long as its polypeptide extinction coefficient is known. Equation 9 is the actual "three-detector method" equation used in our study. The instrument calibration constant, $K_{RI}^2 / (K_{LS}K_{UV})$, can be obtained by running protein standards.

Next, we want to use equation 9 to determine the stoichiometry of a protein-protein complex containing carbohydrates. In most common situations only the polypeptide extinction coefficients of each protein in a complex are known, and thus we must to calculate the polypeptide extinction coefficient of the complex as a whole. The polypeptide extinction coefficient of a complex, ε_p, with a known stoichiometry (A_mB_n) can be calculated using the following equation:

$$\varepsilon_p = (m\varepsilon_A M_A + n\varepsilon_B M_B) / (mM_A + nM_B) \tag{10}$$

where ε_A, ε_B, M_A and M_B are the polypeptide extinction coefficients and molecular weights of either proteins A or B. After obtaining the polypeptide extinction coefficient of a complex, we can calculate the polypeptide molecular weight by equation 9 and, hence, determine the stoichiometry.

It is obvious that there is a circular argument here. On the one hand, we want to use equation 10 to calculate the polypeptide extinction coefficient of a complex and then equation 9 to determine the corresponding molecular weight and the stoichiometry; on the other hand, the polypeptide extinction coefficient of a complex cannot be calculated from equation 10 until the stoichiometry of the complex is known. In order to solve this conundrum, a self-consistent method has been developed. In this method we assume various possibilities for the stoichiometry of a complex. For each assumed stoichiometry we then calculate its corresponding theoretical molecular weight from those of its components and its experimental molecular weight from equations 9 and 10. Finally, we select the stoichiometry with the best consistency between the experimental and theoretical molecular weights as the correct stoichiometry for the complex.

B. Experimental Details

A laser light scattering detector (Wyatt Minidawn), a uv absorbance monitor at 280 nm (Knauer A293), and a refractive index detector (Polymer Laboratories PL-RI) were used in series for the on-line light scattering/SEC system. There are three light scattering signals from three different angles in the Wyatt Minidawn. Only the 90^0 light scattering signal was used in these experiments because it provides the best signal/noise. It should be pointed out that the proteins/complexes studied here are too small to have detectable angular dependence of their light scattering. A Superdex 200 (Pharmacia) 10/30 SEC column and Dulbecco's phosphate buffered saline were used for studying the interactions of sTrkB/BDNF and sHer2/mAbs. A TSK G2000 GW (TosoHaas) column and 0.1 M sodium phosphate plus 0.5 M NaCl were used for sTrkC/NT-3. The system was operated at a flow rate of 0.5 ml/min with a 100 μl sample loop. Ribonuclease (Calbiochem), ovalbumin (Sigma) and bovine serum albumin monomer (Sigma) were used as calibration standards to obtain the instrument constant of equation 9.

BDNF, NT-3, sTrkB, and sTrkC were prepared as described previously (9). sTrkB/BDNF and sTrkC/NT-3 complexes were made by mixing 93.3 µl of 1.41 mg/ml sTrkB with 26.7 µl of 1.5 mg/ml BDNF and 47.7 µl of 2.5 mg/ml sTrkC with 52.3 µl of 0.84 mg/ml NT-3, respectively.

The complexes of mAbs/sHer2 were made by mixing 55 µl of 1.5 mg/ml mAb35, 0.8 mg/ml mAb52, 1.2 mg/ml mAb58, and 1.6 mg/ml mAb42b with 55 µl of 2.0, 2.0, 1.3, and 2.0 mg/ml sHer2, respectively.

III.   Results and Discussion

A. Stoichiometry of sTrkB/BDNF and sTrkC/NT-3

BDNF and NT-3 are highly associated homodimers (10). Before measuring protein complexes, we usually first measure each component by itself. However, because BDNF alone does not elute from a Superdex 200 column when using PBS elution buffer, only a sTrkB control is shown in Figure 1A.

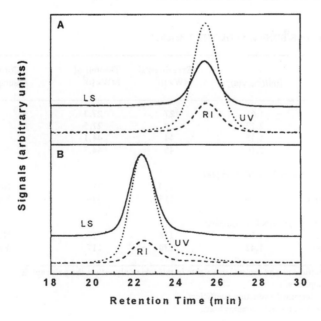

Figure 1. Chromatograms of sTrkB (panel A) and the mixture of sTrkB and BDNF (panel B). LS: light scattering signal (solid line); RI: refractive index signal (dash line); UV: uv absorbance at 280 (dot line).

sTrkB is a glycoprotein and therefore the three-detector method was used. The 44-kDa polypeptide molecular mass thus determined for sTrkB (Table 1) agrees well with the molecular weight calculated from its sequence, indicating that the extinction coefficient was adequately estimated. The chromatogram of a sTrkB/BDNF mixture made at a molar ratio of two TrkB per BDNF dimer is shown in Figure 1B. In order to calculate the molecular weight and stoichiometry of the sTrkB/BDNF complex, it is necessary to calculate its polypeptide extinction coefficient as discussed in *Materials and Methods*. However, the extinction coefficient cannot be calculated until the stoichiometry of the complex is known. Therefore, we assume either one or two sTrkB bind to one BDNF dimer and calculate the corresponding experimental and theoretical molecular weights. The results are summarized in Table I. Experimental molecular weights were calculated from equations 9 and 10, and theoretical molecular weights were calculated from the sequence of each component under each assumed stoichiometry. Obviously, the experimental molecular weight agrees with the theoretical one under the assumption that two sTrkB bind to one BDNF dimer. Therefore, we conclude that the stoichiometry, 2sTrkB:1BDNF dimer, is the correct stoichiometry for the sTrkB/BDNF complex.

Using the same method, sTrkC, NT-3 and the mixture of sTrkC/NT-3 were studied, and the results are summarized in the second part of Table 1. The results clearly indicate that NT-3 can also dimerize sTrkC.

Table I. Summary of sTrkB/BDNF and sTrkC/NT-3 Results [a]

Proteins or Complexes	ε [ml/(mg·cm)]	Experimental MW×10³	Theoretical MW×10³	Correct Assumption?
BDNF dimer [b]	1.6	27	27.3	
NT-3 dimer [b]	2.17	28	27.5	
sTrkB [c]	1.15	44	44.1	
sTrkC [c]	1.19	45	44.7	
Assumed Stoichiometry of sTrkB/BDNF Complex:				
1sTrkB : 1BDNF dimer	1.32	110	71	No
2sTrkB : 1BDNF dimer	1.26	115	116	Yes
Assumed Stoichiometry of sTrkB/NT-3 Complex:				
1sTrkC : 1NT-3 dimer	1.56	118	72	No
2sTrkC : 1NT-3 dimer	1.42	130	117	Yes

[a]All extinction coefficients and molecular weights in this table only reflect the polypeptide components of the complexes.
[b]Experimentally determined extinction coefficient.
[c]Calculated from the amino acid compositions.

B. Stoichiometry of sHer2 and its Monoclonal Antibodies

Samples of sHer2, mAb35, mAb52, mAb58, and mAb42b controls, as well as mixtures of sHer2 with each mAb, were analyzed by light scattering/SEC. The results are summarized in Table II and the chromatograms of one set of samples (sHer2 and mAb35) are shown in Figure 2. The experimental molecular weights for the complexes are most consistent with the theoretical ones under the assumption that two sHer2 bind one mAb for each mAb tested. This shows that these antibodies do dimerize Her2 even when it is not bound to a membrane.

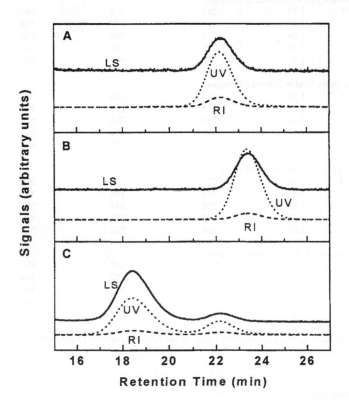

Figure 2. Chromatograms of mAb35 (panel A), sHer2 (panel B), and the mixture of sHer2 and mAb35 (panel C). Lines are the same as defined in Figure 1.

Table II. Summary of sHer2 and mAbs' Light Scattering Results [a]

Proteins or Complexes	ε [ml/(mg·cm)]	Experimental MW × 10^{-3}	Theoretical MW × 10^{-3}	Correct Assumption?
sHer2	0.85	69		
mAb 35	1.4	139		
mAb 52	1.4	151		
mAb 58	1.4	142		
mAb 42b	1.4	136		
Assumed Stoichiometry of sHer2/mAb35 Complex:				
1sHer2 : 1mAb35	1.24	237	208	No
2 : 1	**1.14**	**261**	**277**	**Yes**
3 : 1	1.08	275	346	No
1 : 2	1.31	226	347	No
1 : 3	1.41	208	486	No
Assumed Stoichiometry of sHer2/mAb52 Complex:				
1sHer2 : 1mAb52	1.24	252	220	No
2 : 1	**1.14**	**275**	**289**	**Yes**
3 : 1	1.08	289	358	No
1 : 2	1.31	240	371	No
1 : 3	1.41	223	522	No
Assumed Stoichiometry of sHer2/mAb58 Complex:				
1sHer2 : 1mAb58	1.24	252	211	No
2 : 1	**1.14**	**272**	**280**	**Yes**
3 : 1	1.08	289	348	No
1 : 2	1.31	237	353	No
1 : 3	1.41	220	522	No
Assumed Stoichiometry of sHer2/mAb42b Complex:				
1sHer2 : 1mAb42b	1.24	246	205	No
2 : 1	**1.14**	**266**	**274**	**Yes**
3 : 1	1.08	281	343	No
1 : 2	1.31	232	341	No
1 : 3	1.41	214	477	No

[a]All extinction coefficients and molecular weights in this table only reflect the polypeptide components of the complexes.

IV. Conclusion

With the introduction of the self-consistent "three-detector method" light scattering/SEC has become a very accurate and reliable technique for studying the stoichiometry of a protein-protein interaction. Using this technique, BDNF and NT-3 were shown to dimerize their soluble receptors, sTrkB and sTrkC, respectively, and anti-Her2 monoclonal antibodies were also shown to dimerize sHer2.

References

1. Blume-Jensen, P., Claesson-Welsh, L., Sieghahn, A., Zsebo, K.M., Westermark, B., and Heldin, C.-H. (1991) *EMBO J.* **10**, 4121-4128.

2. Li, W. and Stanley, E.R. (1991) *EMBO J.* **10**, 277-288.

3. Glass, D.J., Nye, S.H., Hantzopoulos, P., Macchi, M.J., Squinto, S.P., Goldfarb, M., and Yancopoulos G. D. (1991) *Cell* **66**, 405-413.

4. Lamballe, F., Klein, R., and Barbacid, M. (1991) *Cell* **66**, 967-979.

5. Harwerth, I. -M., Wels, W., Schlegel, J., Muller, M., and Hynes, N. E. (1993) *Br. J. Cancer* **68**, 1140-1145.

6. Srinivas, U., Tagliabue, E., Campiglio, M., Menard, S., and Colnaghi, M. I. (1993) *Cancer Immunol Immunother* **36**, 397-402.

7. Takagi, T. (1990) *J. Chromatogr.* **506**, 409-416.

8. Gill, S. C. and von Hippel, P. H. (1989) *Anal. Biochem.* **182**, 319-326.

9. Philo, J., Talvenheimo, J., Wen, J., Rosenfeld, R., Welcher, A., and Arakawa, T. (1994) *J. Biol. Chem.* **269**, 27840-27846.

10. Arakawa, T., Yphantis, D.A., Lary, J.W., Narhi, L.O., Lu, H.S., Prestrelski, S.J., Clogston, C.L., Zsebo, K.M., Mendiaz, E.A., Wypych, J., and Langley, K.E. (1991) *J. Biol. Chem.* **266**, 18942-18948.

Measurements of Protein-Protein Interaction by Isothermal Titration Calorimetry with Applications to the Bacterial Chemotaxis System

Jiayin Li and Robert M. Weis

Department of Chemistry and Program in Molecular and Cellular Biology, University of Massachusetts, Amherst, Massachusetts

I. Introduction

Specific macromolecule-macromolecule and ligand-macromolecule interactions are involved in nearly every biological process including signal transduction, gene expression, protein folding, cell differentiation and development. The strength of macromolecule interactions is often regulated physiologically by covalent modification and noncovalent interactions with small molecules (as well as temperature, solution ionic strength and pH), so that an accurate characterization of the interactions between macromolecules in their various functional states can help determine mechanistic aspects of the aforementioned cellular processes. As a consequence of their importance, numerous methods have been developed that quantify an observable property (e.g. gravimetric, hydrodynamic or spectroscopic) of the system which changes as a function of the extent of the interaction (reviewed in Connors, 1987 and Phizicky and Fields, 1995). Most methods either require labeled molecules or are time-consuming; features which reduce their generality and ease of use. Surface plasmon resonance (SPR) intruments and similar refractive-index-based sensors have found recent widespread use in the rapid characterization of macromolecule interactions (Chaiken et al., 1991; Panayotou et al., 1993; Davies et al., 1994, and references therein). The SPR method detects binding interactions when freely soluble macromolecule binds to its partner which is covalently attached to the sensor surface. In some cases the surface immobilization results in unwanted steric effects and may also lead to multivalent binding interactions with oligomeric systems, which can influence the value of K_a (Ladbury et al., 1995).

In contrast to other methods isothermal titration calorimetry (ITC) can determine in a single experiment all of the thermodynamic parameters of the binding process (the association constant, K_a, changes in enthalpy, ΔH, and entropy, ΔS; and the binding ratio, n) with unlabeled macromolecules in solution. If titrations are carried out at more than one temperature then changes in the heat capacity due to binding (ΔC_p) can also be determined. Although ITC is not a new technique (see for example Langerman and Biltonen, 1979, and references therein), the availability of a high-quality commercial instrument manufactured by Microcal Inc. (Northampton, MA) has made ITC a widely-used technique for fundamental studies (Wiseman et al., 1989; Brandts et al., 1990).

II. The ITC Method

An isothermal titration calorimeter measures the temperature difference between a reaction and a reference cell (ca. 1.5 mL ea.) using a thermopile, and nulls any difference detected through feedback control of the cell heating elements (Figure 1). A titration consists of automated injections of a ligand solution into a stirred solution of macromolecule, and the heat released (or absorbed) after an injection is recorded as the perturbation in

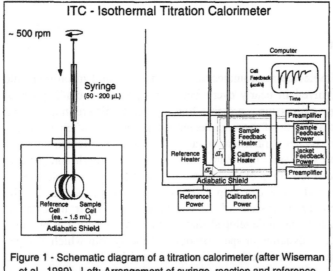

Figure 1 - Schematic diagram of a titration calorimeter (after Wiseman et al., 1989). Left: Arrangement of syringe, reaction and reference cells. Right: Block diagram of control system.

the heat flow (μcal/s) to the reaction cell from its baseline value. Integration of the area under the curve gives the injection heat. Since the Omega® ITC was introduced (Wiseman, 1989), improvements in instrumental design have lowered the absolute and cell-volume-normalized detection sensitivities to ca. 0.1 μcal and 0.07 μcal/mL under optimum conditions. Considering these detection sensitivities, and the maximum macromolecule concentration that can feasibly used in the reaction cell (ca. 1 mM), the range of K_as that can be determined in simple titrations is 10^2 to 10^9 M^{-1}. (In principle it may be possible to extend the K-range with competitive binding experiments.) The product of the macromolecule concentration in the cell (M_t) and K_a is the parameter that determines the shape of the titration curve and should have a value between 10 and 500 for accurate fits of the data (Wiseman et al., 1989). More generally the binding reaction should be driven to completion by the end of the titration. For tight binding interactions, e.g. $K_a \sim 10^6$ M^{-1}, a two-fold molar excess of the injectant relative to the number of sites in the reaction cell is typical, for weak interactions (10^2 M^{-1}), a larger excess (~ 10) is required to drive the reaction to completion.

Provided that the modest solubility requirements can be met, it is probable that an interaction can be studied. Both species must either be soluble or form suspensions so that one macromolecule (or ligand, typically 100 μL of a 1 mM solution) can be injected into the a solution of the other (1.5 mL of a 50 μM solution). However under optimum conditions the association constant can be determined from a titration that altogether evolves (or absorbs) *ca.* 50 μcal. Thus for a typical molar binding enthalpy of 10 kcal only 5 nanomoles of macromolecule is required. As described below it is possible to carry out titrations on a variety of systems including soluble macromolecules, cell membrane suspensions and protein-detergent mixed micelles.

All titrations in this study were carried out using a MicroCal MCS ultrasensitive titration calorimeter (MicroCal Inc., Northampton, MA) with Observer™ software for instrument control and data acquisition (Wiseman et al., 1989; Brandts et al., 1990; Lin et al., 1994). Titrations were carried out in a buffer of 20 mM sodium phosphate, 20 mM NaCl, 1 mM EDTA, 1 mM PMSF, 10%(v/v) glycerol, pH 7.4, using a syringe stirring speed of 500 rpm.

III. Results and Discussion

A. *Applications to Bacterial Chemotaxis*

The chemosensory signal transduction pathways in *Escherichia coli* and *Salmonella typhimurium* are well-studied examples of 'two-component' signaling pathways, which are widely distributed among prokaryotes (Kofoid and Parkinson, 1992) and are also found in eukaryotes (Swanson et al., 1994). The components involved in signaling pathway have been identified and the major biochemical features of the pathway have been determined (reviewed in Bourret et al., 1991; Stock et al., 1992). Attention has shifted toward characterizing the numerous interactions among the signaling proteins and ligand binding interactions of the transmembrane receptors since this information may help determine molecular mechanisms of transmembrane signaling and signal amplification. The signaling pathway starts with ligand binding; *E. coli* is sensitive to a wide variety of environmental cues, some of these are small molecule attractants detected by transmembrane receptors located in the inner-membrane (the aspartate, Tar; serine, Tsr; dipeptide and ribose/galactose receptors). Attractant binding/unbinding triggers excitation and adaptation responses in the cytoplasm by regulating the activity of a 73 kDa autophosphorylating kinase (CheA). CheA amplifies and propagates the excitation signal by a phosphotransfer reaction to the response regulator CheY (14 kDa), which is active in modulating the reversal frequency of the flagellar motor through direct binding interactions. Adaptation attenuates the excitation response to a steady stimulant concentration through a negative feedback loop involving the (reversible) methylation/demethylation and (irreversible) deamidation reactions of the receptor mediated by a methyltransferase, CheR (32 kDa) and a methylesterase (which is also the deamidase, 37 kDa), CheB (Springer et al., 1979). The esterase is also a substrate of CheA-mediated phosphorylation. CheB catalytic activity is increased by phosphorylation.

ITC facilitates the exploration of factors that regulate ligand binding and protein-protein interactions. For example does receptor covalent modification have any effect on transferase-receptor interactions? Do CheY and CheB bind to independent or overlapping sites on CheA?, and what effect does CheY phosphorylation have on CheY-CheA interactions? These and other examples are described below and serve to illustrate the utility of the ITC method for characterizing protein-protein interactions.

1. Protein-protein interactions

a. Protein-Protein Interaction Between a Cytoplasmic Signaling Protein and a Transmembrane Receptor. These experiments demonstrated the feasibility of studying interactions between a soluble protein and a membrane-bound receptor by ITC. The methyltransferase (32 kDa) is a soluble cytoplasmic protein that catalyzes methylation of glutamyl groups on the cytoplasmic domain of the receptor with the methyl group donor s-adenosyl-L-methionine (Springer et al., 1979). Data that measured the binding of the transferase to E. coli whole membranes samples containing the 60 kDa serine receptor (Tsr) are presented in Figure 2. The middle and lower traces of Figure 2a are typical of a specific stoichiometric binding. The first several injections resulted in the evolution of heat (indicated by the decrease in the heat flow to the reaction cell) that decreased in proportion to the remaining available receptor binding sites in the reaction cell. Once the receptor was saturated with transferase (at molar ratios > 1.5), the residual heats were steady and could be attributed to buffer mismatch. Small heats were observed in a control experiment (top trace) where 0.54 mM CheR was injected into a solution of E. coli membrane vesicles that lacked receptor (Tsr⁻). Taken together the data strongly suggested that the saturation observed in the middle and lower traces was due to Tsr-transferase interactions. The integrated heats, plotted in Figure 2b, were fit to a single-set-of-sites model where a one-to-one transferase-receptor binding interaction was indicated by the n-value returned in the fit. Unexpectedly the presence of the extracellular ligand serine at saturating concentrations (1 mM) had no effect on transferase binding. The n-values and values of K_a and ΔH for the transferase-receptor interactions are reported in Table I (expts. #5 & 6).

The Tsr in the membrane samples of Figure 2 were genetically engineered to mimic a high level of methylation, four glutamines and one glutamate (4Q1E) at the five sites of methylation on the Tsr molecule (Rice and Dahlquist, 1991). Receptors with low (5E, Table I expts. # 1 & 2) and intermediate (QEQEE = wt, expts. #3 & 4) levels of modification were also titrated with transferase (in the absence and presence of 1 mM serine), and resulted in values for n, K_a and ΔH that were the same within experimental error, demonstrating that the level of covalent modification on the receptor did not influence the transferase binding significantly. The lack of an influence of covalent modification on serine binding was

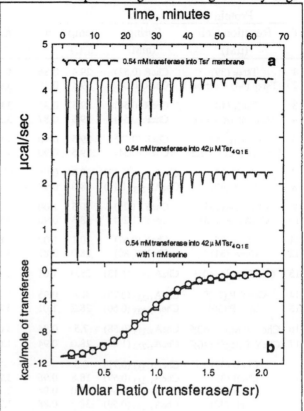

FIGURE 2: Titration calorimetry of Tsr with transferase. a) 10-μL injections of 0.54 mM transferase into Tsr-minus control membranes (top), 42 μM Tsr$_{4Q1E}$ (middle, Table I # 5), and 42 μM Tsr$_{4Q1E}$ in the presence of 1 mM serine (bottom, Table I # 6). b) Integrated, background-subtracted data for the middle (O) and the bottom curves (□) of panel a. Solid lines are the best-fit curves using a single-set-of-sites model.

demonstrated in a previous ITC study (Lin et al., 1994).

Comparing experiments 1 to 6 in Table I, receptor with or without bound serine, or in various levels of covalent modification: demethylated (5E), wildtype (QEQEE), and amidated (4Q1E) all bind to the transferase in a 1 to 1 ratio characterized by a K_a of 4×10^5 M^{-1} and a ΔH of ca. 11 kcal/mol. It appears that the ability of the receptor to serve as a substrate for the methylation reaction is not regulated by varying the strength of transferase-receptor interactions, as neither the number of modifiable sites on the receptor nor the presence of the attractant in the ligand binding pocket had any effect.

Table I. Thermodynamic Parameters for Titrations involving Chemotaxis Proteins: Tsr with CheR; CheA and CheA$_{1-233}$ with CheY and CheB

#	Protein (concentration)		temp ($^\circ$C)	n	K_a ($M^{-1} \times 10^{-5}$)	ΔH (kcal/mol)	ΔC_p (kcal/mol·$^\circ$C)
	Reaction Cell (μM)	Syringe (mM)					
1	Tsr$_{5E}$ (39)	CheR (0.52)	29.4	0.86	4.1	-10.8	
2	Tsr$_{5E}$ (44) + serine	CheR (0.54)	29.4	0.99	3.9	-10.8	
3	Tsr$_{wt}$ (40)	CheR (0.54)	29.5	0.85	3.9	-10.3	
4	Tsr$_{wt}$ (40) + serine	CheR (0.54)	29.5	0.87	3.3	-10.4	
5	Tsr$_{401E}$ (42)	CheR (0.54)	29.5	0.90	4.8	-11.8	
6	Tsr$_{401E}$ (42) + serine	CheR (0.54)	29.5	0.90	4.1	-12.0	
7	CheY (37)	CheA$_{1-233}$ (1.13)	10.3	0.95	17.8	-8.4	-0.22
8	CheY (40)	CheA$_{1-233}$ (1.13)	28.0	0.98	7.4	-12.4	
9	CheA$_{1-233}$ (35)	CheY (1.29)	10.5	1.00	19.5	-9.2	-0.20
10	CheA$_{1-233}$ (33)	CheY (1.29)	28.5	0.90	8.8	-12.8	
11	CheA (25)	CheY (0.67)	10.1	1.02	8.8	-7.2	-0.40
12	CheA (31)	CheY (0.80)	28.0	0.87	5.1	-14.3	
13	CheY·Mg^{2+} (40)[a]	CheA$_{1-233}$ (1.13)	28.0	1.01	6.9	-12.2	
14	CheY-P (21)[b]	CheA$_{1-233}$ (0.70)	8.2	0.95	6.4	-7.1	-0.29
15	CheY-P (50)[b]	CheA$_{1-233}$ (0.60)	28.2	1.01	3.8	-12.9	
16	CheY-P·Mg^{2+} (60)[c]	CheA$_{1-233}$ (0.58)	7.5	0.94	3.0	-6.9	-0.063
17	CheY-P·Mg^{2+} (40)[c]	CheA$_{1-233}$ (1.05)	28.0	0.94	1.6	-8.2	
18	CheB (59)	CheA$_{1-233}$ (0.71)	8.5	-	-	0	
19	CheB (38)	CheA$_{1-233}$ (0.59)	18.8	0.96	3.2	-5.3	-0.45
20	CheB (36)	CheA$_{1-233}$ (0.59)	28.0	0.94	2.7	-10.1	
21	CheB (30)	CheA$_{1-233}$ (0.59)	38.2	0.88	1.7	-13.2	

[a] 6 mM MgCl$_2$ was added to the titration buffer (20 mM sodium phosphate, 20 mM NaCl, 1 mM EDTA, 1 mM PMSF, 10%(v/v) glycerol, pH 7.4.

[b] CheY-P: CheY was phosphorylated with 20 mM acetyl phosphate in the titration buffer plus 6 mM MgCl$_2$ and quenched with 20 mM EDTA.

[c] CheY-P·Mg^{2+}: 6 mM MgCl$_2$ and 20 mM acetyl phosphate were added to the buffer. Under these conditions CheY was ~ 90 % phosphorylated. CheY solution was added to both the reaction and reference cells to balance the heat effects produced by the CheY phosphorylation and dephosphorylation reactions.

b. Single-site *versus* multivalent interactions. Titrations were carried out between intact CheA and CheY to determine what effect, if any, the deletion of amino acids 234 to 654 had on CheA-CheY and CheA-CheB interactions. The CheY binding site had been previously localized to the region between amino acids 98 and 233 of the CheA molecule using SPR (Swanson et al., 1993). Titrations of CheA$_{1-233}$ and *intact*-CheA with CheY produced similar binding curves (Figure 3), and were adequately fit using a

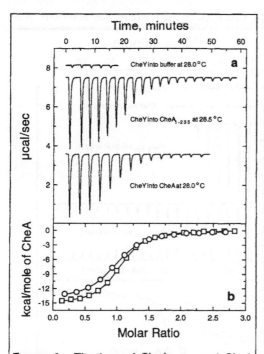

FIGURE 3: Titrations of CheA$_{1-233}$ and CheA with CheY. (a) ITC traces of 10-μL injections of 0.8 mM CheY into buffer (top), and CheY into CheA$_{1-233}$ (middle, Table I #10), and CheA (bottom, Table I #12). (b) Integrated areas & fits of CheY into CheA (O) and CheA$_{1-233}$ (□).

single-set-of-sites model. The modest difference in K_a measured by ITC (Table I, expts. #7-12) were in sharp contrast to the large differences observed by SPR (Schuster et al., 1993; Swanson et al., 1993), where the *intact*-CheA-CheY interaction was measured to be more than 10-fold stronger than the CheA$_{1-233}$-CheY interaction. In the SPR experiments the opportunity for multivalent interactions to form between intact (dimeric) CheA and (surface-attached) CheY but not with (monomeric) CheA$_{1-233}$ may account for the stronger apparent *intact*-CheA-CheY interactions. Since ITC measures the intrinsic single-site interaction between freely dissolved CheA and CheY, the values of K_a are expected to be similar in the absence of cooperative interactions between the CheY binding sites on (intact, dimeric) CheA.

c. CheA Binds to CheY and CheB with Comparable Affinity. To locate the CheB binding site and to measure K_a precisely, titrations were performed between CheA$_{1-233}$, CheY, and CheB. As expected for a simple reversible binding process, titrations of CheY with CheA$_{1-233}$ (Figure 4) resulted in fit parameters identical to those obtained from the titrations of CheA$_{1-233}$ with CheY reported above (Figure 3). The binding interaction between CheA$_{1-233}$ and CheB produced K_a values comparable to the corresponding values of the CheY-CheA$_{1-233}$ interaction (by a factor of 2 to 3), but the enthalpic (and the thus entropic) contribution(s) were significantly different in the two cases. The values of ΔC_p (determined from the temperature-dependence of ΔH) were also significantly different. These data are diagnostic of differences in the molecular details of the binding interaction.

d. Effects of Covalent Modification on CheA-CheY Interactions. Table I summarizes a series of experiments that evaluated the effect of CheY phosphorylation on CheA-CheY interactions (Li et al., *submitted*). Titrations were carried with phosphorylated CheY (CheY-P) where the reactions was quenched with excess EDTA, and also in the presence of excess Mg^{2+} and acetyl phosphate (a low molecular weight phosphate donor that specifically modifies Asp-57 on CheY, Lukat et al., 1991). K_a values of the EDTA-quenched CheY-P/CheA$_{1-233}$ interactions were *ca.* 50% of the unphosphorylated CheY values. Titration experiments performed with CheA$_{1-233}$ and CheY in the presence of acetyl phosphate and Mg^{2+} produced values of K_a that were *ca.* five-fold lower than unphosphorylated CheY. Mg^{2+} had no influence on CheA$_{1-233}$-CheY binding (Table I, # 13-17). From these results it appears that CheY

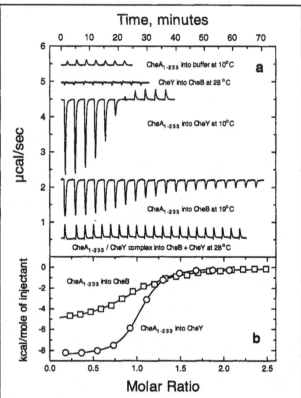

FIGURE 4: (a) ITC data of CheA$_{1-233}$, CheY and CheB. Inj. vols. were 10 μL unless noted. From top: 0.85 mM CheA$_{1-233}$ → buffer; 0.17 mM CheY → 10 μM CheB, CheA$_{1-233}$ → CheY (Table I # 7); CheA$_{1-233}$ → CheB (Table I # 19), & a competitive titration of CheB with CheA$_{1-233}$/CheY complex consisting of 5-μL inj. of 0.33 mM CheA$_{1-233}$ and 0.87 mM CheY → 17 μM CheB in the presence of 36 μM CheY. (b) Integrated areas and fits for CheY (□) and CheB (O) titrated with CheA$_{1-233}$.

phosphorylation and Mg^{2+} binding together produce a significant reduction in the affinity of CheY to CheA interactions, and may be important to the signal amplification process *in vivo*.

2. Competitive titration experiments

CheY and CheB both bind to CheA in the first 233 residues of CheA. Additional relevant information on the organization of the two binding sites can be obtained through competitive binding experiments. CheY and the regulatory domain of CheB are structurally homologous but have a comparatively low sequence identity (23%), so ITC can provide corroborating evidence for competitive interactions. ITC measures the fractional saturation of sites in the reaction cell (Θ) as a function of ligand concentration $[X_1]$. The association constant for ligand, K_1, is related to Θ (which is proportional to the specific enthalpy change of binding) in a single-set-of-sites model by

$$K_1 = \frac{\Theta}{(1-\Theta)[X_1]} \tag{1}$$

When a second competitively binding molecule (X_2) is present in the reaction cell, then apparent value of K_1 (K_{app}) is approximately given by

$$K_{app} = \frac{K_1}{1 + K_2\lfloor X_2 \rfloor} \tag{2}$$

where K_2 is the true association constant of the macromolecule for X_2 and $\lfloor X_2 \rfloor$ is the average free concentration of X_2 during the course of the titration (Li et al., *submitted*). K_{app} and the difference in the binding enthalpy of ligand and the competitor, $\Delta\Delta H$, are obtained when the data are fit to a single-set-of-sites model. One type of competitive binding experiment is plotted in Figure 4a (bottom curve). A solution 0.33 mM in CheA$_{1-233}$ and 0.87 mM in CheY was injected into a solution 17 µM in CheB and 36 µM in CheY. Thus CheB was titrated with CheA$_{1-233}$ in the presence of saturating concentrations of CheY (in both the syringe and the reaction cell). The results were consistent with CheY competitively blocking the binding of CheA$_{1-233}$ to CheB. When compared to the simple titration of CheB with CheA$_{1-233}$ (4th trace from top), the competitive titration showed a only small *decrease* in the endothermicity of the injection heats which was consistent with a weakly endothermic, saturative process. A weakly endothermic process was expected from the sign of $\Delta\Delta H$, the difference in the ΔHs of CheY and CheB binding (cf. Table I). Simulations of trends in the injection heats were also found to be quantitatively consistent with thermodynamic parameters obtained from the simple titrations of CheY and CheB with CheA$_{1-233}$ (Li et al., *submitted*).

To establish competitive binding interactions, ITC can thus provide strong supporting data. Once the competitive nature of the interaction has been established, ITC may be used to extend the range of measurable binding constants. The effect of a competitor on a tight ligand binding interaction (e.g. for $K_a > 10^{10}$ M^{-1}) will be to reduce the apparent interaction according to equation 2. Provided that $\Delta\Delta H$ is nonzero, and the association constant for competitor binding has been measured, then it should be possible to measure K_a for the tight-binding ligand.

3. Dissociation of oligomeric proteins

In the examples given above the ITC experiment yields information about the interactions between different macromolecules, but ITC can also characterize the thermodynamic properties of self-associating macromolecules. In our experience, proteins that have no tendency to self associate have yielded small and nearly constant heats when injected into buffer (See the protein-into-buffer experiments in Figures 2 - 4). For example the CheA$_{1-233}$ fragment was found to be monomeric

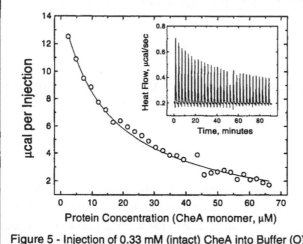

Figure 5 - Injection of 0.33 mM (intact) CheA into Buffer (O). The fit line is to a dimer dissociation model. Inset: Raw data.

by gel filtration chromatography (Swanson et al., 1993), and no significant trends in the injection heats were detected when CheA$_{1-233}$ was diluted into buffer (Figure 4a). In contrast, steadily decreasing heats were observed when 10 μL aliquots of a solution 0.33 mM in *intact*-CheA were injected into buffer (open circles in Figure 5). This trend represented an endothermic process that diminished as the concentration of CheA in the reaction cell increased. These data are plausibly explained by a reversible dissociation process involving an oligomeric species (which would be a larger fraction of the protein in the syringe than in the reaction cell), that dissociates in an endothermic process when it is diluted as a result of the injection process (in this example over 100-fold for the 1st injection). As the macromolecule concentration in the cell increases, the difference between the fraction of oligomeric protein in the syringe and reaction cell decreases, which results in a decrease of the observed heat. CheA has been shown to be dimeric (Gegner and Dahlquist, 1991). Also, the autophosphorylation of CheA (known to be an intermolecular process) exhibits second order kinetics (Ninfa, 1992), which provided an estimate for K_a of 2×10^6 M^{-1}. A fit to the data in Figure 5 to a monomer-dimer model resulted in estimates for K_a and $\Delta H_{dissociation}$ of *ca.* 1.5×10^4 M^{-1} and 6 kcal/mol of CheA monomer. This large difference in the values of K_a warrants further experimentation to determine the origin of the effect seen in the ITC, especially since it was not possible to distinguish between a monomer-dimer and a dimer-tetramer equilibrium by the ITC data. The data could also be fit to a dimer-tetramer process (where the total CheA concentration in the syringe, *as dimer,* was 0.165 mM) and resulted in values for K_a and $\Delta H_{dissociation}$ of 1×10^4 M^{-1} and 14 kcal/mol of CheA dimer. Clearly an independent method is required to establish the existence and molecularity of the putative association process detected in this preliminary ITC experiment. Nevertheless these data illustrate the potential for rapidly determining the thermodynamic parameters of a clustering process.

B. Applications to Other Systems

In the examples cited above ITC has been shown to be useful for studying ligand-protein and protein-protein interactions in a variety of ways. Since 1989 over 100 experimental studies using ITC have been published. ITC has proven to be useful in providing detailed thermodynamic information in a variety of systems including binding interactions between DNA and repressor proteins (Jin et al., 1993; Ladbury et al., 1994); phosphopeptide- and peptide-protein interactions (Lemmon and Ladbury, 1994; Varadarajin, 1992); peptides and membranes (Montich et al., 1993); ligand-receptor interactions (Cunningham et al., 1991; Lin et al., 1994), antigen-antibody interactions (Hibbits et al., 1994), and metal ion-protein interactions (Lin et al., 1991).

IV. Conclusion

As a result of continued improvements in instrument design the measurable range of K_a values should continue to increase, which in some circumstances may be aided through the use of competitive binding reactions. Because of the generality of the thermal response in chemical processes, it is likely that the technique will find additional use in the characterization of a variety of biochemical processes (e.g. enzyme-catalyzed reactions) in addition to measurements of macromolecular interactions.

References

Bourret, R. B., Borkovich, K. A., and Simon, M. I. (1991) *Ann. Rev. Biochem.* **60,** 401-441.

Brandts, J. F., Lin, L.-N., Wiseman, T., Williston, S., and Yang, C. P. (1990) *International Laboratory* **20,** 29-35.

Connors, K. A. (1987) "Binding Constants: The measurement of molecular complex stability" Wiley-Interscience, New York.

Chaiken, I., Rose, S., and Karlsson, R. (1991) *Anal. Bioch.* **201,** 197-210.

Cunningham, B. C., Ultsch, M., de Vos, A. M., Mulkerrin, M. G., Clauser, K. R., and Wells, J. A. (1991) *Science* **254,** 821-825.

Davies, R. J., Edwards, P. R., Watts, H. J., Lowe, C. R., Buckle, P. E., Yeung, D., Kinning, T. M., and Pollard-Knight, D. V. (1994) *in* "Techniques in Protein Chemistry V" (Crabb, J. W. ed), p. 285-292.

Panayotou, G., Waterfield, M. D., and End, P. (1993) *Curr. Biol.* **3**, 913-915.

Hibbits, K. A., Gill, D. S., & Willson, R. C. (1994) *Biochemistry* **33**, 3584-3590.

Jin, L., Yang, J., and Carey, J. (1993) *Biochemistry* **32**, 7302-7309.

Kofoid, E. C., and Parkinson, J. S. (1992) *Ann. Rev. Genet.* **26**, 71-112.

Lemmon, M. A., and Ladbury, J. E. (1994) *Biochemistry* **33**, 5070-5076.

Ladbury, J. E., Wright, J. G., Sturtevant, J. M., and Sigler, P. B. (1994) *J. Mol. Biol.* **238**, 669-681.

Ladbury, J. E., Lemmon, M. A., Zhou, M., Green, J., Botfield, M. C., and Schlessinger, J. (1995) *Proc. Natl. Acad. Sci. USA* **92**, 3199-3203.

Langerman, N., and Biltonen, R. L. (1979) *Meth. Enzymol.* **61**, 261-286.

Li, J., Swanson, R. V., Simon, M. I., and Weis, R. M. *(submitted)*

Lin, L. -N., Mason, A. B., Woodworth, R. C., and Brandts, J. F. (1991) *Biochemistry* **30**, 11660-11669.

Lin, L.-N., Li, J., Brandts, J. F., and Weis, R. M. (1994) *Biochemistry* **33**, 6564-6570.

Lukat, G. S., McCleary, W. R., Stock, A. M., and Stock, J. B. (1992) *Proc. Natl. Acad. Sci. USA* **89**, 718-722.

Montich, G., Scarlata, S, McLaughlin, S., Lehrmann, R., and Seelig, J. (1993) *Biochim. Biophys. Acta* **1146**, 17-24.

Ninfa, E. G. (1992) Doctoral Dissertation. (Princeton University, Princeton, NJ).

Phizicky, E. M., and Fields, S. (1995) *Microbiol. Rev.* **59**, 94-123.

Rice, M. S., and Dahlquist, F. W. (1991) *J. Biol. Chem.* **266**, 9746-9753.

Schuster, S. C., Swanson, R. V., Alex, L. A., Bourret, R. B., and Simon, M. I. (1993) *Nature* **365**, 343-347.

Springer, M. S., Goy, M. F., and Adler, J. (1979) *Nature* **280**, 279-284.

Stock, J. B., Surette, M. G., McCleary, W. R., and Stock, A. M. (1992) *J. Biol. Chem.* **267**, 19753-19756.

Swanson, R. V., Schuster, S. C., and Simon, M. I. (1993) *Biochemistry* **32**, 7623-7629.

Swanson, R. V., Alex, L. A., and Simon, M. I. (1994) *Trends Bioch. Sci.* **19**, 439-518.

Varadarajan, R., Connelly, P. R., Sturtevant, J. M., and Richards, F. M. (1992) *Biochemistry* **31**, 1421-1426.

Wiseman, T., Williston, S., Brandts, J. F., and Lin, L.-N. (1989) *Anal. Biochem.* **179**, 131-137.

Solution Assembly of Cytokine Receptor Ectodomain Complexes

Zining Wu

Department of Pharmacology and Toxicology
Dartmouth Medical School
Hanover, NH 03755 and
The Veterans Administrations Hospital
White River Junction, VT 05009

Kirk W. Johnson

Chiron Corporation
Emeryville, CA 94608-2916

Byron Goldstein

Theoretical Biology and Biophysics
Los Alamos National Laboratory
Los Alamos, NM 87545

Thomas M. Laue

Department of Biochemistry
University of New Hampshire
Durham, NH 03867

Thomas L. Ciardelli

Department of Pharmacology and Toxicology
Dartmouth Medical School
Hanover, NH 03755 and
The Veterans Administrations Hospital
White River Junction, VT 05009

TECHNIQUES IN PROTEIN CHEMISTRY VII

I.   Introduction

For the majority of single transmembrane-spanning cell surface receptors, signal transmission across the lipid bilayer barrier involves several discrete components of molecular recognition. The interaction between ligand and the extracellular segment of its cognate receptor (ectodomain) initiates either homomeric or heteromeric association of receptor subunits. Specific recognition among these subunits may then occur between ectodomain regions (1) within the membrane by interhelical contact (2) or inside the cell between cytoplasmic domains (3, 4). Any or all of these interactions may contribute to the stability of the signaling complex. It is the characteristics of ligand binding by the ectodomains of these receptors, however, that controls the heteromeric or homomeric nature and the stoichiometry of the complex.

Cytokines and their receptors belong to a growing family of macromolecular systems that exhibit these functional features and share many structural similarities as well (5). Interleukin-2 is a multifunctional cytokine (6) that represents, perhaps, the most complex example to date of ligand recognition among the hematopoietin receptor family. The high affinity interleukin-2 receptor (IL-2R, $K_d = 1 \times 10^{-11}$ M) is composed of three different cell surface subunits, each of which participates in ligand recognition (7). The 55 kDa α-subunit binds IL-2 with a K_d of 1×10^{-8} M. The larger 75 kDa β-subunit and the recently discovered 64 kDa γ-subunit combine to form the intermediate affinity site ($K_d = 1 \times 10^{-9}$ M). It is the cooperative binding of IL-2 by all three proteins on the surface of activated T- lymphocytes, however, that ultimately results in crosslinking of the β- and γ-subunits and signaling via association of their cytoplasmic domains (4). Although the high-affinity IL-2R functions as a heterotrimer, heterodimers of the receptor subunits are also physiologically important. The α/β heterodimer or "pseudo-high affinity" receptor captures IL-2 as a preformed cell surface complex (8) while the β/γ intermediate affinity site exists, in the absence of the α subunit, on the majority of natural killer cells (8).

In an effort to facilitate structure-function analysis of IL-2 and its receptor, we have begun to employ a strategy of directed and stable solution assembly of receptor ectodomains (9, 10). The goal of these studies is to prepare stable complexes of cytokine receptor ectodomains of defined composition and stoichiometry and that mimic the ligand binding characteristics of the equivalent cell surface receptor sites. We have chosen coiled-coil molecular recognition to direct the formation of these complexes. As result of more than 20 years of study (11-13), the noncovalent coiled - coil interaction is now well understood. Recent reports stimulated by the rediscovery of this motif as the "leucine zipper" in eukaryotic transcription factors (14) have defined the critical parameters controlling both stability and stoichiometry of complex formation (15 -17). In fact, others have previously employed sequences based on transcription factor leucine zippers to direct complexes of antibody segments (18, 19). We chose idealized coiled-coil motifs due to their inherently greater stability, thus eliminating the need for additional covalent linking.

II.   Coiled-Coil Designs

The design of our hydrophobic heptad repeats employed Leu residues at both the first and fourth (**a** and **d** positions) of the heptad. Leucines at these positions favor the formation of trimeric complexes (16, 20). Therefore we employed seven repeats of the heptad **LEALEKK** fused to the IL-2Rβ subunit ectodomain in a prototype complex to test the feasibility of this approach (9). Displayed in a helical wheel representation (Fig.1, left), the **e** and **g** positions of the heptad are available for favorable electrostatic interactions within a homomeric complex.

For the preparation of heteromeric complexes, we altered the sequences at the **e** and **g** positions (**LEALKEK** for IL-Rα and **LKALEKE** for IL-2Rβ) in order to disfavor homomeric association based on presumed electrostatic interactions (Fig. 1, center and right). These sequences were fused to their respective IL-2R ectodomains to generate a heteromeric complex (10). Using these recognition sequences, three IL-2R complexes were constructed (Fig.2).

III.   Protein Expression and Purification

To express large quantities of properly folded and glycosylated fusion proteins, we utilized baculovirus mediated insect cell expression (21). This method has been employed for expression of other hematopoietin receptor ectodomains (22) and several versions of baculovirus expression vectors are

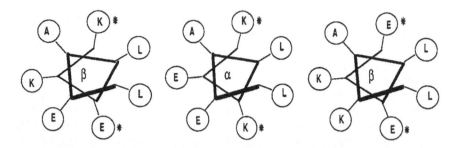

Figure 1. Helical wheel representations of the coiled-coil heptad repeat designs employed, (left) for the IL-2Rβcc homotrimer, (center) for the IL-2Rαcc homotrimer and IL-2Rαβcc heterotrimer and (right) for the IL-2Rαβcc heterotrimer. A single heptad is depicted with positions **a - g** viewed along the helical axis from the top. * indicates residues that may participate in interhelical electrostatic interactions.

commercially available. cDNA's encoding the desired fusion proteins were inserted into pBlueBac II (Invitrogen) and recombinant baculovirus were conveniently identified by color selection. High titer virus was produced in *Sf9* insect cells, while protein expression was carried out by infection of Trichoplusia ni (*High-Five™*) cells in protein free media (10). The preparation of homomeric complexes was achieved by infection of a single virus while simultaneous co-infection by viruses harboring the IL-2Rα and IL-2Rβ coiled-coil fusion proteins generated a heteromeric complex.

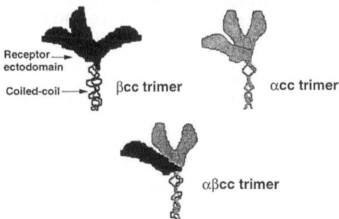

Figure 2. Schematic representations of the three coiled-coil IL-2 receptor ectodomain complexes examined in this study.

Protein purification was carried out by immunoaffinity chromatography over IL-2R subunit specific affinity columns. For homomeric complexes (IL-2Rαcc and IL-2Rβcc), this single step procedure yielded homogenous products. To isolate the heteromeric complex (IL-2Rαβcc), the fraction bound to the IL-2Rα specific immunoaffinity column was eluted and applied to the IL-2Rβ specific column. The fraction that bound to both columns represented the IL-2Rαβcc protein complex. Reverse phase HPLC confirmed the presence of both subunit fusion proteins (Fig. 3).

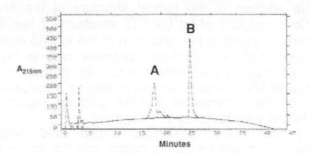

Figure 3. Reverse phase HPLC analysis of the IL-2Rαβcc complex demonstrating the presence of the α-subunit fusion protein (**A**) and the β-subunit fusion protein (**B**). Column: Dynamax 300A C18(4.6 x 250 mm) column (Rainin). Buffers: A, 0.1% TFA in water; B, 0.1% TFA in acetonitrile; gradient, 25-70 % solvent B over 45 min; flow rate,1 ml/min; absorbance was monitored at 215 nm.

IV. Protein Characterization

In addition to reverse phase HPLC, the complexes were analyzed for homogeneity by SDS-PAGE. The proteins migrated at the expected molecular weights corresponding to the IL-2R ectodomains fused to seven coiled-coil heptad repeats (9, 10). The IL-2Rαβcc complex provided bands corresponding to both fusion proteins. Although confirming the homogeneity of the preparations, these analyses did not reveal the stoichiometry of the complexes. Non-denaturing gels as well as gel filtration experiments were also not definitive in this regard due to the expected non-globular nature of the complexes. If fully formed, the coiled-coil segment would extend as a rod-like structure approximately 75 Å in length. To confirm the size of the complexes analytical ultracentrifugation was performed. Equilibrium sedimentation analysis carried out at three to four different protein concentrations and several rotor speeds revealed molecular weights in the 105 - 120 kDa range for all of the complexes (9, 10).   These values were consistent with the trimeric stoichiometry expected from the heptad designs and were further supported by sedimentation velocity experiments.   Sedimentation coefficients of approximately 4.5s were obtained in addition to axial ratios (assuming a hydrated prolate ellipsoid) of 15 suggesting an elongated structure.

     Since the secondary structure of the receptor ectodomains is primarily beta sheet (1, 23), circular dichroism proved to be useful in revealing the addition of the helical structure of the coiled-coil segment.   Deconvolution of far UV CD spectrum of the IL-2Rβcc complex when compared to the spectrum of the IL-2Rβ ectodomain itself suggested that the coiled-coil segment was at least 75% formed (9).   Attempts to determine the stability of the complex by monitoring the far UV CD spectrum while adding denaturants were complicated by the observation that the globular regions of the complex unfolded prior to coiled-coil segment.

     Finally, automated N-terminal amino acid sequencing confirmed that the complex subunits began with the amino terminal residues predicted from the cDNA sequences and were not unexpectedly truncated. Furthermore, sequencing of the IL-2Rαβcc protein predicted a subunit ratio of two IL-2Rα to one IL-2Rβ subunits (10).

V.   Ligand Binding

The ability of the complexes to interact with ligand is a critical factor for the success of this approach. To determine whether the addition of the coiled-coil complex perturbed ligand binding, we compared the ability of IL-2Rβcc complex and the β-subunit ectodomain to compete the binding of ^{125}I-IL-2 to cell surface receptors in a solution binding assay. In this assay, the amount of labeled ligand specifically bound to cell surface receptors from a constant ^{125}I-IL-2 concentration (500 pM) is determined as a function of increasing concentrations of competitor (either receptor complexes, ectodomains or cold IL-2 as a control). From this data, a dissociation constant for the soluble receptor may be calculated (24). The K_d for the IL-2Rβcc complex was determined to be 100 nM in this assay (9). Since this complex contains three β-subunit ectodomains, the apparent K_d for each subunit was ≈ 300 nM, a value that corresponds closely to that determined for the individual β-ectodomain (24). Thus, ligand recognition was not perturbed by the process of coiled-coil complexation.

A second aspect of ligand binding is the extent of cooperativity that heteromeric complexes would exhibit. The α/β "pseudo high affinity" cell surface receptor displays a high degree of cooperativity in ligand capture. The affinity of this site (150 - 600 pM) is ~ 100 fold and ~ 1000 fold higher than the dissociation constants of the α- and β-subunits, respectively. To examine this, we compared the ability of the IL-2Rαβcc complex with a mixture of uncomplexed IL-2R α- and β-ectodomains to compete ligand binding to cell surface receptors (Fig. 4).

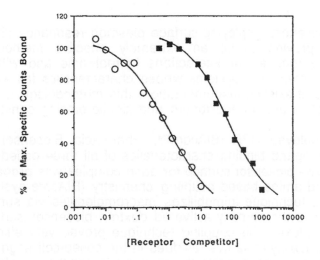

[Receptor Competitor]

Figure 4. Competitive displacement of ^{125}I-IL-2 on MT-1 cells (10). % Of maximum specific radioactivity bound is plotted versus concentration (nM) of soluble receptor for (O) the αβcc complex and an equimolar mixture of α- and β-ectodomains (■). Each point represents the mean of triplicate determinations. The assay was performed at 37° C under conditions where the depletion of free ^{125}I-IL-2 is negligible.

As can be seen from this data, the heteromeric complex is ~ 100 fold more potent in competing cell surface receptor binding than the simple mixture. This observation suggests that ligand binding by the complex is highly cooperative. Quantitative determination of the effective K_d for the IL-2Rαβcc complex in repeated assays provided values in the 300 - 500 pM range (10). The apparent dissociation constant for the α- and β-ectodomain mixture was ~ 40 nM. This value is similar to the K_d s obtained for the IL-2Rα ectodomain and the IL-2Rαcc complex (10). These data, then, not only confirm the cooperative nature of ligand binding by the heteromeric complex, but also suggest that there is no tendency of the individual α- and β-ectodomains to associate at the concentrations examined either in the presence or absence of ligand.

VI.   Surface Plasmon Resonance

The use of biosenors employing surface plasmon resonance (SPR) detection is proving to be an extremely effective method of examining macromolecular interactions in real time and without labeling.   For interacting systems whose characteristics fall within the limitations of existing instrumentation, this methodology allows the determination of both equilibrium and kinetic binding constants (25).

We employed SPR (BIAcore™, Pharmacia Biosensor) to determine the ligand binding characteristics of all three coiled-coil complexes. The biosensor surface for each complex was prepared using standard amine based coupling chemistry (BIAcore systems manual). This technique immobilizes macromolecules via surface amino groups to the carboxy-activated dextran biosensor surface. For these complexes, this coupling technique proved very efficient due to the availability of 42 Lys residues in the coiled-coil segment. Attachment to the surface in this region also oriented the complex in a manner the minimized steric interference with ligand binding.

By employing multiple surface densities of each complex, we were able to examine the binding behavior of wild type IL-2 in addition to two analogs possessing receptor subunit specific mutations (α-subunit, **T41P**; β-subunit, **D20K**). Typical SPR sensorgrams for these ligands are shown in Fig. 5.

Analysis of the binding data over an extended concentration range for these ligands on surfaces of each complex by nonlinear least squares curve fitting (BIAevaluation 2.0 software, Pharmacia) provided estimates of the kinetic and equilibrium binding parameters. Table I lists the dissociation constants obtained from these measurements.

The values obtained for the dissociation constants from Scatchard analysis of the equilibrium bound values compared favorably with the values obtained from the ratio of the kinetic binding constants (k_{off}/k_{on}). These constants were also similar to those obtained in previous solution and cell surface receptor binding assays, where available (26). Of note, these data confirm that the **D20K** mutation eliminates binding to the β-subunit (Fig. 5B) but has no influence on interaction with the α-subunit, while **T41P** is primarily α-subunit directed (Fig. 5A).

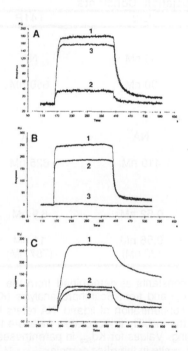

Figure 5. BIAcore™ SPR sensorgrams (relative response in RU after background subtraction vs. time in sec.) of **IL-2 (1)** and analogs **T41P (2)** and **D20K (3)** on biosensor surfaces prepared from: **A**, the IL-2Rαcc complex (ligand concs. = 200nM); **B**, the IL-2Rβcc complex (ligand concs. = 1000 nM) and **C**, IL-2Rαβcc complex (ligand concs. = 50nM). All reagents were dialyzed into standard HBS buffer. Biosensor surfaces were regenerated with 10 mM HCl. The flow rate was 8 μl/min.

Values for the kinetic binding constants were obtained for most of the analogs (26). In some cases, however, the rate constants were too rapid to be measured by the instrument. Of note were the kinetic constants calculated for binding of IL-2 to the IL-2R$\alpha\beta$cc surface (k_{on}= 3.9 X 10^6 $M^{-1}s^{-1}$, k_{off} = 2.0 X 10^{-3} s^{-1}). These values were surprisingly similar to those reported (27) for the cell surface "pseudo high affinity" site (k_{on}= 1.7 X 10^6 $M^{-1}s^{-1}$, k_{off} = 1.1 X 10^{-3} s^{-1}).

Table I. SPR Dissociation Constants[a]

	IL-2	T41P	D20K
IL-2Rαcc			
Kd_{kin}	33 nM	NA	34 nM
Kd_{Sca}	30 nM	650 nM	40 nM
IL-2Rβcc			
Kd_{kin}	NA	NA	
Kd_{Sca}	410 nM	625 nM	NA
IL-2R$\alpha\beta$cc			
Kd_{kin}	0.52 nM	6.9 nM	38 nM
Kd_{Sca}	0.56 nM (30 nM)	11 nM (787 nM)	108 nM

[a] Dissociation constants as determined from the ratio of the kinetic rate constants (Kd_{kin} = k_{off} /k_{on}) and from Scatchard analysis (Kd_{Sca}) of the equilibrium bound values for each concentration of ligand. NA refers to values that could not be determined due to the inability to accurately measure the kinetic constants or lack of detectable binding. Values for Kd_{Sca} in parentheses are those determined for the extra low affinity α site in the $\alpha\beta$cc complex.

VII. Conclusion

This study has demonstrated that it is possible to direct the assembly of cytokine receptor subunits in solution by employing coiled-coil molecular recognition. Furthermore, heteromeric complexes may be prepared in which the subunits function cooperatively to bind ligand. The affinity of the IL-2Rαβ site thus formed was significantly greater than either of the individual sites. The free energies of binding to each separate subunit were not additive in the formation of the IL-2Rαβ site, however, as the observed dissociation constant (K_d = 3-5 X 10^{-10} M) was much less than the product of the binding constants for each subunit ($K_d \approx 10^{-17}$ M). This is also the case for the cell surface receptors and reflects, at least in part, non-additivity in the entropy contributions to the binding energy form each site (28).

The use of idealized coiled-coil designs rather that naturally occurring sequences resulted in subnanomolar complex stabilities. It is unlikely, however, that use of electrostatic interactions to direct heteromeric assembly had a dominant influence in the formation of these complexes, since stable homomeric assemblies (e.g. IL-2Rαcc) were also isolated using "heteromeric" designs. It now appears that the nature of the residues at the **a** and **d** positions, not only control stoichiometry but may control heterogeneity and helix orientation as well (29,30). The techniques of analytical ultracentrifugation and surface plasmon resonance proved critical in quantitatively characterizing these complexes. Using these approaches and the present knowledge of coiled-coil molecular recognition, it should be possible to prepare soluble receptor complexes of defined stoichiometries and subunit composition.

Acknowledgments

This work was supported by grants form the Hitchcock Foundation, the National Institues Health (T.L.C., AI34331) (B. G., GM35556), the American Cancer Society (T.L.C., FRA-385), The National Science Foundation (T.M.L., DIR-9002027). Partial support from the Veterans Administration, the Norris Cotton Cancer Center, the United States Department of Energy and Chiron Corp. is gratefully acknowledged.

References

1.   A. M. De Vos, M. Ultsch and A. A. Kossiakoff, *Science* **255**, 306-312 (1992).

2.   B. J. Bormann and D. M. Engelman, *Annu. Rev. Biophys. Biomol. Struct.*
       21, 223-242 (1992).
3.   B. H. Nelson, J. D. Lord and P. D. Greenberg, *Nature* **369**, 333-336
       (1994).
4.   Y. Nakamura, S. M. Russell, S. A. Mess, M. Friedmann, M. Erdos, C.
       Francois, C. Jacques, S. Adelstein and W. Leonard, *Nature* **369**, 330-331
       (1994).
5.   D. Cosman, *Cytokine* **5**, 95-106 (1993).
6.   K. A. Smith, *Science* **240**, 1169-1176 (1988).
7.   T. Taniguchi and Y. Minami, *Cell* **73**, 5-8 (1993).
8.   B. E. Landgraf, B. Goldstein, D. P. Williams, J. R. Murphy, T. R. Sana, K. A.
       Smith and T. L. Ciardelli, *J. Biol. Chem.* **267**, 18511-18519 (1992).
9.   Z. Wu, S. F. Eaton, T. M. Laue, K. W. Johnson, T. R. Sana and T. L.
       Ciardelli, *Protein Eng.* **7**, 1137-1144 (1994).
10.   Z. Wu, K. W. Johnson, B. Goldstein, Y. Choi, S. F. Eaton, T. M. Laue and
       T. L. Ciardelli, *J. Biol. Chem.* **270**, 16039-16044 (1995).
11.   R. S. Hodges, J. Sodek, L. B. Smillie and L. Jurasek, *Cold Spring Harbor
       Symp. Quant. Biol.* **37**, 299-310 (1972).
12.   R. S. Hodges, A. K. Saund, P. C. S. Chang, S. A. St. Pierre and R. E. Reid,
       J. Biol. Chem. **256**, 1214-1224 (1981).
13.   B.-Y. Zhu, N. E. Zhou, P. D. Semchuck, C. M. Kay and R. S. Hodges, *Int. J.
       Prot. Pep. Res.* **40**, 171-192 (1992).
14.   T. Alber, *Curr. Opin. Gen. Dev.* **2**, 205-210 (1992).
15.   E. O'Shea, R. Rutkowski and P. S. Kim, *Cell* **68**, 699-708 (1992).
16.   E. O'Shea, K. J. Lumb and P. S. Kim, *Curr. Biol.* **3**, 658-667 (1993).
17.   B.-Y. Zhu, N. E. Zhou, C. M. Kay and R. S. Hodges, *Prot. Sci.* **2**, 383-
       394 (1993).
18.   S. A. Kostelny, M. S. Cole and J. Y. Tso, *J. Immunol.* **148**, 1547-1553
       (1992).
19.   P. Pack and A. Plückthun, *Biochemistry* **31**, 1579-1584 (1992).
20.   B.-Y. Zhu, N. E. Zhou, C. M. Kay and R. S. Hodges, *Prot. Sci.* **2**, 383-394
       (1993).
21.   M. D. Summers and G. E. Smith, *Texas Agricultural Experiment Station
       Bulletin No.1555.College Station,Texas.* (1987).
22.   Y. Ota, A. Asakura, Y. Matsuura, H. Kondo, A. Hitoshio, A. Iwane, T.
       Tanaka, M. Kikuchi and M. Ikehara, *Gene* **106**, 159-164 (1991).
23.   J. F. Bazan, *Immun. Today* **11**, 350-354 (1990).
24.   T. Sana, Z. Wu, K. Smith and T. Ciardelli, *Biochemistry* **33**, 5838-5845
       (1994).
25.   D. J. O'Shannessy, M. Brigham-Burke, K. K. Soneson, P. Hensley and I.
       Brooks, *Anal. Biochem.* **212**, 457-468 (1993).
26.   Z. Wu, K. W. Johnson, Y. Choi and T. L. Ciardelli, *J. Biol. Chem.* **270**,
       16044-16051 (1995).
27.   M. Matsuoka, T. Takeshita, N. Ishii, M. Nakamura, T. Ohkubo and K.
       Sugamura, *Eur. J. Immunol.* **23**, 2472-2476 (1993).
28.   W.P. Jencks, *Proc. Natl. Acad. of Sci, USA* **78**, 4046 - 4050 (1981).
29.   O.D. Monera, N.E. Zhou, C.M. Kay and R.S. Hodges, *J. Biol. Chem.* **268**
       19218-19227 (1995).
30.   K.J. Lumb and P.S. Kim, *Biochemistry* **34**, 8642-8648 (1995).

Use of a Peptide Library to Characterize Differential Peptide Binding Specificities of Bacterial and Mammalian Hsp70

K.P. Williams, D.M. Evans, S. Rosenberg and S. Jindal

PerSeptive Biosystems Inc., Framingham, MA

I. Introduction

Screening of libraries of peptides and small molecules for development of drugs has become a major focus in the pharmaceutical industry (1). Peptide-based libraries are generally   prepared by synthetic means (either free in solution, ref. 2 or immobilized on solid beads or pins, ref. 3) or by display of peptides on the surface of filamentous phage (4). Current methods of screening these libraries are labor-intensive, slow and often lead to many hits with low affinity and poor specificity. We are developing automated chromatographic techniques for high throughput screening of soluble libraries in which selection is based upon high affinity of the ligand for a specific target molecule, which depends upon a number of factors, including wash volumes and target density.

Heat shock proteins (hsps), a family of abundant, ubiquitous and highly conserved proteins, are essential for normal cell growth, but their concentration in the cell further increases under stress conditions, such as heat shock, nutrient deprivation and infections (5). Hsp70 family members are among the best characterized hsps and have been shown to play essential functions in the cell, such as protein folding, assembly, translocation and dissolution of insoluble protein aggregates (6). Hsp70, similar to other hsps, is highly conserved: mammalian hsp70 and its bacterial counterpart, Dna K,   share approximately 50% sequence identity (7), which may also suggest that these proteins are performing similar functions. Due to their functions, these proteins have also been termed polypeptide binding proteins or molecular chaperones (8). Both hsp70 and Dna K have been demonstrated to bind to certain linear

TECHNIQUES IN PROTEIN CHEMISTRY VII

57

peptides (9,10), but it is unclear whether these proteins bind to a similar set of peptides, both in terms of length as well as sequence.

In this report, we have screened a peptide library containing a random mixture of peptides of different lengths and sequences and have selected peptides by virtue of their affinity for mammalian hsp70 or its bacterial counterpart, Dna K. The results show that although mammalian and bacterial hsp70 are highly conserved proteins, they differ in their specificity for binding peptides. Such a screening approach should be useful for obtaining ligands that differentiate between closely related targets.

II. Materials and Methods

Target proteins

The gene encoding the stress-inducible member of human heat shock protein, hsp70, was expressed in *E. coli* as described (11). Recombinant hsp70 (rhsp70) was purified using a combination of ATP agarose affinity chromatography and ion exchange chromatography, and was demonstrated to be functional by its ability to bind to certain known peptides and to ATP (11). Purified DnaK was obtained from StressGen Biotechnologies Corp. (Victoria, Canada).

Preparation of target columns

Hsp70 and DnaK were biotinylated with NHS-(long chain)-biotin (Vector Labs.) and were immobilized to streptavidin-POROS (PerSeptive Biosystems Inc.). A "control" column consisting of biotin immobilized to streptavidin-POROS was prepared in a similar manner.

Automated multi-column chromatographic system

Chromatographic-based peptide library screening procedures were carried out on a BioCAD workstation or on an Integral™ micro-analytical workstation (PerSeptive Biosystems Inc.). The BioCAD is equipped with three computer controlled six port biocompatible valves and the Integral™ is equipped with an autosampler and three computer controlled ten port biocompatible valves.

Screening using immobilized target proteins

A BioCAD workstation was configured in two-column mode, with the first column plumbed in-line to a reversed-phase column. The first column consists of either immobilized human hsp70 or immobilized Dna K, and the second column was a POROS R2 reversed-phase (RP) column (PerSeptive Biosystems, Framingham, MA). A peptide library consisting of a random mixture of various lengths and sequences was injected onto the first column, and after washing, the peptides bound to the target proteins were eluted, using either 12 mM HCl or 3 mM ATP, directly onto the RP column. The RP column was later developed with a linear acetonitrile gradient (0-80% acetonitrile in 0.1% TFA).

Screening using targets free in solution

Hsp 70 or Dna K, 50 mg each, were incubated with 100 μg of peptide library, at 4°C for 16 hours. The target-bound peptides were separated from unbound peptides using a size exclusion column (TSK 2000, TosoHass, Japan). The SEC was plumbed in tandem mode to a RP column on a BioCAD workstation. The target peak was identified and the target with any bound peptides was transferred directly onto the RP column, which was developed as above.

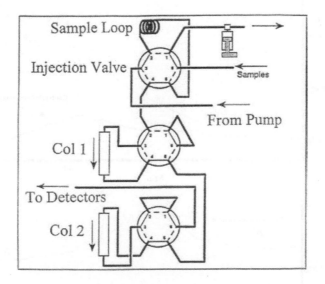

Fig. 1. Configuration of two-column method on a BioCAD workstation.

III. Results

Screening using immobilized target

A BioCAD workstation was configured in a two column configuration (Fig. 1). Hsp70 when immobilzed on POROS retained its ability to bind to a peptide, previously identified to bind to this protein (11). The peptide library was injected onto the hsp70 column, and after washing, the hsp70-bound peptides were either acid-eluted or eluted with ATP (ATP binding to hsp70 has been shown to promote release of polypeptides, ref. 12) and captured directly on the in-line RP column. Employing identical conditions, screening was also performed using a "control column." Overlay analysis of the RP chromatograms obtained using the hsp70 target column and the "control column" indicated that with acid elution (Fig 2A) there are at least two major peptide peaks unique to the chromatogram obtained when using hsp70 column, suggesting that these peaks consisted of peptides that bind specifically to hsp70. When peptides were eluted with ATP instead of 12 mM HCl, there were two hsp70-specific peptide peaks observed (Fig 2B). The hsp70-speciifc peptide peaks were collected and subjected to Edman sequencing. Preliminary results demonstrated that sequences of hsp70-specific peptides obtained using acid elution or ATP elution were different (data not shown).

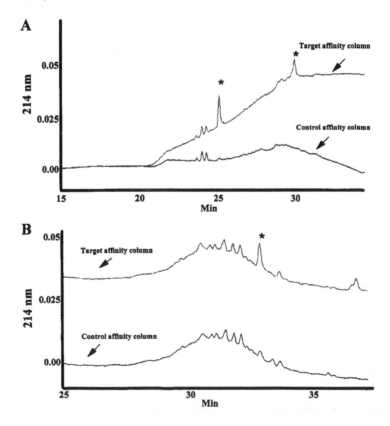

Fig. 2. RP chromatogram of peptides eluted from hsp70 column or control column by 12 mM Hcl (A) or 3 mM ATP (B).

   Because a direct correlation between the number of wash volumes and dissociation rates of target-bound ligands had been earlier demonstrated (13), we next investigated the effect of increasing wash volumes on the recovery of hsp70-bound peptides. For this purpose, the peptide library was injected onto the hsp70 column, and the column was washed using 5, 10, 20 or 40 column volumes (cvs) of washing buffer (25 mM HEPES, pH 7.4, 2mM magnesium acetate). The peptides remaining bound to the hsp70 were recovered by acid elution onto the RP column. Increasing the wash volume from 5 to 40 cv resulted in decreased recovery of the main binding peak (Fig 3, **), however, the peak is still present at 40 cv wash, suggesting that it has affinity in the low μM range (13). A second peak (*) present at 5 cv wash is absent in later washes suggesting that this binder has a much weaker affinity for hsp70.

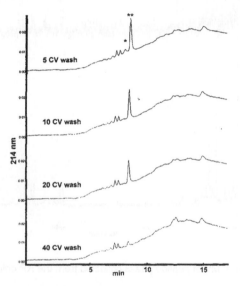

Fig. 3 Effect of wash volumes (CVs) on recovery of peptides bound to hsp70 after elution with 12 mM HCl. The eluted peptides were capture on the RP column for resolution.

Screening using targets in solution

Further investigation of peptide binding specificities of hsp70 and Dna K was performed by screening target proteins free in solution. Hsp70 or Dna K were incubated with the peptide library, and the protein and bound peptides were separated from unbound peptides using size exclusion chromatography. The protein peak was transferred directly to the RP column in-line. The presence of 0.1% TFA results in the release of peptides from the target, which are resolved from the protein by acetonitrile gradient elution. The peptide peaks of interest were

analyzed by MALDI-TOF MS using α-cyanohydroxycinnamic acid as
the matrix (Fig. 4). MALDI-TOF MS suggested that hsp70 and Dna K
bind to different sets of peptides, though a few are in common (Table 1).
In addition, Dna K seems to bind to peptides ranging in mass from 525
to 2109 Da, whereas hsp70 binds predominantly to longer peptides
(1493 to 1748 Da).

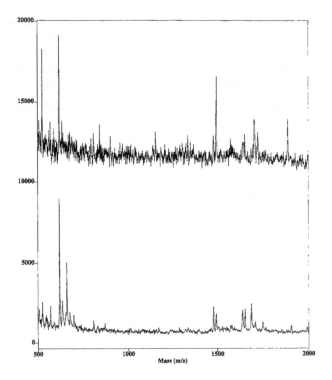

Fig. 4. MALDI-TOF MS of peptide peaks collected from the RP column.

IV. Discussion

Although there is extensive research implicating hsp70 proteins in many
diverse cellular functions, such as protein folding, assembly/disassembly
of oligomeric complexes, translocation of polypeptides across cell
membranes and protein synthesis, degradation and secretion (6), the
mechanism of their action is still poorly understood. Hsp70 proteins
have been suggested to have two domains (12): the N terminal domain
that binds ADP and ATP, and the C-terminal domain that binds short
segments of extended polypeptides. Hsp70 has been demonstrated to
associate with polypeptide substrates in the presence of ADP, whereas
ATP binding promotes release of polypeptides. The peptide binding
specificities of certain hsp70 members, such as BiP (endoplasmic
reticulum member of hsp70) and a constitutive member, hsc 70, have

been earlier investigated employing synthetic peptide libraries and phage display libraries (9,10,14,15). In both cases, the libraries, though diverse, contained peptides of only one length. Moreover, it has also been observed that hsp70 members may have more than one peptide binding site, and the nature of the peptides selected depends upon whether release of peptides was achieved by treatment with acid or by ATP during the screening process.

Stress-inducible members of the hsp70 family (designated as hsp70) and its bacterial counterpart, Dna K, exhibit 50 % identity in amino acid sequence, based upon which it has been suggested that they may play similar roles in cells. It is not clear whether these proteins differ in their peptide binding specificity, both in terms of length and sequence. To characterize the peptide binding specificities of these proteins, we prepared a library of peptides with random sequences and varying lengths. By using the chromatographic approaches, with either target immobilized to column or free in solution, we demonstrated that hsp 70 and Dna K can differ in their binding of peptides. Based upon the mode of screening described herein, it seems that hsp70 binds predominantly to longer peptides (1493 to 1748 Da) wheras Dna K binds peptides of a much wider mass range (525-2109 Da).

Table I. m/z values of peptide peaks selected by hsp70 and Dna K

Hsp70	Dna K
478	
	525
	618
	840
	901
	1154
	1339
	1480
1493	1497
	1574
1637	
1651	1652
1685	
	1705
	1722
1748	
	1886
	2109

The chromatographic approach of screening libraries of biomolecules described in this report has several advantages. 1) The technique is rapid and can be applied for high throughput screening. 2) Based upon wash volumes and target density, one can select ligands with predetermined affinity while screening a library. 3) The technique can be applied for selection of ligands that differentiate between closely related targets.

References

1. Ecker, D.J. and Crooke, S.T. (1995) *Bio/Technology* **13**, 351.
2. Houghten, R.A., Pinilla, C., Blondelle, S.E., Appel, J.R., Dooley, C.T. and Cuervo, J.H. (1991). *Nature* **354**, 84
3. Lam, K.S., Salmon, S.E. , Hersh, E.M., Hurby, V.J. , Kazmierski, W.M. and Knapp, R.L. (1991). *Nature* **354,** 82
4. Scott, J.K. and Craig, L. (1994). *Curr. Biol.* **5**, 40
5. Morimoto, R. I., Tissieres, A. and Georgopoulos, C. (1994). "The Biology of Heat Shock Proteins and Molecular Chaperones". Cold Spring Harbor Laboratory Press, New York.
6. Frydman, J. and Hartl, F.-U. (1994). *In* "The Biology of Heat Shock Proteins and Molecular Chaperones" (R.I Morimoto, A. Tissieres, and C. Georgopoulous, eds.) . p. 251. Cold Spring Harbor Laboratory Press, New York.
7. Hunt, C., and Morimoto, R. I. (1985). *Proc. Natl. Acad. Sci. USA* **82**, 6455.
8. Gething, M.-J. and Sambrook, J. (1992). *Nature* **355**, 33.
9. Flynn, G.C., Pohl, J., Folocco, M.T. and Rothman, J.E. (1991). *Nature* **353**,726.
10. Gragerov, A., Zeng, Li, Zhao, X., Burkholder, W. and Gottesman, M.E. (1994). *J. Mol. Biol.* **235**, 848.
11. Jindal, S., Murray, P., Rosenberg, S., Young, R.A. and Williams, K.P. (1995). *Bio/Technology*, In press.
12. Glick, B.S. (1995). *Cell* **80**, 11.
13. Evans, D.M., Williams, K.P., Parsons, G. and Jindal, S. (1995). *Anal. Biochem.* **229**, 42.
14. Hightower, L.E., Sadis, S.E., and Tanaka, I.M. (1994) *In* "The Biology of Heat Shock Proteins and Molecular Chaperones" (R.I Morimoto, A. Tissieres, and C. Georgopoulous, eds.), p. 179, Cold spring Harbor Press, New York.
15. Blond-Elguindi, S., Cwirla, S.E., Dower, S.E., Lipshutz, R.J., Sprang, S.R., Sambrook, J.F., and Gething, M.J. H. (1993). *Cell* **75**, 717.

The Design of Self-Assembling Peptide Complexes Using Conformationally Defined Libraries

Enrique Pérez-Payá[1], Richard A. Houghten and Sylvie E. Blondelle
Torrey Pines Institute for Molecular Studies
San Diego, CA 92121

I. Introduction

In order to generate diverse planar or topographical landscapes, we have developed a novel approach that involves soluble synthetic combinatorial libraries [SCLs, (1-3)]. The generation of molecular diversities based on defined structural motifs is expected to broaden the use of SCLs for those applications that require the occurrence of a well-defined secondary and/or tertiary structure. For instance, the *de novo* design of artificial receptors and catalysts requires the generation of protein-like molecules having the general structural and functional properties found in natural receptors or enzymes. We are particularly interested in the design of basic self-associating peptides that are able to adopt a highly α-helical conformation in physiological buffered conditions. In our initial study, conformationally defined SCLs were designed and constructed based upon an 18-residue peptide composed solely of leucine and lysine residues. In this report, the structural aspects and the catalytic activity of these libraries on the decarboxylation of oxaloacetate are described.

II. Materials and Methods

A. Preparation of Conformationally Defined SCLs

The conformationally defined SCLs were prepared by simultaneous multiple peptide synthesis using t-Boc chemistry as described elsewhere (4). The

[1]Permanent address: Departament de Bioquimica i Biologia Molecular, Universitat de Valencia, E-46100 Burjassot, Valencia, Spain

mixture positions were obtained using a mixture of 19 L-amino acids (cysteine was omitted) based on a predefined chemical ratio (2,5) at each coupling step. Final cleavage and deprotection steps were carried out using a "low-high" hydrogen fluoride procedure (6,7) and a 120-vessel cleavage apparatus. Individual peptides were purified by preparative reversed phase-high performance liquid chromatography (RP-HPLC) using a DeltaPrep 3000 RP-HPLC (Millipore, Waters Division, San Francisco, CA). Analytical RP-HPLC and laser desorption time-of-flight mass spectroscopy (Kompact Maldi-Tof mass spectrometer - Kratos, Ramsey, NJ) were used to determine the purity and identity of the individual peptides.

B. Circular Dichroism Measurements

All measurements were carried out on a Jasco J-720 circular dichroism spectropolarimeter (CD - Eaton, MD) in conjunction with a Neslab RTE 110 waterbath and temperature controller (Dublin, CA). CD spectra were the average of a series of three to seven scans made at 0.2nm intervals. Ellipticity is reported as the mean residue ellipticity [θ] (deg cm^2 dmol^{-1}). Peptide concentrations were determined by UV spectrophotometry at 276nm using $\epsilon = 1420M^{-1}cm^{-1}$ for tyrosine (8) and $\epsilon = 5570M^{-1}cm^{-1}$ for tryptophan (9).

C. Decarboxylation of Oxaloacetate

The kinetic parameters for decarboxylation of oxaloacetate (Sigma, St Louis, MO) with the different peptide mixtures were determined in 5mM MOPS/NaOH buffer 200mM NaCl, pH 7.0 at a peptide or peptide mixture concentration of 0.2mM, and oxaloacetate concentrations varying from 1.9 up to 11.4mM. The specific activity as well as the initial rates were measured by following the loss in absorbance at 280nm arising from the enol of oxaloacetate as described by Johnsson et al. (10).

III. Results and Discussion

The conformationally defined SCLs were based upon an 18-residue peptide composed solely of leucine and lysine residues ([YLK] - Table 1), which was found in earlier studies to adopt a random conformation in aqueous buffer and a monomeric amphipathic α-helical conformation in trifluoroethanol [TFE - (11)]. The [YLK] peptide was found to self-assemble in a tetrameric aggregate at high ionic strength and high peptide concentration. While many different approaches can be envisioned that encompass combinatorial libraries based on such structural motifs, the initial SCL constructed here was in a positional scanning format [PS-SCL (2)] utilizing the hydrophilic face of the α-helix. The hydrophobic face was kept constant to allow hydrophobic packing to occur and to promote self-association (Table 1). A PS-SCL consists of individual positional SCLs in which a given position is defined with a single amino acid, while the remaining positions are composed of mixtures of amino acids. Each single positional SCL differs from the others only in the location of the defined position. The conformationally defined PS-SCL prepared in this study was built with one position (the "O" position) defined with one of the 19 L-amino acids (cysteine was initially omitted to prevent the formation of dimers or polymers), and three mixture positions (the "X" positions), which were made up of a close to equimolar mixture (2,5) of the same 19 L-amino acids. These mixture positions replaced lysine-6, -9, -13 and -16 (Table I). Each peptide mixture is thus composed of 6,859 (19^3) individual peptides, for a total of 130,321 (19 x 19^3) peptides within the library. Each positional SCL addresses a separate position of the sequence, and, when used in concert, the data derived from the four individual positional SCLs yield information about the relative importance of the amino acids at each position. This approach allows information to be obtained in a single screening assay, which is then used to synthesize a series of individual peptides directly from the screening results.

To evaluate the structural behavior of this library, the apparent helical contents were measured for each peptide mixture using CD spectroscopy at 25°C in MOPS buffer at neutral pH. The CD spectra of all of the peptide mixtures displayed α-helical features with varying helical content. This apparent helical content for a given defined amino acid was found to vary according to its location within the sequence (i.e., which library the corresponding peptide mixture belongs to), and the degree of variability was found to be amino acid-dependent. For instance, a leucine at position 9, (located in the middle of the hydrophilic face of the helix) led to a higher helical content than when it was located close to the N-terminus (at position 6 Figure 1A). In contrast, a proline at position 6 resulted in a higher helical content than when proline was located in the middle of the helix or close to the C-terminus (Figure 1B). Other amino acids such as glutamine could be located at almost any position without significant variations in the resulting helical content (Figure 1C). In addition, the helical content was found to be concentration-dependent, with a tetramer aggregate form found to best fit the concentration dependence curve.

The ability of the peptide mixtures to adopt an α-helical conformation in buffer made them good candidates as structure-related catalysts. For instance, Johnsson et al. (10) recently reported the catalytic activity of a 14-mer peptide, called oxaldie-1, for the oxaloacetate decarboxylation reaction. Such activity was postulated to be closely related with the ability of this peptide to adopt a partial α-helical conformation in aqueous solution which lowers the pK_a of the

Table I. Positional Scanning Conformationally Defined Peptide SCLs

Sequence	# mixtures	"O"substitution	"X"substitution
Parent sequence [YLK] YKLLKKLLKKLKKLLKKL-NH$_2$			
YKLLKOLLXKLKXLLXKL-NH$_2$	19	K-6	K-9, K-13, K-16
YKLLKXLLOKLKXLLXKL-NH$_2$	19	K-9	K-6, K-13, K-16
YKLLKXLLXKLKOLLXKL-NH$_2$	19	K-13	K-6, K-9, K-16
YKLLKXLLXKLKXLLOKL-NH$_2$	19	K-16	K-6, K-9, K-13

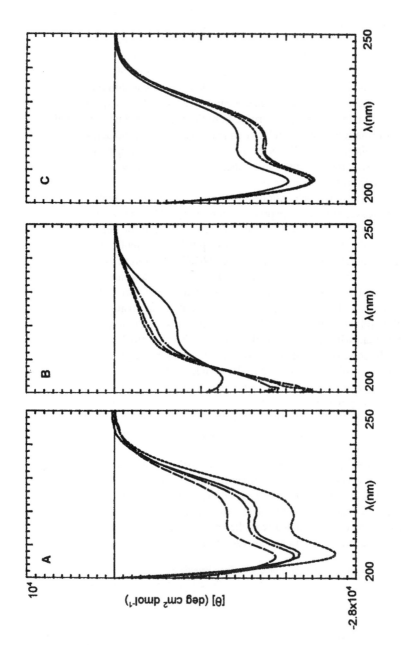

Figure 1. CD spectra of conformationally defined PS-SCLs. The CD spectra were recorded at 25°C in 5mM MOPS/NaOH (pH 7.0), 200mM NaCl. The peptide mixture concentrations were 150μM. The different CD spectra are shown for three different amino acids at the defined "O" position: **(A)** leucine, **(B)** proline, and **(C)** glutamine, when the defined amino acid is at the 6th position (solid line), the 9th position (dotted line), the 13th position (dashed line), and the 16th position (center line).

reactive N-terminal amino group, and to electrostatically attract the substrate through lysine residues. Therefore, this catalytic system is a good model for validation studies of the use of a conformationally defined SCL based upon an α-helical motif for the identification of new catalytic compounds. Thus, each of the 76 separate peptide mixtures of the conformationally defined PS-SCL described above was assayed for its ability to enhance the decarboxylation of oxaloacetate. The peptide oxaldie-1 was synthesized separately to serve as a reference in the assay used in this study. The screening of each peptide mixture, as well as [YLK] and oxaldie-1, showed that all of the peptide mixtures, except when proline was at the defined position, have equal or higher specific activity than the original [YLK] and oxaldie-1 (Figure 2). Good correlations were observed between the specific activity and the ability of the mixtures to adopt an α-helical conformation in buffer (Figures 1 and 2). These results support the reported role of this conformation in the catalytic activity of such peptides.

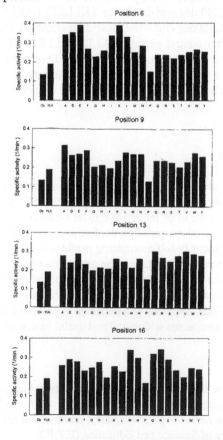

Figure 2. Catalytic activity of a conformationally defined PS-SCL for the oxaloacetate decarboxylation reaction. The specific activity values were determined as described in the text. Each panel represents a single positional SCL (i.e., 20 peptide mixtures having a common defined residue) and each bar represents the specific activity for each peptide mixture, with the x axis corresponding to the defined residue.

The identification of active sequences from the screening of an SCL in a positional scanning format consists of selecting the most active peptide mixtures from each individual library and combining their defined amino acids in all possible combinations to generate individual peptides (2). Thus, 144 individual peptides were prepared, representing all possible combinations of the amino acids defining the most active peptide mixtures identified in the screening assay, and were tested for catalytic activity in a manner similar to the libraries. Among the most active sequences, the peptide YKLLKELLAKLKWLLRKL-NH$_2$ was found to catalyze the decarboxylation of oxaloacetate following Michaelis-Menten saturation kinetics with k_{cat}/K_m= 2.66M^{-1}s^{-1} (versus k_{cat}/K_m= 0.72M^{-1}s^{-1} found for oxaldie-1 in our assay conditions). This activity represents a three to four orders of magnitude enhancement as compared to simple amine catalysts. The activity was also directly related to the peptide's ability to fold into an α-helical conformation in buffer, as shown by CD spectroscopy analysis. The molar ellipticity value at 222nm was $[\theta]$=-28,300 deg cm^2 dmol^{-1} for YKLLKELLAKLKWLLRKL-NH$_2$ versus $[\theta]_{222}$=-4,460deg cm^2 dmol^{-1} found for oxaldie-1 in our assay conditions. A negative control peptide, YKLLKLLLPKLKPLLPKL-NH$_2$, was also tested in order to confirm the poor activity observed for those peptide mixtures with proline as the defining amino acid. As anticipated, this peptide showed very low activity (k_{cat}/K_m<0.003M^{-1}s^{-1}), and a random structure in buffer.

IV. Conclusions

We have demonstrated the ability to generate self-assembling peptide complexes through the use of conformationally defined SCLs. Using a scaffold sequence known to be inducible into an α-helical conformation in the presence of TFE, but to adopt a random conformation in aqueous buffer, we were able to rapidly identify self-associating peptides that folded into a highly α-helical conformation in mild buffer. The generation of conformationally defined diversities can successfully lead to the identification of peptides or protein-like complexes having specific binding activities, and ultimately catalytic activities. This approach is anticipated to provide a means for developing peptide complexes with novel and useful biological properties.

Acknowledgments

The authors wish to thank Ema Takahashi for her technical assistance, and Eileen Silva for editing this manuscript. This work was funded by Houghten Pharmaceuticals, Inc., San Diego, Calif. and a postdoctoral fellowship from the Spanish Ministry of Science and Education (E.P.P.).

References

1. Houghten, R.A., Pinilla, C., Blondelle, S.E., Appel, J.R., Dooley, C.T. and Cuervo, J.H. (1991) *Nature* **354**, 84-86.
2. Pinilla, C., Appel, J.R., Blanc, P. and Houghten, R.A. (1992) *Biotechniques* **13**, 901-905.
3. Pinilla, C., Appel, J., Blondelle, S.E., Dooley, C.T., Dörner, B., Eichler, J., Ostresh, J.M. and Houghten, R.A. (1995) *Biopolymers (Peptide Science)* **37**, 221-240.
4. Houghten, R.A. (1985) *Proc. Natl. Acad. Sci. USA* **82**, 5131-5135.
5. Ostresh, J.M., Winkle, J.H., Hamashin, V.T. and Houghten, R.A. (1994) *Biopolymers* **34**, 1681-1689.
6. Tam, J.P., Heath, W.F. and Merrifield, R.B. (1983) *J. Am. Chem. Soc.* **105**, 6442-6455
7. Houghten, R.A., Bray, M.K., De Graw, S.T. and Kirby, C.J. (1986) *Int. J. Pept. Protein Res.* **27**, 673-678.
8. Marqusee, S., Robbins, V.H. and Baldwin, R.L. (1989) *Proc. Natl. Acad. Sci. USA* **86**, 5286-5290.
9. Quay, S.C., Condie, C.C. and Minton, K.W. (1985) *Biochim. Biophys. Acta* **831**, 22-29
10. Johnsson, K., Allemann, R.K., Widmer, H. and Benner, S.A. (1993) *Nature* **365**, 530-532.
11. Blondelle, S.E., Ostresh, J.M., Houghten, R.A. and Pérez-Payá, E. (1995) *Biophys. J.* **68**, 351-359.

"Fishing Out" Ligand Binding Proteins from Protein Mixtures by a Two-Dimensional Gel Electrophoresis System Utilizing Native PAGE Followed by SDS-PAGE

Kurt Hollfelder,
Feng Wang, and
Yu-Ching E. Pan

Department of Biotechnology
Hoffmann-La Roche Inc.
Nutley, New Jersey

I.   Introduction

The ability to identify a ligand binding protein in a protein mixture without going through tedious purification steps would be desirable to biochemists. To achieve this goal, a preliminary study was carried out   with the use of a new two-dimensional (2-D) gel electrophoresis protocol involving native polyacrylamide gel electrophoresis (PAGE) and sodium dodecyl sulfate (SDS) PAGE. In the first dimension, native PAGE is used for initial separation of the protein/ligand complex from other non-complexed proteins (1).   In the second dimension, SDS-PAGE is used to further resolve the proteins based on their molecular weights (2). The proteins involved in complex formation can be unambiguously identified by comparing the 2-D profiles of the following: (i) the protein mixture with ligand added, (ii) the protein mixture without ligand added, and (iii) the non-complexed ligand control. Any new protein bands detected in (i), that are not present in (ii) or (iii), must be involved in a complex. The bands of interest can be transferred from the second dimension gel onto PVDF membrane for further analyses. The use of sequential native and SDS-PAGE has been previously reported for protein concentration and purification (3). The application described here exploits the non-denaturing environment of native PAGE to maintain protein/ligand complexes. A ligand binding protein present in a mixture can be identified at the micro-scale level with minimal sample manipulation.

II.   Materials and Methods

A.   Protein and Reagents

The following recombinant proteins used in this study were provided by colleagues at the Roche Research Center: human interleukin-2 (IL-2), human IL-2 receptor α chain (IL-2Rα), murine interleukin-1α (IL-1α), murine IL-1 receptor (IL-1R) and human interferon-α (IFN-α)(4-6). Acrylamide, N,N'-methylene-bis-acrylamide, ammonium persulfate, riboflavin-5'-phosphate, N,N,N',N'-tetramethylethenediamine (TEMED), and Tris-Gly SDS 10x running buffer were purchased from Bio-Rad (Hercules, CA). Low Molecular Weight Range Rainbow Markers were purchased from Amersham (Arlington Heights, IL). Other chemicals and solvents were of the highest purity available.

B.   *Apparatus*

An X-cell apparatus (Novex, San Diego, CA) was employed for both native and SDS-PAGE equipped with a Bio-Rad Model 1000/500 power supply.

C.   *Preparation of Sample Mixtures*

Each study involved the side-by-side analysis of two mixtures containing known proteins. These mixtures differed only by the presence or absence of one known ligand. The proteins ranged in concentration from 0.5 to 3.0 µg/µl. Approximately 200 to 500 picomoles of each component was added sequentially to phosphate buffered saline (PBS) while maintaining a neutral pH, vortexed, and allowed to stand for at least 15 min. Samples were then adjusted to a final concentration of 10% sucrose with 0.05% Bromophenol Blue marker in a total volume of 15-30 µl, and immediately subjected to native PAGE.

D.   *Native PAGE*

Analyses were performed in a discontinuous buffer system using published procedures with some modifications (1,2,7). The compositions of the stacking and resolving gels are listed in Table 1. The gels were freshly prepared in Novex cassettes (8.5 cm x 7.5 cm x 1.0 mm) and acrylamide polymerization was allowed to proceed at room temperature. Both anode and cathode reservoirs contained 37 mM Tris-Gly, pH 8.9. The protein samples were electrophoresed toward the anode (+) at 25°C for 30 min at 60 V followed by 110 min at 120 V until the dye front reached the bottom of the gel. The portion of the gel not used for the second dimension was placed in a fix solution (12% Trichloroacetic Acid) for 1/2 hr before staining with 0.25% Coomassie Blue G-250.

Table 1. Native gel composition

	Resolving gel	Stacking gel
Acrylamide	7.5%	3%
N,N'-Methylenebisacrylamide	0.19%	0.08%
Tris-HCl, pH 8.5	240 mM	
Tris-phosphate, pH 6.9		40 mM
Ammonium persulfate	0.0075%	0.03%
Riboflavin	0.00025%	0.001%
TEMED	0.2%	0.1%

E.   *SDS-PAGE*

A large sample well (6.5 cm x 1 cm x 1.0 mm), along with a regular sample well for the molecular weight standard, was prepared. Each entire native gel lane selected for 2-D was excised without staining, wetted with a few drops of water, and carefully loaded into the large sample well. Handling of these native gel slices must be rapid (within 5 min) to prevent diffusion of bands and loss of resolution. A solution of 0.1% agarose (containing 0.1M Tris, 0.1% SDS, and tracking dye) is then poured around the native gel slice and all air bubbles are removed. SDS-PAGE was run (90 V for 1/2 hr followed by 160 V for 2 hrs) until the dye front reached the bottom of the gel. For this 2-D analysis the SDS-PAGE maps of both sample mixtures are run simultaneously, preferably in the same apparatus. After electrophoresis, the gel was stained by Coomassie blue R-250.

III.   Results and Discussion

To demonstrate the usefulness of this 2-D approach, the results of two model systems are presented here. The first system involves the use of IL-2Rα as the "bait" to detect IL-2 in a mixture (designated as M_1) containing IL-2, IFN-α, and IL-1α. The 2-D maps of M_1 in the presence (Fig. 1A) and absence (Fig. 1B) of IL-2Rα are shown. At the top of each 2-D pattern is the stained equivalent of the native PAGE lane

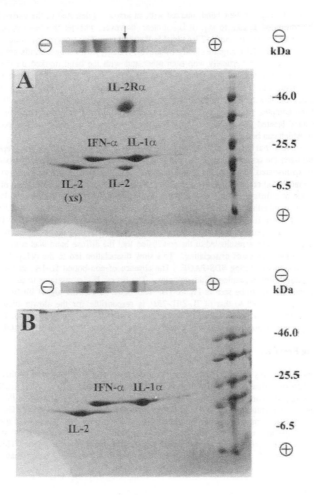

Figure 1. Two-dimensional maps of protein mixture M_1 in the presence (A) and absence (B) of IL-2Rα. At the top of the 2-D maps are the stained equivalents of native gel slices used to generate the 2-D map. (xs, excess)

used to generate the 2-D map. A new band, marked with an arrow, is detected in the native PAGE lane of Fig. 1A upon the addition of IL-2Rα to M_1. It is not clear, however, whether this band is comprised of the IL-2/IL-2Rα complex or the non-bound IL-2Rα added to the mixture. This uncertainty is due to the fact that IL-2/IL-2Rα complex and IL-2Rα migrate closely in native PAGE (1). Two new bands are detected in the 2-D map (Fig. 1A) which align vertically with each other and with the band marked on the native PAGE lane. They are not present in the 2-D map of M_1 without IL-2Rα added (Fig. 1B). These two new bands match the molecular weight of IL-2 and IL-2Rα in the second dimension SDS-PAGE. These observations led to the conclusion that the new bands originated from an IL-2/IL-2Rα complex. The excess IL-2 not used in the formation of the complex (Fig. 1A) migrates to the same position as the non-complexed protein (Fig. 1B). The upper band labeled IL-2Rα (Fig. 1A) is located at a slightly different vertical position than the non-bound protein which can be seen in Fig. 2B.

It should be noted that dissociation of the complex during the second dimension depends upon the binding affinity between the components. A case in which the complex does not fully dissociate during the second dimension is illustrated in the second model system. In this system IL-1α was used to detect IL-1R in a mixture (designated M_2) containing IL-1R, IL-2Rα, and IFN-α (Fig. 2). A comparison of the 2-D maps revealed the presence of a diffuse band in Fig. 2A that is not in Fig. 2B. This diffuse band has a molecular weight higher than that of either IL-1R or IL-1α. The pattern has proven to be reproducible. This band appeared to be related to the IL-1α/IL-1R complex. The diffuse band was transferred to PVDF membrane, divided into top and bottom portions and sequenced. The top portion yielded the sequence of IL-1R and the bottom portion of IL-1α. These results led to the conclusion that the diffuse band was comprised of the IL-1α/IL-1R complex in the process of dissociating. This slow dissociation led to the delayed mobility of the IL-1α and IL-1R components during SDS-PAGE. The absence of non-bound IL-1R in Fig. 2A indicates that it was all used during the formation of the complex. The excess IL-1α migrates to the same position as the non-bound protein which can be seen in Figure 1B. It is believed that the higher binding affinity of IL-1α/IL-1R complex as compared to that of IL-2/IL-2Rα is responsible for the slower dissociation of the former during SDS-PAGE analysis. The SDS concentration surrounding the proteins, at the start of the second dimension gel run, is presumably low as a result of slow diffusion into the native PAGE lane.

IV.   Concluding Remarks

The two model systems shown in this communication demonstrate the potential of this 2-D approach for the identification of a ligand binding protein in a mixture. The 1-D native PAGE resolution may be adequate for identifying the complex forming species in the two protein mixtures used for this investigation. In the case of a more complicated protein mixture, however, the resolution of the native PAGE alone will not be sufficient to separate all components from the protein/ligand complex. This 2-D method greatly improves the resolution of all species, while establishing the binding ability for the components of interest. As mentioned previously, these proteins can be recovered onto PVDF for further analyses. This method does not require large amounts of sample (just enough for Coomassie blue staining) and may be capable of capturing a binding protein out of an unpurified protein mixture.

Acknowledgments

We thank F. Khan, P. Bailon, and S. Roy for providing purified proteins along with Hanspeter Michel and Doreen Ciolek for reviewing the manuscript and Kimberly Mitchell for her assistance in its preparation.

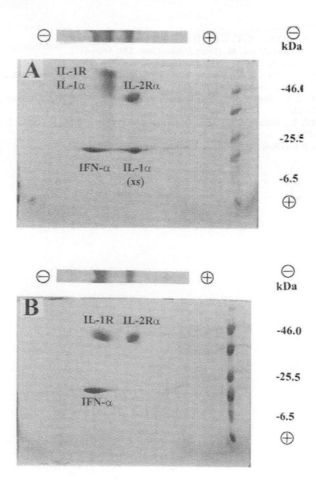

Figure 2. Two-dimensional maps of protein mixture M$_2$ in the presence (A) and absence (B) of IL-1α. At the top of the 2-D maps are the stained equivalents of native gel slices used to generate the 2-D map. (xs, excess)

References

(1)   Wang, F., and Pan, Y.-C. E (1991) *Anal. Biochem.* **198** 285-291.
(2)   Weber, K., and Osborn, M. (1969) *J. Biol. Chem.* **244**, 4406-4412.
(3)   Trudel, J. and Asselin, A. (1994) *Anal. Biochem.* **221**, 214-216.
(4)   Bailon, P., Weber, D. V., Keeney, R. F., Fredericks, J. E., Smith, C. Familletti, P. C. and Smart, J. E. (1987) *Bio/Technol.* **5**, 1195-1198.
(5)   Weber, D. V., Keeney, R. F., Familletti, P. C. and Bailon, P. (1988) *J. Chromatogr. Biomed. Appl.* **431**, 55-63.
(6)   Steophil, T., Hobbs, D. S., Kung, H.-F., Lai, C.-Y., and Pestka, S. (1981) *J. Biol. Chem.* **256**, 9750-9754.
(7)   Davis, B.J. (1964) *Ann. NY Acad. Sci.* **121**, 404-427.

Mass Spectrometric Analyses of the Activation Products of the Third Component of Complement

William T. Moore and John D. Lambris
The Protein Chemistry Laboratory, Department of Pathology and Laboratory Medicine
The Medical School, University of Pennsylvania, Philadelphia, PA 19104

I. Introduction

The complement system constitutes an immune defense mechanism that originated early in evolutionary history. The system is comprised of a collection of effector and control proteins that interact in a highly intricate and specific manner that leads to both proteolytic fragmentation and association of proteins that provide known effector functions of humoral immunity and inflammation in addition to other not yet defined biological functions. Of the 25 or so proteins that comprise the complement system , one protein in particular plays a central role, the third component identified known as C3, a 187 kD glycoprotein (for reviews, see Refs. 1 and 2). This protein is synthesized as a single chain and is posttranslationally cleaved into two chains, alpha (α) and beta (β), that are disulfide bonded through a single linkage between Cys^{816} of the α and Cys^{559} of the β. In connection with the immunobiological role of C3, the protein is remarkable in that the 187 kD form is relatively inactive and acquires a diverse multifuntionality expressed through specific ligand binding upon further specific fragmentation by proteolysis. C3 also contains a unique internal thioester bond that becomes involved with covalent "fixations" to other molecules often resulting in "surface bound" C3 - derived species.

Specific proteolytic events and subsequent conformational change in the resulting C3-derived fragments appear to be the mechanism for the tight regulatory control that is necessary for this system to function properly. Biophysical methods employing spectral and scattering techniques using various chemical probes have revealed in low-resolution various fragment specific conformational states in the fluid phase (1). Immunological probing with monoclonal antibodies has been employed to probe conformational states in both surface bound fragment and fluid phase situations and has identified areas in higher resolution if the epitope for the monoclonal antibody has been defined (1). Although much information is known at the protein level about C3 and its fragments (1), a fine detail map of the structure at the protein level using protein chemistry techniques is not yet available because of its large molecular size. The deduced amino acid sequence of human C3 has been obtained (3) and has only been partially verified with protein chemistry methods. We have begun to employ mass spectrometry to explore the structures comprising the functional life-cycle of C3 at the protein level. The goal is to fully characterize the C3 molecule and its fragments at the protein level cataloging any post-translational covalent modifications using mass spectrometric methods. Here we reveal the results of our initial attempts to characterize C3 and its activation fragments. The initial endeavor also indicates that the mass spectrometric approach is also yielding information about the profound conformational changes that exist in this family of protein derivatives.

II. Experimental

A. *Preparation of C3 and fragments*

C3 was isolated from EDTA-plasma using precipitation and chromatographic methods as previously described (4). The final step of purification employs ion exchange chromatography on Mono-Q HR10. C3 fragments were generated "experimentally" (in vitro) using protease treatment: a mild trypsin treatment for the generation of C3b form C3 and an elastase treatment of purified C3 to generate C3c . C3dg was derived as described previously (5) from chromatographic isolation employing ion exchange and gel filtration from whole plasma that had been incubated for 6 days at 37°C in taking steps to eliminate plasmin activity. The final purification step employed reverse phase HPLC using a PE-ABI Model 130 microbore system and a Brownlee Aquapore BU-300 C4 (2 X 220 mm) column. Acetonitrile gradients in approximately 0.1% TFA were used for elution of components. Typically fragments are initially evaluated by SDS-PAGE and immunoblotting and bioassayed by methods described previously (6).

B. *Disulfide Reductions*

Reductions were performed with tributylphosphine (TBP) (7). In connection with MALDI-MS analysis proteins in the 10 to 50 μM range were diluted 1/10 in 0.05% ammonium bicarbonate. TBP dissolved in MeOH was added to either a final 200 μM or 2 mM concentration. Molar ratios of TBP to disulfide ranged from 3 to 300. When the molar ratio was increased several-fold using 2 mM TBP concentrations, no further changes in reduction patterns were observed. Reductions were allowed to proceed for at least 30 min at room temperature unless specified otherwise. When full and complete reduction was desired for isolation of the separate chains heat (100 °C for 3 min.) was applied.

C. *Mass Spectrometry*

Matrix-assisted laser desorption ionization mass spectrometry (MALDI-MS)

(8) was performed on a Fisons Instruments (Beverly, MA) VG TofSpec time-of-flight mass spectrometer (0.6 m flight tube) outfitted with a N_2 (337 nm) laser. Mass calibrations were established using BSA. Data were analyzed using Fisons Instruments OPUS software. Two matrices were used. For C3 and the C3b fragments 4 to 20 µM solutions of polypeptide in PBS were diluted in 1/5 in sinapic acid so that each target site contained sample in the 1 to 5 pmole range. For reduction studies and for higher sensitivity when needed 2-(4-hydroxyphenylazo)benzoic acid (HABA) (9) was the matrix of choice. Acceleration voltage was set to 29000 volts and the multichannel plate detector was set to 1900 volts.

Electrospray ionization mass spectrometry (ESI-MS) (10) was performed on a Fisons Instruments (Beverly, MA) VG Quattro triple quadrupole mass spectrometer outfitted with the manufacturer's electrospray source. Samples were diluted into 50% acetonitrile 1% formate in water and injected as a 10 µL bolus into a stream of matrix consisting of 50% acetonitrile, 1 or 5 mM in triethylamine (TEA) to permit a family of lower charge state distributions that is essential for analysis of large molecular weight proteins. Matrix flow was established at 12 µL/min using a microliter syringe pump. In most cases 10 scans were acquired and averaged. The instrument was calibrated using equine myoglobin and data was processed using Fisons Instruments MassLynx software.

III. Results

A. Mass Spectrometric Analysis of C3 and the activation fragments

The C3 molecule and various preparations of activation fragments were analyzed by MALDI-MS to survey whether the observed masses would approximate theoretical masses. Typical spectra obtained for the larger fragments C3 and C3b are shown in Fig. 1. The upper panel of Fig.1 is the mass spectral result showing the ion signals for +1, +2 and +3 charge states of C3. These charge states suggest an average mass of 186400 for C3 which is 0.5% lower than the theoretical mass of 187443 based on amino acid sequence deduced from cDNA sequence and average carbohydrate composition (11). The carbohydrate micro-heterogeneity has the potential to be an issue in the mass analysis of these proteins even though the percent carbohydrate is low 1.7 %. The observed mass is also within the experimental error arising from external calibration using BSA. Given this moderate level of uncertainty, we conclude cautiously that this preparation of C3 approximates the theoretical construct. It is of interest to note that an ion signal having a mass assignment of 8980 was observed in the low mass end of the mass spectrum as shown in the upper panel of Fig. 1. This ion signal could represent a minor amount of C3a that has been

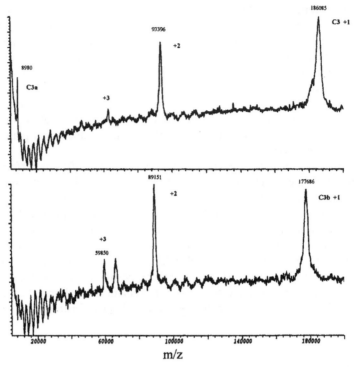

Figure 1. MALDI-MS analysis of C3 (upper panel) and C3b (lower panel)

generated by proteolysis of this intact C3 preparation. It should be noted that minor amounts of smaller molecular weight peptides may present as exaggerated signals in the MALDI-MS analyses of proteins and that the signal heights of the peptide signals do not quantitatively reflect the actual content. The lower panel of Fig. 1 shows the mass spectrum of C3b, the C3 activation fragment that arises from the proteolytic removal of the 9 kD C3a peptide from the N-terminal region of the alpha subunit of C3. The average mass for C3b deduced from the three charge states shown in the lower Panel of Fig. 1 is 178500 a value that agrees very well with that for the theoretical construct (178510.5). Again these results superficially suggest that the theoretical model of C3 developed over the years may be close to reality.

B. Disulfide reduction studies

To check the theoretical construct more stringently we chose to mass analyze the individual chains of C3 (2 interchain disulfide bonded polypeptides) and the subsequent activation products: C3c (3 interchain disulfide bonded polypeptides) and the single chain C3dg product. The rationale was that mass analysis of this collection of fragments would permit a more fine-detailed evaluation of most of the molecule. We chose tributylphosphine (TBP) because it is a potent reducing agent working at nearly equimolar concentrations of disulfide. Micro-molar concentrations of TBP reagent were thought to minimally interfere with the MALDI-MS ionization process. It was feared that the mmolar concentrations of other reducing reagents such as DTT and β–mercaptoethanol would possibly interfere with MALDI-MS analysis and require a greater dilution of sample and subsequent loss of sensitivity. Before proceeding to C3 and C3 fragment analysis, the TBP reagent was used on purified rabbit IgG and bovine insulin and was found to be completely effective in reducing the disulfides that held the polypeptide chains and permitted MALDI-MS visualization of the respective chains (data not shown). The MALDI-MS analysis of the TBP reduction of C3 is shown in Fig. 2. The upper panel of Fig. 2 is the nonreduced control and the lower panel is the result for reduced material. TBP treatment resulted in highly modified spectrum that yielded ion signals that are consistent with the visualization of the two disulfide α and β bonded chains of C3. The predominant ion signals observed in the reduced material (Fig. 2, lower panel) were for the multi-charge state species for the β chain. An average mass of 72544 for the β chain was derived from the multicharge state family shown in the lower panel of Fig. 2. This value is within 0.1% of the theoretically derived mass (72630.8). The other major but relatively weaker ion signals in the spectrum are derived from the α chain, the +1 of which was mass assigned at 113959 (theoretical = 114813) and some

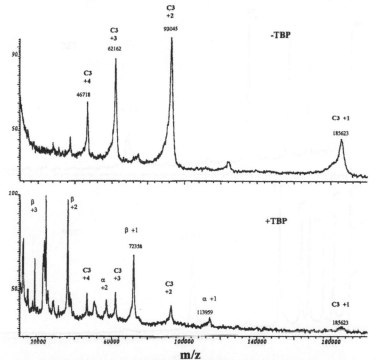

Figure 2. Reduction of C3 with Tributylphosphine (TBP). MALDI-MS analysis of nonreduced control (upper panel) and MALDI-MS analysis of TBP-reduced C3 (Lower panel).

unreduced C3 mass assigned at 186233. The poor peak shapes for both of these ions led to appropriate but less accurate mass assignments for these fragments.

To achieve an additional refinement in the mass analysis, an elastase-generated C3c preparation that had been characterized by N-terminal sequencing and subjected to an identical TBP treatment was examined with the expectation that the three chains would be observed. C3c is essentially a C3b molecule with the middle segment of the alpha chain (C3dg) removed. The three chains are held together by two disulfides. An N-terminal region of the alpha chain designated α_1 having a theoretical mass of 24533 is bonded to the C-terminal region of the alpha chain designated α_2 having a theoretical mass of 36305. The beta chain having a theoretical mass of 72630 is bonded to the α_1 chain through the disulfide formed between αCys^{816} and βCys^{559} major ion. The MALDI-MS results for the reduction of C3c are shown in Fig. 3. The non reduced control is presented in the upper panel of Fig. 3 and the spectrum for the TBP-treated material is presented in the lower panel. An unexpected and paradoxical result was achieved. No beta chain signal was obtained and signals having masses at 96 kD and 36 kD were observed indicating that the reduction was partial in this case. This result was the same for C3c treated with TBP at a low concentration of TBP (data not shown) as used in the C3 experiment (Fig. 2) and for a reduction experiment where nearly a hundred fold of the amount of TBP was employed as shown in the lower panel of Fig. 3. Only two ion signal

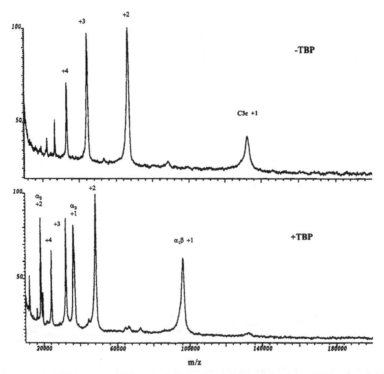

Figure 3. Reduction of C3c with Tributylphosphine (TBP). MALDI-MS analysis of nonreduced control (upper panel) and MALDI-MS analysis of TBP-reduced C3c (lower panel).

multicharge state families are evident; one for the 96188 representing the disulfide bonded $\alpha_1\beta$ species and the other 36430 species representing the α_2 species. The conclusion of these results are schematically portrayed in Fig. 4. There is a differential reactivity of the disulfide formed between αCys^{816} and βCys^{559} depending whether it exists in the C3 molecule or the C3c molecule. The suggestion is that there are extreme differences in conformation that exist in C3 and C3c that affect reagent access to this region of the molecule surrounding the αCys^{816}- βCys^{559} disulfide. We propose that coupling reduction with a mass spectrometric evaluation may be an additional approach to probing conformational states in this family of molecules as well as validating theoretically proposed structure at the protein level.

In order to test the resistance of this disulfide resistance further and to rule out a MALDI-MS related artifact, we analyzed C3c by ESI-MS. Although C3c is at the upper mass limit of this mass spectrometric method, the resulting reduced fragments would fall within. The ESI-MS analysis of the C3c reduction is shown in Fig. 5. Similar conclusions can be drawn from the ESI-MS data shown in Fig. 5. The ESI-MS pattern for untreated C3 as shown in the upper panel of Fig. 5 yielded a low quality multicharge state ion pattern that appears to arise from an ion having a mass of 138123, a value around that for C3c but much higher than expected and observed in MALDI-MS analysis (Fig. 3, upper panel). We believe this inaccuracy is related to the molecular weight of the protein being at or over the upper limit for this ionization technique. Treatment of this C3c preparation with TBP at room temperature resulted in a distinct change in the electrospray ion pattern as shown in the middle panel of Fig. 5. This family of charge states transformed or deconvoluted to a mass of 96606 a

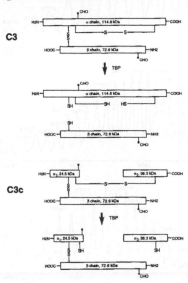

Figure 4. Schematic presentation of the differential reduction of αCys^{816}-βCys^{559} in C3 and C3c by TBP.

value consistent with that for the $\alpha_1\beta$ species observed in MALDI-MS (Fig. 2, lower panel). It is of interest to note that the 96.6 kD ion pattern is the dominant signal in the spectrum and that the 36kD signal MALDI-MS detectable (Fig. 2, lower panel) was not observed in the ESI-MS analysis (Fig. 5, middle panel). If the TBP reduction is carried out in the presence of heat (100°C for 3 min) and the protein is electrosprayed (Fig. 5, bottom panel) another distinct multicharge state ion pattern is observed that yields a mass of 72697 upon transformation or deconvolution of the multicharge state data. Again the observed mass is the dominant species and is in very good agreement with the theoretical mass of the β chain (72630). This result indicates that with the addition of heat and subsequent thermal denaturation of the protein exposes the αCys^{816}- βCys^{559} disulfide to reagent and complete reduction is achieved as has been indicated by SDS-PAGE analysis after reduction with β–mercaptoethanol in the presence of SDS and heat (5). It is of interest to note that the MALDI-MS analysis gave a more complete picture of the reduction results than the electrospray analysis. However the electrospray result, although incomplete, fortuitously yielded spectra for pertinent fragments that permitted interpretation of the reduction of C3c at the two different temperatures. These results also represent a case where complex mixtures of proteins are more readily assessed by MALDI-MS than ESI-MS.

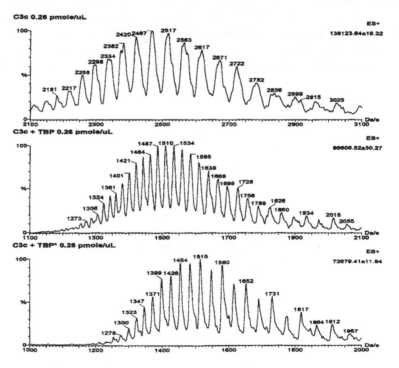

Figure 5. ESI-MS analysis of the TBP reduction of the C3c molecule. Nonreduced C3c control (upper panel); Room temperature TBP treatment of C3c (middle panel) and TBP treatment at 100°C/3 min (lower panel).

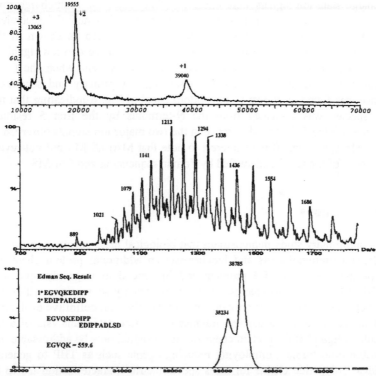

Figure 6. Mass spectrometric analysis of C3dg. MALDI-MS analysis of C3dg (upper panel); ESI-MS raw data for C3dg (middle panel) and deconvoluted or transformed raw data spectrum for C3dg (lower panel).

C. C3dg characterization

Our ultimate goal is to mass spectrometrically fine-map each fragment of C3 to survey and clarify the status of post translational modifications if any and to verify the theoretically proposed primary and secondary structure of the complete C3 molecule. In addition the physiological proteolytic processing sites need clarification as well as those created in the laboratory by enzyme treatment. Clarification of these are essential for structure-function relationship studies employing fragments. Along these lines we have characterized a "physiological" C3dg fragment that has been generated in plasma and isolated from plasma treated as described previously (5). The final product was obtained by fractionation over reverse phase column as described in the Experimental section. MALDI-MS and ESI-MS analyses of the RP-HPLC derived product are shown in Fig. 6. The upper panel of Fig. 6 indicates the MALDI-MS analysis of the C3dg preparation. The observed major ion signal mass assigns at 39000 in close agreement with the theoretical proposed mass of 38906. In the MALDI-MS analysis another minor signal was evident at 36000 (possibly C3d) that has not yet been identified or characterized. When this preparation was analyzed by ESI-MS as shown in the middle panel of Fig. 6, two sets of multicharge state ion signals were observed. The major ion multicharge state pattern deconvoluted or transformed to a mass of 38785 and the secondary multicharge state pattern deconvoluted to a mass of 38235 as shown in the bottom panel of Fig. 6. The mass difference between these two ion signals is 547 u. The basis to this mass difference became apparent when the C3dg preparation was subjected to Edman sequencing. A primary and secondary sequence was found as shown in the bottom panel of Fig. 6 indicating that the minor species was truncated from the N-terminus by the first 5 residues accounting for the 547 mass difference in the two major ion signals observed in the ESI-MS spectrum. It is of interest to note that MALDI-MS did not reveal the presence of the species identified by Edman sequencing and ESI-MS.

IV. Conclusions

This preliminary study demonstrates that biological mass spectrometry offers complement immunologists and biochemists an additional protein chemsitry level perspective on the C3 problem and in general to large glycoproteins. MALDI-MS and ESI-MS analyses in a complementary sense along with other protein chemistry techniques provide a comprehensive and definitive evaluation of C3 molecules and the derived fragments. A fine-detail analysis of a C3 molecule or ligand at the protein chemistry level should now be addressable in a reasonable time frame. Employing reducing agents such as TBP to generate individual polypetide chains for mass spectrometric characterization led to an interesting observation concerning one particular disulfide in the C3 molecule formed between αCys^{816} and βCys^{559}. The use of this reagent in conjunction with mass spectrometric techniques allows a probing of the conformational structure around this particular region in the molecule. The present data suggest that profound differences must exist around this disulfide in the C3 and C3c fragments.

Acknowledgments

The technical support from Yang Wang and James Nicoludis are greatly appreciated. This work was supported by grants awarded to JDL from the NSF(MCB93-19111) and the NIH(AI-30040). The Protein Chemistry Laboratory also receives support from core grants awarded to the University of Pennsylvania Cancer Center (NCI CA-16520), and the University of Pennsylvania Diabetes Research Center (DK-19525)

References

1. Lambris, J. D. Ed. (1990) The monograph - *The Third Component of Complement, Current Topics in Microbiology and Immunology* Vol. 153. Springer-Verlag, New York.

2. Müller-Eberhard, H.J. & Miescher, P.A. Eds.(1984) The monograph-*Complement*, Springer-Verlag, New York.
3. De Bruijn, M.H.L. and Fey, G.H. (1985) *PANS* 82, 708-712.
4. Lambris, J.D., Dobson, N.J., Ross, G.D., (1980) *J. Exp. Med.* 152,1625-1644.
5. Vik, D.P. and Fearon, D.T. (1985) *J. of Immunol.* 134, 2571-2579.
6. Becherer,J.D. and Lambris, J.D. (1988) *JBC* 263, 14586-14591.
7. Ruegg, U.T. and Rudinger, J. (1977) *Methods Enzymol.* 47, 111-116.
8. Hillenkamp, F., Karas, M., Beavis, R.C. and Chait, B.T. (1991) *Anal. Chem.* 63, 1193A-1203A.
9. Juhasz, P. Costello, C.E., and Biemann, K (1993) *JASMS* 4, 399-116.
10. Meng, C.K., Mann, M. and Fenn, J.B. (1988) *Zeitschrift fur Physik D(Z. Phys. D)* - Atoms, Molecules and Clusters 10, 361-368.
11. Hirani, S. Lambris, J.D. and Müller-Eberhard, H.J. (1986) *Biochem J.* 233, 613-616.

Site-specific Reversible Conjugation: A Novel Concept in Peptide-DNA Interactions

Dusan Stanojevic and Gregory L. Verdine

Department of Chemistry
Harvard University
Cambridge, Massachusetts 02138

I. Introduction

Sequence-specific binding of polypeptides to DNA results in a large reduction in the translational entropy of the system. Transcription factors must compensate for this unfavorable energetic cost in order to bind DNA. In nature, the most common strategy for overcoming this problem employs multimerization of either DNA-binding motifs or protein subunits (1-3). We reasoned that the same effect could be achieved artificially by linking a peptide to DNA with a flexible tether. In this manner, the interaction of a tethered peptide with DNA should cause no major change in the translational entropy of the system.

Our experimental system is based on the yeast transcriptional activator GCN4. This leucine zipper homodimeric protein binds DNA sequence element 5'-ATGA(G/C)TCAT through the basic region containing ~20 amino acid residues (Figure 1A)(4-5). Peptides derived from the basic region of GCN4 can bind DNA sequence-specifically only after being dimerized by various methods (6-9). We have synthesized several wild-type and mutant GCN4-derived peptides and DNA duplexes (Figure 2). A disulfide tether was introduced into the DNA to link the C-terminus of peptides to an adenine residue (**A**) immediately 5' to the GCN4 half-site: 5'-AGTCAT (Figures 1B, 1C, 2A and 2B). Various combinations of peptides and DNA were linked together by the formation of a disulfide bond.

II. Materials and methods

A. Synthesis of disulfide-linked peptide-DNA conjugates

All modified DNA oligonucleotides were synthesized and purified as described previously (10-11). Peptides were synthesized on an Applied Biosystems Model 431A synthesizer using Fmoc amino acids and following standard procedures. The peptides were purified by reverse-phase high-performance liquid chromatography with a semi-preparative C18 column (Beckman) and a linear gradient of acetonitrile-water containing 0.1% trifluoroacetic acid. The expected molecular weights and homogeneity of the peptides were confirmed by fast atom bombardment and electrospray ionization mass spectrometry. The formation of a disulfide bond between peptides and DNA was accomplished by thiol-disulfide exchange reaction. The modified single-stranded oligonucleotides having the thiol group activated as the mixed 2-thioelthylamine disulfide form were mixed with peptides in a buffer containing 20 mM Tris (pH 7.5), 5 mM KCl, 2 mM EDTA, 2 mM MgCl2, 0.1% NP-40, 100 mM oligonucleotide and 0.5 mM

TECHNIQUES IN PROTEIN CHEMISTRY VII

peptide. After incubation at room temperature for 8 hours, the reaction mixture was resolved by high-performance liquid chromatography using a FAX anion-exchange column (Millipore). Under these conditions, the efficiency of conjugation ranged from 60% to 90%. The fractions corresponding to the disulfide-linked peptide-DNA complex were collected, desalted by ethanol precipitation and redissolved in water.

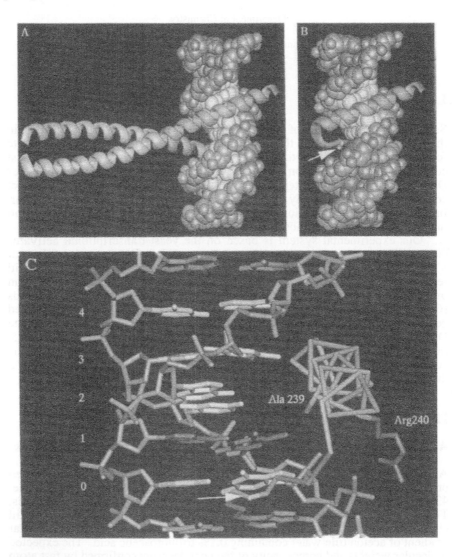

Figure 1. (A) Structure of the homodimeric GCN4 bZip bound to DNA, as determined by Ellenberger et al. (4). The bases shown in white constitute the GCN4 recognition sequence. (B and C) Model of the monomeric disulfide-crosslinked complex formed between a 24-mer GCN4 basic region peptide and DNA containing a consensus GCN4 half-site (white). Disulfide bond is marked by arrow. Residues Ala 239 and Arg 240 as well as the base pair at position 1 were subjected to mutational analysis. Numbers on the left (C) correspond to the bases that form the GCN4 consensus half-site (see Figure 2). Reprinted with the permission from *Nature Structural Biology*.

(A) GCN4 recognition element (GCRE):

non-consensus half-site

consensus half-site

```
         0 1 2 3 4
5' - ATGAGTCAT - 3'
3' - TACTCAGTA - 5'
         4 3 2 1
```

(B) Oligonucleotides:

```
              HS
                0 1 2 3 4
wt   5' - ... TGCAAGTCATCGG ... - 3'
     3' - ... ACGTTCAGTAGCC ... - 5'
```

```
              HS
                1
M3   5' - ... TGCAAGgCATCGG ... - 3'
     3' - ... ACGTTCcGTAGCC ... - 5'
```

(C) Peptides:

wild-type 24-mer (Gcn)

H_2N-DPAALKRARNTEAARRSRARKGGC-CO_2H
   226   239   243 246   SH

A239L mutant (Leu)

H_2N-DPAALKRARNTEALRRSRARKGGC-CO_2H
                      239   SH

R240P mutant (Pro)

H_2N-DPAALKRARNTEAAPRSRARKGGC-CO_2H
                       240   SH

wild-type 19-mer (19)

H_2N-KRARNTEAARRSRARKGGC-CO_2H
        231   246   SH

GCN4 basic region   linker

Figure 2. (A) DNA sequence of the 9 base-pair GCN4 recognition element (GCRE). The non-consensus half-site (on the left) is shaded, and the dot denotes the center of dyad pseudosymmetry. Numbering refers to the position of each base-pair. (B) Sequences of thiol-bearing DNA duplexes linked to peptides; wt: wild-type consensus GCN4 half-site; M3: mutant half-site having a T•A to G•C substitution at position 1; nc: non-consensus GCN4 half-site. Open and shaded boxes are as in (A), except the mutated base-pair in M3, which is unboxed and in lower-case lettering. A refers to the thiol-tethered adenine residue, N^6-(2-thioethyl)adenine. The flanking sequences present in the 39-mer oligonucleotides used in DNase I protection assays (top strand only) are 5'-AAGGTTAAACG......TATAGGTCGAGAAGT-3', and in the 20-mers used in reduction assays are 5'-CG...CGCGT-3'. (C) Sequences of wild-type and mutant peptides, with mutated residues enclosed in a shaded box. Numbering refers to wild-type GCN4 (4). The wild-type 24-mer peptide contains 21 residues derived from the GCN4 basic region as well as a Gly-Gly-Cys linker (Gcn, Figure 2C). The A239L (Leu) mutation in the peptide and an T•A→G•C mutation at position 1 (M3) have been reported to abrogate binding in the native GCN4 system (12). The R240P mutant replaces a phosphate contact residue with one that destabilizes the a-helix. The truncated 19-mer peptide lacks five N-terminal residues that do not contact DNA, but are essential for specific DNA binding by an artificially linked peptide dimer (7). Reprinted with the permission from *Nature Structural Biology*.

B. DNAse protection assays

The peptide-DNA complexes for DNase protection analysis were formed by mixing 0.1 pmol of the purified single-stranded DNA-peptide complex with the same amount of ^{32}P end-labeled complementary strand in incubation buffer containing 20 mM Tris (pH 7.5), 2 mM EDTA, 0.1% NP-40, 100 mM KCl and 2 mM MgCl$_2$ in a total volume of 50 ml. After a 20 min. incubation at the appropriate temperature the digestion was initiated by the addition of 50 ml of a 10 mM MgCl$_2$/5 mM CaCl$_2$ solution followed by the 2 ml of freshly diluted (1:1000) DNAse I stock (1 mg/ml, BRL). The reactions were incubated for 6 min. at 0°C, 3 min at 10°C and 1 min. at 20°C and stopped by the addition of 90 ml of 1% SDS, 20 mM EDTA, 200 mM KCl and 200 mg/ml glycogen (Boehringer Mannheim). The samples were extracted with phenol/chloroform (1:1), chloroform and then ethanol precipitated and electrophoresed on 20% sequencing (7.5 M urea) polyacrylamide gel.

C. Reduction assays

The peptide-DNA complexes for the reduction assays were prepared as described in the legend of Figure 3. 10 pmol of each complex was end-labeled with ^{32}P in a buffer containing 70 mM Tris-HCl (pH 7.5), 10 mM MgCl$_2$, 100 nM $\gamma^{32}P$-ATP (DuPont) and 0.1 unit of T4 polynucleotide kinase (New England Biolabs). After incubation for 15 min. at room temperature, the labeling reactions were extracted with phenol/chloroform (1:1), chloroform, and precipitated with ethanol 2 times. The labeled peptide-DNA complexes were redissolved in 100 ml of buffer (20 mM Tris (pH 7.5), 2 mM EDTA, 2 mM MgCl$_2$, 0.1% NP-40 and 100 mM KCl). A 4 ml aliquot of each complex, containing 10,000 counts per minute, was diluted into a 50 ml of the same buffer and the solutions were incubated at 10°C for 20 min. The reduction of the disulfide bond was started by the addition of the DTT or 2-mercaptoethanol (2-ME) to 50 mM. After 1.5 hour incubation at 10°C, glycerol was added to 5% along with a trace amount of bromphenol blue dye, and the reactions were loaded immediately on 20% native polyacrylamide gel.

III.   Results and Discussion

The specificity of the interaction between the tethered peptide and the adjacent GCN4 half-site was analyzed by DNase I protection assays (Figure 3). At 0°C, the wild-type 24-mer peptide (Gcn) protects the GCN4 half-site (Figure 3A). The protection remains visible even at elevated temperatures. The tethered Leu peptide (Figure 1C) gives consistently weaker protection, especially at temperatures above 0°C (Figure 3A, Leu). The tethered Pro peptide exhibits no significant protection at any temperature (Figure 3A, Pro). The tethered 19-mer peptide protects from DNase I attack at 0°C but not at higher temperatures (Figure 3A, lane labeled 19). None of the peptides protects a sequence that contains M3 mutation in the GCN4 half-site (Figure 3B). These results demonstrate that tethered peptides interact with DNA in a sequence-specific manner.

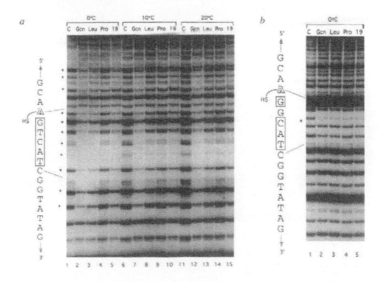

Figure 3. DNase I protection assays of tethered peptide-DNA complexes. (A) Complexes of various peptides with wt DNA, the sequence of which is shown on the left. Lanes 1, 6, 11: control lanes representing the DNase I digestion pattern of wt DNA bearing the tethered adenine residue but no peptide. Lane 2: digestion pattern of the wild-type complex (Gcn/wt) at 0°C; asterisks to the left of the control lane denote positions that are more than 50% protected in Gcn/wt at 0°C. Strong protection is also observed for the Gcn/wt complex at 10°C (lanes 7), but is noticeably weaker at 20°C (lane 12). The Leu/wt complex shows strong protection at 0°C (lane 3), only weak protection at 10°C (lane 8) and little or no protection at 20°C (lane 13). Lanes 4, 9 and 14: the Pro/wt complex exhibits no protection at any temperature. Lanes 5, 10 and 15: the 19-mer/wt complex reveals protection of wt GCN4 half-site only at 0°C. The arrow denotes a band that is particularly suitable for comparisons among various lanes. (B) Complexes of various peptides with the M3 mutant DNA, the sequence of which is shown on the left. None of the peptides shows significant binding at 0°C, as evidenced by the lack of a zone of protection from DNase I. A single band (marked by an asterisk) shows a uniformly decreased intensity in all of the peptide-DNA complexes relative to the control, and can therefore be ascribed to a non-specific effect caused by the presence of the tethered peptide. Reprinted with the permission from *Nature Structural Biology*.

Our model implies that the reduction potential of the disulfide bond linking a peptide to DNA is dependent on the non-covalent peptide-DNA interaction. The rationale for this novel phenomenon is described in Figure 4A. This concept was tested by the incubation of peptide-DNA complexes in the presence of disulfide reducing reagents. The products of the reaction were analyzed by native polyacrylamide gel electrophoresis (Figure 4B). These results indicate that the complex formed between the wild-type 24-mer peptide and wild-type half-site was consistently more resistant to reduction when compared with mutant complexes. The unexpected difference in gel mobility of various complexes could be ascribed to the average conformational state of each complex.

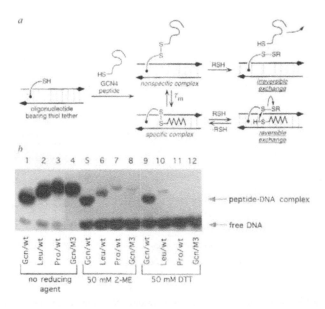

Figure 4. (A) A model illustrating the relationship between the strength of noncovalent peptide-DNA interaction and reducibility of the disulfide bond through which the two are linked. The nonspecific complex represents the state of the system in which there is no sequence-specific interaction between the peptide and DNA (the predominant state of Gcn/M3 or Pro/wt); the specific complex represents the state of the system in which the peptide and DNA make sequence-specific contacts with each other, and the peptide possesses an ordered secondary structure (the predominant state of Gcn/wt at 0°C); T_m corresponds to the temperature at which the nonspecific and specific complexes are present in equal amounts. Addition of a thiol-containing reagent (RSH) results in breakage of the disulfide bond. In the case of a specific complex (e. g. Gcn/wt), the peptide remains transiently bound to DNA, and can thus participate in the reverse reaction, resulting in reformation of the disulfide bond; In the case of a nonspecific complex, the peptide is instantaneously released into solution, and hence is essentially incapable of undergoing the reverse reaction. Consequently, peptides that establish stronger non-covalent interactions with DNA should have disulfide bonds that are less susceptible to reduction. (B) An experimental test of the proposed relationship between the strength of tethered peptide-DNA interactions and disulfide bond strength. Lanes 1-4 show intact peptide-DNA complexes migrating through native 20% polyacrylamide gel. Treatment with reducing agents 2-mercaptoethanol (2-ME) or dithiothreitol (DTT) results in a breakage of the disulfide bond and the appearance of the band corresponding to free DNA (lower bands, lanes 5-12). The amount of intact complex left after reduction represents a measure of the reducibility of the corresponding disulfide bond (upper bands, lanes 5-12). Of all four complexes the Gcn/wt is consistently most resistant toward reduction, followed by the Leu/wt. The notable differences in mobility may be due to differences in the average conformational state of each peptide-DNA complex. Reprinted with the permission from *Nature Structural Biology*.

Here we present an experimental system that is based on a novel concept of intramolecular DNA-peptide interactions. The entropic cost of peptide-DNA association is reduced by covalent conjugation, thus allowing exceptionally short peptides to bind DNA sequence-specifically. The specificity of peptide-DNA interaction is preserved even though 3-dimensional diffusion has been largely eliminated from the system. This is evident from the fact that the same alterations in nucleotide and amino acid sequence that abolish native GCN4-DNA interactions have a significant effect on the ability of the tethered peptide to protect the DNA from the digestion by DNAse I. This system substantially reduces the size of specific protein-DNA complexes and therefore has a potential to facilitate high-resolution structural analysis by X-ray and NMR. A disulfide linkage also provides a novel method to measure the strength of the peptide-DNA interaction through the reduction assay (Figure 4B). The exquisite selectivity of this assay holds promise for the development of new strategies for the isolation of novel sequence-specific DNA-binding molecules from various kinds of peptidic or non-peptidic combinatorial libraries.

Acknowledgments

We are grateful to T. Ellenberger and S. Harrison for sharing the data prior to publication and J. Green and N. Sinitskaya for peptide synthesis. D.S. is a Scholar of the American Foundation for AIDS Research. This work was supported by grants from the Chicago Community Trust (Searle Scholars Program), National Science Foundation (Presidential Young Investigator Award) and Ariad Pharmaceuticals.

References

1. Pabo, C. O. & Sauer, R. T. *A. Rev. Biochem.* **61**, 1053-1095 (1992).
2. Harrison, S. C. *Nature* **353**, 715-719 (1991).
3. Steitz, T. A. *Q. Rev. Biophys.* **23**, 205-280 (1990).
4. Ellenberger, T. E., Brandl, C. J., Struhl, K. & Harrison, S. C. *Cell* **71**, 1223-1237 (1992).
5. König, P. & Richmond, T. J. *J. Mol. Biol.* **233**, 139-154 (1993).
6. Talanian, R. V., McKnight, C. J. & Kim, P. S. *Science* **249**, 769-778 (1990).
7. Talanian, R. V., McKnight, C. J., Rutkowski, R. & Kim, P. S. *Biochemistry* **31**, 6871-6875 (1992).
8. Cuenoud, B. & Schepartz, A. *Science* **259**, 510-513 (1993).
9. Park, C., Campbell, J. L. & Goddard, W. A.,III *Proc. Natl. Acad. Sci. USA* **90**, 4892-4895 (1993).
10. Ferentz, A. E. & Verdine, G. L. *J. Am. Chem. Soc.* **113**, 4000-4002 (1991).
11. Ferentz, A. E., Keating, T. A. & Verdine, G. L. *J. Am. Chem. Soc.* **115**, 9006-9014 (1993).
12. Suckow, M., Von Wilcken-Bergmann, B. & Müller-Hill, B. *Nucl. Acids. Res.* **21**, 2081-2086 (1993).
13. Stanojevic, D. & Verdine, G. L. *Nature Structural Biology* **2**, 450-458 (1995).

NOTE: Figures and methods were reprinted with the permission from *Nature Structural Biology*. For more details about the system please see reference 13.

SECTION II

Interactions of Proteins with Ligands

Scintillation Proximity Assay to Measure the Binding of Ras-GTP to the Ras-Binding Domain of c-Raf-1

Julie E. Scheffler, Susan E. Kiefer, Kathleen Prinzo and Eva Bekesi

Department of Metabolic Diseases, Hoffmann-La Roche Inc.
Nutley, New Jersey 07110

I. Introduction

The protein product of the c-Raf-1 protooncogene is a 74 kDa kinase that specifically phosphorylates MEK (MAP kinase kinase) in response to growth factor stimulation of the Ras signal transduction pathway (1-2). The catalytic activity of Raf is associated with a 40 kDa domain that includes the C-terminus. The Ras-binding domain of Raf is a 78-residue fragment (RafRBD; Raf_{55-132}) that binds with high affinity to the effector domain of activated Ras-GTP (3). RafRBD constitutes a structural domain as well. The tertiary folding of the protein was recently shown by NMR methods to closely resemble ubiquitin, having a ubiquitin $\alpha\beta$ roll superfold structure (4-5). To fully integate the emerging structural information into a model describing the interaction of Ras-GTP with RafRBD, it is desirable to measure the contribution of individual amino acid residues to the binding. Previously reported assay methods for the interaction of Ras-GTP with RafRBD are either indirect or require a high affinity interaction. These include: (i) competitive inhibition of GAP-stimulated Ras GTPase activity by Raf (6); (ii) the inhibition of nucleotide dissociation from Ras by Raf (7); (iii) capture of the protein complex with an affinity resin directed toward one partner (8-9), and; (iv) an enzyme-linked immunosorbant assay (10). A scintillation proximity assay (SPA) is reported here that allows the direct measurement of both low and high affinity interactions between Ras and Raf.

II. Materials and Methods

Protein Constructs. During the development of the direct binding assay for Ras and Raf, Raf proteins were assayed as competitive inhibitors in the GAP assay which measures enhancement of Ras GTPase activity (6). The RafRBD protein construct used in the SPA was a fusion between maltose-binding protein (MBP) and RafRBD that contained a signal sequence for biotinylation at the N-terminus (11). Biotinylated MBP-RafRBD was co-expressed in *E. coli* with biotin ligase and purified by chromatography on amylose-derivitized resin as previously described (6). The biotinylated portion (80%) was isolated on monomeric avidin-modified resin (Promega Corp.) Purification methods used to isolate Harvey Ras with a Q61L mutation and GAP334, both from *E. coli*, were described previously (6).

Pre-Binding of [^{3}H]-GTP to [Q61L]-Ras. Ras co-purifies with a stoichiometric equivalent of GDP bound to it. To exchange [^{3}H]GTP for GDP, 20 μM Ras was incubated with 40 μM [^{3}H]GTP (5-10 Ci/mmole; Amersham) in Nucleotide Exchange Buffer (100 mM sodium phosphate, pH 6.8, 0.5 mM EDTA, 0.005% sodium deoxycholate, 0.1 mM DTT) at room temperature for 10 minutes. In theory, 67% of the Ras should contain bound [^{3}H]GTP. The actual efficiency of radioactive GTP exchange was determined by separating Ras-bound and free nucleotide by gel filtration (PD-10 column, Pharmacia) and measuring the radioactivity in each fraction. By this method, 50-67% GTP-labeling of Ras was routinely achieved. Ras-GTP concentrations in this manuscript refer specifically to the labeled fraction. Binding of Ras-GDP to RafRBD in the SPA was shown to be negligible at the concentrations present in the labeled mixture. The Ras-GMPPNP complex was similarly prepared by incubating Ras-GDP in the presence of a 10-fold molar excess of GMPPNP in Nucleotide Exchange Buffer followed by dialysis to remove excess nucleotide.

Scintillation Proximity Assay. The standard scintillation proximity assay contained 67 μg of scintillant-impregnated microspheres (US Patent #4568649, Amersham) biotin-MBP-RafRBD and [Q61L]-Ras-[^{3}H]GTP in 50 mM Hepes (pH 7.5), 1 mM MgCl$_2$, 0.1 mM DTT a total volume of 100 μl. Samples were counted in a standard liquid scintillation counter (Packard) and corrected for non-specific background counts measured in control assays lacking RafRBD. 80-100% maximal binding was observed within 2 hours at 25 °C, after which time the binding equilibrium was stable for over 20 hours. An increase in the actual signal was observed during this time period due to settling of the beads from the assay solution. Binding capacity of the beads and counting efficiency of the bound ^{3}H-labeled ligand was determined by measuring the binding of [^{3}H]-biotin (Amersham). 50 μg of beads had a maximal binding capacity of approximately 7.5-10 pmoles biotin and 20% counting efficiency (dpm/cpm) when measured in microfuge tubes in a standard scintillation counter. In a 96-well microtiter plate the counting efficiency was 7% (Packard

TopCount). The maximal bead capacity for RafRBD was 80% of the binding capacity observed with biotin. Assays to examine competitive inhibitors contained 50 nM biotin-MBP-RafRBD and 148 μM Ras-[^{3}H]GTP, to produce half-maximal binding.

III. Results

Scintillation Proximity Assay. An equilibrium binding assay was developed to directly measure the interaction between Ras-GTP and the Ras-binding domain of c-Raf-1, as shown in Figure 1. The assay strategy employs streptavidin-coated microspheres, that are impregnated with scintillant, to capture a biotinylated fusion protein containing the RafRBD. The subsequent binding of Ras-[^{3}H]GTP is detected with a standard scintillation counter to measure light emitted as a result of the close proximity of the tritium β-emission and the scintillant in the bead. Hence, this method allows the measurement of bound ligand in the presence of free ligand (12).

Biotinylation of MBP-RafRBD using chemical reagents with specificity for lysine, histidine or tyrosine residues inactivated the protein, as measured by the ability to competitvely inhibit GAP-enhanced Ras activity. Therefore, the MBP-RafRBD construct was modified to include a signal sequence for *in vivo* biotinylation at the amino terminus and co-expressed in *E. coli* with biotin holoenzyme synthetase to improve the efficiency of the *in vivo* biotinylation (11). This material was isolated by affinity chromatography, shown to be biotinylated, and found to have full activity as an inhibitor of GAP-catalyzed stimulation of Ras-associated GTPase activity (6).

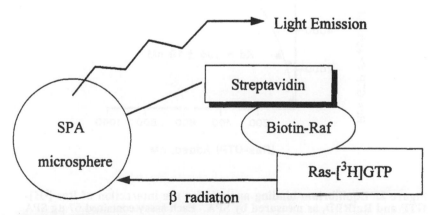

Figure 1. Schematic representation of the scintillation proximity assay to measure Ras-GTP binding to RafRBD. The SPA protocol employs streptavidin-coated, scintillant-impregnated beads to capture a biotinylated fusion of maltose binding protein with the Ras-binding domain of c-Raf-1.

Initial work in developing the SPA employed Ras with an activating Asp[12] mutation. However, this mutant protein retains sufficient GTPase activity that a loss of signal in the SPA was observed with time (approximately 5% per hour). [Q61L]Ras, a different activating mutant, that is also present in human tumors, was subsequently tested. Signal output in the SPA was stable for at least 24 hours with [Q61L]Ras-GTP.

Equilibrium Binding of the RafRBD to Ras-GTP. A typical equilibrium binding curve obtained by SPA for the binding of [Q61L]Ras-[3H]GTP to biotin-MBP-RafRBD is shown in Figure 2. Samples were corrected for non-specific background counts (10% of the specific binding) measured in control assays lacking RafRBD. Analysis of the data yielded a Kd = 105 nM for a single site at 25 °C. The same RafRBD construct produced 50% inhibition at 100 nM concentration when it was examined for binding to Ras-GTP indirectly as a competitive inhibitor of GAP, also at 25 °C. In a previous study, the MBP fusion protein and the isolated 9 kDa RafRBD domain were equally active inhibitors of GAP (6). Therefore, it appears that the SPA provides a reliable measurement for the binding of RafRBD and Ras-GTP. Ras-GDP was inactive in the SPA when tested at the same concentrations as Ras-GTP, as expected.

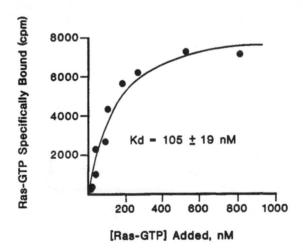

Figure 2. Equilibrium binding analysis for the interaction of Ras-[3H]-GTP and RafRBD, as measured by SPA. Each assay contained 67 µg SPA beads (Amersham), 50 nM biotinylated RafRBD and the indicated concentrations of [Q61L]Ras-[3H]GTP. The binding curve was obtained by non-linear regression analysis (Enzfitter) for a single binding site (Kd = 105 nM; Bound max = 8730 cpm).

Competitive Displacement. A number of proteins were tested as competitive inhibitors in the SPA for Ras/Raf binding to both validate and establish the sensitivity of the assay method. The results are shown in Figure 3. The commonly employed antibody Y13-259 inhibits Ras signaling *in vivo* and as previously reported has an IC_{50} of 150 nM in the GAP assay (6). The IC_{50} for Y13-259 was comparable when measured by SPA (IC_{50} = 100 nM). In contrast, the catalytic domain of the GAP protein, GAP334, has a weak affinity for Ras-GTP and exhibited an IC_{50} in the SPA of approximately 20 μM. These results demonstrate the broad range of binding affinities that are measurable by the SPA protocol. Furthermore, Ras and Raf protein constructs can be evaluated directly by tritium or biotin labeling, respectively, or they can be evaluated as competive inhibitors without labeling. In competition experiments the use of non-hydrolyzable GMPPNP to maintain conformational activation of Ras allows binding measurements for catalytically active Ras proteins.

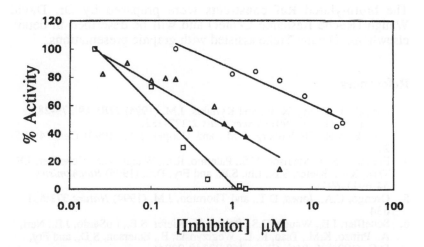

Figure 3. Competitive Displacement of Ras-GTP from RafRBD. Inhibition of Ras-[³H]GTP binding to RafRBD by antibody Y13-259 (square), [Q61L]Ras-GMPPNP (triangle) and GAP334 (circle). Each assay contained 50 nM RafRBD protein construct and 148 μM Ras-[³H]GTP, to achieve half-maximal binding in the absence of competing ligand. GAP334 is the catalytic domain of the GTPase Activating Protein (GAP120). Y13-259 antibody inhibits Ras signaling *in vivo* and has previously been shown to inhibit the Ras/Raf interaction (6).

IV. Discussion

The general strategy employed here, capturing protein-bound [^{3}H]GTP on SPA beads, lends itself to a number of applications. For example, it is possible to measure the binding of Ras to other effectors, like neurofibromin (13) or to measure the interaction of other G-proteins with their respective effector molecules. A similar format would also be useful to follow the exchange of GTP for GDP on Ras that is catalyzed by guanine nucleotide exchange factors.

The advantages of the scintillation proximity assay method (12) include: (i) stability of the signal; (ii) amenability to a high-flux format requiring the addition of pre-mixed buffer and proteins to microtiter plate wells to facilitate inhibitor screening, and; (iii) an equilibrium binding format that allows the determination of dissociation constants for both weak and high-affinity ligands and inhibitors. Therefore, the binding energy contributed by individual protein residues can be analyzed using the SPA to quantitate the effects of amino acid substitutions in Ras or Raf.

Acknowledgments

The biotinylated Raf constructs were prepared by Dr. David Waugh (Roche Research Center) and will be described in detail, elsewhere. Dennis Tighe assisted with graphic presentations.

References

1. Avruch, J., Zhang, X.-F., and Kyriakis, J.M. (1994) *TIBS* **19**:279-283.
2. Morrison, D.K. (1990) *Cancer Cells* **2**:377-382.
3. Vojtek, A.B., Hollenberg, S.M., and Cooper, J.A. (1993) *Cell* **74**:205-214.
4. Emerson, S.D., Madison, V.S., Palermo, R.E., Waugh, D.S., Scheffler, J.E., Tsao, K-L., Kiefer, S.E., Liu, S.P., and Fry, D.C. (1995) *Biochemistry* **34**:6911-6918.
5. Orengo, C.A., Jones, D.T., and Thornton, J.M. (1994) *Nature* **372**:631-634.
6. Scheffler, J.E., Waugh, D.S., Bekesi, E., Kiefer, S.E., LoSardo, J.E., Neri, A., Prinzo, K.M., Tsao, K.-L., Wegrzynski, B., Emerson, S.D., and Fry, D.C. (1994) *J. Biol. Chem.* **269**:22340-22346.
7. Herrmann, C., Martin, G. A., and Wittinghofer, A. (1995) *J. Biol.Chem.* **270:** 2901-2905.
8. Warne, P.H., Viciana, P.R. and Downward, J. (1993) *Nature*, **364**:352-355.
9. Fabian, J.R., Vojtek, A.B., Cooper, J.A., and Morrison, D.K. (1994) *Proc. Natl. Acad. Sci. U.S.A.* **91**:5982-5986.
10. Brtva, T.R., Drugan, J.K., Ghosh, S., Terrell, R.S., Campbell-Burk, S. Bell, R.M., and Der, C.J. (1995) *J. Biol. Chem.* **270**:9809-9812.
11. Schatz, P.J. (1993) *Biotechnology* **11**:1138-1143.
12. Bosworth, N., and Towers, P. (1989) *Nature* **341**:167-168.
13. Skinner, R.H., Picardo, M., Gane, N.M., Cook, N.D., Morgan, L., Rowedder, J., and Lowe, P.N. (1994) *Anal. Biochem.* **223**:259-265.

SECTION III

Behavior of Proteins at Surfaces

Asparagine Rearrangements in Two Isoforms Of Erythropoietin Receptor

Patricia L. Derby, Kennneth H. Aoki, Viswanatham Katta, and Michael F. Rohde
Amgen, Inc., Thousand Oaks, CA

Introduction

Erythropoietin (EPO), a serum glycoprotein and a member of the cytokine family, regulates the growth and differentiation of erythroid hematopoietic cells through its interaction with the cellular transmembrane EPO receptor (EPOr). Two features on the cytoplasmic domain of the EPOr are shared with other members of the cytokine family: a set of four cysteine residues and a five-residue motif close to the transmembrane domain, Trp-Ser-X-Trp-Ser (1). This motif is thought to be involved in the binding of EPO to the receptor and leading to receptor homodimerization (2,3). Since normal erythroid progenitor cells exhibit less than 1000 EPOr per cell (4), the cloning of the EPOr in 1989 has provided greater opportunity to study this receptor (5).

Soluble, recombinant EPOr (226 amino acids) was purified in our laboratory and shown to be comprised of two isoforms that behaved differently when run on a native gel. One (EPOr-2) migrated further in the gel than the other (EPOr-1), indicating that EPOr-2 was either processed at the amino or carboxy terminus or modified in some way that would make it more acidic. Liquid chromatographic mass spectrometry (LC/MS) of the two forms did not show any mass difference. Interestingly, over time, EPOr-1 was becoming a mixture of EPOr-1 and a small amount of EPOr-2.

It is known that asparagine, followed by a glycyl residue, may form a succinimide or cyclic imide at a high rate (6,8); EPOr contains one Asn-Gly combination at residues 163-164. To discern if modification of Asn-163 was the difference between these two isoforms, both EPOr-1 and EPOr-2 were examined by hydroxylamine cleavage using conditions that would cleave a cyclic imide only (7). A portion of EPOr-1 was cleaved, but EPOr-2 alone was not cleaved, indicating the presence of a cyclic imide in EPOr-1. Tryptic peptide mapping on reverse phase HPLC was followed by Matrix assisted laser desorption/ ionization mass spectrometry (MALDI/MS) of the peptides. A peptide from each isoform with a mass consistent with the mass of residues 155-170 was identified. N-terminal sequencing of these two peptides revealed that the major part of EPOr-1 had an Asn at residue 163. This residue in EPOr-2 was not detected by N-terminal sequencing, indicating that a modification of residue 163 had occurred, which would block further sequencing. Rearrangement of asparagine to an isoaspartate would be such a modification.

Materials and Methods

1. Tryptic digests and mapping:
   a. Digests were conducted using an enzyme to protein ratio of 1:50 and incubation at 37° C for 18 hours.
   b. Mapping was performed by reverse phase HPLC (RP-HPLC) on a Hewlett Packard 1090M using a Vydac C18 column (4.6 X 250mm) with a gradient of 0-90% acetonitrile in 0.1% TFA in 75 minutes. Peaks were manually collected.

2. N-terminal sequencing:
   Performed on ABI 477A Protein Sequencer.

3. MALDI mass spectrometry:
   Performed on a Kratos Kompact Maldi III using the linear and positive ion modes with laser settings that were at the threshold of peak detection. The matrix used was a 33mM solution of alpha-cyano-4-hydroxycinnamic acid in acetonitrile and methanol (Biomolecular Laboratories). One to twenty pmoles of sample was used with a matrix to sample ratio of 1:1 (vol/vol).

4. LC/Mass Spectrometry:
   Performed on a Sciex API III on-line to a Hewlett Packard 1090 using a PLRP-S column 1.0x50 mm, 8u, 4000A; the gradient was 4-90% acecetonitrile in 0.1% TFA in 6 minutes. A stepped orifice potential was used; 50 pmoles of sample was injected.

5. Gel Electrophoresis:
   a. Native Gel electrophoresis (N/SDS) of the two isoforms was performed on an 8% Tris-Glycine gel using a Novex system.
   b. Post hydroxylamine treatment was assessed by electrophoresis of 4µg of each sample on a 14% Tris-glycine gel using a Novex system.

6. Hydroxylamine Treatment:
16µg of EPOr-1 and 16µg of EPOr-2 were incubated in 19 volumes of 2M hydroxylamine, 0.2 M Tris buffer, pH 9, for 2 hours at 45°C.

Results and Discussion

The EPO receptor is a 507 amino acid transmembrane protein. Soluble recombinant EPOr was purified from Chinese hamster ovary (CHO) cells. It was comprised of 225 amino acids, and two disulfide linkages (2). Glycosylation of this molecule consisted of one N-site at asparagine 51; no O-sites were present. Isoelectric focusing gel electrophoresis of the receptor after the purification process revealed several isoforms of the EPOr. After treatment with neuraminadase and N-glycanase, two forms were present. When they were subsequently run on a native gel after neuraminadase and N-glycanase treatment, it was observed that one of these isoforms (EPOr-2) migrated slightly further in the gel than the other isoform (EPOr-1) (Fig. 1.) indicating the possibility that: 1) the protein was modified at the amino or carboxy terminal or 2) the protein was modified in a way that would result in a more positively charged, hence more acidic, species. All of the subsequent work on the EPOr isoforms was performed on the unglycosylated protein.

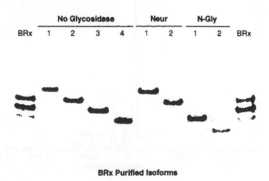

Fig. 1. Gel Electrophoresis of the Erythropoietin Receptor Isoforms Before and After Glycosidase Treatment.

LC/MS was performed on both EPOr forms as seen in
Figure 2. The average mass for the unglycosylated protein is
24756.1694 atomic mass units. The molecular weights attained for
EPOr-1 and EPOr-2, respectively, of 24,755 and 24,756 daltons(D)
demonstrated that no modification of the amino or carboxy termini
had occurred.

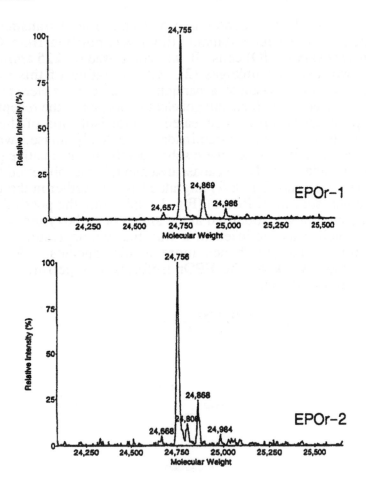

Fig. 2. LC/MS of EPOr-1 and EPOr-2.

Deamidation of asparagine and formation of isoaspartate from aspartic acid are protein degradation processes that involve succinimide (cyclic imide) intermediates (6) (Fig. 3). This degradation would convey an extra positive charge to a protein. Small hydrophilic residues such as glycine, serine, or threonine are often found adjacent to the Asp/Asn residues that form cyclic imides (8). The EPO receptor contains one sequence that meets this criteria: residues 163-164 are Asn-Gly. Hydroxylamine, under the conditions in Materials and Methods, cleaves on the C-terminal side of a cyclic imide without cleaving the uncyclized Asn-Gly bond (7). This cleavage process was performed on the unfractionated isoforms and on EPOr-1 and EPOr-2. Results of hydroxylamine treatment were assessed by gel electrophoresis (Fig.4). If a cyclic imide had formed and cleavage had occurred there, the resulting fragments would be approximately 18kD and 6kD. These fragments were detected in the unfractionated EPOr and in EPOr-1, however, the main portion of these two samples was not cleaved. EPOr-2 had no cleavage products. This demonstrated that a portion of EPOr-1 had indeed formed a cyclic imide.

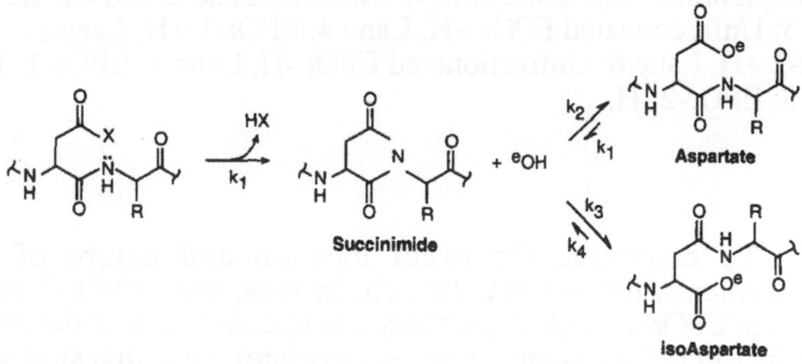

X	R
–OH (Asp)	–H (Gly)
–NH$_2$ (Asn)	–CH$_3$ (Ala)
–OCH$_3$ (Asp β-Me ester)	–CH$_2$OH (Ser)

Fig. 3. The Reactions of Conversion of Asparagine to a Cyclic Imide Intermediate and Subsequent Conversion to Isoaspartate or Aspartate.

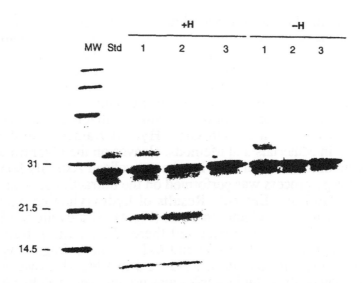

Fig. 4. Gel Electrophoresis of Unfractionated EPOr, EPOr-1, and EPOr-2 with Hydroxylamine treatment (+H) and without Hydroxylamine (-H). Lane 1: MW Markers, Lane 2: Std EPOr, Lane 3: Unfractionated EPOr +H, Lane 4: EPOr-1 +H, Lane 5: EPOr-2 +H, Lane 6: Unfractionated EPOr -H, Lane 7: EPOr-1 -H, Lane 8: EPOr-2 -H.

To determine the exact location and nature of the rearrangements, the isoforms (which, by now, were EPOr-2 and a mixture of EPOr-1/EPOr-2, probably due to conversion of the cyclic imide in EPOr-1 to aspartate or isoaspartate) were digested with trypsin. The peptides were separated and collected on RP-HPLC (Fig. 5 and 6). MALDI/MS of these peptides from each isoform identified a peptide from each that had a mass consistent with the mass of residues 155-170: 1609.7 D (Fig. 7 and 8). (There was another peptide present, with a mass of 974 daltons, that had coeluted with the peptide of interest; this was determined to be residues 1-9.) The mass for residues 155-170 obtained for EPOr-1/EPOr-2 was 1609.0 and the mass obtained for EPOr-2 was 1611.4. These two peptides were then subjected to N-terminal sequencing. Results of that sequencing are seen in Table I.

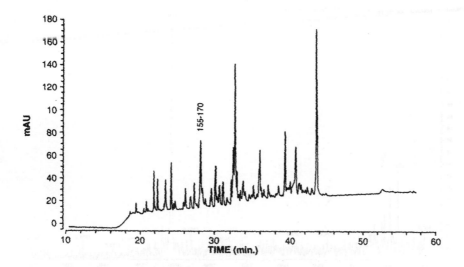

Fig.5. Tryptic Peptide Map of EPOr-1/EPOr-2 Showing the Peak that Contains Residues 155-170.

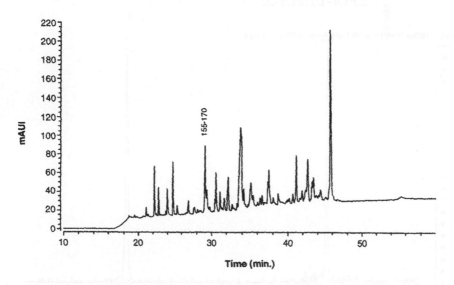

Fig. 6. Tryptic Peptide Map of EPOr-2 Showing the Peak that Contains Residues 155-170.

Patricia L. Derby *et al.*

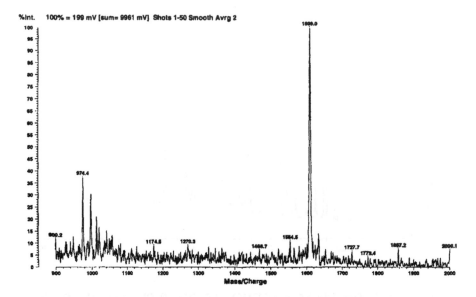

Fig. 7. MALDI/MS Spectrum of the Tryptic Peptide from
EPOr-1/EPOr-2.

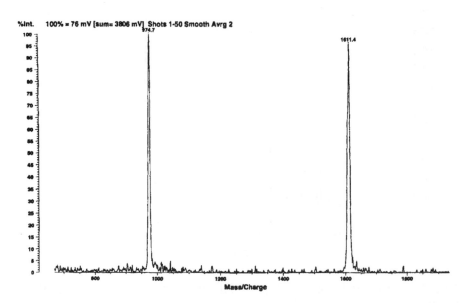

Fig. 8. MALDI/MS Spectrum of the Tryptic Peptide from EPOr-2.

TABLE I

Amino   Acids(155-170)

N-Terminal Sequencing EPOr-1/EPOr-2	Yield(pmoles) EPOr-2
Tyr	15.87	27.18
Glu	11.64	16.80
Val	28.80	29.00
Asp	18.80	21.22
Val	28.82	25.98
Ser	20.49	6.24
Ala	16.21	20.00
Gly	11.30	12.08
Asn 163	6.63	ND*
Gly 164	7.38	ND
Ala	7.21	ND
Gly	5.81	ND
Ser	1.21	ND
Val	6.23	ND
Gln	2.87	ND
Arg	ND	ND

*ND = Not Detected.

There was no detectable sequence in EPOr-2 after Gly-162. Since the MALDI/MS results were consistent with the presence of the entire peptide from residues 155 to 170, the fact that the sequence ended at Gly-162 in EPOr-2 was an indication that Asn-163 had undergone some modification that would block sequencing from that point. Conversion of asparagine to isoaspartate might be such a modification.

EPOr-1/EPOr-2 sequence revealed an Asn at residue 163; no Asp was detected. Sequencing continued until the end of the peptide, but the yield dropped 41% at residue 163. The drop in yield at this residue in EPOr-1/EPOr-2 is consistent with the results of the hydroxylamine treatment showing that a portion of EPOr-1 had formed a cyclic imide at Asn-163. The presence of a cyclic imide would block sequencing. In addition, this cyclic imide was then converting to isoaspartate, which would result in EPOr-1 becoming a combination of EPOr-1 and EPOr-2. The portion of the molecule that had an isoaspartate at site 163 would not sequence beyond that residue, adding to the overall drop in yield.

Conclusions

There are two isoforms of the EPO receptor (EPOr-1 and EPOr-2) which differ in charge from each other due to a modification in EPOr-2 of Asn-163 to isoaspartate. The nature of this modification was detected after hydroxylamine treatment demonstrated that a small portion of EPOr-1 contained a cyclic imide. Cyclic imides form at a high rate on asparagines followed by glycine and the EPOr had one Asn-Gly combination at residues 162-163. Tryptic digestion of the two isoforms and examination of the peptides by MALDI/MS identified the two peptides consisting of residues 155-170. N-terminal sequencing of these peptides revealed that the major portion of EPOr-1 had an asparagine at residue 163. Over time, this asparagine appeared to be converting to a cyclic imide intermediate which then converted further to an isoaspartate.

EPOr-2, however, did not cleave with hydroxylamine treatment, indicating the absence of a cyclic imide in this isoform. N-terminal sequencing of the peptide containing residues 155-170 stopped sequencing at Asn-163, inferring that a modification of Asn-163 to isoaspartate had occurred. This would account for the observation that EPOr-2 was more acidic than EPOr-1. The same modification was in process in EPOr-1 and, if left to completion, there would be no difference between EPOr-1 and EPOr-2; both would have isoaspartate at residue 163 instead of an asparagine.

Acknowledgment: I would like to thank Thomas W. Strickland for the helpful discussions.

References

1. Mayeaux, P., Pallu, S., Gobert, S., Lacombe, C. and Gisselbrecht, S. (1994). Structure of the erythropoietin receptor. *Proc. Soc. Exp. Biol. Med.* **206**: 200-204.
2. Youssoufian, H., Longmore, G., Neumann, D., Yoshimura, A., and Lodish, H.F. (1993). Structure, function, and activation of the erythropoietin receptor. *Blood* **81**: 2223-2236.
3. Longmore, G.D., Watowich, S.S., Hilton, D.J., and Lodish, H.F. (1993). The erythropoietin receptor: Its role in hematopoiesis and myeloproliferative diseases. *J. Cell Biol.* **123**: 1305-1308.
4. Mayeaux, P., Billat, C., and Jacquot, R. (1987). The erythropoietin receptor of the rat erythroid progenitor cells: Characterization and affinity cross-linking. *J.Biol. Chem.* **262**: 13985-13990.
5. D'Andrea, A.D., Lodish, H.F., and Wong, G.G. (1989). Expression cloning of the erythropoietin receptor. *Cell* **52**: 277-285.
6. Stephenson, R.C., and Clarke, S. (1989). Succinimide formation from aspartyl and asparaginyl peptides as a model for the spontaneous degradation of proteins. *J. Biol. Chem.* **264**: 6164-6170.
7. Kwong, M.Y. and Harris, R. J. (1994). Identification of succinimide sites in proteins by N-terminal sequence analysis after alkaline hydroxylamine cleavage. *Protein Sci.* **3**: 147-149.
8. Tyler-Cross, R. and Schirch, V. (1991). Effects of amino acid sequence, buffers and ionic strength on the rate and mechanism of deamidation of asparagine residues in small peptides. *J. Biol. Chem.* **266**: 22549-22556.

Initial Characterization of a Peptide Epitope Displayed on the Surface of fd Bacteriophage by Solution and Solid-State NMR Spectroscopy

Martine Monette
Holly Gratkowski
Stanley J. Opella[1]
Department of Chemistry, University of Pennsylvania
Philadelphia, Pennsylvania, 19104

Judith Greenwood[2]
Anne E. Willis[3]
Richard N. Perham[4]
Cambridge Centre for Molecular Recognition
Department of Biochemistry, University of Cambridge
Tennis Court Road, Cambridge, CB2 1QW, England

I. Introduction

The circumsporozoite (CS) protein of the human malaria parasite *Plasmodium falciparum* provides one of the most promising targets for the development of a vaccine (1). The main antigenic determinant of the CS protein contains several repetitions of the tetrapeptide NANP, and proteins containing several units of this epitope have been successfully used to raise a high titer of antibodies (1). When the epitope is displayed on the surface of the filamentous bacteriophage fd, an even higher immunogenic response is observed (2,3). Bacteriophage display of peptide epitopes can enhance their immunogenicity by providing a protein environment that facilitates and stabilizes their folding, and by

[1]Supported by grants No. R376M24266 and No. P41RR09793 from the National Institute of Health

[2]Present address: Institute of Biotechnology, University of Cambridge, Tennis Court Road, Cambridge, CB2 1QT, England

[3]Present address: Department of Biochemistry, University of Leicester, Adrian Building, Leicester, LE1 7RH, England

[4]Supported by a grant from the Wellcome trust

attaching them to already highly immunogenic particles. While peptides with up to six residues can be displayed close to the N-terminus of all 2700 copies of the major coat protein of fd, generating a recombinant virion, epitopes containing more than 6 residues generally require hybrid phages in which the insert containing coat proteins are interspersed among wild type proteins (2,4).

In this paper, we describe spectroscopic methods and preliminary results from structural studies of the epitope (NANP)₃ displayed close to the N-terminus of the major coat protein in hybrid phage. First, multidimensional solution NMR is used to study the three-dimensional structure of the epitope in isolated coat proteins solubilized in micelles. Second, solid-state NMR is applied to intact phage particles; the spontaneous orientation of the virion in the presence of strong magnetic fields enables the epitope to be characterized on the surface of the bacteriophage.

It is generally accepted that the coat protein has a structure containing more than 90% α-helix, as determined by solid-state NMR spectroscopy (5) and X-ray fiber diffraction (6). The structure and the dynamics of the fd coat protein have also been characterized by solution NMR studies in micelles (7,8). In addition, a recent solid-state NMR study has shown that the membrane bound form of fd coat protein is mainly helical, having one hydrophobic membrane spanning helix and a shorter amphipathic helix parallel to the plane of the bilayer (9). These studies provide an excellent background for the structural study of coat proteins with peptide inserts, such as epitopes. Solving the three-dimensional structure of epitope inserts has great potential in structural biology since their immunogenicity is likely to be related to their folding in the context of the coat protein.

II. Experimental

A. Sample Preparation

The modified coat protein containing the sequence (NANP)₃ is expressed from the plasmid PKfdMAL1 (2). *Escherichia coli* TG1 or JM101 cells containing the PKfdMAL1 plasmid are grown in minimal media containing (^{15}NH$_4$)$_2$SO$_4$ (Isotec) as the only source of nitrogen, in the case of uniformly labeled samples. For selective labeling, ^{15}N labeled alanine (Isotec) was added to the growth media, together with (^{14}NH$_4$)$_2$SO$_4$, unlabeled valine, threonine, leucine and isoleucine. In the presence of the inducer (isopropyl-β-D-thiogalactopyranoside, IPTG), large amounts of the mutant coat protein are produced. At the mid-log phase of the growth, the cells are infected with wild-type fd for 7 hours. Isolation by polyethylene glycol (PEG) precipitation and purification of the phage particles by

KBr gradient results in 5-20 mg of phage per liter of growth medium. The ratio of wild-type:mutant coat protein as determined by SDS-PAGE (2), is approximately 4:1. The solution NMR spectra were obtained from uniformly or selectively labeled fd/fdMAL1 coat proteins (2 mM) as a mixture in ^{2}H-SDS micelles (480 mM), at pH=4, in 40mM NaCl. The solid-state spectra were obtained on solutions of magnetically oriented fd/fdMAL1 hybrid bacteriophages (25-50 mg/ml) at 25°C, pH=8.

B. Solution NMR Spectroscopy

The two-dimensional HMQC (Heteronuclear Multiple Quantum Coherence) spectra were acquired on Bruker DMX 500 and DMX 750 NMR spectrometers using 5-mm triple resonance probes. Water suppression was achieved either by selective presaturation for 1.5 s or by the application of WATERGATE gradient pulses (10) prior to data acquisition. Typically, 128 increments in the ^{15}N dimension and 16 transients in the proton dimension were acquired using the TPPI method. The data were processed with the UXNMR program with sine bell multiplication in the ^{15}N dimension and an exponential multiplication (line broadening of 5 Hz) in the ^{1}H dimension. The size of the final transformed matrices were 2048 X 256 points.

The three-dimensional HMQC-TOCSY (total correlation spectroscopy) and 3D-HMQC-NOESY (nuclear Overhauser enhancement spectroscopy) experiments were performed on a Bruker DMX 750 NMR spectrometer. Water suppression was achieved by the application of WATERGATE gradient pulses prior to the acquisition. The mixing time was set to 55 ms for the TOCSY experiment and 150 ms for the NOESY experiment. The spectra were acquired with 64 increments in the ^{15}N dimension, 128 increments in the ^{1}H indirect dimension, and 16 transients were collected using the states-TPPI method (11). The data were processed with the UXNMR program with sine bell multiplication in the ^{15}N dimension, as well as in the indirect proton dimension, and an exponential multiplication (line broadening of 5 Hz) in the direct ^{1}H dimension. The size of the final transformed matrices were 1024 X 256 X 128 points. Chemical shifts were referenced internally to the residual ^{1}H$_2$O peak (4.75 ppm) and externally to liquid ammonia (0 ppm).

C. Solid-state NMR Spectroscopy

The solid-state NMR experiments were performed on Chemagnetics CMX-400 and home-built 360 MHz spectrometers using home-built double resonance probes (^{1}H and ^{15}N) equipped

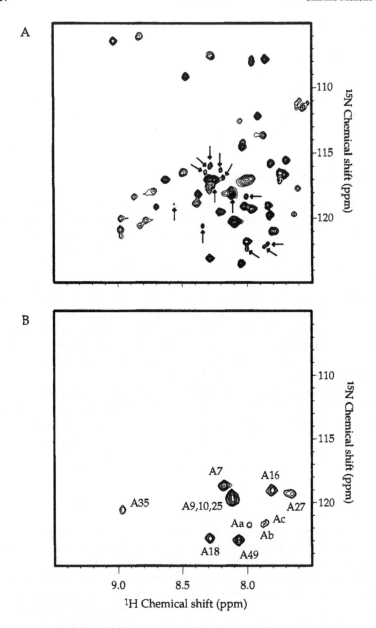

Figure 1: a) ^{1}H-^{15}N 2D-HMQC spectrum of the ^{15}N uniformly labeled fd/ fdMAL1 hybrid phage coat proteins, at 50°C. The spectrum was acquired on a Bruker DMX 750 NMR spectrometer, at the ^{1}H resonance frequency of 750 MHz; b) ^{1}H-^{15}N 2D-HMQC spectrum of the ^{15}N alanine labeled fd/fdMAL1 hybrid phage coat proteins, at 35°C. The spectrum was acquired on a Bruker DMX 500 NMR spectrometer, at the ^{1}H resonance frequency of 500 MHz.

with 5 or 7-mm solenoid coils. The ^{15}N spectra were acquired with a phase cycled cross polarization sequence (12) in which the 90° pulse was 4 μs and the mixing time was 3 ms. Typically, 256 points were acquired with a spectral width of 50 kHz. The recycle delay was 8 s and 10^3-10^4 transients were signal averaged for each spectrum. Exponential multiplication (line broadening of 100 Hz) and zero filling to 1024 points were performed prior to Fourier transformation.

III. Results and Discussion

Two-dimensional [1]H-^{15}N HMQC spectra of the mixture of fd/fdMAL1 coat proteins (4:1 ratio) solubilized in detergent micelles are shown on figure 1. The spectrum on figure 1A was obtained with a uniformly ^{15}N labeled sample, whereas the spectrum in figure 1B is of a selectively ^{15}N alanine labeled sample. The most intense peaks belong to the wild type coat protein and were assigned by direct comparison with the spectrum of wild type fd (9). The marked smaller extra peaks represent resonances from the epitope (NANP)3, inserted between amino acids 3 and 4 in the wild type coat protein. In the modified coat protein, residue 3 is changed from Gly to Val, and residue 4 from Asp to Asn, as a result of the creation of the *HpaI* restriction site in the gene that enables the oligonucleotide specifying the peptide epitope to be inserted (2). The amino acids E2' and D5' in the mutant coat protein show different chemical shifts than those of the wild type (E2 and D5). A total of 13 extra peaks are present on the HMQC spectrum, shown in figure 1A (resonances from prolines do not appear in this type of spectrum). The fd/fdMAL1 hybrid phage was also selectively labeled with ^{15}N alanine, and the HMQC spectrum is shown in figure 1B. The most intense peaks represent the alanines of the wild type coat protein and the three smaller extra peaks are from residues in the epitope. The resonances in figure 1B are superimposable to those assigned to alanine in figure 1A, and no other peaks are present, even in the baseline noise level. These spectra clearly indicate that the selective labeling is very efficient and isotopic scrambling did not occur during the incorporation of ^{15}N alanine.

Because the resonances in figure 1A are well resolved, and the extra peaks are readily identified, additional information on the residues in the epitope can be obtained from three-dimensional experiments. Figure 2A contains spectral strips extracted from a 3D-HMQC-TOCSY experiment. In this experiment, correlations through bonds are observed between the chemical shifts of both ^{15}N and ^{1}H of the amide group and the chemical shifts of nearby protons. Therefore, the peaks in the strips correspond to the resonance frequencies of the amide proton (around 8 ppm), α protons (from 4 to 5 ppm) and β protons (from 1 to 3 ppm) of each

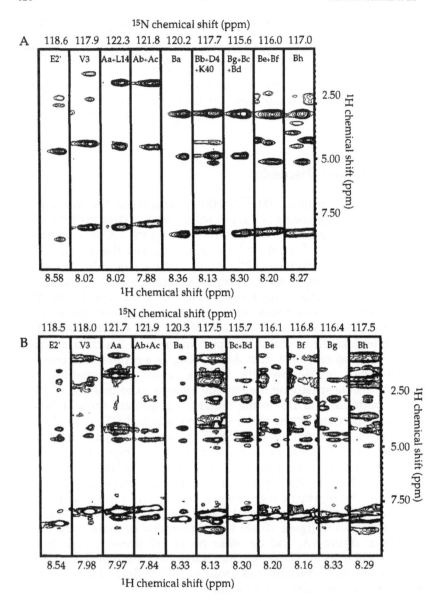

Figure 2: A) Homonuclear ^{1}H-^{1}H TOCSY spectral strips taken from the ^{1}H-^{15}N 3D-HMQC-TOCSY spectrum of the ^{15}N uniformly labeled fd/fdMAL1 hybrid phage coat proteins, at 50°C. B) Homonuclear ^{1}H-^{1}H NOESY spectral strips taken from the ^{1}H-^{15}N-HMQC-NOESY spectrum of the ^{15}N uniformly labeled fd/fdMAL1 hybrid phage coat proteins, at 35°C.

residue. In the case of residues E2' and V3, correlation to γ protons are also apparent. On the basis of the different chemical shifts of α and β proton resonances, assignment of the different types of amino acids can be made, and they are in agreement with the assignments obtained from the selective labeling (figure 1B). The three alanines are marked Aa, Ab and Ac. The six subsequent strips represent the asparagine residues contained in the epitope as well as D5'. On the sole basis of the chemical shifts, N and D cannot be differentiated and the spins systems are labeled Ba through Bh in figure 2A. In order to determine the three dimensional structure of the peptide epitope, geometry and distance constraints have to be established. NOEs between protons provide information on inter nuclear distances. Figure 2B shows spectral strips extracted from a 3D-HMQC-NOESY experiment. Correlations through space are established between the chemical shifts of both ^{15}N and ^{1}H of the amide group and the chemical shifts of nearby protons. The strips contain both intra-residue and inter-residue correlations. With the help of the HMQC-TOCSY data presented in figure 2A, it is possible to identify the intra-residue connectivities, and assign the remaining peaks to correlations of other residues of the epitope. As shown by the TOCSY strips, the lack of chemical shift dispersion among the resonances from the α protons of the alanines, the β protons of the alanines, the α protons of asparagines and the β protons of asparagines makes the total assignment of the peptide resonances difficult on the basis of these data alone. Other types of multidimensional experiments involving double labeling (^{13}C and ^{15}N) will be used to complete the assignments.

The coat proteins were also studied on fd/fdMAL1 oriented phage by solid-state NMR spectroscopy. Figure 3A shows the solid-state NMR spectrum of uniformly ^{15}N labeled oriented wild-type fd. The resonance near 190 ppm indicates that most of the peptide planes in the backbone are oriented with their N-H bond parallel to the phage axis (13). Three resolved resonances around the isotropic chemical shift of the amide tensor (80 ppm) are assigned to the motionally averaged residues 2-4. The terminal amino groups of the coat protein contribute to the peak at 12 ppm. The preliminary solid-state NMR spectrum of the ^{15}N uniformly labeled fd/fdMAL1 hybrid phage in figure 3B displays the same features as the fd spectrum in figure 3A, as well as some additional peaks (arrows), attributed to some residues of the epitope in the coat protein. The peptide planes of these residues have different orientations with respect to the magnetic field, leading to different chemical shifts. The measurement of the ^{1}H chemical shift, ^{1}H-^{15}N dipolar coupling, and ^{15}N chemical shift for each site will provide information on the orientation of the peptide planes with respect to the magnetic field, leading to structural information on the peptide epitope displayed on the surface of the bacteriophage.

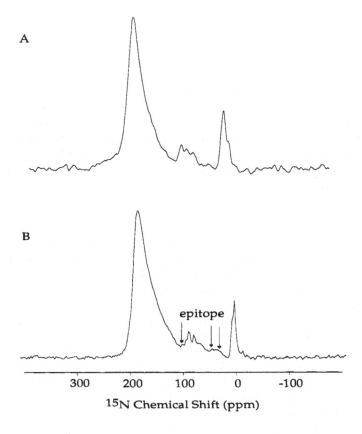

Figure 3: a)^{15}N solid-state spectrum of oriented wild type fd phage. The data was acquired on a Chemagnetics CMX-400 spectrometer, at the ^{15}N resonance frequency of 40 MHz; b)^{15}N solid-state spectrum of oriented fd/fdMAL1 hybrid phage. The data was acquired on a home-built spectrometer at the resonance frequency of 36 MHz.

IV. Conclusion

We have obtained large amounts of uniformly ^{15}N and selectively ^{15}N alanine labeled fd/fdMAL1 hybrid phage displaying the epitope insert (NANP)$_3$. Preliminary results obtained by solution and solid-state NMR spectroscopy show that the epitope is folded in the context of the phage coat protein, and its structure in micelles as well as in the virion particles can be determined by NMR spectroscopy.

Acknowledgments

M.M. is grateful to the Medical Research Council of Canada (MRC) for a post-doctoral fellowship. J.G. acknowlegdes the support of a fellowship from the Lloyds of London Tercentenary Foundation.

References

1. Young, J.F., Hockmeyer, W.T., Gross, M., Ballou, W.R., Wirtz, R.A., Trosper, H., Beaudoin, R.L., Hollingdale, M.R., Miller, L.H., Diggs, C.L., and Rosenberg, M., (1985). *Science* **228**, 958-962.
2. Greenwood, J., Willis, A.E., and Perham, R.N. (1991). *J. Mol. Biol.* **220**, 821-827.
3. Willis, A.E., Perham, R.N., and Wraith, D. (1993). *Gene* **128**, 79-83.
4. Iannolo, G., Minenkova, O., Petruzzelli, R., and Cesareni, G. (1995). *J. Mol. Biol.* **243**, 835-844.
5. Opella, S.J., Stewart, P.L., and Valentine, K.G. (1987). *Q. Rev. Biophys.* **19**, 7-49.
6. Marvin, D.A., Hale, R.D., Nave, C., and Citterich, M.H. (1994). *J. Mol. Biol.* **235**, 260-286.
7. Bogusky, M.J., Tsang, P., and Opella, S.J. (1985). *Biochem. Biophys. Res. Commun.* **127**, 540-545.
8. Bogusky, M.J., Leo, G.C., and Opella, S.J. (1988). *Proteins: Struct. Funct. Genet.* **4**, 123-130.
9. McDonnell, P.A., Shon, K., Kim, Y., and Opella, S.J. (1993). *J. Mol. Biol.* **233**, 447-463.
10. Skelnar, V., Piotto, M., Leppik, R., and Saudek, V. (1993). *J. Magn. Res. Ser A* **102**, 241-245.
11. Bax, A., and Ikura, M. (1991). *J. Biomol. NMR* **1**, 99-104.
12. Pines, A., Gibby, M.G., and Waugh J.S. (1973). *J. Chem. Phys.* **59**, 569-590.
13. Cross, T.A., Tsang, P., and Opella, S.J. (1983). *Biochemistry* **22**, 721-726.

NMR Structures for the Membrane Binding Gla Domain of Blood Coagulation Factor IX

James D. Baleja, Steven J. Freedman[a], Barbara C. Furie[a], and Bruce Furie[a]

Departments of Biochemistry and [a]Medicine
Tufts University School of Medicine, Boston, MA 02111
and
[a]Center for Hemostasis and Thrombosis Research, Division of Hematology-Oncology
New England Medical Center, Boston MA 02111

I. Introduction

Factor IX is a vitamin K-dependent protein whose critical role in blood coagulation is exemplified by hemophilia B, a bleeding disorder that results from a deficiency of biologically active Factor IX. The activity of Factor IXa, the enzyme form of the Factor IX zymogen, is greatly enhanced on binding anionic phospholipids, Factor VIII, and calcium ions (1). At the amino terminus of Factor IX, the Gla domain mediates phospholipid recognition by liganding calcium ions and the consequent formation of a phospholipid binding site. In contrast to Ca(II), other divalent metal ions induce conformers that do not bind phospholipids (2).

A two-step sequential conformational transition model for the metal ion-induced structural changes has been proposed based on observations using conformation-specific antibodies (3,4). In the first step, the apo Gla domain may bind a variety of divalent metal ions, all of which result in a conformational change with the expression of a common antigen. In the second step, calcium ions induce a specific conformer that presents an additional antigenic determinant. Antibodies and their Fab fragments directed against the Ca(II)-stabilized epitope block phospholipid binding (3,4).

We have synthesized a peptide of 47 residues (Figure 1) that contains the γ-

```
       1         5          10        15        20
FIX    Y N S G K L γ γ F V Q G N L γ R γ C M γ γ K C
PT     A N K G F L γ γ V R K G N L γ R γ C L γ γ P C

           25        30        35        40        45
FIX    S F γ γ A R γ V F γ N T γ K T T γ F W K Q Y V
PT     S R γ γ A F γ R L γ S L S A T D A F W A K Y T
```

Figure 1. Sequences of the N-terminal domains of blood coagulation proteins human Factor IX (FIX) and bovine prothrombin (PT). A disulfide bond connects cysteines 18 and 23. The Gla domain consists of residues 1 to 38 and the aromatic amino acid stack domain consists of residues 39 to 46. The γ-carboxyglutamic acid residues are designated as γ. The two sequences shown here are 54% identical. The sequences of other homologous proteins (Factor X, protein C, Factor VII, protein S, protein Z) are given elsewhere (9). To facilitate its synthesis, the Factor IX peptide also contained a C-terminal Asp residue, which is the first residue of the next domain (an EGF-like domain).

carboxyglutamic acid-rich (Gla) domain, which has metal-binding and membrane-binding affinities similar to full length Factor IX (5). In this report, we have investigated the apo, Mg(II), and Ca(II) forms of the peptide using nuclear magnetic resonance (NMR) spectroscopy (6-8). The structures of the Gla domain delineate the region of Factor IX that interacts with a lipid bilayer. They also explain the sequential conformational transition model proposed using conformation-specific antibodies.

II. Sample Conditioning and NMR Spectroscopy

The peptide representing the membrane-binding portion of Factor IX was synthesized by FMOC chemistry (5). At pH values greater than 4, the apo peptide was soluble to >3 mM. Collection of ^{1}H NMR data presented no special problems (7).

The situation for the Ca(II) bound peptide was quite different—this form readily precipitated, which was not surprising given that the calcium bound 1-47 peptide was known to be membrane-binding competent with a hydrophobic contribution to the interaction. The same hydrophobic forces presumably cause peptide aggregation.

On the other hand, initial NMR spectra on a 0.2 mM solution of the peptide with 8 equivalents of calcium (chloride) were very encouraging. An amide proton was observed at 10.5 ppm and, at the other end of the spectrum, methyl groups were seen near 0.4 ppm. However, close inspection revealed metal-free peptide (0.05 mM) still present, in slow exchange, with the calcium-bound form (0.15 mM). Addition of two equivalents of calcium (10 equivalents in total) caused a loss in signal preferentially for the calcium form. Two more equivalents (12 equivalents total) caused macroscopic precipitation, which was not immediate, but occurred within an hour or so. The time-dependent phenomenon appears similar to a slow-folding calcium-induced membrane-binding transition observed for the homologous protein prothrombin fragment 1 (2). Precipitation by calcium could be reversed by adding EDTA or Chelex-100 beads.

A range of pH, temperature, and ionic strengths was used in an attempt to increase the solubility of the peptide. Other than causing nearly complete peptide precipitation below pH 4, higher pH (to pH 8) had no discernible effect on peptide solubility. Likewise, less than 0.15 mM Ca(II)-bound peptide was soluble at any temperature between 5 °C and 35 °C. However, addition of salts to 1M increased the solubility of the Ca(II)-bound peptide to about 0.25 mM, with LiCl>NaCl>KCl in solubilization ability. Higher concentrations of calcium were required to obtain the fully Ca-bound peptide NMR spectrum (about 40 equivalents) than in the low ionic strength solution, suggesting the competition of monovalent cations for calcium ions, with calcium-induced aggregation being affected more than calcium-induced folding of the peptide.

Detergents and organic solvents were then tried to increase the solubility the calcium-bound peptide to greater than 0.25 mM. Perdeuterated dodecylphosphocholine (DPC) did not improve solubility at concentrations between 20 and 300 mM, even when including 20% dilauroylphosphatidylserine (Factor IX has a requirement for negatively charged phospholipids). However, the combination of 1M NaCl and 50 mM DPC did give a soluble complex (0.5 mM peptide, 10 equivalents calcium), although the NMR spectrum was broad and resembled the apo peptide. Additional calcium, or reduction of the salt or lipid concentration, decreased solubility of the peptide. We also tried to solubilize 0.25 mM peptide with 10 equivalents calcium in several separate solutions containing 1 mM CHAPS detergent, 20% trifluoroethanol (TFE), 20% dimethylsulfoxide (DMSO), 30% methanol, acetonitrile concentrations between 5 and 50%, or 3 mM phospholipid vesicles. In addition, most of these mixed solvent systems were tried in combination with alterations in temperature and ionic strength, all with little success. The addition of glycerol (> 30%) to a high salt (0.5M NaCl) solution reduced calcium-induced precipitation, but increased NMR linewidths considerably. Lesser amounts of glycerol caused precipitation.

Recent gel filtration studies on a similarly sized fragment (1-45) proteolytically derived from Factor IX, indicated a stable, compact Ca(II)-bound structure in the presence of 5 M urea (10). We likewise found that urea solubilized our 1-47 peptide, but only to about 0.5 mM. However, the spectrum was of high quality and progress clearly had been made. The addition of NaCl or DPC did not help appreciably, nor did variation of temperature. However, the addition of guanidine hydrochloride to the urea solution gave a remarkably clean spectrum, and solubilized the peptide to greater than 1 mM, with saturating amounts of calcium (Figure 2). An unusual feature of the calcium-bound peptide in guanidine/urea solution was its stability. The NMR spectra remained essentially unchanged over a wide range of temperatures (15 to 55 °C). The differences between the spectra were only trivial—higher temperatures sharpened linewidths (presumably by increasing the tumbling rate of the molecules) and decreased amide proton resonance intensities (presumably by increasing proton exchange). The spectrum of 1.5 mM peptide in guanidine/urea solution was very similar to the low concentration peptide collected in low salt buffer (6). Incidentally, solubilization of the peptide in 3 M guanidine alone, or in combination with 0.5 M NaCl, did not result in greater than 0.25 mM solubility for Ca(II)-bound peptide.

An additional problem for the sample in guanidine/urea solution at pH 5.9 was that the pH would slowly rise to above 7, which thereby broadened resonances. This was likely caused by the breakdown of urea to ammonia and bicarbonate, which is a process catalyzed by high pH. The addition of 0.5 M perdeuterated acetate buffer held the pH constant near 5.2, and yielded a sample that was stable for many days at 35 °C.

If presaturation was used to suppress the H_2O resonance, the protons from the urea and guanidine did not appear in the spectrum, because of rapid exchange of their nitrogen-bound protons with solvent at pH 5.2 and 35 °C. At lower pH (or temperature), however, the guanidine peak near 6.8 ppm became large, consis

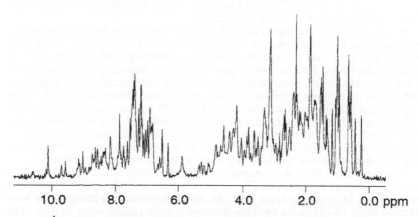

Figure 2. ^{1}H NMR spectrum of Factor IX (1-47) peptide. The sample contained 1.5 mM peptide, 90 mM $CaCl_2$, 0.5 M sodium acetate-d$_4$, pH 5.0, 2.5 M guanidine•HCl, and 3 M urea. 64 scans were co-added and the solvent was suppressed by presaturation.

tent with the pH minimum of guanidino groups being around pH 3. In turn, the urea peak near 5.8 ppm, became large at higher pH, especially around pH 6, nearer to its exchange minimum. The protons of urea and guanidine prevented the collection of NOESY data using binomial H_2O suppression methods, which would yield full intensity for some solvent-exposed amide protons of the peptide. Nonetheless, cross peaks to every amide proton of the peptide were observed using presaturation for H_2O suppression.

Collection of the NMR data in the high salt samples presented no insurmountable problems, although the 90° pulse width was 13 μs, which led to considerable loss in signal intensity, as compared to the situation in low salt solution (90° pulse width of 8 μs). However, the inversion pulses were clean and the two-dimensional spectra were of high quality. The suggested use of a 4 mm tube to shield the probe from the high-salt sample (11) did restore the pulse width to about 9 μs, but the simultaneous loss in sample volume canceled any appreciable signal-to-noise gains. Because of convenience, we used the standard 5 mm NMR tubes.

The Mg(II) bound peptide behaved differently from both the apo and Ca(II) forms. Although the Mg(II) peptide was soluble to about 1 mM, some NMR resonance lines were broad at 35 °C. Lower temperatures narrowed resonances slightly, suggesting exchange broadening. However, lower peptide concentrations or the addition of guanidine and urea, as described above, also reduced linewidths, pointing to aggregation as the more likely explanation.

Homonuclear NOESY and TOCSY data, and especially DQF-COSY data, provided sufficient resolution at 500 MHz for resonance assignment and for obtaining conformational data. Two-dimensional spectra of the Factor IX 1-47 peptide and the procedures used for cross-peak assignment have been published elsewhere (6-8), and were quite standard (12).

III. Structure Determination

After completion of resonance assignment, differences in chemical shift between the metal-free, Mg(II)-bound, and Ca(II)-bound forms of the peptide were apparent (Figure 3). In the N-terminal 14 amino acids, the Mg(II)-bound form appeared similar to the apo peptide, and in the C-terminal 33 amino acids, it was similar to the Ca(II)-bound peptide. These chemical shift differences indicate differences in structure (see below).

Conformational data were gathered and structures were determined for the peptide in the apo, Mg(II)-bound, and Ca(II)-bound forms (Table I). Additional details can be found in separate publications dealing with these structures individually (6-8). Structures were generated using distance geometry (Biosym Technologies). Here, distance-restrained energy minimization was applied to the geometric average of the metal-bound structure to restore the regions of the peptide distorted by the averaging process (mainly solvent-exposed amino acid side-chains). The final energies were -235 kcal/mole and -418 kcal/mole for the Mg(II) and Ca(II)-bound structures, respectively. Since the metal-free structure had extensive unordered regions (7), energy minimization was performed for one member of the structure set and yielded a final energy of -95 kcal/mole.

There are three structural elements in the apo peptide which are linked by a flexible polypeptide backbone (7). These elements include a short N-terminal tetrapeptide loop (amino acids 6-9), the disulfide-containing hexapeptide loop (amino acids 18-23), and a carboxy-terminal α helix (amino acids 37-46). The structured regions in the apo peptide are insufficient to support phospholipid binding, indicating the importance of additional structural features in the Ca(II)-stabilized conformer (13).

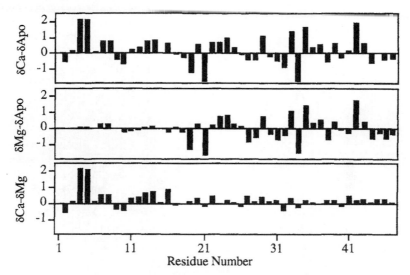

Figure 3. Chemical shift comparison of amide protons for the metal-free, Mg(II)-bound, and Ca(II)-bound forms of Factor IX(1-47) peptide. Chemical shift comparisons for the α protons have been made (8). Most noteworthy are the small chemical shift differences (<0.2 ppm) between the apo and Mg (II) forms for residues 1-15, and between the Ca(II) and Mg(II) forms for residues 16-47.

The Ca(II) bound peptide contains these ordered structural elements of the metal-free form plus other ordered regions (6) (Figure 4). The domain is mainly helical, except for an N-terminal loop. Here, Gla residues are oriented to the interior of the protein, suggesting an internal Ca(II) binding pocket, in a manner similar to that of prothrombin (14). The majority of hydrophobic residues in the carboxyl-terminal three-quarters of the molecule stabilize a globular core. However, an N-terminal hydrophobic surface patch (residues 6-10) appears to represent a component of the phospholipid binding site in Factor IX (bottom of Figure 4). The conformation of these residues in the Mg(II) bound peptide, a form that does not bind phospholipids, was therefore of high interest.

Table I. Structure Determination Statistics for the Factor IX Peptide

	Apo	Mg(II)	Ca(II)
# of distance restraints	600	710	851
# of med. + long-range NOEs	142	330	427
# of torsion measurements	20	0	42
# of structures calculated	15	15	20
residues 6-9 r.m.s.d.[a]	1.0±0.3	1.4±0.4	0.5±0.1
residues 18-23 r.m.s.d.	1.6±0.7	0.9±0.2	0.7±0.1
residues 35-46 r.m.s.d.	1.5±0.6	1.0±0.2	0.7±0.1
residues 12-46 r.m.s.d.	7.0±2.1	1.5±0.3	1.0±0.1
residues 2-46 r.m.s.d.	9.2±2.6	3.5±1.3	1.3±0.2

[a]Root-mean-square deviation (in Å) of backbone atoms between structures superimposed to the energy-minimized conformer. R.m.s.d. values to the geometric average are about 40% lower, and have been reported elsewhere (6,7).

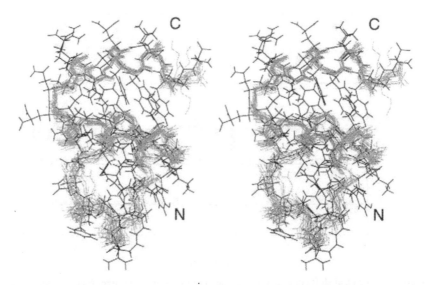

Figure 4. Stereoview of the Factor IX (1-47):Ca(II) structures. The backbone atoms of calculated structures are shown in gray superimposed using the energy-minimized average structure, shown in black. Side-chains of the average structure are also shown.

Despite the impediment of concentration-dependent aggregation to the acquisition of high quality spectra, two-dimensional ^{1}H NMR studies were performed on the Mg(II)-bound peptide. Resonance cross peaks were assigned using the Ca(II) peptide cross-peak patterns as a guide (8). The NOE contacts for the carboxyl-terminal 36 residues were nearly identical to those of the Ca(II)-bound peptide. In contrast, there was little similarity in NOE contacts within the amino-terminal 11 residues between the two metal forms. All NOE contacts confined to this peptide region were instead similar to those found in the apo Factor IX (1-47) spectra.

Mg(II)-bound structures were determined using a total of 710 distance restraints. As a consequence of cross-peak broadening, torsion angle information was not available for structure calculations. The Mg(II)-bound peptide had defined structure from residues 12-46 that was very similar to the Ca(II)-bound structure, with a r.m.s. deviation of 1.2 Å for backbone atoms. In contrast, the amino-terminal 11 residues lacked defined structure in Mg(II)-bound Factor IX (1-47), with the exception of a short loop from residues 6 to 9, which was also found in the apo Factor IX (1-47) structure (7).

The spectrum of a 1.5 mM sample of Mg(II)-Factor IX (1-47) had a small subclass of 22 NOE contacts between residues 1-11 and residues 12-47 that could not be identified in either the apo or Ca(II)-bound peptide spectra (6,7). These cross peaks, some of which were strong, disappeared at lower peptide concentrations and in the solution containing 3 M urea and 2.5 M guanidine HCl, suggesting the presence of intermolecular contacts produced through aggregation.

Incidentally, data recently collected for the Cd(II) peptide shows narrow NMR resonance lines at chemical shift positions similar to the Mg(II)-bound form. It is predicted that the structure of the Cd(II)-bound peptide will resemble the Mg(II) form.

IV. Discussion

The Ca(II)-induced conformational transition that leads to phospholipid binding is between a relatively disordered metal-free structure and a compact Ca(II)-bound structure. The structure of the Mg(II) bound peptide is nearly the same as the Ca(II) bound form, except for residues 1-11, which are disordered (Figure 5). Comparison of the apo, Mg(II) and Ca(II) structures reveals a feature required for phospholipid binding by Factor IX. In the apo and Mg(II) forms, the hydrophobic patch defined by residues 6-10 is unstructured relative to the rest of the protein. In the Ca(II)-bound form, however, the hydrophobic patch is held by the C-terminal part of the domain in a specific conformation for presentation on the surface. The exposed hydrophobic side-chains may interact with the acyl chains of phospholipids.

Populations of conformation-specific polyclonal antibodies, anti-Factor IX:Mg(II), can be isolated that detect an epitope common to Gla domain structures bound to Ca(II) and Mg(II), but do not detect an epitope in the metal-free Gla domain. Other anti-Factor IX:Ca(II)-specific antibodies detect an epitope specific to the Ca(II)-bound structures (3,4). The structure in common between the Ca(II) and Mg(II)-bound Factor IX Gla domains resides within the region from residues 12-47. Furthermore, the Ca(II)-bound, Mg(II)-bound, and metal-free Factor IX (1-47) structures all share a structured α helix from residues 37-46. These observations suggest that the conformation-specific anti-Factor IX:Mg(II) antibodies are directed at determinants between residues 12 and 36. In contrast, the anti-Factor IX:Ca(II)-specific antibodies must be directed, at least partially, toward determinants between residues 1 and 11, since this is the only region that differs between the Ca(II)-bound and Mg(II)-bound structures.

Comparison of Ca(II)-bound Factor IX with the analogous region of the bovine prothrombin fragment I (14) reveals a strikingly similar global fold, particularly for residues 12-47 (r.m.s.d. is ~1.2Å). However, the backbone conformations deviate in the first 11 residues, and most prominently between residues 3 and 6. For example, the ϕ angle of residue 4 is positive in Factor IX

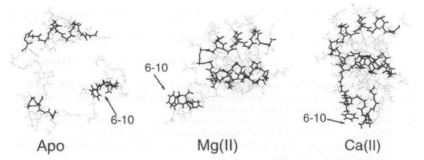

Apo   Mg(II)   Ca(II)

Figure 5. Comparison of structures for the apo, Mg(II)-bound, and Ca(II)-bound forms of the Factor IX(1-47) peptide. Mobile backbone regions are shown in gray. The well-ordered backbone atoms are shown in black, with the corresponding sidechains in gray. In addition, the sidechains of residues 6, 9, and 10 are in black. These residues protrude from the surface of the Ca(II)-bound structure to form a hydrophobic patch that is likely part of the phospholipid binding site of Factor IX.

(84°) and negative in prothrombin (-120°), whereas the ϕ angles of residues 5 and 6 are negative in Factor IX (-99°, -126°) and positive in prothrombin (175°, 74°). There are three possibilities for this observed conformational difference. One, this may represent a true difference in conformation between the native proteins, which reflects their differences in function. It is known through mutagenesis studies that Gla residues at identical positions in vitamin K-dependent proteins play different roles (15,16). In addition, the phospholipid binding affinity and specificity varies among vitamin K-dependent proteins (17). Two, residues 4 and 5 of the prothrombin crystal structure were associated with poor electron density and residues 5 and 6 were reported to be involved in intermolecular van der Waals interactions between adjacent molecules (14). Three, the guanidine/urea system used to solubilize the Ca(II)-bound Factor IX (1-47) peptide may affect this region of the peptide. (Although the chemical shifts and NOE contacts in this region are identical, within experimental error, to the data collected without guanidine and urea.) In sum, it is not clear whether the conformational difference at the amino terminus between prothrombin and Factor IX is significant; the question warrants investigation of the prothrombin peptide using NMR spectroscopy.

Acknowledgments

This work was supported by NIH grant HL42443 and the NMR spectrometer was acquired with NIH grant RR06282.

References

1. Furie, B., and Furie, B. C. (1988). *Cell* **53**, 505-518.
2. Nelsestuen, G. L. (1976). *J. Biol. Chem.* **251**, 5648-5656.
3. Borowski, M., Furie, B. C., Bauminger, S., and Furie, B. (1986). *J. Biol. Chem.* **261**, 14969-14975.
4. Liebman, H. A., Furie, B. C., and Furie, B. (1987). *J. Biol. Chem.* **262**, 7605-7612.
5. Jacobs, M., Freedman, S. J., Furie, B. C., and Furie, B. (1994). *J. Biol. Chem.* **269**, 25494-25501.
6. Freedman, S. J., Furie, B. C., Furie, C., and Baleja, J. D. (1995). *Biochemistry* **34**, (in press).
7. Freedman, S. J., Furie, B. C., Furie, C., and Baleja, J. D. (1995). *J. Biol. Chem.* **270**, 7980-7987.
8. Freedman, S. J., Furie, B., Furie, B. C., and Baleja, J. D. (1995). (Manuscript in preparation).
9. Tulinsky, A., Park, C. H., and Skrzypczak-Jankun, E. (1988). *J. Mol. Biol.* **202**, 885-901.
10. Medved, L. V., Vysotchin, A., and Ingham, K. C. (1994). *Biochemistry* **33**, 478-485.
11. Dykstra, R. W. (1989). *J. Magn. Reson.* **84**, 388-391.
12. Wüthrich, K. (1986) "NMR of Proteins and Nucleic Acids." Wiley, New York.
13. Teleman, O., and Stenflo, J. (1995). *Nature Structural Biology* **2**, 504-509.
14. Soriano-Garcia, M., Padmanabhan, K., de Vos, A. M., and Tulinsky, A. (1992). *Biochemistry* **31**, 2554-2566.
15. Zhang, L., Jhingan, A., and Castellino, F. J. (1992). *Blood* **80**, 942-952.
16. Ratcliffe, J. V., Furie, B., and Furie, B. C. (1993). *J. Biol. Chem.* **268**, 24339-24345.
17. Mann, K. G., Nesheim, M. E., Church, W. R., Haley, P., and Krishnaswamy, S. (1990). *Blood* **76**, 1-16.

Chemical Kinetic Investigations of Neurotransmitter Receptors on a Cell Surface in the μs Time Region[1]

Li Niu, Christof Grewer[2], and George P. Hess[3]

Section of Biochemistry, Molecular and Cell Biology, Division of Biological Sciences, 217 Biotechnology Building, Cornell University, Ithaca, NY 14853-2703

INTRODUCTION

Neurotransmitter receptors are large (~ 200 kD) proteins embedded in the membrane of nerve and muscle cells. Amino-acid sequences suggest that the neurotransmitter receptors are structurally related (Betz, 1990; Stroud et al., 1990). Three-dimensional structural information is so far available only at 9 Å resolution and only for the nicotinic acetylcholine receptor (Unwin, 1995). Upon binding a specific neurotransmitter, excitatory receptors transiently allow the passage of inorganic cations, and inhibitory receptors the passage of inorganic anions. The resulting voltage change across the membrane of a cell determines whether a signal is transmitted to another cell (Kandel et al., 1995). Regulation of signal transmission by these receptor proteins is believed to be involved in various neurological functions such as learning and memory. Receptor malfunction has been indicated in many disorders, including stroke and Parkinson's syndrome (Jessell & Kandel, 1993). Procaine and QX-222, local anesthetics, MK-801, an anticonvulsant, and cocaine, a powerful drug of abuse, all inhibit the nicotinic acetylcholine receptor (Ramoa et al., 1990; Gilman et al., 1992). Because the function of these proteins must be investigated in living cells or tightly sealed membrane vesicles, information about the mechanism of receptor function has come mainly from electrophysiological measurements, which give reliable

---

[1]This research was supported by grants (NS08527 and GM04842) from the National Institutes of Health

[2]CG is grateful for a Feodor Lynen Fellowship of the Alexander von Humboldt Foundation

[3]Corresponding author

information only about the life time and conductance of the open channel (Neher & Sakmann, 1976; Sakmann & Neher, 1983).

Here we describe a rapid chemical kinetic technique using laser-pulse photolysis suitable for measuring the kinetic constants of the receptor-mediated reactions, with a μs time resolution (Milburn et al., 1989; Billington et al., 1992; Matsubara et al., 1992; Niu & Hess, 1993; Ramesh et al., 1993; Gee et al., 1994; Wieboldt et al., 1994a, b). This approach utilizes the rapid release of biologically active molecules from photolabile precursors (Kaplan, 1990; Adams & Tsien, 1993; Corrie & Trentham, 1993; Hess, 1993; Hess et al., 1995). We use procaine, an inhibitor of the muscle acetylcholine receptor, as an example to illustrate how the mechanism of inhibition was investigated using the laser-pulse photolysis technique. The new information and its implications in understanding the structural and functional relationships of the receptors and in rational design of potential therapeutic agents are discussed.

EXPERIMENTAL PROCEDURES

Although many neuronal cells expressing a variety of receptors are now available (Yakel et al., 1990; Tygesen et al., 1994; Tyndale et al., 1994), to illustrate the technique the BC_3H1 cell line, which expresses the muscle acetylcholine receptor (Schubert et al., 1974), was used.

The original cell-flow device, consisting of a U-tube (Krishtal & Pidoplichko, 1980) used to equilibrate the receptors on a cell surface with neurotransmitters in the ms time region, was modified (Figure 1). The new device allows one to preincubate, if necessary, a solution of inhibitor with a cell before a solution containing both inhibitor and activating ligand flows over the cell.

Caged neurotransmitters are equilibrated with the receptors. A laser pulse liberates the neurotransmitter and the resulting current, due to the opening of the receptor-channels, is recorded. A Candela UV 500 flash-lamp-pumped dye laser (dye: oxazine 720 perchlorate) is used. Laser pulses at 343 nm (pulse length: 600 ns) tuned with a secondary harmonic generator are adjusted to give 200-500 μJ from the optical fiber of 200 μm core diameter (Niu & Hess, 1993). The distance between the cell and the end face of the fiber is about 400 μm (Figure 2); the diameter of the illuminated area around the cell is thus about 400 μm. This orientation ensures optimum results: lack of cell damage and high yield of released neurotransmitter. Figure 2 also illustrates a whole-cell current trace induced by 100-μM released carbamoylcholine using laser-pulse photolysis. From time zero where a laser pulse was fired to 3 ms where the current reached a maximum, the kinetic parameters that control the channel opening (see Figure 3) can be obtained, including the channel-opening (k_{op}) and -closing (k_{cl}) rate constants, the receptor:neurotransmitter dissociation constant (K_1), and the equilibrium constant of channel opening ($\Phi = k_{cl}/k_{op}$) (Matsubara et al., 1992). The falling phase of the current reflects receptor desensitization and

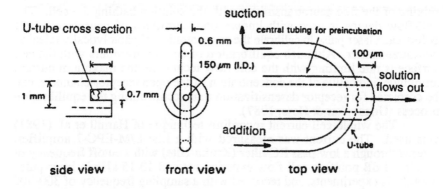

FIGURE 1. Schematic drawing (not to scale) of the device used for rapid equilibration of receptors on a cell surface with neurotransmitters, caged neurotransmitters, and inhibitors. In the modified device was stainless steel tubing with a cut into which the U-tube of the cell-flow device was fitted, having a shape like "ψ". Shown in the side view of this Figure is the cross section of the U-tube (with its hole of 150 μm diameter facing towards right) fitted into the cut. The central tubing and the U-tube were sealed together with epoxy, leaving an opening that surrounds the U-tube hole (for details of the construction of the device, contact the corresponding author). Before measurements, the central tubing is prefilled with the preincubation solution. Suction is applied to the U-tube until the device is activated by a solenoid valve; in this way, any preincubation solution that leaks by diffusion when the device is not in use is removed by suctioning. A BC₃H1 cell is about 15 μm in diameter. The cell suspended from the current-recording electrode is approximately 100 μm away from the hole. The linear flow rate of solution used was 1 cm/s. The direction of the solution application (addition, suction, and solution flowing out when the solenoid valve is activated) is indicated in the drawing.

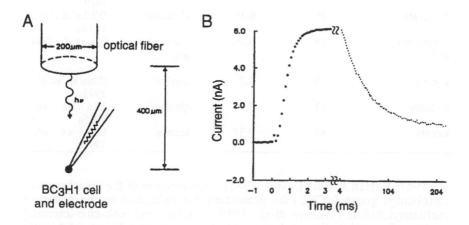

FIGURE 2. (A) Laser-pulse photolysis setup. (B). Whole-cell current induced by 100 μM carbamoylcholine using the laser-pulse photolysis technique on a single BC₃H1 cell (Matsubara et al., 1992).

mixing of the free neurotransmitter with the solution bathing the cell. The cell-flow device is used with known concentrations of neurotransmitter before each laser pulse to calibrate the concentration of photolytically released neurotransmitter. Although equilibration of neurotransmitters with the cell surface is rate limiting with the cell-flow technique, the maximum current amplitude, indicative of the concentration of the open receptor-channels, can be corrected for receptor desensitization that occurs during the equilibration process (Udgaonkar & Hess, 1987).

The whole-cell current recording technique of Hamill et al. (1981) was used. The current was measured with a List L/M-EPC-7 amplifier, filtered through a low-pass RC filter (Krohn-Hite) with a cutoff frequency of 2 kHz (-3-dB point) in cell-flow experiments and 10-15 kHz in laser-pulse photolysis experiments, and recorded with a sampling frequency of 300-500 Hz in cell-flow experiments and 50 kHz in laser-pulse photolysis experiments by a Labmaster DMA digitizer (Scientific Solutions) driven by the PClamp program (Axon).

Several photolabile biologically inert neurotransmitter precursors are now available for chemical kinetic measurements and neuronal mapping. The Table below lists the $t_{1/2}$ values of the photolysis reaction and the product quantum yield for some compounds. *o*-Nitrobenzyl derivatives have been used as photoremovable blocking reagents for many functional groups in organic chemistry (De Mayo, 1960; Barltrop et al., 1966; Patchornik et al., 1970; Kaplan et al., 1978; McCray et al., 1980; Walker et

Caged neurotransmitter	Photolysis lifetime $(t_{1/2} / \mu s)$	Product quantum yield	Target receptor	Reference
carbamoylcholine	45	0.8	acetylcholine	Milburn et al., 1989
glutamate	21	0.14	glutamate	Wieboldt et al., 1994b
γ-aminobutyric acid	19	0.16	γ-aminobutyric acid	Gee et al., 1994
glycine	<3	0.2	glycine	Ramesh et al., 1993
β-alanine	<3	0.2	glycine	Niu et al., 1995a
kainate	45	0.37	kainate	Niu et al., 1995b

al., 1987; Kaplan & Ellis-Davies, 1988). Introduction of the α-carboxyl-*o*-nitrobenzyl group (αCNB) for protecting the carbamate amino group of carbamoylcholine (Milburn et al., 1989), a stable and well-characterized analog of acetylcholine, led to a suitable compound for rapid kinetic investigations of acetylcholine receptors (Matsubara et al., 1992; Niu & Hess, 1993; Niu et al., 1995c), and for neuronal mapping (Callaway & Katz, 1993; Avery et al., 1994; Dalva & Katz, 1994; Denk, 1994). The αCNB group has also been used to prepare caged derivatives of the excitatory neurotransmitters glutamate and kainate and the inhibitory neurotransmitter γ-aminobutyric acid. More recently, we used a new caging group, 2-methoxy-5-nitrophenyl (MNP), which in the case of caged glycine and β-alanine is photolyzed with a $t_{1/2}$ value of < 3 μs. MNP-β-alanine, which upon photolysis activates glycine channels in hippocampal neurons, is stable at neutral pH (Niu et al., 1995a), unlike MNP-glycine which decomposes with a $t_{1/2}$ value of ~7 minutes (Ramesh et al., 1993).

When both inhibitors and caged neurotransmitters are present in investigations of the mechanism of receptor inhibition (see Figure 4A, B, and C), control experiments must be performed to ascertain that the reduction of rate and amplitude is solely attributable to inhibition and not to either attenuation of the laser-pulse energy to which the caged neurotransmitter is exposed in the presence of the inhibitor, thus releasing less neurotransmitter, or damage to the receptor or cell upon repetitive laser pulses. Procaine is used here as an example to illustrate the following control experiments. The wavelength of 343 nm was chosen because at this wavelength caged carbamoylcholine still absorbs strongly but procaine does not. The photolysis products from 750-μM caged carbamoylcholine in the absence and presence of 500-μM procaine (the highest concentration used in the cell experiments) were analyzed using reverse-phase HPLC chromatography. Neither the amount of carbamoylcholine released in the presence of procaine nor the concentration of procaine after photolysis was affected. An independent method, a cell-flow technique (Udgaonkar & Hess, 1987), was used to determine the whole-cell current amplitude obtained at a known concentration of carbamoylcholine in conjunction with the laser-pulse photolysis experiments. This technique was also used to measure the current amplitudes with known concentrations of both carbamoylcholine and procaine. The two independent techniques gave similar results, confirming that the reduction of both rate and amplitude in the presence of procaine determined in laser-pulse photolysis experiments was indeed due to receptor inhibition. To avoid experimental errors in either the laser-pulse photolysis experiments or the cell-flow measurements with preincubation of inhibitors, one must be certain that the concentration of the inhibitor solution is not diluted by the surrounding bath solution prior to data acquisition.

RESULTS

To illustrate the power of the chemical kinetic approach, we shall now consider the mechanism of inhibition of the nicotinic acetylcholine receptor. A prevailing model of inhibition, based upon extensive electrophysiological and chemical labeling experiments, suggested that these inhibitors enter the receptor-channels only after the channels open and sterically plug the channels, thus blocking signal transmission (Adams, 1976; Neher & Steinbach, 1978; Galzi et al., 1991). Deviations from the open-channel blockade model have been observed in electrophysiological experiments (Adams, 1977; Neher, 1983; Papke & Oswald, 1989) and chemical kinetic experiments with a 5-ms time resolution (Shiono et al., 1984; Karpen & Hess, 1986). A regulatory site mechanism was proposed in which inhibitors can bind to a regulatory site of the receptor thus inhibiting both the closed- and open-channels forms (Shiono et al., 1984; Karpen & Hess, 1986). Examination of the mechanisms in Figure 3 led to a decisive experiment (Figure 4) (Niu & Hess, 1993). The regulatory site mechanism predicts that both k_{cl} and k_{op} will be affected in the presence of an inhibitor, whereas the open-channel blockade predicts that only k_{cl}, but not k_{op}, will be

$$I + A + L \xrightleftharpoons{K_1} AL \xrightleftharpoons{K_1} AL_2 \underset{k_{cl}}{\overset{k_{op}}{\rightleftharpoons}} \overline{AL}_2$$

$$IA \xrightleftharpoons{K_1} IAL \xrightleftharpoons{K_1} IAL_2 \cdots \overline{IAL}_2$$

FIGURE 3. Inhibition of acetylcholine receptor. L represents the channel-activating ligand, A the active nondesensitizing receptors, K_1 the dissociation constant for the receptor:ligand complexes, and $\overline{AL}_2$ the open-channel form of the receptor. I represents an inhibitor, and K_I and $\overline{K}_I$ the dissociation constants of the receptor:inhibitor complexes of the closed- and open-channel forms of the receptor respectively. The open-channel blockade mechanism is shown by solid lines, whereas the regulatory site mechanism is shown by dashed lines.

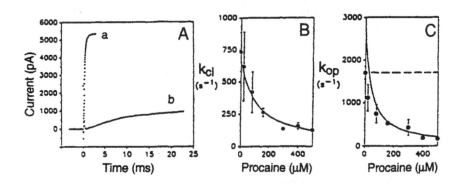

FIGURE 4. (A) Laser-pulse photolysis of 750-μM caged carbamoylcholine with BC3H1 cells at pH 7.4, 22 °C, and -60 mV transmembrane potential. The carbamoylcholine released was 115 μM in the absence (a) and presence (b) of 500-μM procaine. The observed first order rate constants and the maximum current amplitudes for a and b are 2400 and 200 s^{-1}, and 5.4 and 0.8 nA, respectively. (B) Effect of procaine on k_{cl}. (C) Effect of procaine on k_{op}.

affected. The effect of inhibitors on k_{op} has not previously been determined, but can now be measured with the laser-pulse photolysis technique.

Figure 4A shows representative whole-cell current traces induced by 115-µM released carbamoylcholine in the absence (a) and presence (b) of 500-µM procaine. Notice that both the observed channel-opening rate constant (k_{obs}) and the amplitude in the presence of procaine are reduced. The effects of procaine on k_{cl} and k_{op} were measured (Figure 4B and C) (Niu & Hess, 1993). Two concentrations of released carbamoylcholine (Figure 4: B, 20 µM; C, 115 µM) were used. In Figure 4B, the value of k_{obs} reflecting k_{cl} (Matsubara et al., 1992; Niu & Hess, 1993) decreased with increasing procaine concentration. In Figure 4C, the effect of procaine on k_{op} was measured. The effects of procaine on both k_{op} and k_{cl} suggest that procaine interacts with both open- and closed-channel forms (Niu & Hess, 1993). If procaine binds only to the open channel and then blocks it, no effect on k_{op} is expected, indicated by a dotted line in Figure 4C. A $\overline{K_I}$ value of 110 ± 50 µM, associated with the open-channel form of the receptor, suggests the affinity of procaine for this form is somewhat weaker than for the closed forms (K_I of 40 ± 60 µM). Several other inhibitors have been studied using the same technique and cell line. Cocaine binds to the closed-channel forms of the receptor much more strongly than to the open-channel form (Niu et al., 1995c), while QX-222 and MK-801 bind more strongly to the open-channel form than to the closed-channel forms. It was also found that rates of inhibition by cocaine and MK-801 are much slower than the observed rate of channel opening ($t_{1/2} = 2$ ms). These results suggest that cocaine and MK-801 first bind to the receptor rapidly and form inhibitor:receptor intermediates which can still form an open channel, and then slowly form inhibited receptors.

DISCUSSION

Several precautions need to be taken in using the laser-pulse photolysis technique. Among the laser sources, including XeCl excimer (308 nm, 8 ns pulse length) and N_2 lasers (337 nm, 4 ns pulse length), the tunable flash-lamp-pumped dye laser used in the experiments described above is probably the best choice at present. This is based upon the following observations. The longer the lasing wavelength used for photolyzing caged compounds [343 nm is the longest wavelength we used (Niu & Hess, 1993)], the longer each cell can be used in an experiment. The longer the pulse length, the less the damage to the cell or the receptor. The dye laser also offers tunability of the wavelength, which is an important option to have in studying any inhibitor which may absorb light in part of the UV region. Use of high energy (e.g., 1 mJ in the UV region with a dye laser) laser light is not necessarily beneficial. To the contrary, the higher the energy used, the more likely it is that the seal between the cell and electrode breaks and the experiment has to be abandoned. The caged compound stock can be contaminated by some residual amount of free neurotransmitter; the purity can be improved, if necessary, with HPLC (Milburn et al., 1989). In the case of caged carbamoylcholine, a purity of ~99.5% can be obtained. Using higher than necessary concentrations of caged carbamoylcholine can be counterproductive. Even a minute amount of contamination of the caged carbamoylcholine with free carbamoylcholine (~0.5%) can cause problems, e.g., detectable desensitization during the equilibration of caged compound with the cell. This situation becomes worse when a higher concentration of caged carbamoylcholine and a longer equilibration are required (the minimal time of equilibration of caged carbamoylcholine with a BC$_3$H1 cell is 200 ms, but, when inhibitors are used, the equilibration time may become longer depending on how fast the inhibition or binding of the inhibitors is). The desensitization has to be taken into account when the concentration of liberated carbamoylcholine is calibrated using a known concentration of carbamoylcholine with the cell-flow technique. If a cell is equilibrated with caged carbamoylcholine for less than 500 ms prior to photolysis, the receptor desensitization due to the free carbamoylcholine from the caged carbamoylcholine solution may be neglected in the calibration. In kinetic studies, a high concentration of the caged compound can cause lower concentrations of released neurotransmitter. This is likely due to the fact that caged compounds in a solution layer removed from the cell surface absorb the laser light delivered to the cell surface.

A newly developed rapid chemical kinetic technique, laser-pulse photolysis, is described, and the usefulness of this technique for investigation of mechanisms of receptor-channel formation and inhibition in the μs-to-ms time region is demonstrated. The technique is suitable for kinetic investigations not only of the excitatory acetylcholine and glutamate receptors but also of the inhibitory glycine and γ-aminobutyric acid receptors. With it one can determine (i) the rate constants of channel opening and closing separately, (ii) the dissociation constant of the neurotransmitter from the receptor site controlling channel opening, (iii) the apparent dissociation constants of the inhibitor from the closed and open channels separately, and (iv) the rates with which the inhibitor binds to the closed and open channels separately.

The results obtained in our experiments are consistent with the existence of a regulatory (inhibitory) site in the nicotinic acetylcholine receptor. The new information obtained has some implications for not only the nicotinic acetylcholine receptor but also other receptors. (i) Structural studies using inhibitors, chemical labels, and cDNA technology, directed towards delineating the channel pore, are based primarily on the assumption that the inhibitor-binding sites are located inside the open channel (Galzi et al., 1991). Knowledge of the site of action is thus critical in such studies. (ii) A mechanism in which an inhibitor binds to a regulatory site rather than inside the open channel suggests a strategy for the design of therapeutic agents that can be used to alleviate the effects of a drug like cocaine. Some therapeutic agents may be found that, for instance, prevent cocaine from binding to its regulatory site without affecting signal transmission. With the laser-pulse photolysis technique and various caged neurotransmitters, kinetic investigations of how the receptor proteins are activated, regulated, and inhibited can now be conducted in a µs time resolution on the surface of a single cell. Because of the spatial resolution of the technique, caged neurotransmitters can also be used in mapping neuronal circuits (Callaway & Katz, 1993; Avery et al., 1994; Dalva & Katz, 1994; Denk, 1994).

ACKNOWLEDGMENT

We thank Susan Coombs for technical editing.

REFERENCES

Adams, P. R. (1976) *J. Physiol. (London)* **260**, 531-532.

Adams, P. R. (1977) *J. Physiol. (London)* **268**, 291-318.

Adams, S. R., & Tsien, R. Y. (1993) *Annu. Rev. Physiol.* **55**, 755-783.

Avery, L., Davis, M. W., Denk, W., Dent, J., & Hess, G. P. (1994) *Worms Breeders Gazette* **13** (4), 72.

Betz, H. (1990) *Neuron* **5**, 383-392.

Barltrop, J. A., Plant, P. J., & Schofield, P. (1966) *J. Chem. Soc. Chem. Commun.*, 822-823.

Billington, A. P., Matsubara, N., Webb, W. W., & Hess, G. P. (1992) in *Adv. Prot. Chem. III* (Angeletti, R. H., ed.) pp 417-427, Academic Press, New York.

Callaway, E. M., & Katz, L. C. (1993) *Proc. Natl. Acad. Sci. USA* **90**, 7661-7665.

Corrie, J. E. T., & Trentham, D. R. (1993) in *Bioorganic Photochemistry* (Morrison, H., ed.) Vol. 2, pp 243-305, Wiley, New York.

Dalva, M. B., & Katz, L. C. (1994) *Science* **265**, 255-258.

De Mayo, P. (1960) *Adv. Org. Chem.* **2**, 367-425.

Denk, W. (1994) *Proc. Natl. Acad. Sci. USA* **91**, 6629-6633.

Galzi, J.-L., Revah, F., Bessis, A., & Changeux, J.-P.(1991) *Annu. Rev. Pharmacol.* **31**, 37-72.

Gee, K. R., Wieboldt, R., & Hess, G. P. (1994) *J. Am. Chem. Soc.* **116**, 8366-8367.

Gilman, A. G., Rall, T. W., Nies, A. S., & Taylor, P. (eds.) (1992) *The Pharmacological Basis of Therapeutics*, 8th Ed., Pergamon Press, New York.

Hamill, O. P., Marty, A., Neher, E., Sakmann, E., & Sigworth, F. J. (1981) *Pflügers Arch.* **391**, 85-100.

Hess, G. P. (1993) *Biochem.* **32**, 989-1000.

Hess, G. P., Niu, L., & Wieboldt, R. (1995) in *Ann. N.Y. Acad. Sci.* (Abood, L. G., & Lajtha, A., eds.) Vol 757, pp 23-39.

Jessell, T. M., & Kandel, E. R. (1993) *Cell* 72/*Neuron* 10, 1-30.

Kaplan, J. H. (1990) *Ann. Rev. Physiol.* 52, 897-914.

Kaplan, J. H., Forbush, B., & Hoffman, J. F. (1978) *Biochem.* 17, 1929-1935.

Kaplan, J. H., & Ellis-Davies, G. C. R. (1988) *Proc. Natl. Acad. Sci. USA* 85, 6571-6575.

Karpen, J. W., & Hess, G. P. (1986) *Biochem.* 25, 1777-1785.

Kendal, E. R., Schwartz, J. H., & Jessell, T. M. (eds.) (1995) *Essentials of Neural Science and Behavior*, Appleton & Lauge, Norwalk, Connecticut.

Krishtal, O. A., & Pidoplichko, V. I. (1980) *Neurosci.* 5, 2325-2327.

McCray, J. A., Hérbette, L., Kihara, T., & Trentham, D. R. (1980) *Proc. Natl. Acad. Sci. USA* 77, 7237-7241.

Matsubara, N., Billington, A. P., & Hess, G. P. (1992) *Biochem.* 31, 5507-5514.

Milburn, T., Matsubara, N., Billington, A. P., Udgaonkar, J. B., Walker, J. W., Carpenter, B. K., Webb, W. W., Marque, J., Denk, W., McCray, J. A., & Hess, G. P. (1989) *Biochem.* 29, 49-55.

Neher, E. (1983) *J. Physiol. (London)* 339, 663-678.

Neher, E., & Sakmann, B. (1976) *Nature* 260, 799-802.

Neher, E., & Steinbach, J. H. (1978) *J. Physiol. (London)* 277, 153-176.

Niu, L., & Hess, G. P. (1993) *Biochem.* 32, 3831-3835.

Niu, L., Wieboldt, R., Ramesh, D., Carpenter, B. K., & Hess, G. P. (1995a) *Biochem.*, submitted.

Niu, L., Gee, K. R., Schaper, K., & Hess, G. P. (1995b) *Biochem.*, submitted.

Niu, L., Abood, L. O., & Hess, G. P. (1995c) *Proc. Natl. Acad. Sci. USA* (in press).

Papke, R. L., & Oswald, R. E. (1989) *J. Gen. Physiol.* 93, 785-811.

Patchornik, A., Amit, B., & Woodward, R. B. (1970) *J. Am. Chem. Soc.* 92, 6333-6335.

Ramoa, A. S., Alkondon, M., Aracava, Y., Irons, J., Lunt, G. G., Deshpande, S. S., Wonnacott, S., Aronstam, R. S., & Albuquerque, E. X. (1990) *J. Pharmacol. Exp. The.* 254, 71-82.

Ramesh, D., Wieboldt, R., Niu, L., Carpenter, B. K., & Hess, G. P. (1993) *Proc. Natl. Acad. Sci. USA* 90, 11074-11078.

Sakmann, B., & Neher, E. (eds.) (1983) *Single-Channel Recording*, Plenum Press, New York.

Schubert, D., Harris, A. J., Devine, E. E., & Heinemann, S. (1974) *J. Cell Biol.* 61, 398-402.

Shiono, S., Takeyasu, K., Udgaonkar, J. B., Delcour, A. H., Fujita, N., & Hess, G. P. (1984) *Biochem.* 23, 6889-6893.

Stroud, R. M., McCarthy, M. P., & Shuster, M. (1990) *Biochem.* 29, 11009-11023.

Tygesen, C. K., Rasmussen, J. S., Jones, S. V. P., Hansen, A., Hansen, K., & Andersen, P. H. (1994) *Proc. Natl. Acad. Sci. USA* **91**, 13018-13022.

Tyndale, R. F., Hales, T. G., Olsen, R. W., & Tobin, A. J. (1994) *J. Neurosci.* **14**, 5417-5428.

Udgaonkar, J. B., & Hess, G. P. (1987) *Proc. Natl. Acad. Sci. USA* **84**, 8758-8762.

Unwin, N. (1995) *Nature* **373**, 37-43.

Walker, J. W., Somlyo, A. V., Goldman, Y. E., Somlyo, A. P., & Trentham, D. R. (1987) *Nature* **327**, 249-252.

Wieboldt, R., Ramesh, D., Carpenter, B. K., & Hess, G. P. (1994a) *Biochem.* **33**, 1526-1533.

Wieboldt, R., Gee, K. R., Niu, L., Ramesh, D., Carpenter, B. K., & Hess, G. P. (1994b) *Proc. Natl. Acad. Sci. USA* **91**, 8752-8756.

Yakel., J. L., Shao, X. M., & Jackson, M. B. (1990) *Brain Res.* **533**, 46-52.

Site-Directed Isotope Labeling of Membrane Proteins: A New Tool for Spectroscopists

Sanjay Sonar[‡], Chan-Ping Lee[§1], Cheryl F.C. Ludlam[‡],
Xiao-Mei Liu[‡], Matthew Coleman[‡], Thomas Marti[#2],
Uttam L. RajBhandary[§] and Kenneth J. Rothschild[‡*]

[‡]Physics Department and Molecular Biophysics Laboratory, Boston University, Boston MA 02215;
[§]Department of Biology, and [#] Department of Chemistry and Biology, Massachusetts Institute of
Technology, Cambridge, MA 02139

I. Introduction

A key goal in understanding how an integral membrane protein works is to obtain a detailed picture of the structural changes which occur during function. For this purpose, Fourier Transform infrared (FTIR) difference spectroscopy has been increasingly used in conjunction with a variety of methods in biochemistry and molecular biology (for a recent review see (1) and references therein). The technique derives its usefulness from the ability of FTIR-difference spectroscopy to detect bands arising due to the vibrations of individual residues and chemical groups in a protein which undergo conformational and/or chemical change. However, a key problem in the use of FTIR-difference spectroscopy is the assignment of bands to individual amino acid residues. Only after such assignments are made can information be derived about changes in specific components of the protein.

In the past, vibrational assignments have been made on the basis of two different approaches: uniform isotope labeling of amino acids and site-directed mutagenesis. While these methods led to the assignment of several specific bands (2-5), many of these assignments are still tentative, while most bands have not yet been assigned. The essential problems are: *i*) uniform isotope labeling can assign bands to particular amino acids but not to specific residues; and *ii*) site-directed mutagenesis, especially of those amino acid residues in the active site of a protein, can alter the structure and function of a protein, thereby preventing unambiguous assignments of individual bands.

We recently introduced a third method for assigning bands in FTIR-difference spectra (6) based on a technique which we have termed site-directed isotope labeling (SDIL). As summarized in Figure 1, a key element in SDIL is the use of a suppressor tRNA

---

[1] Present Address: Tzu Chi College of Medicine, Hualien, Taiwan
[2] Present Address: Bernard Nocht Institute for Tropical Medicine, 20359 Hamburg, Germany
[*] Address correspondence to this author.

TECHNIQUES IN PROTEIN CHEMISTRY VII

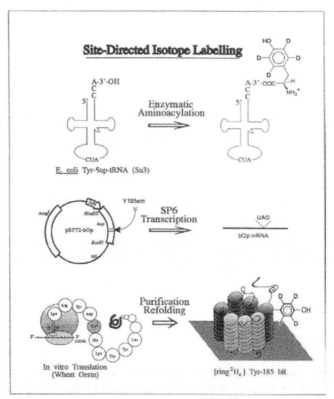

Figure 1: Major steps in site-directed isotope labeling of a protein: i) *E. coli* Tyr sup-tRNA is enzymatically aminoacylated with a deuterated tyrosine; ii) an amber codon is inserted in the gene at the position of the residue to be isotope labeled and iii) the mRNA is translated in a cell-free system in the presence of aminoacylated Tyr sup-tRNA. The resulting SDIL-analog is then purified and refolded. This illustration is reprinted from ref (6) with permission.

aminoacylated with an isotopically labeled amino acid. This tRNA is targeted to insert the isotopic amino acid at the proper position in the nascent protein by using an amber codon at the corresponding position in the gene. Cell-free synthesis (*in vitro* translation) and exogenous addition of the aminoacylated suppressor tRNA prevent aminoacylation of non-suppressor tRNAs with the isotopic amino acid, similar to the approach used for site-directed non-native amino acid replacement (SNAAR) (7-9).

We have chosen the integral membrane protein bacteriorhodopsin as a model system for demonstrating the application of SDIL-FTIR. Bacteriorhodopsin (bR) is a light-activatable proton pump containing an all-*trans* retinylidene chromophore and is found in the purple membrane of the extremely halophilic bacterium *H. salinarium* (10, 11). Due to its unusual properties, including a native 2-D crystalline lattice which has led to a 3-D structural model (12), stability at very high temperature (13), and a photocycle with distinct spectral intermediates ($bR_{570} \rightarrow K_{630} \rightarrow L_{550} \rightarrow M_{412} \rightarrow N_{550} \rightarrow O_{640} \rightarrow bR_{570}$) (14), bR has become an important model of energy transduction and proton transport in membrane proteins. Several recent studies on bacteriorhodopsin have conclusively demonstrated that the FTIR-SDIL approach can probe the *local* environment and structural changes of *specific* residues and backbone carbonyl groups in a protein (6, 15, 16, 24).

II. Materials and Methods

Site-directed Mutagensis and In vitro Transcription

The amber codon (TAG) was introduced into positions Tyr-57, Tyr-83, Tyr-147 and Tyr-185 in the synthetic *bop* gene using either cassette mutagenesis (17) or PCR based mutagenesis (18). mRNAs were obtained by *in vitro* transcription by placing these genes under control of an SP6 or T7 promoter (19).

Amber Suppression in Cell-free Protein Synthesis System

Cell-free synthesis of bacterioopsin and its amber mutants were carried out in an mRNA-dependent wheat germ translation system as described (19) using ^{35}S-methionine as the radioactive label. Aminoacylated *E. coli* tyrosine suppressor tRNA ($[ring\text{-}^{2}H_{4}]$ or $[1\text{-}^{13}C]$Tyr-sup tRNA) was used for suppression of amber codons introduced in the *bop* gene at position Tyr-57, Tyr-83, Tyr-147 and Tyr-185. The suppressor tRNA was overexpressed in *E. coli* and isolated as total tRNA (20). Aminoacylation of the total tRNA was carried out using *E. coli* S-100 extract prepared as described (21). Synthesis of bOp and amber suppression was optimized with respect to the concentration of Mg^{2+}, K^{+} and aminoacylated suppressor tRNA concentrations. The reaction mixtures were analyzed by gel electrophoresis and the extent of suppression was calculated using phosphorimager measurements (Molecular Dynamics) by comparing the band intensities of full length bOp and the truncated fragment (see Figure 2, Panel A, lanes 5 and 8). The amber suppressions were also carried out where the proteins synthesized were labeled with ^{3}H-Tyr (100 mCi/ml; Amersham) present in the incubation mixture as free amino acid or as aminoacylated suppressor tRNA. SDS-PAGE was carried out in a 15% gel.

Functional Refolding and Regeneration of SDIL-analogs

Cell-free expressed bacterioopsin and its SDIL mutants were purified using procedures described earlier (19). Refolding in halobacterial lipids along with regeneration of bOp by exogenous addition of all-*trans* retinal results in formation of a purple complex absorbing at 560 nm, characteristic of dark-adapted bR (bR$_{560}$). Light adaptation was carried out as described previously (22). Time-resolved absorption measurements of SDIL-analogs was carried out as described previously (19, 22).

FTIR-difference spectroscopy of SDIL-analogs

FTIR-difference spectra were recorded at 2 cm^{-1} on a 740 Nicolet spectrometer equipped with an MCT-B detector. Approximately 25-30 micrograms of purple membrane or SDIL sample suspended in distilled water was placed on an AgCl window and was allowed to dry overnight in a dry-box with a -150 °C dewpoint. The window with the deposited sample was then placed in a specially designed cell, light-adapted and FTIR difference spectra for the bR→K and bR→M transitions measured at 80 K and 250 K, respectively, and at 2 cm^{-1} resolution using methods previously reported (23, 25).

ATR-FTIR difference spectroscopy of SDIL-analogs

Sample solutions of SDIL-analogs of bR were prepared (1 mM NaPi, 1 mM KCl, 0.4 mM MgCl$_2$, 0.6 mM CaCl$_2$, and 50 mM NaCl), and 25 µl of each of these samples (2 µM bR) was dried onto a Ge-crystal internal reflection element as described (15). The sample film was equilibrated with a pH 9.0 solution (25 mM NaPi titrated with 0.5 M NaOH) at 3 °C, and Attenuted total reflection (ATR)-FTIR difference spectra were recorded as described earlier (15).

III. Results and Discussion

SDIL Analogs of Bacteriorhodopsin

Seven SDIL analogs of bacteriorhodopsin, [*ring*-^{2}H$_4$] Tyr-57, [*ring*-^{2}H$_4$] Tyr-147, [*ring*-^{2}H$_4$] Tyr-185, [1-^{13}C] Tyr-57, [1-^{13}C] Tyr-83, [1-^{13}C] Tyr-147 and [1-^{13}C] Tyr-185 have been produced (6, 15, 16, 24). In contrast to earlier studies aimed at site-specific non-native amino acid replacement (SNAAR) in proteins (7-9), which relied on chemical aminoacylation of the suppressor tRNA involving several synthetic steps, we developed a highly efficient method employing a single step enzymatic aminoacylation for site-directed isotope labeling.

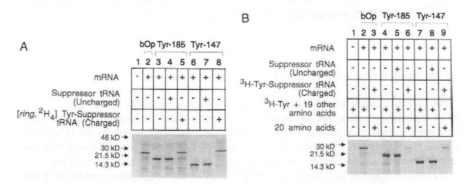

Figure 2: Assays for amber suppression and site-directed isotope labeling using SDS-PAGE followed by autoradiography using ³⁵S-methionine (A) and ³H-Tyr (B). The reaction composition of samples loaded in each lane is as indicated. The reaction mixtures were analyzed by SDS-PAGE in a 15% polyacrylamide gel. For (B), the reactions were similar as in (A) except that the proteins synthesized were labeled with ³H-Tyr (100 mCi/ml; Amersham) present in the incubation mixture as a free amino acid or as aminoacylated suppressor tRNA. Reprinted from ref. 6 with permission.

The results of cell-free expression of bR shown in Figures 2A and B establish that the aminoacylated *E. coli* Tyr suppressor tRNA efficiently suppresses the amber mutations and site-directed isotope labeling does not cause significant "scrambling" of the isotope label. Figure 2A and 2B show that: *i*) Cell-free translation of Y185am and Y147am mRNAs resulted in incomplete translation in the absence of the aminoacylated suppressor tRNAs, whereas full-length bacterioopsin was produced only in the presence of suppressor tRNAs aminoacylated with [*ring*-²H₄] Tyr (Figure 2A). *ii*) Aminoacyl-linkage between the isotopic amino acid and the suppressor tRNA was not hydrolyzed and "scrambling" of [*ring*-²H₄] Tyr did not take place (lane 2, Fig 2B). *iii*) After incorporating [*ring*-²H₄] Tyr at the amber codon, the free suppressor tRNA was not aminoacylated by the aminoacyl synthetases present in the cell-free extract, and thus neither "dilution" of the isotopic label nor insertion of normal amino acids occurred. *iv*) Efficiency of amber suppression is very high, typically 80-90%.

Functional Characterization

After cell-free synthesis, the SDIL proteins were purified, refolded in halobacterial lipids and regenerated with exogenously added all-*trans* retinal using the methods reported recently (19). Typically, 20-25 µg of refolded proteins were obtained per ml of cell-free translation reaction. We have recently reported various vector constructs that hold potential for further increasing the yields (18). Importantly, both SDIL proteins as well as the unlabeled cell-free expressed bR (cf-bR) exhibit a normal visible absorption spectra in the light and dark as compared to purple membrane (6, 22). Static and time-resolved visible absorption spectra indicate that the labeling, expression and reconstitution procedures yield SDIL analogs which are functionally similar to bR (6).

FTIR Analysis of SDIL Analogs

An essential requirement for the successful application of SDIL is that high quality FTIR-difference spectra can be obtained that allow band assignments to particular molecular groups in a protein. This was demonstrated conclusively by us in several reports (6, 15, 16). As a first example, we focused on tyrosine residues, although with the help of chemical aminoacylation, this approach can be extended to any residues in bR.

As seen in Figure 3, high signal-to-noise FTIR difference spectra can be obtained from SDIL samples. Isotope labeling facilitates band assignment by producing frequency shifts in bands that arise from vibrations of those chemical groups which contain the isotope. Thus, for example, perdeuteration of the tyrosine ring causes a downshift in the frequency of a tyrosinate mode near 1277 cm^{-1} which involves C-O$^-$ stretching (23). Figure 3A shows the bR→M difference spectra of bR, cell-free expressed bR (cf-bR), the three SDIL analogs of bR and bR containing uniform labeling at all tyrosines (L-[ring-^{2}H$_4$] all-Tyr). In the case of L-[ring-^{2}H$_4$] Tyr 57 and L-[ring-^{2}H$_4$] Tyr 147, all of the bands previously assigned to tyrosine vibrational modes on the basis of uniform isotope labeling remain unaltered. For example, a negative band at 833 cm^{-1} previously assigned to a characteristic Fermi resonance of tyrosinate (23), negative/positive bands at 1277/1271 cm^{-1} assigned to the CO$^-$ stretch of tyrosinate, a positive band at 1456 cm^{-1} assigned to a tyrosine vibrational mode and a negative band at 1590 cm^{-1} possibly due to a tyrosinate ring mode all appear (23, 25). In contrast to L-[ring-^{2}H$_4$] Tyr 57 and L-[ring-^{2}H$_4$] Tyr 147, the bR→M difference spectrum of L-[ring-^{2}H$_4$] Tyr 185 exhibits significant changes compared to unlabeled bR (Figure 3A). Furthermore, these changes are very similar to those produced by uniform L-[ring-^{2}H$_4$] labeling of bR. In particular, bands characteristic of tyrosine or tyrosinate vibrations including the 1590, 1456, 1277 and 833 cm^{-1} bands disappear (Figures 3 and 4), whereas new bands appear in both spectra at 1575 cm^{-1}, 1416 cm^{-1}, and 1240-1250 cm^{-1}, which were tentatively assigned on the basis of model compound studies to L-[ring-^{2}H$_4$] Tyr (or tyrosinate) vibrational modes (23, 25). This is especially clear in the case of the negative/positive pair at 1277/1271 cm^{-1} (Figure 4) previously attributed to a change in the environment of a tyrosinate residue. This demonstrates that all of the spectral changes due to uniform [ring-^{2}H$_4$] Tyr labeling arise exclusively from the isotope labeling of Tyr-185. Thus, we conclude that Tyr-185 is the only tyrosine residue which is structurally active during the bR→M transition of the bR photocycle. Since similar results were also obtained for the bR→K (6) bR→L and bR→N difference spectra (unpublished results, data not shown), it appears likely that Tyr-185 is the only tyrosine which is structurally active during the entire bR photocycle.

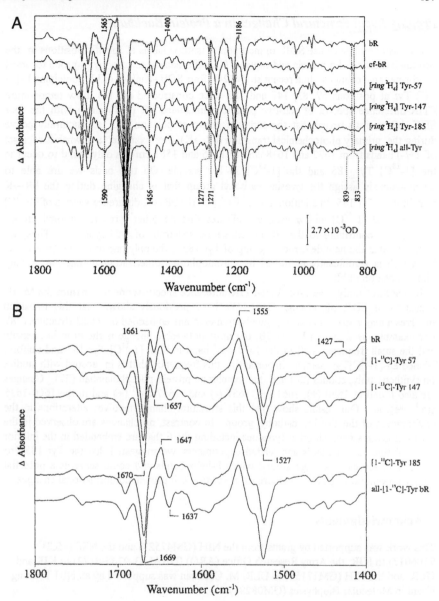

Figure 3: FTIR Difference spectra of SDIL-analogs of bR: Top: FTIR Spectra corresponding to bR→M transition recorded at 250 K (16). Bottom: ATR-FTIR spectra corresponding to the bR→N transition at 276 K (15).

Probing Local Structural Changes in a Protein Backbone

It is also possible to use SDIL to incorporate isotope labels at specific positions in the peptide backbone of a protein and to use FTIR difference spectroscopy to probe for local conformational changes. In a recent study (24), four SDIL-analogs of bR containing L-[1-^{13}C] Tyr at positions 57, 83, 147 and 185 were constructed (6, 15). Low temperature FTIR-difference spectra of cell-free produced unlabeled bR (WT) along with the four SDIL bR analogs reveal that the carbonyl group at position 185 is structurally active during the primary photochemical transition (bR→K). For example, a downshift of a set of (+/-) bands from 1622 and 1618 cm^{-1} to 1586 and 1581 cm^{-1} is only found to occur in the [1-^{13}C] Tyr 185 and the [1-^{13}C] all-Tyr sample. On this basis we are able to unambiguously assign the tyrosine carbonyl group that is changing during the bR→K transition to Tyr 185. In addition, the similarity of the bR→K difference spectra of [1-^{13}C] Tyr 185 and [1-^{13}C] all-Tyr samples indicates that no other tyrosine carbonyl groups significantly contribute to the bR→K difference spectrum of wild type bR. Thus, we conclude that the peptide carbonyl group of Tyr 185 is the only one of the 11 tyrosines in bR which is significantly perturbed by chromophore isomerization during the primary phototransition of bR.

In a second study, we used ATR-FTIR difference spectroscopy to measure the M→N transition of SDIL-bR analogs (15). We have previously shown that this transition involves a major structural change involving membrane embedded α-helical structure (28). For example, as shown in Figure 3B, the most noticeable change is the drop in intensity and the small frequency shift of the 1670 cm^{-1} band in both [1-^{13}C] Tyr 185 and uniform labeled all-[1-^{13}C]Tyr sample. Importantly, this frequency shift is in accord with studies on synthetic polypeptides (26) and the membrane protein phospholamban (27). Changes are also found at 1680 (+), 1661 (+), 1657 (-) and 1647 cm^{-1} (+) and in the 1600-1635 cm^{-1} region. Our results show that this structural change involves alterations in the environment of the Tyr 185 carbonyl group. In contrast, no changes are observed in the carbonyl groups from all other tyrosines including the 6 that are embedded in the interior of membrane. On this basis and related experiments, we suggested that the Tyr 185-Pro 186 region serves as a hinge around which F-helix bends and could serve as a potential coupling point between chromophore isomerization and protein conformational changes.

Acknowledgments

This work was supported by grants from the NIH (GM47527) and the NSF (MCB 9106017) to KJR, the Army Research Office (ARO) (DAAL03-92-G-0172) to KJR and ULR and the NIH (GM17151) to ULR. M. Coleman was supported by an NIH Training Grant in Molecular Biophysics (GM08291).

References

1. Rothschild, K. J. (1992). *J. Bioenerg. Biomembr.* **24**, 147-167.
2. Braiman, M. S., Mogi, T., Marti, T., Stern, L. J., Khorana, H. G. & Rothschild, K. J. (1988). *Biochemistry* **27**, 8516-8520.
3. Braiman, M. S., Mogi, T., Stern, L. J., Hackett, N. R., Chao, B. H., Khorana, H. G. & Rothschild, K. J. (1988). *Proteins: Struct., Funct., Genet.* **3**, 219-229.
4. Gerwert, K., Hess, B., Soppa, J. & Oesterhelt, D. (1989). *Proc. Natl. Acad. Sci. U. S. A.* **86**, 4943-4947.
5. Maeda, A., Sasaki, J., Shichida, Y., Yoshizawa, T., Chang, M., Ni, B., Needleman, R. & Lanyi, J. K. (1992). *Biochemistry* **31**, 4684-4690.
6. Sonar, S., Lee, C.-P., Coleman, M., Marti, T., Patel, N., Liu, X., Khorana, H. G., RajBhandary, U. L. & Rothschild, K. J. (1994). *Nature Struct. Biol.* **1**, 512-517.
7. Noren, C. J., Anthony-Cahill, S. J., Griffith, M. C. & Schultz, P. G. (1989). *Science* **244**, 182-188.
8. Ellman, J. A., Mendel, D. & Schultz, P. G. (1992). *Science* **255**, 197-200.
9. Bain, J. D., Diala, E. S., Glabe, C. G., Wacker, D. A., Lyttle, M. H., Dix, T. A. & Chamberlin, A. R. (1991). *Biochemistry* **30**, 5411-5421.
10. Stoeckenius, W. & Bogomolni, R. A. (1982). *Annu. Rev. Biochem.* **51**, 587-616.
11. Rothschild, K. J. & Sonar, S. (1995) In *CRC Handbook of Organic Photochemistry and Photobiology*: Horspool, W. M. & Song, P-S., Eds., CRC Press Inc.: London, pp 1521-1544.
12. Henderson, R., Baldwin, J. M., Ceska, T. A., Zemlin, F., Beckmann, E. & Downing, K. H. (1990). *J. Mol. Biol.* **213**, 899-929.
13. Shen, Y., Safinya, C. R., Liang, K. S., Ruppert, A. F. & Rothschild, K. J. (1993). *Nature* **366**, 48-50.
14. Lozier, R. H., Xie, A., Hofrichter, J. & Clore, G. M. (1992). *Proc. Natl. Acad. Sci. U. S. A.* **89**, 3610-3614.
15. Ludlam, C. F. C., Sonar, S., Lee, C.-P., Coleman, M., Herzfeld, J., RajBhandary, U. L. & Rothschild, K. J. (1995). *Biochemistry* **34**, 2-6.
16. Liu, X.-M., Sonar, S., Lee, C.-P., Coleman, M., RajBhandary, U. L. & Rothschild, K. J. (1995). *Biophys. Chem.* **56**, 63-70.
17. Hackett, N. R., Stern, L. J., Chao, B. H., Kronis, K. A. & Khorana, H. G. (1987). *J. Biol. Chem.* **262**, 9277-9284.
18. Coleman, M., Sonar, S., Patel, N. & Rothschild, K. J. (1994). in *Dual-Use Technologies and Applications Conference: 4th Annual 1994 IEEE Mohawk Valley Section* (SUNY Institute of Technology at Utica/Rome,)., May 23-26 Ed., Vol. 1, pp. 326-331.
19. Sonar, S. M., Patel, N. P., Fischer, W. & Rothschild, K. J. (1993). *Biochemistry* **32**, 13777-13781.
20. Lee, C. P., Seong, B. L. & RajBhandary, U. L. (1991). *J. Biol. Chem.* **266**, 18012-18018.
21. RajBhandary, U. L. & Ghosh, H. P. (1969). *J. Biol. Chem.* **244**, 1104-1113.
22. Sonar, S., Krebs, M. P., Khorana, H. G. & Rothschild, K. J. (1993). *Biochemistry* **32**, 2263-2271.
23. Rothschild, K. J., Roepe, P., Ahl, P. L., Earnest, T. N., Bogomolni, R. A., Das Gupta, S. K., Mulliken, C. M. & Herzfeld, J. (1986). *Proc. Natl. Acad. Sci. U. S. A.* **83**, 347-351.
24. Sonar, S., Liu, X.-M., Lee, C.-P., Coleman, M., He, Y.-W., Herzfeld, J., RajBhandary, U. L. & Rothschild, K. J. (1995). *J. Amer. Chem. Soc.* **117**, In Press.
25. Roepe, P., Ahl, P. L., Das Gupta, S. K., Herzfeld, J. & Rothschild, K. J. (1987). Biochemistry 26, 6696-6707.
26. Martinez, G. V., Fiori, W. R. & Millhauser, G. (1994). *Biophys. J.* 66, A65.
27. Arkin, I. T., Rothman, M. S., Ludlam, C. F. C., Aimoto, S., Engelman, D. M., Rothschild, K. J. & Smith, S. O. (1995). *J. Mol. Biol.* 248, 824-834.
28. Rothschild, K. J., Marti, T., Sonar, S., He, Y. W., Rath, P., Fischer, W., Bousche, O. & Khorana, H. (1993). *J. Biol. Chem.* 268, 27046-27052.

SECTION IV

Modifications to Proteins *in Vivo*

PROTEIN C-GLYCOSYLATION

Jan Hofsteenge, Andreas Löffler
Friedrich Miescher-Institut, CH-4002 Basel, Switzerland

Dieter R. Müller, W.J. Richter
Ciba-Geigy Ltd., CH-4002, Basel, Switzerland

Tonny de Beer, Johannes F.G. Vliegenthart
Department of Bio-organic Chemistry, Utrecht University, NL-3584 CH Utrecht,
the Netherlands

I. INTRODUCTION

Glycosylation of amino acid side chains of proteins is a common and widespread modification. Two types of carbohydrate-protein linkages have been known for a long time, i.e. the N- and O-glycosidic bonds to Asn, and hydroxy amino acids, respectively (1). Recently, we have reported a third type of linkage, the C-glycosidic attachment of an α-mannopyranosyl moiety to the indolic C-2 atom of Trp-7 in human RNase U_s (2,3):

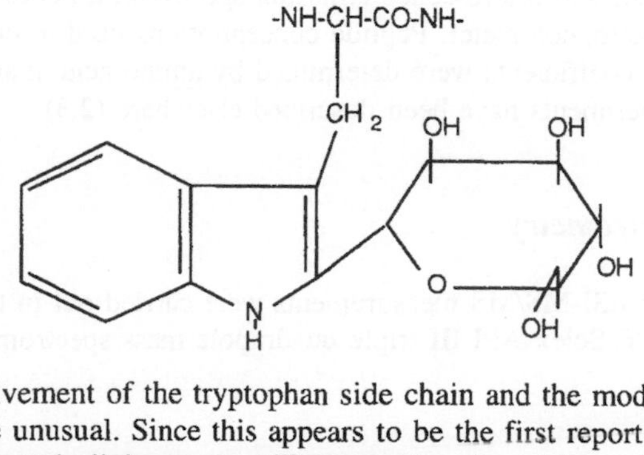

Both the involvement of the tryptophan side chain and the mode of attachment are unusual. Since this appears to be the first report on such a carbohydrate-protein linkage, we will summarize the evidence for this modification, and describe a number of its properties that may be useful for its detection in other proteins.

TECHNIQUES IN PROTEIN CHEMISTRY VII

II. MATERIALS AND METHODS

A. *Protein chemistry*

RNase U_s was isolated from human urine or erythrocytes as described (2). Proteolytic digestion of the enzyme with thermolysin required denaturation by reduction and carboxymethylation (2). Purification of peptides was performed by C_{18} (Vydac, Hispania,CA) reversed-phase HPLC using the standard 0.1% TFA/acetonitrile system. Peptides were demonstrated to be pure by amino acid analysis and Edman degradation using an Applied Biosystems model 470A or 477A sequencer. Protein and peptides were sequenced in the absorptive mode with polybrene, using the standard 03CPTH cycle. Phenylthiohydantoin-amino acids (PTH-amino acids) were separated on-line with a model 120A chromatograph equipped with a Brownlee C_{18} column (2.1x220 mm; Applied Biosystems) operated at 55°C. Buffer A consisted of 7 ml premix buffer per liter of 3% THF in H_2O (Perkin Elmer), 130 µl formic acid, and eluent B was CH_3CN, containing 12% (v/v) isopropanol. The flow rate during separation was 210 µl/min. The gradient used is specified in the "Results" section. PTH-Trp^{Man} was prepared as described previously (2), and its concentration was determined from A_{269} ($\varepsilon = 19'700$ $M^{-1}cm^{-1}$).

Proteolytic digestions of the tetrapeptides with aminopeptidase M (10^{-2} units/nmol, Boehringer, Mannheim, FRG) and carboxypeptidase Y (Boehringer, enzyme/substrate ratio : 1/80, for 4 or 18 h at 37°C) were carried out in 50 mM NH_4HCO_3 and 100 mM Na-citrate, pH 5.0, respectively.

Modification of tryptophan residues with o-nitrophenylsulfenylchloride (NPS) was done in glacial acetic acid for 30 min in the dark. The concentrations of peptide and NPS were 0.2 and 3.3 mM, respectively.

B. *Spectroscopy*

Absorption spectra were recorded on a Hewlett Packard 8452 A spectrophotometer, and fluorescence emission spectra on a Perkin Elmer LS-3 fluorescence spectrophotometer. Peptide concentrations used to calculate molar extinction coefficients were determined by amino acid analysis. Details of the NMR experiments have been described elsewhere (2,3).

C. *Mass spectrometry*

ESI-MS and ESI-MS/MS measurements were carried out in the positive ion mode on a PE Sciex API III triple quadrupole mass spectrometer.

III. RESULTS AND DISCUSSION

A. Summary of the approaches to identify Trp^Man (2,3)

The chemically determined primary structure of RNase U$_s$ from human urine (4), and the one deduced from the cDNA coding for eosinophil derived neurotoxin (EDN) were identical (5), except for residue 7. The predicted tryptophan residue could not be identified by Edman degradation. The latter was also true for the enzyme isolated from a variety of human tissues or cells (see summary in (2)), suggesting a modification of this residue. A number of different peptides containing Trp-7 were found by ESI-MS to be 162 Da heavier than expected, and to have properties in ESI-MS/MS experiments typical for aromatic C-glycosides (see below).

Since C-glycosidic bonds are resistant to acid hydrolysis, and do not yield a "sugar" ion in MS experiments, the identification of the monosaccharide had to be achieved by NMR.

B. Edman degradation of peptides containing Trp^Man

Automated Edman degradation in the absorptive mode of RNase U$_s$ and its N-terminal peptides yielded one major, and two minor PTH-derivatives at position 7 (2). Although it was initially reported that the major derivative eluted shortly after PTH-Tyr, further experiments revealed that most of the time it co-eluted. This could result from changes in the buffer composition, or variations in the batches of column material. To separate PTH-Trp^Man from PTH-Tyr, the conditions shown in Fig. 1 were used.

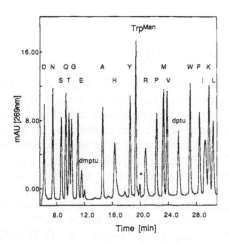

time [min]	%B
0	10
20	36
28	36
29	90
35	90

Figure 1 Improved separation of PTH-Trp^Man and PTH-Tyr
The major PTH-derivative of Trp^Man can be separated from PTH-Tyr by including 12% (v/v) isopropanol in eluent B. The minor PTH-Trp^Man derivative, most likely a diastereomer (2), has been indicated with an asterisk.

Another variation during Edman degradation concerns the yield of PTH-Trp^{Man}. During sequencing of F-T-W^{Man}-A the ratio PTH-Trp^{Man}/PTH-Thr varied between 0.2 and 0.9. This variability did not result from the HPLC separation, but rather from differences in the extraction conditions of the anilinothiazolinone. Probably, a more reliable identification of Trp^{Man} could be achieved by solid phase sequencing, and extraction with more polar solvents (6).

C. Mass spectrometry of peptides containing Trp^Man

Glycosylation of tryptophan can be most sensitively detected by MS. The most salient feature of Trp^{Man}-containing peptides was their increased mass by 162 Da. In addition, multiple losses of 18 Da from various ions (" $\subsetneq$ " in Fig. 2) were consistently observed in MS/MS experiments of small peptides containing Trp^{Man}. ESI-MS and -MS/MS experiments on (perdeuterio)-acetylated peptides turned out to be very informative: they demonstrated the incorporation of four acyl groups into the mannopyranosyl moiety, and the conspicuous absence of a "sugar ion" (m/z 343), which would be expected for N- or O-glycosidically linked sugars. Finally, the loss of 120 Da (formally four times CH_2=O) observed in the unmodified peptides (Fig. 2) can be taken as evidence for a C-glycosidically linked carbohydrate, as such a loss was also observed with low molecular weight flavone C-glycopyranosides (7).

It is of interest to note that Gäde et al. (8) have identified a hexosylated tryptophan in a neuropeptide from the insect *Carausius morosus*, based on an increased mass of 162 Da. These investigators proposed an N-glycosidic linkage, but a C-glycosidic bond seems to be very well compatible with the FAB-MS/MS spectra (8), in that they also showed the loss of 120 Da from $[M + H]^+$.

D. NMR of peptides containing Trp^Man

^{1}H-NMR spectroscopy of the peptide F-T-W^{Man}-A-Q-W established the residue at position 7 to be a modified tryptophan. All protons of the indole side chain were observed, except for the one at position 2, thus demonstrating the carbohydrate attachment site to be C-2. The combination of ^{1}H- and ^{13}C-NMR data unequivocally showed the presence of an aldohexopyranose, as well as the C-C link to C-2 (2). From vicinal proton-proton coupling constants and ROE intensities it was concluded that the carbohydrate was an α-manno-pyranosyl moiety (3). The monosaccharide did not adopt the usual chair conformation, but appeared to exist in a number of (yet unidentified) different conformations. The availability of NMR data on Trp^{Man} will greatly facilitate the detection of this feature in other peptides. A 1D ^{1}H-NMR spectrum, which can be recorded on amounts as low as 10 nmol of peptide material, should display very similar to identical chemical shifts and coupling constants as observed for Trp^{Man} in F-T-W^{Man}-A-Q-W.

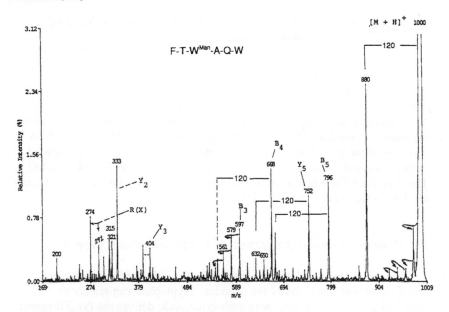

Figure 2   ESI-MS/MS of the hexapeptide F-T-W^{Man}-A-Q-W
Fragments have been labeled using the standard nomenclature. The multiple losses of 18 Da (H_2O) have been indicated by " 〈 ". The loss of 120 Da from fragments containing Trp^{Man} has been indicated, and is typical for aromatic C-glycosidic compounds.

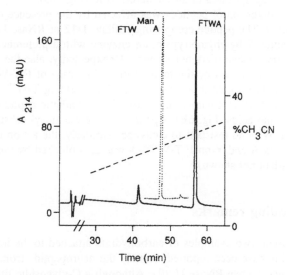

Figure 3   Chromatographic properties of the peptide F-T-W^{Man}-A
Equal amounts of F-T-W-A (full line), or F-T-W^{Man}-A (stippled line) were injected onto a C_{18} reversed phase column equilibrated in 0.1% TFA. The peptides were eluted with a linear gradient of acetonitrile (broken line).

E. Effect of mannosylation of Trp on chromatographic and spectroscopic properties

Fig.3 shows the chromatographic properties of the N-terminal peptide F-T-W^{Man}-A and the synthetic peptide F-T-W-A. The modified peptide eluted considerably earlier from a C$_{18}$ reversed-phase column.

Also the spectroscopic properties of the indole moiety were influenced by the mannosyl residue. The shape of the absorption spectrum of F-T-W^{Man}-A was typical for that of an indole, but its maximum was shifted 2 nm toward longer wavelength and was 1.6-times higher than that of F-T-W-A (Fig.4A). The maximum fluorescence emission increased 2.8-fold, and was shifted 2 nm toward shorter wavelength (Fig. 4B).

F. Protein chemical properties of C-mannosylated peptides

The mannopyranosyl moiety attached to the indole influenced several protein chemical properties of the peptides. Tryptophan can readily be modified at the C-2 position with sulfenylchloride derivatives (9). Treatment of F-T-W-A with o-nitrophenylsulfenylchloride resulted in the conversion of 98% of the peptide into the expected product with a higher retention time on HPLC (Fig.5A), and the characteristic absorbance maximum at 365 nm (data not shown). In contrast, no such product was found with F-T-W^{Man}-A, and 91% of the peptide appeared in an unaltered form (Fig. 5B).

Proteolytic digestion of peptides was affected by the presence of the mannosyl moiety. The peptide comprising residue 1-12 of RNase U$_s$ was not cleaved at position 7 by chymotrypsin, an enzyme with a preference for cleaving after tryptophan (data not shown). Unexpectedly, also the action of exopeptidases was affected by the modification. Cleavage of F-T-W^{Man}-A with Aminopeptidase M resulted in a limited digestion, yielding T-W^{Man}-A as the end product (Fig.6A). In contrast, digestion of the synthetic peptide F-T-W-A yielded free amino acids (Fig. 6B). Apparently, the mannosyl substituent rendered the T-W bond resistant to cleavage. Similarly, the action of carboxypepidases A and Y on F-T-W^{Man}-A was also blocked by the modification (data not shown).

IV. Concluding remarks

At present, two examples of carbohydrate attached to the indole of Trp in a polypeptide have been reported, namely the neuropeptide from *C. morosus* (7), and human RNase U$_s$ (2). Although a C-glycosidic linkage has only been established for the latter, it seems likely that the same occurs in the neuropeptide as well (see above). It remains to be seen how general this kind of modification is, but its occurrence in man and insects suggests a wide phylogenetic distribution. Other questions to be answered are whether other monosaccharides are C-glycosidically linked to Trp, and how the biosynthesis of C-glycosylated Trp takes place. It may be of interest to point out that only

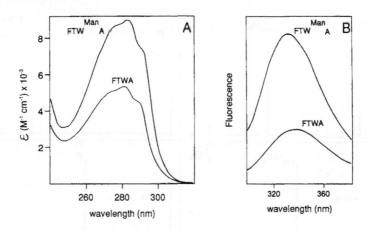

Figure 4   Spectroscopic properties of the peptide F-T-W^Man-A
A. Absorbance spectra of the peptides F-T-W-A and F-T-W^Man-A.
B. Emission fluorescence spectra of the same peptides; excitation was at 285 nm.

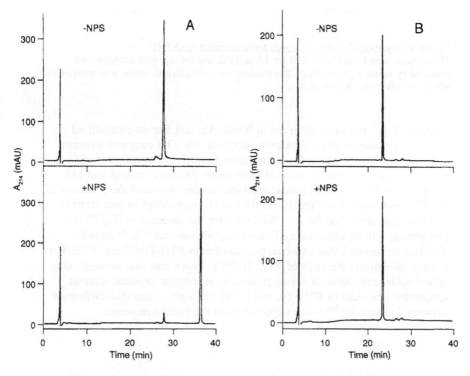

Figure 5   Reaction of the peptide F-T-W^Man-A with o-nitrophenyl sulfenylchloride (NPS)
The peptides were treated with NPS as described in "Experimental". The reaction mixture was
analysed by reversed phase HPLC. In order to identify the reaction product, the eluate was
also monitored at 365 nm (data not shown).
A: F-T-W-A;   **B**: F-T-W^Man-A.

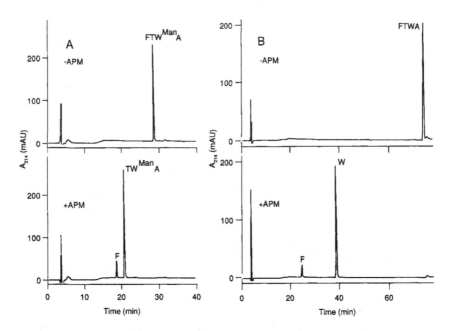

Figure 6 <u>Digestion of F-T-W^{Man}-A with aminopeptidase M (APM)</u>
The peptides were treated with APM for 2 h at 37°C, and the digestion products were
separated by reversed phase HPLC. The products were identified by amino acid analysis and
MS. <u>A</u>: F-T-W^{Man}-A; <u>B</u>: F-T-W-A

modified Trp-7 has been observed in RNase U_s, and that an unmodified
tryptophan occurs in close proximity (position 10). This suggests a certain
degree of specificity of the C-glycosylation reaction. In man this modification
is not restricted to proteins isolated from urine. Protein chemical and MS
analysis of RNase U_s from human erythrocytes demonstrated the presence of
Trp^{Man}-7 and unmodified Trp-10 (Löffler et al., manuscript in preparation).

It seems likely that the first indication for the presence of Trp^{Man} in a
polypeptide will be obtained by Edman degradation and MS. It should
therefore be stressed that under certain conditions PTH-Trp^{Man} and PTH-Tyr
coelute. Moreover, the yield of PTH-Trp^{Man} is lower than that of most other
amino acids, and varies. It seems prudent to reexamine peptides with an
apparently low yield of PTH-Tyr and a 185 Da higher mass (the difference
between Tyr and Trp^{Man}) than expected from the protein sequence.

ACKNOWLEDGMENTS

We would like to thank Renate Matthies for help with the protein chemistry, Drs. J. Krieg and D. Hess for reading the manuscript, and the latter also for preparing Figure 1.

REFERENCES

1. Lis, H. and Sharon, N. (1993) Eur. J. Biochem. **218** , 1-27.
2. Hofsteenge, J., Müller, D.R., de Beer, T., Löffler, A., Richter, W.J. and Vliegenhart, J.F.G. (1994) Biochemistry **33**, 13524-13530.
3. de Beer, T., Vliegenthart, J.F.G., Löffler, A., and Hofsteenge, J. Biochemistry, in press.
4. Beintema, J.J., Hofsteenge, J., Iwama, M., Morita, T., Ohgi, K., Irie, M., Sugiyama, R.H., Schieven, G.L., Dekker, C.A. and Glitz, D.G. (1988) Biochem. **27**, 4530-4538.
5. Hamman, K.J., Ten, R.M., Loegering, D.A., Jenkins, R.B., Heise, M.T., Schad, C.R., Pease, L.R., Gleich, G.J. and Barker, R.L. (1990) Genomics **7**, 535-546.
6. Gooley, A.A., Classon, B.J., Marschalek, R. and Williams, K.L. (1991) Biochem. Biophys. Res. Commun. **178**, 1194-1201.
7. Becchi, M, and Fraisse, D. (1989) Biomed Environ Mass Spectrom **18**, 122-130.
8. Gäde, G., Kellner, R., Rinehart, K.L. and Proefke, M.L. (1992) Biochem. Biophys. Res. Comm. **189**, 1303-1309.
9. Fontana, A. and Scoffone, E. (1972) Methods Enzymology **25**, 482-494

Characterization of the *N*- and *O*-Linked Oligosaccharides from Glucoamylase E4 from *Monascus Rubignosus* using Electrospray Mass Spectrometry and Glycosidase Digestion

Christine A. Settineri, Yu Dong Chen, Ke Jiang[§], Shui Zheng Zhang[†]
and Alma L. Burlingame

Department of Pharmaceutical Chemistry and Mass Spectrometry Facility,
University of California, San Francisco, CA 94143-0446

[§]Research Center for Eco-Environmental Sciences. [†]Institute of Microbiology,
Academia Sinica, Beijing, China 100085

I. Introduction

Glucoamylases are enzymes which catalyze the release of β-D-glucose (primarily 1,4 linked) from the nonreducing end of oligosaccharide chains such as starch (1). They are used industrially in the production of glucose syrup, which is then used in the production of ethanol or high fructose corn sweeteners (2). The most highly characterized glucoamylases are from *Aspergillus niger* and *Aspergillus awamori*. These enzymes have identical primary sequences and contain a large catalytic domain of about 440 amino acids, as well as a much smaller starch binding domain of about 25 amino acids (3, 4). These two domains are linked by a sequence of about 72 amino acids which is known to have extensive *O*-linked glycosylation containing primarily mannose residues (5). The catalytic domain also contains *N*-linked glycosylation (6). Recent experiments with glucoamylase from *Aspergillus awamori* indicated that more highly glycosylated forms of the protein have significantly improved thermal and pH stabilities.

Glucoamylase E_4 (E_4) from the fungus *Monascus rubiginosus* is a novel microbial glucoamylase of unknown sequence. Glycopeptide mapping experiments were performed to obtain detailed structural information on the carbohydrates including where they were attached to the protein, so that it could be determined if this relates to the stability of this particular glucoamylase.

Tryptic and endoproteinase glu-C mapping experiments using HPLC/electrospray ionization mass spectrometry (HPLC/ESIMS) with collision induced dissociation (CID) and selected ion monitoring (SIM) of diagnostic carbohydrate fragment ions (7, 8) on a single quadrupole instrument (HPLC/ESI/CID/MS) were used to identify and partially characterize several *N*- and *O*-linked glycopeptides. To further characterize the individual oligosaccharide structures, several glycosidase digestions were performed on either the isolated glycopeptides or peptide/glycopeptide mixtures, followed by HPLC/ESIMS (8). Methylation analysis of the β-eliminated *O*-linked structures was also performed to obtain linkage information on the *O*-linked glycosylation. Edman degradation on the two isolated *N*-linked glycopeptides revealed sequences spanning the same glycosylation site in E_4. These sequences are highly homologous to other fungal glucoamylases and they contain a conserved *N*-linked glycosylation site (4).

II. Materials and Methods

Carbohydrate Composition Analysis. Carbohydrate analyses were performed on acid hydrolyzed, reduced and S-carboxymethylated E_4. Approximately one to two nmole of protein was hydrolyzed in each of two different solutions: 2 M TFA (5 hours) or 6 N HCl (3 hours) for quantitation of amino sugars and neutral sugars, respectively. Analyses were performed using a Dionex Bio-LC high performance anion exchange chromatography system consisting of a gradient pump, and a pulsed amperometric detector (PAD) with a gold working electrode. Samples were injected using a Spectra Physics SP8780 autosampler onto a CarboPac PA-1 column (4.6 x 250 mm). Eluant 1 was 200 mM NaOH, eluant 2 was 1 M NaOAc and eluant 3 was H_2O. Quantitation was determined by comparing the results to injection of a standard monosaccharide mixture.

Enzymatic Digestions. Glucoamylase E_4 was reduced in 0.1 M ammonium bicarbonate, 6 M guanidine-HCl, pH 8.0 with a 50-fold excess of dithiothreitol (DTT) at 50°C for 1.5 hr under argon. A 20-fold excess of iodoacetic acid was then used for cysteine alkylation in the dark at room temperature for 1.5 hr. E_4 was proteolytically cleaved with trypsin and portions of the tryptic digest were then subdigested with endoproteinase glu-C. Trypsin digestions were performed in 0.1 M ammonium bicarbonate (pH 8.3), at a weight ratio of 4% (2 X 2%) for 20-26 hr at 37°C. Endoproteinase glu-C digestion conditions were the same as for trypsin digestion except at a weight ratio of 8% (2 X 4%). Peptide N-glycosidase F (PNGase F, Genzyme) digestions were performed at 37°C in 0.1 M ammonium bicarbonate (pH 8.3), using 1.0-1.5 units of enzyme for 1.0 nmole of E_4 for 20-26 hr. Endoglycosidase H (Endo-H, Boehringer Mannheim) digestions were performed in 50 mM sodium acetate (pH 5.0) at 37°C for 20-24 hr. Aliquots of 1 mU were added to 1 nmole of Glucoamylase E_4 at 0 and 4 hours. Endo-α-N-acetylgalactosaminidase from *Dipplococcus pneumoniae* (O-glycosidase) digestions were performed at 37°C in 100 ml 50 mM sodium acetate (pH 6.0), using 2 mU enzyme for 600 pmole of E_4 for 20 hr. The HPLC purified *N*-linked glycopeptide was digested with α 1→2 mannosidase from *Aspergillus saitoi* (Oxford GlycoSystems) at a concentration of 1.0 U/μl in 15 μl 100 mM sodium acetate (pH 5.0) containing 14 μg BSA for 18 hr at 37°C.

Chromatography. HPLC-Electrospray ionization mass spectrometric (HPLC/ESIMS) analyses were performed using an Applied Biosystems 140A gradient HPLC system and a C_{18} Vydac microbore column (250 X 1.0 mm, 5 μ particle size). The mobile phases used were (solvent A) 0.1% TFA, 98% H_2O, 2% acetonitrile and (solvent B) 0.08% TFA 98% acetonitrile, 2% H_2O, and the flow rate was 50 μL/min. Typical gradient conditions were as follows: 2%B to 50%B linearly in 60min; then linearly to 98%B in 15 min. Using an Isco μLC-500 micro flow pump, a 1:1 mixture of isopropanol:2-methoxyethanol was mixed with the column effluent post-column (after UV detection at 215 nm on an Applied Biosystems UV detector model 204 with a U-shaped capillary flow cell) at a flow of 30 μl/min. This combined flow of 80 μL/min was split at a ratio of 4:1 post-column, so that only 20 μL/min entered the mass spectrometer.

Mass Spectrometry. Electrospray ionization (ESI) spectra were acquired on-line, after microbore HPLC, on a VG Biotech/Fisons Bio-Q or VG Platform mass spectrometer equipped with an atmospheric pressure electrostatic spray ion source and a quadrupole mass analyzer (maximum m/z=3000). The scan time in non-continuum mode from m/z= 350 to m/z= 2200 was 5 seconds. Typical operating voltages for the Bio-Q were as follows: probe tip 3.6 kV; counter electrode 1.5 kV; sampling orifice 60 V; SIM sampling orifice 160-180 V. Typical operating voltages for the Platform were as follows: probe tip 4.2 kV; counter electrode 550 V; sampling orifice 40-50 V. Nebulizing gas flow rate=150 mL/min.; Drying gas flow rate=4 L/min.; Source temperature=60°C.

Edman Degradation. Microsequencing was performed on an Applied Biosystems 470A gas phase protein sequencer, with an Applied Biosystems 120A phenylthiohydantoin analyzer (9).

III. Results and Discussion

A. Carbohydrate Composition Analysis

Carbohydrate composition analysis on the oligosaccharides hydrolyzed from intact, reduced and alkylated E$_4$ using high pH anion exchange chromatography (10) indicated the presence of mannose (Man, 43.9 mole/mole E$_4$), N-acetyl glucosamine (GlcNAc, 12.2 mole/mole) and galactose (Gal, 7.9 mole/mole), indicating the possible presence of high mannose N-linked oligosaccharides as well as some O-linked oligosaccharides.

B. Identification of Glycopeptides

Approximately 300 pmole E$_4$ was first digested with trypsin and the resulting digest analyzed by microbore reversed phase HPLC/ESIMS and HPLC/ESI/CID/MS. By inducing fragmentation as the eluant entered the mass spectrometer and monitoring a variety of carbohydrate-specific fragment ions (SIM), the glycopeptide-containing fractions were identified in the mixture of peptides and glycopeptides. The HPLC/ESIMS and HPLC/ESI/CID/MS with the SIM data for m/z=204 is shown in Figure 1. The same HPLC/ESIMS conditions were used to identify and distinguish the N- linked glycopeptides from the O-linked glycopeptides by further digestion of the E$_4$ tryptic digest with PNGase F, an endoglycosidase which removes all classes of N-linked oligosaccharides from proteins and glycopeptides, converting the attached Asn to an Asp (13). The SIM at m/z=204 chromatogram for the PNGase F digest is also shown in Figure 1 (bottom panel).

C. Characterization of N-linked Glycopeptides

As shown in Figure 1, two groups of ions are present in the SIM at m/z=204 chromatogram of the tryptic digest, and the later eluting group of ions disappeared after PNGase F digestion, indicating the presence of N-linked glycopeptides in this region of the tryptic digest. However, the ions identified in this region were in the m/z=1800-2100 range, which deconvoluted to molecular weights around 8000 Da. These ions were quite weak, presumably due to the fact that the transmission of ions and therefore sensitivity of the quadrupole mass analyzer drops significantly above mass 1800. Therefore, in order to obtain smaller glycopeptides, the fraction of glycopeptides and peptides in this section of the tryptic digest (marked "E" fraction in Figure 1) were collected by microbore HPLC and subdigested with endoproteinase glu-C and then reanalyzed by HPLC/ESIMS and HPLC/ESI/CID/MS. The HPLC/ESIMS with SIM data at m/z= 204 for the glu-C digest of the E fraction indicated the presence of two major and four minor glycopeptide components as shown in Figure 2. These were confirmed as N-linked glycopeptides by treatment of this glu-C digest with PNGase F. As shown in Figure 2 (bottom panel) the SIM data at m/z= 204 shows no significant peaks, indicating that no glycopeptides were present after PNGase F digestion.

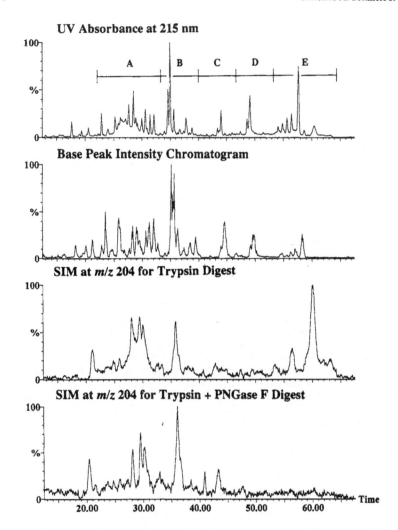

Figure 1. HPLC/ESIMS and HPLC/ESI/CID/MS with the SIM at *m/z*=204 chromatograms for the trypsin and trypsin + PNGase F digests of E_4. The trypsin digest was divided into 5 sections (A-E) for subdigestions with glu-C.

Figure 3A shows an ESIMS spectrum of one of the two major groups of glycopeptides shown in Figure 2 (marked glycopeptide 1). The two identified glycopeptides were then confirmed as high mannose *N*-linked glycopeptides by treatment of the isolated glycopeptides with Endo-H followed by HPLC/ESIMS analysis. The spectrum of the deglycosylated glycopeptide from Figure 3A is shown in Figure 3B. The two isolated glycopeptides marked 1 and 2 in Figure 2 (approximately 40 pmole of glycopeptide 1 and 80 pmole of glycopeptide 2) were sequenced by Edman degradation. The Edman data indicated that the glycopeptides contained the following overlapping sequences:

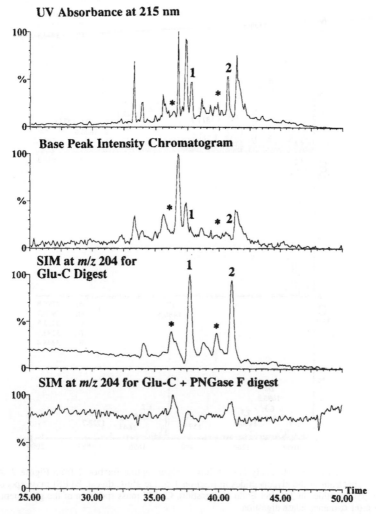

Figure 2. HPLC/ESIMS and HPLC/ESI/CID/MS with the SIM at m/z=204 chromatograms for the glu-C and glu-C + PNGase F digests of fraction E of the E$_4$ tryptic digest. The two starred peaks contain glycopeptides which are the oxidized forms of the glycopeptides found in peaks 1 and 2, respectively.

Glycopeptide 1: Leu-Ser-Tyr-Val-Met-Gln-Tyr-Trp-Xxx-Ser-Ser-Thr-Phe-Asp
Glycopeptide 2: Leu-Ser-Tyr-Val-Met-Gln-Tyr-Trp-Xxx-Ser-Ser-Thr-Phe-Asp-
　　　　　　　　　Leu-Trp-Glu-Glu

Assuming that the blank cycles in both glycopeptides (Xxx) are Asn, the molecular weights of the peptide components based on the sequences obtained by Edman degradation analysis of the two glycopeptides are 1740.8 and 2298.0, respectively. Interestingly, these sequences are highly homologous to other glucoamylases of known sequence such as glucoamylase G1 and G2 from *Aspergillus niger* (4), and the site of the *N*-linked glycosylation is conserved as well, indicating a possible functional significance. Based on the molecular

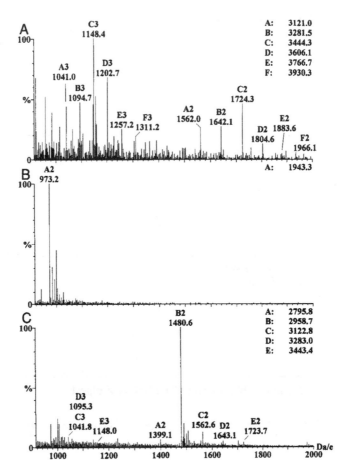

Figure 3. HPLC/ESIMS analysis of *N*-linked glycopeptide fraction 1 from Figure 2. A) ESI mass spectrum of the glycopeptides in fraction 1 of the glu-C digest. B) ESI mass spectrum of the glycopeptides in 3A after Endo-H digestion. C) ESI mass spectrum of the glycopeptides in 3A after α-mannosidase digestion.

weights of the glycopeptides obtained by HPLC/ESIMS (shown in Figure 3A and listed in Table I), combined with the change in molecular weight after Endo-H digestion (shown in Figure 3B), the same high mannose type oligosaccharides are attached to both peptides as follows: $Man_6GlcNAc_2$ (*m/z* 1380), $Man_7GlcNAc_2$ (*m/z* 1542), $Man_8GlcNAc_2$ (*m/z* 1704), $Man_9GlcNAc_2$ (*m/z* 1866), $Glc_1Man_9GlcNAc_2$ (*m/z* 2028), $Glc_2Man_9GlcNAc_2$ (*m/z* 2190). (Note: it is assumed that the species of mass higher than Man_9 are incompletely processed oligosaccharides which still contain glucose units which have not been trimmed from the α1→3 linked mannose branch (14). Based on ion abundance in the mass spectra, the Man_{7-9} are the major oligosaccharides present. The two starred peaks in Figure 2 contain components which are 16 Da higher in mass than glycopeptides 1 and 2, respectively, and are therefore assumed to be the same peptides as glycopeptide 1 and 2 but with oxidized methionines (i. e., the glycopeptide).

To determine the linkages of the major N-linked oligosaccharides, glycopeptide 1 was preparatively purified and digested with an α 1→2 specific mannosidase from *Aspergillus saitoi* (15) and then reanalyzed by HPLC/ESIMS. As shown in Figure 3C, the molecular weight of the major component (B in Figure 3C) resulting from the digest has a molecular weight of 2958.7, which corresponds to a Man₅GlcNAc₂ (*m/z* 1218) oligosaccharide attached to the peptide of mass 1740.8. Therefore, the Man₆₋₉GlcNAc₂ oligosaccharides contained one to four α1→2 linked mannose residues on the nonreducing terminus of the oligosaccharide, respectively. This indicates that the branching is like mammalian high mannose oligosacchrides rather than like those found in yeast (16). Unfortunately, there is no other α mannosidase which is very specific for α1→3 or α1→6 linkages which could be used for further digestion of the Man₅GlcNAc₂ species, so we can only assume that the rest of the N-linked oligosaccharide contains the common high mannose structure and linkages. The oligosaccharides believed to contain the one and two glucose residues could have lost up to two mannose residues after the α mannosidase digestion, leaving glycans with molecular weights of 3445 and 3607, respectively. Although the ions are quite weak, component E in Figure 3C matches very closely to the former mass of 3445. Components C and D are likely due to some incomplete digestion of one or two of the α1→2 linked mannose residues, and component A is likely due to a small percentage of the oligosaccharides which are missing an α1→3 or 6 linked mannose branch.

The mass spectral data determined for the N-linked glycopeptides from E₄ is summarized in Table I. Therefore, the structure of the major N-linked oligosaccharides on glucoamylase E₄ is:

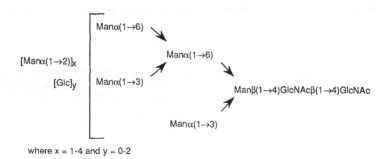

where x = 1-4 and y = 0-2

D. Characterization of O-linked Glycopeptides

Similar procedures using HPLC/ESIMS and HPLC/ESI/CID/MS as described for the N-linked glycopeptides in fraction E above were followed to characterize the O-linked glycopeptides in the fraction marked "A" in the tryptic digest. As shown in Figure 1, this group of glycopeptides was not affected by PNGase F digestion, and therefore assumed to be O-linked glycopeptides. The HPLC/ESIMS and HPLC/ESI/CID/MS data with SIM for *m/z* 163 and *m/z* 204 is shown in Figure 4, indicating the presence of oligosaccharides containing hexoses and N-acetylhexosamines such as Man (and/or Gal) and GlcNAc, respectively. Figure 5 shows the SIM data for *m/z* 163 for the glu-C digest of fraction A (top panel) as well as sialidase + PNGase F (center panel) and sialidase + O-glycosidase (bottom panel) digests of fraction A. This data again demonstrates that the oligosaccharides are not removed by PNGase F and

Table I. Summary of Mass Spectral Data Obtained on the N-linked Glycopeptides from E_4

N-linked Glycopeptides			Mass after Digestion		
Fr. #	Mass	Structure	PNGase F	Endo-H	α-mannosidase
1*	3135.8	$Man_6GN_2 + O$			
	3298.4	$Man_7GN_2 + O$	ND	ND	ND
	3462.2	$Man_8GN_2 + O$			
	3624.3	$Man_9GN_2 + O$			
1	3121.0	Man_6GN_2			
	3281.5	Man_7GN_2	1740.8	1944.4	2958.7
	3444.3	Man_8GN_2			(Man_5GN_2)
	3606.1	Man_9GN_2			
	3766.7	$Glc_1Man_9GN_2$			3443.4
	3930.3	$Glc_2Man_9GN_2$			$(Glc_1Man_7GN_2)$
2*	3694.4	$Man_6GN_2 + O$			
	3856.2	$Man_7GN_2 + O$	2298.0	2502.6	3517.2
	4017.7	$Man_8GN_2 + O$			(Man_5GN_2)
	4179.4	$Man_9GN_2 + O$			
2	3678.4	Man_6GN_2			
	3840.4	Man_7GN_2	2298.0	2502.6	3517.2
	4002.5	Man_8GN_2			(Man_5GN_2)
	4163.5	Man_9GN_2			

therefore must be O-linked glycans. In addition, the glycans are not cleavable by O-glycosidase, as would be expected for non-mammalian derived oligosaccharides, since this enzyme cleaves only Galβ1→3GalNAc structures (12). Interestingly, the SIM chromatograms are very similar for m/z 204 as well. This indicates that the O-linked oligosaccharides contain HexNAc residues, which has not been previously reported for other glucoamylases where the O-linked oligosaccharides have been studied in detail (5).

Figure 6 shows the HPLC/ESIMS spectrum of the major group of O-linked glycopeptides (starred in Figure 5). As shown in Figure 6, at least 11 oligosaccharide components are present in this spectrum, with the masses of each differing by one hexose unit (162 Da). Two other groups of glycopeptides were also identified, which contained much smaller components ranging from 2347-2844 Da. The individual components also differed only by 162 Da. No differences of 204 were identified in any of the O-linked glycopeptides. Unfortunately, efforts to sequence the peptide components of these glycopeptides by Edman degradation were unsuccessful, due to the presence of multiple peptide components. Since the sequence of this glycoprotein is unknown, it is unclear how many different sites are glycosylated and what sites are glycosylated. Since other glucoamylases are known to be multiply O-glycosylated, we assume this is the case with E_4. α-Mannosidase digestion of the O-linked glycopeptides followed by HPLC/ESIMS may give more insight into the size of the structures. The mass spectral data determined for the O-linked glycopeptides from E_4 is summarized in Table II. No glycopeptides could be identified in the HPLC/ESIMS data corresponding to peak 3 in Figure 5, presumably due to the presence of other peptides in the same fraction which suppressed the ionization of the glycopeptides.

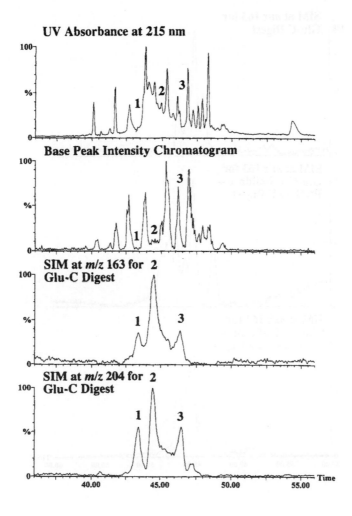

Figure 4. HPLC/ESIMS and HPLC/ESI/CID/MS with the SIM at m/z=163 and 204 chromatograms for the glu-C digest of fraction A from the E$_4$ tryptic digest.

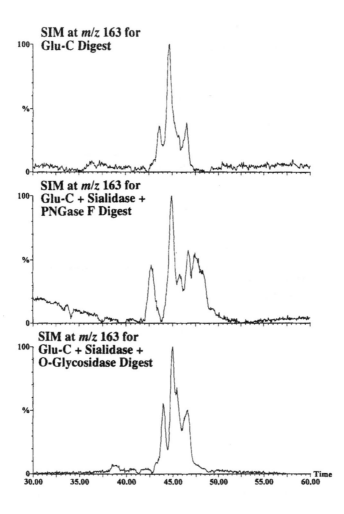

Figure 5. SIM at m/z=163 chromatograms for the glu-C, glu-C + sialidase + PNGase F, and gluc-C + sialidase + O-glycosidase digests of fraction A from the E_4 tryptic digest.

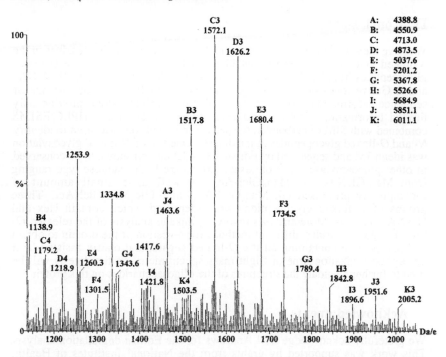

Figure 6. ESI mass spectrum of the major group of O-linked glycopeptides in the glu-C digest of fraction A from the tryptic digest (from Figure 5).

Table II. Summary of Mass Spectral Data Obtained on the O-linked Glycopeptides from E₄

O-linked Glycopeptides

Fr. #	Mass	Comments
1	2364.4	
	2522.1	Oxidized glycopeptides
	2682.5	
	2843.8	
	2347.2	
	2505.5	Masses differ by 162 Da
	2668.3	
	2832.5	
2	4388.8	
	4550.9	
	4713.0	
	4873.5	
	5037.6	Masses differ by 162 Da
	5201.2	
	5367.8	
	5526.6	
	5684.9	
	5851.1	
	6011.1	

IV. Conclusions

We have identified at least 11 hexose and 2 N-acetylglucosamine residues in the
N-linked glycopeptides, as well as 11 to 14 hexose residues in the O-linked
glycopeptides from E_4. This does not account for all of the 36 Man, 10 GlcNAc
and 6 Gal residues determined by composition analysis. In the absence of
sequence information, however, the molecular weights alone provide only
limited information. Protease and glycosidase digestion and HPLC/ESIMS
combined with SIM of carbohydrate-specific fragment ions was used to identify
N- and O-linked glycopeptides in the digests. One site of N-linked glycosylation
was identified and sequenced by Edman degradation and shown to be conserved
in other glucoamylases. The oligosaccharides are high mannose type ranging
from $Man_6GlcNAc_2$ to $Man_9GlcNAc_2$, as well as a small amount of
incompletely processed $Glc_1Man_9GlcNAc_2$ and $Glc_2Man_9GlcNAc_2$. Three
groups of O-linked glycopeptides were identified which contain Hex and
HexNAc residues. More analyses involving linkage analysis of the released O-
linked oligosaccharides, different methods for digestion of the protein to obtain
a single peptide containing all the O-linked sugars, and matrix assisted laser
desorption ionization time of flight mass spectrometry on the digests will be
used to further elucidate the structures of the oligosaccharides on this protein.

Acknowledgments

We gratefully acknowledge Lori Andrews for the Edman degradation analysis.
This work was supported by grants from the National Institutes of Health,
National Center for Research Resources (Grant RR01614), the National
Institute of Environmental Health Sciences (Grant ES04704) (to Alma L.
Burlingame), and the National Natural Science Foundation of China.

References

1.   Pazur, J. H., and Ando, T. (1960) *J. Biol. Chem.* **235**, 297-302.
2.   Saha, B. C., and Zeikus, J. G. (1989) *Starch/Starke* **41**, 57-64.
3.   Boel, E., Hjort, I., Svensson, B., Norris, F., Norris, K. E., and Fiil, N. P. (1984) *EMBO J.*
      3, 1097-1102.
4.   Svensson, B., Larsen, K., and Gunnarsson, A. (1986) *Eur. J. Biochem.* **154**, 497-502.
5.   Gunnarsson, A., Svensson, B., Nilsson, B., and Svensson, S. (1984) *Eur. J. Biochem.*
      145, 463-467.
6.   Aleshin, A., Golubev, A., Firsov, L. M., and Honzatko, R. B. (1992) *J. Biol. Chem.* **267**,
      19291-19298.
7.   Carr, S. A., Huddleston, M. J., and Bean, M. F. (1993) *Protein Sci.* **2**, 183-196.
8.   Schindler, P. A., Settineri, C. A., Collet, X., Fielding, C. J., and Burlingame, A. L. (1995)
      Protein Sci. **4**, 791-803.
9.   Hunkapiller, M. W., Hewick, R. M., Dreyer, W. J., and Hood, L. E. (1983) *Methods
      Enzymol.* **91**, 399-413.
10.   Townsend, R. R., and Hardy, M. R. (1991) *Glycobiology* **1**, 139-147.
11.   Uchida, Y., Tsukada, Y., and Sugimori, T. (1979) *J. Biochem.* **86**, 1573-1585.
12.   Umemoto, J., Bhavanandan, V. P., and Davidson, E. A. (1977) *J. Biol. Chem.* **252**, 8609-
      8614.
13.   Tarentino, A. L., Gómez, C. M., and Plummer, T. H. (1985) *Biochem.* **24**, 4665-4671.
14.   Kornfeld, R. and Kornfeld, S. (1976) *Ann. Rev. Biochem.* **45**, 217-237.
15.   Yamashita, K., Ichishima, E., Arai, M., and Kobata, A. (1980) *Biochem. Biophys. Res.
      Commun.* **96**, 1335-1342.
16.   Hernandez, L. M., Ballou, L., Alvarado, E., Gillece-Castro, B., Burlingame, A. L.,
      Ballou, C. L. (1989) *J. Biol. Chem.* **264**, 11849-11856.

Post-Translational Changes In The Protein Core Of The Proteoglycan, Aggrecan, In Human Articular Cartilage

Frank P. Barry*, Rose Maciewicz[a] and Michael T. Bayliss[b]

Osiris Therapeutics Inc., 2001 Aliceanna Street, Baltimore, MD 21231, USA, [a]ZENECA Pharmaceuticals, Mereside, Alderley Park, Macclesfield, Cheshire SK10 4TG, UK and [b]Biochemistry Division, Kennedy Institute of Rheumatology, Hammersmith, London, UK

I.   Introduction

Aggrecan is present in abundance in the extracellular matrix of cartilage, where it forms a multimolecular complex with hyaluronan and link protein.   The protein core of aggrecan, which is 2,200 amino acid in length, is organized into a series of globular and extended subdomains (1).   Two globular domains at the N-terminus, referred to as G1 and G2 are separated by an extended polypeptide of about 90 residues, known as the interglobular domain (IGD).   G1 mediates the interaction between aggrecan, hyaluronan and link protein.   G2 shows some structural similarity with G1 but does not apparently bind to hyaluronan (2).   A third globular domain, G3, is found at the C-terminus.   The extended region between G2 and G3 is substituted with a large number of keratan and chondroitin sulfate chains.

During normal aging, aggrecan in human articular cartilage undergoes a number of changes in composition.   There are changes in the length, distribution and degree of sulfation of the glycosaminoglycan chains (3, 4).   In addition, cleavage of the protein core of aggrecan occurs in the extracellular matrix (5).   The consequence of these events is an increase in the heterogeneity of aggrecan and an accumulation of the G1 domain (free-G1).   These changes eventually lead to a loss of proteoglycan from the tissue and a reduction in biomechanical strength.   Several enzymes have been shown the cleave the Asn^{341}-Phe^{342} bond in the IGD, including matrix metalloproteinases-2, 3, 7 and 9 (6).   A second cleavage site has been identified in interleukin-1 stimulated explant cultures at the Glu^{373}-Ala^{374} bond (7) but the enzyme(s) responsible for this, termed "aggrecanase" have not been identified.

The objective of this study was to identify the site within the IGD which is proteolytically cleaved to give rise to the free-G1 fragments that accumulate during aging. We used a neocarboxy terminal-specific antibody, RAM 3-2, to address this question.   RAM 3-2 shows reaction with the C-terminal sequence that arises from cleavage at the Glu^{373}-Ala^{374} bond.   It does not react with the new amino terminus or with the uncleaved protein.   Using this approach it is possible to gain information about the degradative process that leads to the accumulation of these fragments in the tissue.   In addition, the level of glycosylation of the free-G1 is compared with that of the G1 peptide isolated from aggrecan after digestion with chondroitinase ABC and trypsin (tryptic G1).

II.   Materials and Methods

A.   *Purification of G1 fragments*

Human articular cartilage, obtained post-operatively, was shredded and extracted with 4 M guanidinium hydrochloride, 50 mM sodium acetate, pH 5.8, containing 0.1 M 6-aminohexanoic acid, 10 mM EDTA, 5 mM benzamidine HCl, 1 mM PMSF and 10 mM N-ethylmaleimide as proteinase inhibitors for 24 h at 4° C.   The extract was filtered, dialyzed to remove the guanidine and solid CsCl was added to give a final concentration of 3.5 M.   Equilibrium density gradient centrifugation was carried out at 100,000 g for 48 h.   The tube was divided into four fractions and the fraction of highest buoyant density (termed A1) was further fractionated by dissociative density gradient centrifugation. The fraction of lowest density (termed A1D6) was dialyzed, dried and applied to a TSK G3000SW column and eluted with 4 M guanidinium hydrochloride, 20 mM Tris-Cl, pH 6.8.   The free-G1 fractions were pooled and rechromatographed on a Superose 6 column using the same eluent.   To prepare tryptic G1 the A1 fraction was dissolved in 0.1 M Tris-acetate, pH 7.4 and digested with chondroitinase ABC and trypsin. The digested material was applied to a column of Sepharose CL6B.   The void fraction was then chromatographed on TSK G3000SW under dissociative conditions followed by Superose 6.

B.   *Preparation of Antisera*

Peptide GCPLPRNITEGE was conjugated to KLH via the cysteine and used to raise a rabbit polyclonal antibody.   By ELISA, the resulting serum was found to specifically bind the peptide-KLH conjugate.   No cross-reaction was observed using a peptide which spans the "aggrecanase" cleavage site (RNITEGEARGSVCL) when it was conjugated to KLH. Further characterization by Western blotting showed that the antibody cross-reacted with the free-G1 of aggrecan when it was cleaved at the "aggrecanase" site, but failed to bind to native bovine or human aggrecan, or bovine aggrecan cleaved by plasmin (Gly^{375}-Ser^{376}), cathepsin B (Gly^{344}-Val^{345}) or cathepsin L (Ala^{690}-Ala^{691}). Additionally, digestion of the "aggrecanase"-generated free-G1 domain with carboxypeptidase B resulted in loss of the signal.   These results indicate that the rabbit polyclonal antiserum specifically recognized the neocarboxy terminal of free-G1 generated by cleavage of aggrecan between Glu^{373} and Ala^{374}.

C.   *Other Methods*

Peptide immunogens were synthesized using an ABI peptide synthesizer.   Purity of each peptide was determined by analysis of the peptides on a Finnigan Matt Laser Desorption Mass spectrometer and by HPLC on Vydac 2.1 mm C18 columns.   Sequence analysis was carried out on an ABI 477 sequencer.   Monosaccharide analysis was carried out on a Dionex LC system with a Carbopac PA1 column and pulsed amperometric detection. Hydrolysis was at 100° C in either 2 M trifluoroacetic acid or 4 M hydrochloric acid for 3 and 5 hours respectively.

III. Results

The A1D6 fraction obtained from mature human articular cartilage was chromatographed on TSK G3000SW to give a free-G1 fraction (Fig. 1). The tryptic-G1 was prepared from the same tissue by digestion of the A1 fraction with chondroitinase ABC and trypsin. Both the free- and tryptic-G1s isolated in this study had the same N-terminal sequence, AVTVET, which is the expected N-terminus for human aggrecan (1). The free-G1 had substantially less glycosaminoglycan (0.35 mg/mg protein) compared to the tryptic G1 (0.86 mg/mg protein), as determined by reaction with dimethylmethylene blue (7). The monosaccharide composition of each peptide was also measured, showing that the tryptic-G1 was 5-6 times more heavily substituted with galactosamine, glucosamine, galactose and mannose (Figure 2).

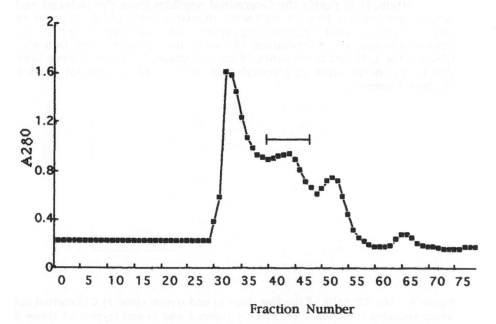

Figure 1. Separation of the A1D6 fraction prepared from a 4M guanidine hydrochloride extract of mature human articular cartilage on a TSK G3000SW column under associative conditions. The horizontal bar marks the fractions containing free-G1

The free-G1 peptide appeared on SDS PAGE as two bands of 65 and 54 kDa after reduction and the tryptic-G1 appeared as a diffuse band of approximately 60-100 kDa (Fig. 3a). After keratanase digestion both fragments were reduced in size indicating that they were both substituted with keratan sulfate (Fig. 3b). Subsequent digestion with stromelysin converted the free-G1 into a product with an apparent molecular weight of 44 kDa (Fig. 3b,). This suggested that the free-G1 did not arise by the *in vivo* action of stromelysin.

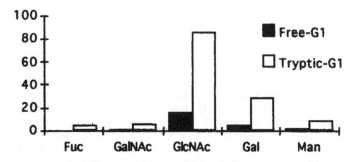

Figure 2. Monosaccharide composition of the free and tryptic G1 peptides isolated from mature human cartilage

Attempts to purify the C-terminal peptides from the reduced and carboxymethylated free-G1 following digestion with either trypsin or chymotrypsin and chromatography on anhydrotrypsin- or anhydrochymotrypsin-Sepharose (8) were unsuccessful and failed to identify the C-terminal sequence of each peptide. This may have been due to a concentration of glycosylation sites at the C-terminal portion of the G1 domain.

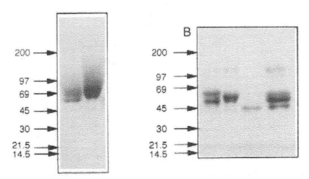

Figure 3. (A) SDS PAGE of the free (lane 1) and tryptic (lane 2) G1s carried out under reducing conditions. (B) Free-G1 (lanes 1 and 3) and tryptic G1 (lanes 2 and 4) after digestion with keratanase (lanes 1 and 2) and stromelysin (lanes 3 and 4). Keratanase-digested free- (lane 1) and tryptic-G1(lane 2).

The identity of the C-terminal sequence was therefore investigated by Western blot analysis using the polyclonal antibody, RAM 3-2, which specifically recognizes the C-terminal neo-epitope generated by "aggrecanase" cleavage of aggrecan. Both of the free-G1 peptides (65 kDa and 54 kDa) were immunolocalized with Ram 3-2 (Fig. 4, lane 1). However there was no immunolocation of the tryptic-G1 (not shown).

Using a polyclonal antiserum raised against trypsin-generated human G1, immunoreactivity was obtained with both free-G1 (Fig. 4, lane 2) and tryptic G1 (not shown).

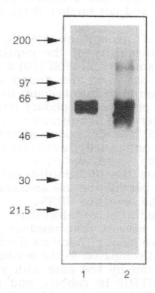

Figure 4.   Western blot of keratanase-digested free G1. Lane 1 was probed with the antibody RAM 3-2 and lane 2 with a polyclonal antiserum recognizing trypsin-generated G1.

IV.   Conclusions

Among the many age-related, post-translational modifications of aggrecan that occur during normal turnover, proteolytic cleavage of the protein core contributes significantly to the heterogeneity in size and composition of the molecule.   Although the matrix metalloproteinases are thought to be involved in aggrecan degradation, it is not clear which of them is primarily responsible or whether several of them work in combination.   For example, *in vitro* studies have shown that a range of metalloproteinases cleave within the IGD of the protein core between residues Asn^{341}-Phe^{342}. The G1 domain which results from this cleavage has been detected in human cartilage, although in low amount (12).   This study shows that the free-G1 peptide that accumulates in normal articular cartilage is generated by cleavage at Glu ^{373}-Ala^{374} by "aggrecanase" and not by stromelysin.   The conclusion is based on (a) the reaction of the free-G1 fragments with the neocarboxy terminal-specific antibody RAM 3.2 and (b) the ability to convert both free-G1 fragments to a single species of smaller apparent molecular weight by cleavage with stromelysin in vitro (Fig. 3b).   The level of glycosylation of the free-G1 is much less than that of the tryptic G1 (Fig. 2) and in fact resembles the glycosylation pattern seen on the G1 domain isolated from newborn cartilage (9). It is possible that those aggrecan molecules that carry low levels of glycosylation on the G1 domain are readily targeted for proteolytic processing in the extracellular matrix. It is also possible that the free-G1 peptide represents a distinct pool of molecules which is retained in the matrix and not further degraded by a cell-mediated process.

The usefulness of neoepitope-specific antibodies is fully dependent upon the antigenicity of the peptides used. Using the algorithm developed by Jameson and Wolf (10) it is possible to calculate the Antigenicity Index, based on the amino acid sequence of the peptide. This calculation takes into account the surface accessibility, backbone flexibility, and predicted secondary structure. On this scale a value close to or greater than 1 indicates that the peptide is highly antigenic, while a value <0 indicates poor antigenicity. For the peptide PLPRNITEGE the Average Antigenicity Index was calculated to be 0.805, and for ARGSVILTVK the value was -0.08. For the former, the antigenicity derives from the hydrophilic nature of the sequence RNITEGE. For the latter peptide the low antigenicity index reflects the hydrophobic nature of the sequence VILTV.

Studies of surface probability of the peptide PLPRNITEGEARGSVILTVK suggest that residues 3 (P) through 14 (S) define an exposed reactive loop. Therefore the neocarboxy segment is more likely to be exposed compared to the neoamino segment. These predictions were borne out by the ease with which antibodies were raised against GCPLPRNITEGE in rabbits, and the inability to raise antibodies against ARGSVILTVKCG in rabbits, mice or sheep. However, Hughes et al. (11) have successfully raised antibodies against this peptide in mice.

It is also necessary to ensure that the sequence of the peptide immunogen is sufficiently unique for it to be antigenic. In the case of PLPRNITEGE a search of the GENBANK and PIR databases revealed no similarities. Furthermore, one must show that the uncleaved peptide has no reaction with the antibody. In this case it was demonstrated by ELISA that there was no reaction with the uncleaved peptide, presumably because it had a different tertiary structure compared to the cleaved product.

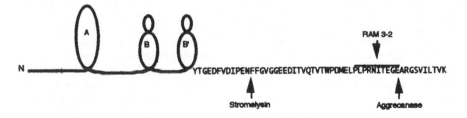

Figure 5. Schematic structure of the G1 domain of aggrecan showing the arrangement of disulfide-bonded loops and the stromlysin and "aggrecanase" cleavage sites. The RAM 3-2 neo-epitope is overlined.

Two important points arise from this study. In the first place, we show that the free-G1 fragments which arise in mature human articular cartilage are generated by cleavage of the protein core of aggrecan at the Glu^{373}-Ala^{374} bond. This proteolytic event is mediated by an uncharacterized enzyme activity termed "aggrecanase". Secondly, we show that the level of glycosylation of the free G1 peptide is substantially less than that of the corresponding G1 isolated from intact aggrecan. This implies that the pattern of glycosylation on these molecules may act as a signal which targets them for degradation in the extracellular matrix. Finally, we demonstrate the usefulness of neoepitope-specific antibodies in studying *in vivo* proteolytic events.

References

1. Doege, K., Sasaki, M., Horigan, E., Hassell, J.R. and Yamada, Y. (1987) *J. Biol. Chem.* 262, 17757-17767
2. Fosang, A.J., and Hardingham, T.E. (1991) *Biochem. J.* 273, 369-373
3. Sweet, M.B.E., Thonar, E.J.-M., and Marsh, J. (1979) *Arch. Biochem. Biophys.* 198, 439-448
4. Front, P., Arprile, F., Mitrovic, D.R. and Swann, D.A. (1989) *Connective Tissue Res.* 19, 121-133
5. Roughley, P. J., White, R., J. and Poole, A. R. (1985) *Biochem. J.* 231, 129-138
6. Fosang, A.J., Neame, P.J., Last, K., Hardingham, T.E., Murphy, G., and Hamilton, J.A. (1992) *J. Biol. Chem.* 267, 18470-19474
7. Farndale, R. W., Sayers, C. A. and Barrett, A. (1982) *Connective Tissue Res.* 9, 247-248
8. Ishii, S., Yokosawa, H., Kumazaki, Y. and Makamura, I. (1983) *Methods Enzymol.* 91, 378
9. Barry, F. P., Rosenberg, L. C., Gaw, J. U., Gaw, J. U., Koob, T. J. and Neame, P. J. (1995) *J. Biol. Chem.* 270, 20516-20524
10. Jameson, B.A., and Wolf, H. (1988) *Comput. Appl. Biosci.* 4, 181-186
11. Hughes, C.E., Caterson, B., Fosang, A.J., Roughley, P.J. and Mort, J.S. (1995) *Biochem. J.* 305, 799-804

In vivo Dipeptidylation of the Amino Terminus of Recombinant Glucagon-like Peptide-1(7-37) Produced in the Yeast *Yarrowia lipolytica* as a Fusion Protein

K. F. Geoghegan, C. A. Strick, S. Guhan, M. E. Kelly, A. J. Lanzetti, K. E. Cole,
S. B. Jones, D. A. Cole, K. J. Rosnack, R. M. Guinn, A. R. Goulet, Ting-Po I,
L. W. Blocker, D. W. Melvin and J. A. Funes

Central Research Division, Pfizer Inc., Eastern Point Road, Groton, Connecticut 06340

I. Introduction

Proteins made in high-yield recombinant expression systems often contain minor components related to the desired product but altered in an unforeseen manner. The reported instances of such variants are too numerous to reference individually, but may be grouped into two major categories; those due to chemical modifications that occur without biocatalysis, such as deamidations and certain types of oxidation, and those unexpectedly generated by the expression system. This paper describes a modification of the second type detected in a 31-residue human peptide hormone secreted from the yeast *Yarrowia lipolytica*.

Among cellular systems used for protein production, yeast and filamentous fungi best combine the benefits of economy with the capacity for extensive post-translational processing proper to eukaryotic cells. Innate protein production and secretion pathways can be recruited to express fusion proteins in which a leader sequence native to the host is followed by a human (or other recombinant) sequence (1,2). The human polypeptide, the desired product, is then cleaved from the fusion protein by endogenous enzymes and secreted from the cell.

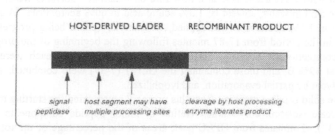

As an example of this approach, *Y. lipolytica* has recently been engineered to produce the insulin-mobilizing peptide hormone glucagon-like peptide-1(7-37) [GLP-1(7-37)], a potential therapeutic agent for type II diabetes (3). This single-chain peptide lacks any disulfide bond or other post-translational modification. (The recombinant form described in this report has a free C-terminus, although the natural form is C-terminally amidated). In the recombinant system, it was derived by proteolytic processing from a fusion protein in which GLP-1(7-37) represented the C-terminal portion and whose N-terminal segment was derived from the host alkaline extracellular protease (AEP), an important secreted activity of the yeast (4-7). Use of the AEP leader allowed GLP-1(7-37) to be made as a secreted product, as this portion of the AEP precursor contains the information that designates the mature form of the protease for export.

While purifying GLP-1(7-37), we characterized a number of minor species that were resolved from it by HPLC. Four of these were shown to be 33-residue N-terminally dipeptidylated forms of GLP-1(7-37) whose amino-terminal sequences were not encoded continuously by the recombinant gene. While these results constituted *prima facie* evidence of genetic instability in the host organism, they are explained here in terms of a post-translational modification of GLP-1(7-37) by a mechanism known to classical enzymology.

II. MATERIALS AND METHODS

Yeast Plasmids and Strains. Construction of GLP-1(7-37) expression plasmids pNB747 and pNB813 and generation and characterization of *Y. lipolytica* GLP-1(7-37) expression strain NBL625 will be described in detail elsewhere (C. A. Strick et al., manuscript in preparation).

Purification of GLP-1 (7-37). Rich broth medium containing yeast extract, peptone and dextrose and conditioned by growth of the *Y. lipolytica* expression strain NBL625 was clarified by centrifugation at 10,000 rpm in a Beckman J2-21 centrifuge followed by filtration through Whatman GF/B glass-fiber paper (1.0 μm retention). The filtrate was fractionated using a 400 ml column of Mitsubishi HP-21, a polymeric reversed-phase resin. Typically, the column was loaded at 1-2 mg GLP-1 (7-37) per ml of column bed volume. The effluent and a wash with one bed volume of 10% methanol were combined and stripped.

The concentrated, combined flow through/wash stream was fed to a 100 ml column packed with Bondapak C-4 (15 μm) packing. After loading, the GLP-1(7-37) was eluted in a linear gradient of acetonitrile and water (containing 0.1% TFA) at a flow rate of 7.2 ml/min. Fractions containing the GLP-1 (7-37) were pooled and this pooled stream was freeze-dried.

The freeze-dried solids were dissolved in 15 ml of 1.5% acetic acid and loaded on a Vydac C18 reversed-phase column (2.2 cm x 25 cm; type 218TP1022) previously equilibrated with 1.5% acetic acid at a flow rate of 9 ml/min. Solvents were: solvent A, 1.5% acetic acid in water; solvent B, 1.5% acetic acid in 60% acetonitrile. A gradient of acetonitrile in 1.5% acetic acid was applied, with the key phase being progression from 45-85% B in the period from 17-87 minutes following the beginning of the program (i.e., after the completion of the sample loading). Fractions of 5 ml each were collected beginning at 49% B, and those containing the GLP-1 (7-37) were combined, stripped of organic solvent by partial evaporation, and lyophilized.

The solid product resulting from this protocol represented the starting material for analytical-scale isolation and characterization of minor peptide components related to GLP-1 (7-37). Estimates from HPLC peak areas of the percentage of the total product represented by any minor component refer to this bulk product, an extensively purified fraction compared to the product as initially resident in yeast-conditioned medium.

Analytical-Scale (Micropreparative) Chromatography. Analytical reversed-phase HPLC was conducted at 25 °C using a Vydac C4 column (4.6 x 250 mm; 5 µm particle size) at a flow rate of 1 ml/min. Solvents were: solvent A, 0.1% TFA in water; solvent B, 0.1% TFA in acetonitrile. From initial conditions of 25% B, a gradient was formed from 25-50% B in the period from 25-35 min after injection of a 200 µl (or smaller) sample. The sample concentration was 1 mg/ml, and the detector was set to 215 nm. Mixed-mode chromatography was performed at 40 °C using an Alltech C18/Cation Exchange column (4.6 x 150 mm; 5 µm particle size) at a flow rate of 1 ml/min. Solvents were: solvent A, 0.1 M KH_2PO_4, pH 3.0; solvent B, acetonitrile. Beginning with initial conditions of 30% B, a gradient was formed from 30-36% B from 25-40 min after injection; and from 36-60% B from 40-50 min after injection. The sample concentration was 0.8 mg/ml, and the detector was set to 215 nm. A sample of not more than 240 µl was normally injected.

Mass Spectrometry and Protein Sequencing. Electrospray ionization MS was performed on a Finnigan TSQ-700 spectrometer using 2-methoxyethanol as the sheath liquid and scanning the *m/z* range from 600-2,200. Automated Edman sequencing was performed with either (i) a Hewlett-Packard G1000S sequencing system consisting of a G1000A sequencer, an on-line 1090M HPLC unit, and associated computer system and software, or (ii) an Applied Biosystems Model 470A gas-phase sequencer with Model 120A on-line PTH Analyzer and a data system based on PE Nelson Model 2600 software.

III. RESULTS

Expression of GLP-1(7-37) in Yeast. GLP-1(7-37) was expressed in the form of a 188-residue fusion protein composed of 157 residues from the N-terminus of the alkaline extracellular protease of *Y. lipolytica* followed by the 31 residues of GLP-1(7-37) (Figure 1). This gene product consisted of several functional segments, including a putative 15-residue signal peptide followed by a region with the sequence APLAAPAPAPDAAPAA-(one-letter amino acid code). GLP-1(7-37) was expected to be liberated from the fusion protein by the Kex2-like host processing enzyme Xpr6, which cleaves following dibasic residue pairs. The leader sequence from the host protease designated the gene product for secretion, allowing GLP-1(7-37) to be recovered in final form from the growth medium. The N-terminal sequence of the desired product was His-Ala-Glu-Gly-Thr- (Figure 1).

Transformation of *Y. lipolytica* to harbor ten stably incorporated copies of the recombinant gene (strain NBL625) and subsequent cell culture to achieve high levels of GLP-1(7-37) expression will be described elsewhere (C. A. Strick et al.; S. E. Lee et al., manuscripts in preparation). Procedures to recover the product in pure form evolved with experiment, but invariably ended with reversed-phase HPLC so that the product could be dried without excess nonvolatile salts. Minor components of the product were purified by analytical-scale HPLC and then investigated by mass spectrometry and automated protein sequencing.

Dipeptidylaminopeptidase processing

MKLATAFTILTAVLAAPLAAPAPAPDAAPAAVPEGPAAAAYSSILSVVAK

Δ

QSKKFKHHKRDLDEKDQFIVVFDSSATVDQIASEIQKLDSLVDEDSSNGI

▲ ▲

TSALDLPVYTDGSGFLGFVGKFNSTIVDKLKESSVLTVEPDTIVSLPEIP

▲

GLP-1(7-37)

ASSNAKRHAEGTFTSDVSSYLEGQAAKEFIAWLVKGRG

▲

Figure 1. Amino acid sequence of the fusion protein encoded by the recombinant gene. The open triangle (Δ) indicates the signal peptidase cleavage site; the closed triangles (▲) indicate demonstrated or putative sites for cleavage by Xpr6, a Kex2-like enzyme of *Y. lipolytica* .

Minor Components of the Product. A number of minor products were copurified at the preparative level with correctly processed GLP-1(7-37) and resolved from it by analytical separations. In several cases, the origin of these was easy to understand. For example, a 6077 Da peptide with the N-terminal sequence of Glu-Ser-Ser-Val-Leu- (see Figure 1) was evidently a fragment of the original gene product that had escaped Xpr6-catalyzed cleavage at the Lys-Arg pair immediately ahead of the GLP-1(7-37) peptide. Likewise, a des-Gly-37 form of the product was isolated, presumably a product of carboxypeptidase-catalyzed C-terminal truncation of GLP-1(7-37). A host enzyme similar to Kex1 of *S. cerevisiae* was a good candidate as the responsible enzyme. Such low-frequency variations of product structure are typical in high-yield protein expression systems.

Another minor component detected was a 29-residue C-terminal fragment of GLP-1(7-37), i.e. the desired product lacking the N-terminal His-Ala dipeptide. This component was presumed to have been generated either by two cycles of aminopeptidase action on GLP-1(7-37) or one cycle of dipeptidylaminopeptidase action. As a dipeptidylaminopeptidase that selectively removes X-Ala or X-Pro dipeptides participates in processing the alkaline extracellular protease of *Y. lipolytica* (8), the same enzyme was probably responsible for forming des-His-Ala-GLP-1(7-37).

N-Terminally dipeptidylated forms of GLP-1(7-37). When the solid product from the pilot-scale purification was resolubilized and analyzed by RP-HPLC, the resulting chromatograms consistently showed a tailing shoulder at the main band (retention time ~23 min) with a maximal area percentage of 0.7% relative to the principal peak (Figure 2a). The components responsible for it were separated from the main product (although not from each other) by a two-step enrichment procedure (Figure 2b,c).

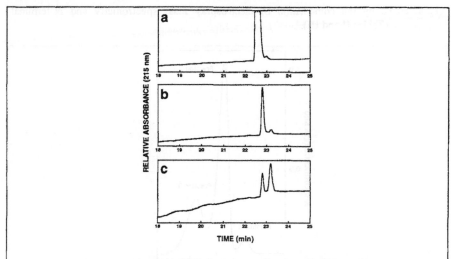

Figure 2. Successive reversed-phase HPLC fractionations yielding a peak (retention time of approximately 23.1 min) containing the minor products whose characterization is reported in Table I.

TABLE I *Mass Spectrometric and Edman Sequencing Data on Impurity Fraction Isolated by RP-HPLC*

Mass Obs.	Identified with Sequence	Identified Peptide	Mass Theor.
3539	LAHAEGTFTSDVSSYLEGQAAKEFIAW...	Leu-Ala-GLP-1(7-37)	3539.9
3355	HAEGTFTSDVSSYLEGQAAKEFIAWLV...	GLP-1(7-37)	3355.7
3297	same as for mass 3355 component	GLP-1(7-36) (lacks Gly-37)	3298.6
6077	EXSVLXVEPXXIXXXPXIPXSXNXK..	Fusion protein res. 132-188	6077.3

After each fraction collection and concentration, the samples were divided and subjected to electrospray ionization mass spectrometry, N-terminal amino acid sequencing, and HPLC. From the mass analyses, it was apparent that the isolated fraction contained GLP-1(7-37) and three minor compounds. As the limited quantities to hand precluded further purification, the whole peptide mixture was subjected to automated N-terminal sequence analysis. Three sequence signals were resolved on the basis of differential level and by reference to mass data from the mixture (Table I). Among the components found in this fraction was Leu-Ala-GLP-1(7-37), the first N-terminally dipeptidylated form of GLP-1(7-37) to be recovered.

Three further dipeptidylated forms of GLP-1(7-37) were identified by mixed-mode HPLC using a fraction from the preparative chromatography which contained higher contents of other components (this fraction was not utilized in the final pool to isolate GLP-1(7-37)). This method separated three peaks (Figure 3) which were then

characterized by the combination of electrospray mass spectrometry and N-terminal
sequencing (Tables II and III).

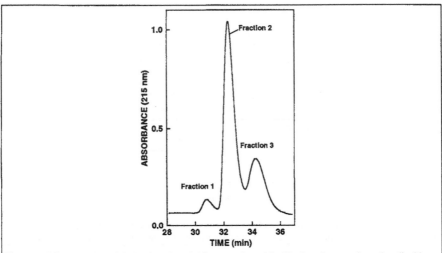

Figure 3. Mixed-mode HPLC fractionation yielding a peak enriched in the minor products described in
Tables II and III.

TABLE II. *Interpretation of Sequencing Data on Impurity Fraction 1 from Mixed-Mode HPLC*

Sequence Observed	Identified Peptide
`APHAEGTFTSDVSSYLEGQAAKEFIAW....`	Ala-Pro-GLP-1(7-37)
`LAHAEGTFTSDVSSYLEGQAAKEFIAW....`	Leu-Ala-GLP-1(7-37)
`AAHAEGTFTSDVSSYLEGQAAKEFIAW....`	Ala-Ala-GLP-1(7-37)

TABLE III. *Mass Spectrometric and Sequence Data on Impurity Fraction 3 from Mixed-Mode HPLC*

Mass Obs.	Sequence Observed	Identified Peptide	Mass Theor.
3355	`HAEGTFTSDVSSYLEGQAAKEFIAW....`	GLP-1(7-37)	3355.7
3542	`DAHAEGTFTSDVSSYLEGQAAKEFIAW...`	Asp-Ala-GLP-1(7-37)	3540.9

　　　　To summarize, analytical fractionations led to the identification of four N-
terminally dipeptidylated forms of GLP-1(7-37). These were: Leu-Ala-GLP-1(7-37); Ala-
Ala-GLP-1(7-37); Asp-Ala-GLP-1(7-37); and Ala-Pro-GLP-1(7-37). The sequence of

the fusion protein (Figure 1) shows that the dipeptides added to the N-terminus of GLP-1(7-37) all occur in even dipeptide phase following the signal cleavage site.

To validate this result further, the four dipeptidylated forms of GLP-1(7-37) were synthesized and chromatographed. Each synthetic peptide behaved identically to the respective anomalous recombinant product (data not shown).

IV. DISCUSSION

While the protein secretion pathway of *Y. lipolytica* has been studied in less detail than that of *S. cerevisiae*, the processing of several proteins secreted from *Y. lipolytica* has been well characterized. One of these is AEP, which has been cloned and sequenced here and elsewhere (6,7). As a result of these studies, homologues of processing enzymes active in *S. cerevisiae* have been demonstrated or implicated to function during the processing of proproteins in *Y. lipolytica* as well. In proAEP, the placement of Lys-Arg immediately prior to the mature AEP protein suggests that processing at this site is effected by a Kex2-like proteinase (8), and such an activity has recently been identified and its gene (*xpr6*) cloned from *Y. lipolytica* (9). The expression/secretion system used for GLP-1(7-37) production relied on an in-frame fusion between the prepro and pro regions of AEP and GLP-1 (7-37) which retained the Lys-Arg sequence between the pro region of AEP and the mature heterologous protein (Figure 1). Cleavage at this site released an authentic protein molecule for secretion.

The expression system is very successful, with correctly processed GLP-1(7-37) released into the medium at a concentration of >0.1 g l^{-1}. In such processes, it is not unusual to detect minor forms of the product differing from the desired material in a manner that can be accounted for by ordinary mechanisms, and several such species appeared in the present work (see Results). Quite unexpected, however, were the four N-terminally dipeptidylated forms of GLP-1(7-37) that have been described.

As these peptides contained amino acid sequences not directly encoded by the source gene, their appearance raised the issue of whether the host system was genetically stable. To monitor genetic stability, we conducted Southern hybridization analyses of the expression units within the host genome and nucleotide sequencing of the fusion gene mRNA population (to be described elsewhere; C. A. Strick et al., in preparation). Results from the expression culture before and after fermentation gave no indication of rearrangements in the recombinant genes. Abundant production of the correct product set aside the possibility that the process was operating in an entirely different manner from that intended. It then remained to explain the origin of the dipeptidylated forms of GLP-1(7-37) in terms of post-translational addition of a dipeptide unit.

A single explanation presented itself, based on what is known of the processing and secretion pathway for *Y. lipolytica* AEP. A dipeptidylaminopeptidase similar to Ste13 of *S. cerevisiae* effects serial removal of dipeptide units from the N-terminus of the fusion protein through the region of sequence APLAAPAPAPDAAPAA-. This type of processing does take place in vivo in *Y. lipolytica* during maturation of the pro-AEP (8), and appears to occur earlier than the Kex2-like cleavage that releases the pro region from the mature protein. Kex2 and Ste13 are proposed to be colocalized (with Kex1) in *S. cerevisiae* to a single late Golgi compartment (10). We assume that colocalization occurs in *Y. lipolytica* as well, so that although the two processing steps may be separated in time, the two late processing proteinases and their products are located in a single late Golgi compartment. Transfer of dipeptidyl units originating from the N-terminal region of the fusion protein to the N-terminus of fully processed GLP-1(7-37) can reasonably be attributed to the activity of an Ste13-like host enzyme.

The Ste13 dipeptidylaminopeptidase of *S. cerevisiae* is an apparent serine proteinase of >90 kDa (11). Its function is thought to be stepwise trimming of dipeptide units from the N-termini of yeast proteins intended for secretion, as in the case of α-factor in *S. cerevisiae* (12,13). The exact importance of this activity is unclear, but there has been speculation that it serves to prevent premature (intracellular) development of activity intended to take place outside the cell.

An extensive literature deals with proteinase-catalyzed formation of peptide bonds, of which a highlight is the review by Fruton (14). Carpenter (15) showed that ionizations of the liberated carboxyl and amino groups are principal energetic factors favoring the hydrolysis of peptide bonds at pH 7. Conditions that moderate the tendency of these groups to be ionized, such as the presence of an organic solvent, favor bond synthesis. The pH range best favoring peptide synthesis is that intermediate between the pK_a values of the two groups, generally pH 5-7. Synthesis is more favored for the joining of two peptides, a description that encompasses N-terminal dipeptidylation, than for the coupling of a single amino acid to a peptide. Even in peptide mapping experiments, where complete hydrolysis of all susceptible peptide bonds is the preferred outcome, unpredicted synthetic reactions can occur in detectable yield (16). We conclude that there is ample precedent and rationale for ascribing formation of N-terminally dipeptidylated GLP-1(7-37) synthesized in *Y. lipolytica* to dipeptidylaminopeptidase-catalyzed, nonribosomal peptide bond formation. Definitive proof of this concept will await further studies.

References

1. Fleer, R., Chen, X. J., Amellal, N., Yeh, P., Fournier, A., Guinet, F., Gault, N., Faucher, D., Folliard, F., Fukuhara, H. and Mayaux, J.-F. (1991) *Gene* 107, 285-295.
2. Ward, P. P., Piddington, C. S., Cunningham, G. A., Zhou, X., Wyatt, R. D. and Conneely, O. M. (1995) *Bio/Technology* 13, 498-503.
3. Holz, G. G., Kühtreiber, W. M. and Habener, J. F. (1993) *Nature* 361, 362-365.
4. Simms, P. C. and Ogrydziak, D. M. (1981) *J. Bacteriol.* 145, 404-409.
5. Ogrydziak, D. M. and Scharf, S. J. (1982) *J Gen. Microbiol.* 128, 1225-1234.
6. Davidow, L. S., O'Donnell, M. M., Kaczmarek, F. S., Pereira, D. A., DeZeeuw, J. R. and Franke A. E. (1987) *J. Bacteriol.* 169, 4621-4629.
7. Cheng, S.-C. and Ogrydziak, D. M. (1987) *J. Bacteriol.* 169, 1433-1440.
8. Matoba, S., Fukayama, J., Wing, R. A. and Ogrydziak, D. M. (1988) *Mol. Cell. Biol.* 8, 4904-4916.
9. Enderlin, C. S. and Ogrydziak, D. M. (1994) *Yeast* 10, 67-79.
10. Bryant, N. J. and Boyd A. (1993) *J. Cell Sci.* 106, 815-822.
11. Anna-Arriola, S. S. and Herskowitz, I. (1994) *Yeast* 10, 801-810.
12. Julius, D., Blair, L., Brake, A., Sprague, G. and Thorner, J. (1983) *Cell* 32, 839-852.
13. Julius, D., Brake, A., Blair, L., Kunisawa, R. and Thorner, J. (1984) *Cell* 37, 1075-1089.
14. Fruton, J. S. (1982) *Adv. Enzymol.* 53, 239-306.
15. Carpenter, F. H. (1960) *J. Am. Chem. Soc.* 82, 1111-1122.
16. Canova-Davis, E., Kessler, T. J. and Ling, V. T. (1991) *Anal. Biochem.* 196, 39-45.

Detection of Acetylated Lysine Residues using Sequencing by Edman Degradation and Mass Spectrometry

Faith E. Ross[*], Tom Zamborelli[#], Alan C. Herman[*], Che-Hung Yeh[*], Nicole I. Tedeschi[*], and Edward S. Luedke[*]

[*]Department of Analytical Research and Development
Amgen Inc., Amgen Ctr, Thousand Oaks, CA 91320-1789
[#]Peptide Synthesis Group
Amgen Boulder Inc., Boulder, CO 80301-2549

I. Introduction

Acetylation of lysine residues is a well known post-translational modification in recombinant proteins expressed in *Escherichia coli* (1,2,3). Acetylation sites have also been used to determine the conservation of sequence in related proteins (4) Acetylation of lysine residues have been found to prevent carbamylation and glycosylation of some proteins (5). Much of the previous research in identifying ε-N-acetyllysines requires *in vivo* labeling with a radioactively labeled acetylating agent followed by amino acid analysis(AAA) or amino acid sequencing to determine the location of the incorporated label (2,4,6-9). Detection of post-translational acetylated lysines is slightly more difficult. Mass spectrometry of digested proteins may indicate a possilbe modification site by the detection of peptide mass changes of +42 amu. Incomplete digestion or shifts in peptide location in chromatographic separations have been noted in peptides and proteins containing acetylated lysines (5,10,11). It is difficult to confirm the acetylation using amino acid analysis because the acetyl group is labile in normal hydrolysis conditions. However, exhaustive enzymatic digestions followed by AAA, can be compared with ε-N-acetyllysine standards to detect and quantify these species (1,8,12,13).

When subjected to N-terminal sequencing acetylated lysine residues commonly do not yield a peak at the lysine retention time. In order to use sequencing as a detection method for unlabeled acetylated lysine, a PTH-derivative standard of the modified amino acid residue can be used (2,6,11).

In our efforts to characterize post-translational modifications in recombinant Neurotrophin-3 we suspected the occurance of acetylated lysines due to mass increases of +42 amu observed on LC-MS (electrospray) and MALDI-TOF data. Endoproteinase Lys-C digestions of modified NT-3 peaks showed incomplete digestion of peptides containing suspected acetylated residues. In addition to the inablity to be enzymatically hydrolyzed, mass increases of +42 amu were observed in the peptides. Edman degradation sequencing did not yield a peak at the normal PTH-lysine retention time. Depending on the sequencer used, peaks eluted near the PTH-alanine or PTH-histidine retention time for the acetylated lysine residues. The mass data did not support a substitution of either amino acid.

TECHNIQUES IN PROTEIN CHEMISTRY VII

Synthetic NT-3 peptides were prepared with and without acetylation of the C-terminal lysine and they were used as PTH-standards for three different sequencers. Chromatograms from the appropriate sequence cycles were compared to standards and the peak location for acetylated lysine determined. The sequencing results supported the mass spectrometry data for post-translational acetylations sites in NT-3.

The acetylation sites were also confirmed from Lys-C digested native NT-3 which had been chemically acetylated. A comparison of retention time and mass of the incompletely digested peptides containing chemically acetylated lysines to the collected, modified NT-3 forms showed similarity. Edman sequencing of these peptides confirmed ε-N-acetylysine residues.

This report describes the application of Edman degradation sequencing in addition to mass spectrometry for the identification of post-tranlational, acetylation of lysine residues in a recimbinant protein expressed in *E. coli*.

II. Methods

A. Protein

r-metHuNT-3 (NT-3) was expressed and purified from insoluble fractions of *E. coli* lysates using a series of chromatographic procedures following oxidation and refolding steps as previously described (14).

B. Reverse-phase HPLC Separation and Collection of r-metHuNT-3 Forms Containing e-N-acetyllysine

Initial separation and purification of the acetylated forms of NT-3 was achieved by dilution of the sample in a denaturing buffer (6M Guanidine-HCl buffered to pH 8.3) and preparative HPLC. Reverse-phase HPLC was performed using a Hewlett Packard (HP) 1090 solvent delivery system and a C4 column (YMC, 5µm, 200Å, 20mmID X 250mm). A linear gradient using standard reverse-phase techniques (water + 0.1% TFA and 90% acetonitrile + 0.1% TFA) was used to separate the modified protein forms. Individual peaks were manually collected, lyophilized, and stored at -20°C for later analysis.

C. Endoproteinase Lys-C Peptide Mapping

Samples were denatured in a Guanidine-HCl solution buffered to pH 8.3 with 0.01M DTT at ~55°C for 30 minutes. 0.02M IAA (iodoacetic acid) was used to alkylate the denatured protein at ~55°C for 1 hour. The samples were dialyzed using a 8,000 mw cut-off membrane into 2M Urea + 0.1M NH_4HCO_3. Endoproteinase Lys-C was added at 1% w/w and samples were incubated at 37°C for 20 hours. An HP 1090 solvent delivery system and a RP-HPLC column (Polymer Labs 5µm, 100Å, 2.0mmID X 250mm) were used to separate the peptides using standard reverse-phase techniques. The endoproteinase Lys-C peptide map was eluted directly into a Finnigan SSQ 710 mass spectrometer (LC-MS) to get accurate mass data on the resolved peptides. Peptide peaks were manually collected from the HPLC or from a post-column in-line split at the LC-MS interface for sequencing analysis.

D. Chemical Acetylation of r-metHuNT-3 using Acetic Anhydride

NT-3 was chemically acetylated with acetic anhydride (7,10). The sample was diluted with 6M Guanidine-HCl + 0.2M NaHCO$_3$, pH 8.7 to a concentration of 1.4mg/ml. 0.5M acetic anhydride was added until a final concentration of 50mM was reached. A pH range of 8.5-8.7 was maintained during the addition of acetic anhydride using 0.5M NaCO$_3$. The solution was passed through a Sephadex G-25 column to terminate the reaction.

E. Synthetic peptide preparation

Amgen Boulder Inc., Peptide Technology Group synthesized the native and acetylated forms of peptides to be used as sequencing and mass spectrometry controls. Peptides were synthesized on a ABI 431 peptide synthesizer by the Fmoc (fluorenylmethoxycarbonyl)/t-butyl based solid phase peptide chemistry method (15). Synthesis of the acetylated peptide was done on PAM resin, which was loaded with Fmoc-Lys(Boc) through the DMAP (4-dimethylaminopyridine)-catalyzed, DIC (diisopropylcarbodiimide)-mediated coubling of Fmoc-Lys(Boc)-COOH. Subsequent amino acids were coupled as HOBT (hydroxybenztriazole) esters. Without removal of the final N-terminal Fmoc, side chains protecting groups were removed by treatment for 4 hours with TFA:thioanisole:β-mercaptoethanol:water:phenol (80:5:5:5:5). Next, the C-terminal lysine was acetylated by mixing the resin for 30 minutes in a solution of 25% acetic anhydride in DMF (dimethyl formamide). The N-terminal Fmoc was then removed, and the peptide was cleaved from the resin by treatment with neat HF (0°C for 1 hour).

The native peptide(without acetylated lysine) was assembled starting with a commercially available Fmoc-Lys(Boc)-HMP (hydroxymethyl-phenylacetyl) derivatized polystyrene resin. The N-terminal Fmoc, side chain protecting groups were removed and the peptide cleaved from the resin by treatment with TFA:thioanisole:β-mercaptoethanol:water:phenol as before. The crude peptides were purified by preparative reverse-phase HPLC (Vydac C4, 2.5cm X 25cm column). The sequence, mass, and amino acid composition of the resulting peptides were confirmed by standard methods.

F. Sequencing of Peptides

The common protein sequencing protocol for Edman degradation and phenylthiohydantoin-derivative identification was used. Automated N-terminal amino acid sequencing of the collected peptides and synthetic peptides was performed on two HP G1000A sequencers and also an ABI 494 sequencer.

III. Results

NT-3 fractions containing amino acid variations were isolated from the native form of the protein by preparative reverse phase HPLC. Acetylation was suspected when endoproteinase Lys-C digested proteins resulted in incomplete digestion of some expected peptides and mass increases of +42 amu were observed. Figure 1 is the chromatographic separation of Lys-C digested standard NT-3 verses a modified NT-3 peak containing a suspected acetylated lysine. Mass data for the peptide containing the acetylated lysine as well as the native peptides is show in Figure 2. Sequencing of the acetylated peptide showed the expected amino acid order, however the signal for the cycle containing the acetylated lysine was low at the standard PTH-lysine(PTH-K) retention time. The Hewlett Packard sequencers consistantly had a noticable increase in a peak eluting near PTH-alanine(PTH-A) while the ABI sequencer had an increase in a peak eluting between the PTH-histidine(PTH-H) and PTH-alanine(PTH-A) retention times. The retention time for the acetylated lysine was slightly later than the expected PTH-A or PTH-H peak and was often distinguished by a doublet at the amino acid location.

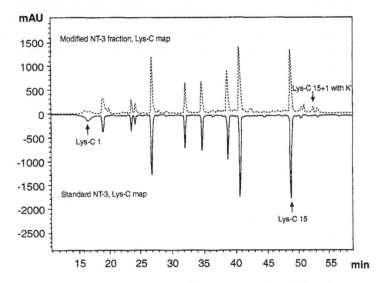

Figure 1. Peptide map of standard NT-3 verses a modified NT-3 fraction. Samples were digested with Lys-C and separated chromatographically. The native peptides are indicated by the arrows in the standard NT-3 chromatogram. An arrow on the chromatogram for the modified NT-3 fraction indicates the resulting peptide possessing the acetylated lysine residue.

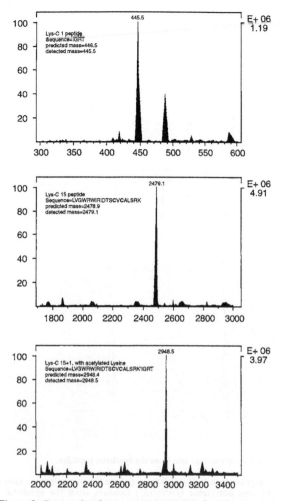

Figure 2. Deconvoluted mass spectometry data for the native and acetylated peptides shown in Figure 1.

The N-terminal peptide for endoproteinase Lys-C digested NT-3 contains the six amino acid sequence (MYAEHK). Synthetic peptide preparations of this peptide with and without acetylating the C-terminal lysine were prepared and used as PTH-standards for the three sequencers. A comparison of the sequence cycles for the acetylated verses native lysines for the three sequencers is shown in Figure 3. Amino acids with similar retention times as observed for the acetylated lysine residues are also compared in these figures. The acetylated lysine (PTH-derivative) sequence chromatgram resembled data which we had observed for the modified NT-3 peptides containing suspected acetylated lysines.

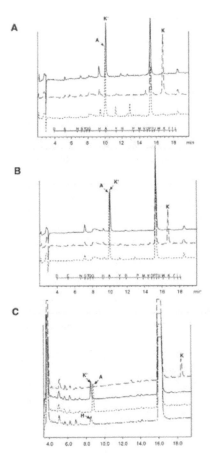

Figure 3 - Sequence cycle chromatograms from the synthetic peptides.

A. Overlay of chromatograms from the H/P G1000A sequencer. The chromatograms show the alanine (A), lysine (K), and acetylated lysine (K') residues from the synthetic peptides.
B. Overlay of chromatograms from the H/P G1000A sequencer. The chromatograms show the alanine (A), lysine (K), and acetylated lysine (K') residues from the synthetic peptides.
C. Overlay of chromatograms from the ABI 494 sequencer. The chromatograms show the alanine (A), histidine (H), lysine (K), and acetylated lysine (K') residues from the synthetic peptides.

To confirm the acetylation sites of the modified NT-3 forms native NT-3 was chemically acetylated using acetic anhydride. The acetylated NT-3 was digested with endoproteinase Lys-C and the peptides were separated in the same manner as the modified NT-3 fractions. The RP-HPLC peptide maps were eluted directly into a mass spectrometer. Mass and retention times of the peptides containing the chemically acetylated lysines was similar to that observed for the modified NT-3 peptides containing acetylated lysine residues. The chemically acetylated peptides were sequenced. Figure 4A is an example of the chromatograms for the acetylated lysine and alanine cycles produced by the H/P sequencer for a chemically acetylated peptide. Figure 4B shows similar chromatograms for cycles generated by the ABI sequencer.

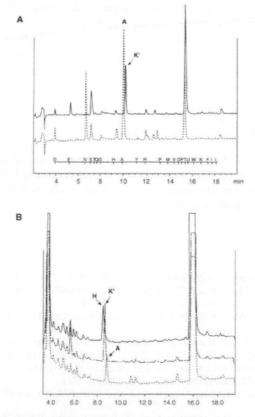

Figure 4 - Sequence cycle chromatograms from a chemically acetylated NT-3 peptide.

A. Overlay of chromatograms from the H/P G1000A sequencer. The chromatograms show the alanine (A) and acetylated lysine (K') residues from a chemically acetylated NT-3 peptide.
B. Overlay of chromatograms from the ABI 494 sequencer. The chromatograms show the alanine (A), histidine (H) and acetylated lysine (K') residues from a chemically acetylated NT-3 peptide.

All peptides in these experiments produced similar results on the sequencers for the acetylated verses native lysine residues. These results indicate that amino acid residues with retention times slightly later than alanine (H/P) or slightly later than histidine (ABI) may be acetylation of a native lysine residue.

IV. Conclusion

Peptide mapping in conjunction with mass spectrometry may indicate acetylation due to incomplete hydrolysis and increases in mass of +42 amu. We were able to use N-terminal amino acid sequencing to confirm lysine acetylation. Sequencer manufacturers use different chemistries and columns. The elution patterns vary and modifications to amino acids produce different results and retention times. The two Hewlett Packard G1000A sequencers produced very similar results. The PTH-acetylated lysine produced a peak with a retention time close to PTH-alanine on the HP sequencers. The ABI 494 sequencer showed a peak with a retention time between the PTH-alanine and PTH-histidine for the PTH-acetylated lysine residue. The retention times for the PTH-acetylated lysine peaks are shifted only slightly from the PTH-alanine or PTH-histidine peaks which may be mistaken for amino acid substitution rather than post-translational modifications. The final confirmation was achieved with chemically modified synthetic standards.

Acknowledgments

We thank S. Burrell, T. Jacobson and D. Lenz of the Amgen Boulder Inc., Peptide Technology Group, for their technical support and preparation of the synthetic peptides used in this research. We thank Kathryn Rubenstein for assistance in preparation of the manuscript figures.

References

1. Han, K.K. and Martinage, A. (1993) *Int. J. Biochem. Vol. 25, No. 7*, 957-70.
2. Eddie, B., Rossier, J., Le Caer, J.P., Berwald-Netter, Y., Koulakoff, A., Gros, F. and Denoulet, P. (1991) *J. Cell. Biochem. Vol. 46, No. 2*, 134-42.
3. Groenen, P. J. T. A., Merck, K.B., De Jong, W.W., and Bloemendal, H. (1994) *Eur. J. Biochem. 225*, 1-19.
4. Waterborg, J.H., Fried S.R., and Matthews, H.R. (1983) *Eur. J. Biochem. 136*, 245-252.
5. Hasan, A., Smith, J.B., Qin, W., and Smith, D.L. (1993) *Exp. Eye Res.*, 57, 29-35.
6. Waterborg, J.H. (1990) *J. Biol. Chem., Vol. 265, No 28*, pp 17157-17161.
7. Waterborg, J.H. (1992) *Biochemistry, Vol.31, No. 27*, 6211-6219.
8. Sterner,R., Vidali, G., and Allfrey, V.G. (1981) *J. Biol. Chem. Vol. 256, No. 17*, pp. 8892-8895.
9. Pesis, K. and Matthews, H.R. (1986) *Arch. Biochem. Biophys., Vol. 251, No. 2*, pp 665-673.
10. Takano, M., Takahashi, M., Oobatake, M. and Asada, K. (1985) *J. Biochem. 98*, 1333-1340.
11. Suckau, D., Mak, M., and Przyblski, M. (1992) *Proc. Natl. Acad.. Sci.* USA, *Vol. 89*, pp 5630-5634.
12. Zhu, D.X., Xu, L.X., Zhu, N.Z., Briand, G., and Han, K.K. (1985) *Int. J. Biochem. Vol. 17, No. 6*, pp.719-721.
13. Zhu, D.X., Zhang, A., Zhu, N.C., Xu, L.X., Deutsch, H.F., and Han, K.K. (1986) *Int. J. Biochem. Vol. 18, No. 5*, pp. 473-476.
14. Philo, J., Talvenheimo, J., Wen, J., Rosenfeld, R., Welcher, A., and Arakawa, T. (1994) *J. Biol. Chem. Vol. 269, No. 45*, 27840-27846.
15. Atherton, E.; Sheppard, R.C. (1985) *J. Chem. Soc., Chem. Commun.*, pp 165-166.

Mechanism of the nucleoside diphosphate kinase reaction: X-ray structure of the phosphohistidine intermediate

Ioan Lascu
Institut de Biochimie et Génétique Cellulaires
33077-Bordeaux, France

Solange Moréra
Mohammed Chiadmi
Jacqueline Cherfils
Joël Janin
Laboratoire de Biologie Structurale
CNRS-Université Paris-Sud
91198-Gif-sur-Yvette, France

Michel Véron
Unité de Biochimie Cellulaire
Institut Pasteur
75724-Paris, France

1. Introduction

Nucleoside diphosphate kinase (NDP kinase, EC 2.7.4.6) catalyses the reversible transfer of the terminal phosphate of nucleoside triphosphates to nucleoside diphosphates:

$$N_1TP + N_2DP \longleftrightarrow N_1DP + N_2TP \qquad (1)$$

NDP kinase is a highly efficient enzyme (k_{cat} higher than 10^3 s^{-1}, k_{cat}/K_m of about 10^7 M^{-1}s^{-1}) with little specificity for the base moiety and for the presence of the OH in position 2' of ribose of the donor and acceptor nucleotides. The activity is high in all cells consistent with a key role for NDP kinase in the biosynthesis of non-adenine nucleoside and deoxynucleoside triphosphate precursors of nucleic acid synthesis, some of which are involved in biosynthetic pathways (UTP, CTP) and in signal transduction (GTP). NDP kinase also plays regulatory functions (reviewed by De la Rosa et al., 1995). It was shown to be involved in differentiation of *Drosophila* larvae, in proliferation and metastasis of human tumors and in transcription of the *c-myc* oncogene. These functions are not necessarily related to the catalytic function discussed here.

This short review will focus on a particular aspect of the NDP kinase structure and mechanism, the properties of the phosphohistidine form, an obligatory intermediate during catalysis. The role of phosphoryl intermediates in catalysis has been reviewed from biochemical and structural point of view (Frey, 1992; Johnson & Barford, 1993). The older biochemical data on NDP kinase have been reviewed by Parks and Agarwal (1973). Since then, our knowledge of NDP kinase structure and catalysis has accumulated. The first sequence was published in 1990 (more than 25 are known now) followed soon after by the first high-resolution X-ray structure in 1992.

2. NDP kinase is phosphorylated on a histidine residue during the catalytic cycle

It is now well-established that reaction (1) proceeds *via* a covalent intermediate, the enzyme being transiently phosphorylated on a histidine residue:

$$N_1TP + E \longleftrightarrow N_1DP + E\text{-}P \qquad (1a)$$
$$N_2DP + E\text{-}P \longleftrightarrow N_2TP + E \qquad (1b)$$

The following biochemical data demonstrate the existence and the competence of this intermediate:

a. All NDP kinases have a ping-pong kinetic mechanism. Inhibition by excess substrate and the exchange N_1TP/N_1DP in the absence of the second substrate have been observed (Garces & Cleland, 1969), as requested by this mechanism.

b. The phosphorylated enzyme can be isolated. However, the equilibrium constant of the reaction (1a) is 0.2-0.25 (Garces & Cleland, 1969; Lascu et al., 1983) and the stoichiometry was found to be less than one phosphate/subunit if the enzyme was incubated with ATP or GTP. Full phosphorylation could be obtained only by rephosphorylating the ADP formed by a regenerative system, like phosphoenolpyruvate/pyruvate kinase (Lascu et al, 1983). This is particularly useful for keeping the enzyme fully phosphorylated during data collection over extended periods of time (like in NMR experiments) since the phosphohistidine intermediate hydrolyzes slowly.

c. Reaction (1) occurs with retention of configuration for the phosphate being transfered. This is compatible with double inversion upon each of the two reactions (1a) and (1b) (Sheu et al., 1979).

d. The rate of both reactions (1a) and (1b) is equal to or faster than the over-all reaction rate (1). This proves that the intermediate is kinetically competent (Walinder et al., 1969).

e. The site of phosphorylation was identified as a histidine by chemical analysis (Edlund et al., 1969), site-directed mutagenesis (Dumas et al., 1992) and direct protein sequencing (Gilles et al., 1991). The phosphorylated position of histidine has been identified as N_δ by alkaline hydrolysis and thin-layer chromatographic separation (Edlund et al., 1969), stability studies (δ-phospho-histidine is more labile than ϵ-phosphohistidine) and recently by ^{31}P-NMR studies (Lecroisey et al, 1995). N_δ and N_ϵ positions have quite different chemical properties: N_δ is a better nucleophile than N_ϵ and N_δ phosphohistidine is more reactive than N_ϵ phosphohistidine (Hultquist et al., 1966; Wei & Matthews, 1991). This difference in reactivity could be relevant to the reaction mechanism of different enzymes having phosphohistidine intermediates.

Auto-phosphorylation of NDP kinase on serine residues has been demonstrated in some cases, but the stoichiometry was low and the functionality is questionable (Bominaar et al, 1994).

Phosphohistidine is labile in acid and has also a limited stability at neutral pH.

$$E\text{-}P + H_2O \longrightarrow E + P_i \qquad (1c)$$

Therefore, an NTPase activity exists as a sum of reaction (1a) and (1c). It is different in nature from the ATPase activity of other kinases, hexokinase for instance, where the reactive water replaces glucose as a second substrate in a in-line mechanism. In NDP kinase, the NTPase activity results from chemical hydrolysis of the reaction intermediate. X-ray structures show that there is no room at the active site for a water molecule to insert between the γ-phosphate of NTP and the histidine. In counterpart, the water attack on the phosphohistidine cannot be prevented by a conformational change of the protein, as in hexokinase.

3. Purification and structural properties of NDP kinases

Eukaryotic NDP kinases are hexamers, and bacterial enzymes are tetramers made of small subunits (16-17 kDa). Purification of recombinant enzymes expressed in *E. coli* is easy for proteins having an alkaline pI, like the *Dictyostelium* and *Drosophila* NDP kinases, or the B isoform of the mammalian enzyme. Passing the bacterial extract through a DEAE column eliminates most of bacterial proteins, and also the NDP kinase of *E. coli* which is acidic. Next, dye-mediated chromatography on Blue Sepharose (or similar preparations) and elution with nucleotides (ATP 1 mM) yields highly purified enzyme (Lascu et al., 1993). This protocol can be easily scaled both up or down (Lascu et al., 1981). For the NDP kinases having an acidic pI, such as the mammalian A isoform, the purification is less straightforward. Ammonium sulfate precipitation, DEAE chromatography can be used for cleaning up the protein solution. Next, hydroxyapatite gives a good purification since NDP kinases bind more tightly than most contaminant proteins. The yield for this step is poor when using a low protein/gel ratio. Gel filtration chromatography is being used routinely in our laboratories for eliminating contaminants like adenylate kinase and low molecular weight non-protein molecules. This is not a purification in terms of increasing specific activity, since the molecular weight of NDP kinase is close to the mean of cellular proteins.

Several crystal structures are now available (Table I). We will focus mostly on *Dictyostelium* NDP kinase, for which we have a detailed atomic description of the free (Moréra et al., 1994), phosphorylated (Moréra et al, 1995) and nucleoside diphosphate complexed (Moréra et al., 1994; Cherfils et al, 1994) enzyme.

Sequence similarity is high between the NDP kinases from various origins (Moréra et al., 1994). Residue numbering from *Dictyostelium* enzyme will be used below. As could be anticipated, the over-all subunit fold is near identical, with differences in the C-terminal region which correlate with the different quaternary structures. The subunits are made of a single structural domain of the $\beta\alpha\beta)(\beta\alpha\beta$ type. It contains about 130 residues and is followed by a C-terminal extension of 15-25 residues. Their structure is described in detail in Moréra et al. (1994). The central β-sheet of the α/β structure is antiparallel with a $\beta2\beta3\beta1\beta4$ strand order; α-helices alternate with β-strands: $\alpha1$ connects $\beta1$ to $\beta2$, $\alpha3$ connects $\beta3$ to $\beta4$ on one side of the β-sheet. The other face is covered by αa and $\alpha2$ inserted between $\beta2$ and $\beta3$, and helix $\alpha4$ following $\beta4$. Helix $\alpha3$ is connected to $\beta4$ by a twenty residue loop, the Kpn loop. This was named from the natural mutation *killer-of-prune* of *Drosophila* which is a point substitution in that loop (P97S). The Kpn loop is essential for hexamer, but not for tetramer formation, and it also contains some active-site residues. The NDP kinase fold is original for a phosphotransferase: α/β structures like p21^{ras} or adenylate kinase have a parallel β-strand and are monomeric.

Table I. X-ray structures of NDP kinases

Protein	Space group	Subunits in asymmetric unit	Resolution (Å)	PDB entry	Reference
Dictyostelium					
form I	P6₃22	2	2.4	*	Moréra et al., 1994b
form II	P6₃22	1	1.8	1npk	"
H122C mutant	P6₃22	1	2.4	1ndk	Dumas et al., 1992
ADP complex	R3	2	2.2	1ndp	Moréra et al., 1994a
dTDP complex	R32	1	2.0	1ndc	Cherfils et al., 1994
phosphorylated	P6₃22	1	2.1	1nsp	Moréra et al., 1995
Drosophila					
native	P3₂21	3	2.4	1ndl	Chiadmi et al., 1993
phosphorylated	P3₂21	3	2.2	1nsq	Moréra et al., 1995
Myxococcus					
ADP complex	P4₃2₁2	2	2.0	1nck	Williams et al., 1993
cAMP complex	P4₃2₁2	2	1.9	1nhk	Strelkov et al., 1995

* Refinement done on even-l reflections only.

NDP kinase subunits form dimers, which assemble in a hexamer in eukaryotic enzymes, or in a tetramer in *Myxococcus* NDP kinase (Williams et al., 1993). Hexamers are formed by the arrangement of three dimers about a three-fold axis, which can also be described as a dimer of trimers. It is unusual to find highly homologous proteins forming oligomers with different symmetries. The regions of the protein surface involved in the assembly of the dimers into the tetramer and hexamer structures are different. The C-terminal part of the polypeptide chains interacts with a neighboring subunit within a dimer in the tetrameric enzyme and within a trimer in the hexameric ones. The oligomeric structure appears necessary for enzymatic activity of the hexameric, but not of the tetrameric NDP kinases (I. Lascu, in preparation). There is no kinetic evidence for allosteric behaviour of NDP kinases.

4. Complexes with nucleotides and the phosphorylated enzyme

Initial attempts to infiltrate crystals of *Dictyostelium* NDP kinase with nucleotides were unsuccessful, which can be explained *a posteriori* by the molecular packing in the crystal. High quality crystals were obtained by co-crystallization with ADP or dTDP with PEG as precipitant, but not with ATP or the analogue AMPPNP.

Essentially identical contacts are made by ADP and dTDP with residues from the active site in *Dictyostelium* NDP kinase, or by ADP in *Myxococcus* NDP kinase, despite the different quaternary structure of these enzymes. A relevant part of the NDP kinase-ADP complex is shown in Fig. 1A. Though ADP and dTDP occupy almost identical positions at the binding site, minor differences could explain the small difference in reactivity. Several reports based on steady-state kinetic data, suggest that purine nucleotides are slightly better substrates than pyrimidine. As this is by a factor of less than 10 for both k_{cat} and K_m, this difference could also be attributed to enzyme inhibition by excess substrate, incorrect estimation of effective substrate concentration (nucleotide protonation, affinity for Mg^{++} ion) or inhibition by the Mg^{++}-free nucleotides.

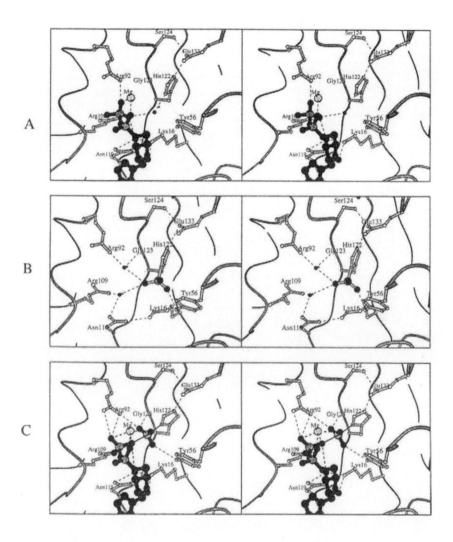

Fig 1. Model of the NDP kinase active site of NDP kinase. A. ADP bound to the *Dictyostelium* enzyme as seen in the X-ray structure of the complex. The adenine base (truncated) points towards the bottom of the figure. His 122 is the active site histidine. A water molecule bridges its N_δ atom to the β-phosphate. B. Structure of the phosphorylated *Dictyostelium* NDP kinase. C. A model for the transition state. The γ-phosphorus atom replaces the water molecule in (A). The penta-coordinated bipyramidal configuration will lead to the inversion of the γ-phosphate as it is transfered onto the histidine.

In ADP and dTDP complexes, the base makes hydrophobic contacts with residues Phe64 and Val116 and no direct hydrogen bond. This explains why NDP kinase is not sensitive to the nature of the base moiety. In addition, both the purine and the pyrimidine form a water-mediated hydrogen bond with the side-chain carboxylate of the C terminal residue Glu 155 from a neighboring subunit in the hexameric enzyme, but not in the *Myxococcus* enzyme. The protein makes several contacts with the sugar. The 2' hydroxyl group (absent in dTDP) is hydrogen-bonded to Lys 16 in the NDP kinase-ADP complex. The 3' hydroxyl is essential for substrate binding and/or catalysis (Lascu & Véron, unpublished). It interacts with Lys 16, Asn 119 and it donates an internal hydrogen-bond to a β-phosphate oxygen. The α-phosphate is accessible to solvent and makes rather long hydrogen-bonds with the side-chains of His 59 and Thr 98. The β-phosphate is buried and hydrogen-bonds to the side-chains of Thr 98, Arg 92 and Arg 109.

We also solved the X-ray structure of the phosphorylated intermediate, a non trivial task as the half-life time of the phosphohistidine is a few hours. Although compatible with X-ray data collection using synchrotron radiation, this is much too short for crystallization. We tried regenerating ATP, but this was not useful since nucleotides cannot penetrate the active site in crystals. We looked for smaller molecular weight molecules which would be able to phosphorylate the active-site histidine. Phosphoramidate ($NH_2PO_3^{2-}$) was shown to phosphorylate nitrogen nucleophiles (Benkovic & Sampson, 1971). It was also found to be effective for phosphorylating the active-site carboxylates in CheY and CheB (Lukat et al., 1992). We found that NDP kinase could be phosphorylated slowly by phosphoramidate, and the product could transfer a phosphate to a nucleoside diphosphate (I.L., unpublished). An enzyme catalyzing the phosphorylation of nucleoside diphosphates by phosphoramidate has been described several years ago in yeast (Dowler & Kakada, 1968). It may well be NDP kinase since it had a ping-pong mechanism and was non-specific for the acceptor nucleotide.

Phosphoramidate was prepared as described by Wei & Matthews (1991) and soaked into crystals of NDP kinase (one hour at 40 mM phosphoramidate for the *Dictyostelium* enzyme; 3 hours at 200 mM phosphoramidate for the *Drosophila* enzyme). X-ray diffraction data were collected at the W32 station in the LURE-DCI synchrotron radiation center (Orsay), using only one crystal for each structure. With both enzymes, the electron density showed the active site histidine to be modified on its N_δ atom. No other modification could be observed: no other histidine and no hydroxyle group was phosphorylated. The environment of the phosphoryl group is shown on Fig. 1B. It receives a hydrogen bond from the phenolic group of Tyr 56 and interacts with Arg 92 and Arg 109 through water molecules. No conformational change was detected in the phosphorylated enzyme compared to the un-phosphorylated form and to the complexes with ADP and dTDP. Only a slight movement of the helices α2 and αa was observed, which may be necessary for nucleotide binding and release.

This is only the second example of a high-resolution crystal structure of a protein phosphorylated on histidine, the first being *E. coli* succinyl CoA synthase (Wolodko et al., 1994). This enzyme is different from NDP kinase in almost all respects. It is a α2β2 heterotetramer, its active sites are shared between two subunits. The phosphohistidine Hisα246 is stabilized by an electrostatic interaction with the dipole of two α-helices and by several potential hydrogen bonds. Phosphorylation can be achieved by inorganic phosphate rather than GTP or ATP. It occurs on the N_ε, N_δ giving a hydrogen-bond to a glutamate carboxylate. Only the later interaction is present in NDP kinase, with the two nitrogen atoms exchanged. The role of the phosphorylated intermediate in succinyl CoA synthase is to convert the energy of GTP into that of the thioester bond of acyl co A through a phosphoramide intermediate and vice-versa. Energy conservation is more important than speed in this case, and the turn-over is only about 35 s[-1] instead of over 1000 s.1 for NDP kinase. In succinyl CoA synthase, the histidine is phosphorylated on the N_ε position (more stable but less reactive). NDP kinase has iso-energetic substrates and products and it is optimized for speed. The few interactions made with the phosphoryl group in the intermediate contribute to it, since stabilization of the intermediate would increase the activation energy of the overall reaction and reduce the rate of overall transfer.

5. Modelling the transition state

NDP kinase has only one nucleotide binding site per subunit. The donor nucleoside triphosphate (N_1TP) binds to the free enzyme and phosphorylates His122. The product N_1DP must dissociate before the acceptor N_2DP binds to the same binding site. This has an important mechanistic consequence: reaction (1a) and (1b) are identical except for being written in opposite directions. Therefore, microscopic reversibility requires that they use the same chemical mechanism and the same transition state.

The model depicted in Fig. 1C describes the transition state for both phosphorylation and dephosphorylation. It was built based on the crystal structure of the NDP kinase-ADP complex. There, the distance between the β-phosphate of ADP and the N_δ atom of histidine 122 is 4.7 Å and a penta-coordinated γ-phosphate can be accommodated in place of a water molecule located between ADP and the histidine. In the transition state, the two apical P_γ-O and P_γ-N_δ bonds should be longer than the equatorial bonds. In the model, they are 2.1-2.2Å rather than the usual 1.5-1.7 Å. This geometry is compatible with the in-line associative mechanism of transfer (Knowles, 1980).

The model identifies residues which may be essential for catalysis. They role was tested by site-directed mutagenesis of *Dictyostelium* NDP kinase (Tepper et al., 1994). Glu 133 is hydrogen-bonded to the N_ε atom of tha active site His 122. This arrangement is reminiscent to the catalytic triad in the serine proteases and raises the same question of whether a proton is transfered from the histidine to the carboxylate during catalysis. The E133Q mutant protein where Glu 133 is substituted by a glutamine has 0.5% residual activity. Glutamine can accept the hydrogen bond, but it can not take the proton. Other residues crucial for the stabilization of the transition state are Tyr 56 and Lys 16. The Y56F mutant enzyme has 2% residual activity suggesting than, rather than being an acid catalyst that gives a proton to the γ-phosphate, the Tyr 56 phenol group hydrogen-bonds to it and acts to lower the energy the transition state relative to the Michaelis complex.

Charged residues Arg 92 and Arg 109 interact with the phosphates both of the substrate and of the transition state. Replacing them with alanines leaves less than 1% residual activity, with lysines, 5 to 10% (Tepper et al., 1994). This indicates a role in neutralizing the negative charges. Lys 16 and Asn 119 are important for substrate binding. In their absence the residual activity is less that 1%. They interact with the sugar hydroxyl rather than with the phosphates in the complexes with ADP and dTDP. Lys16 could have an additional role in catalysis, since, in the modelled transition state, its amino group is at 3.5 Å from the nearest phosphate oxygen.

Biochemical data indicate that substrates for NDP kinase are the Mg^{2+} complexes of both NTP and NDP (Garces & Cleland, 1969). Phosphorylation is observed in the absence of Mg^{++}, but its rate has not been measured and may be orders of magnitudes slower than in its presence. In the X-ray structures, a magnesium ion is seen to bind to the α and β-phosphates of ADP. In the modelled transition state, it also interacts with the γ-phosphate being transfered.

Of particular interest is the potential involvement in catalysis of the 3' hydroxyl of the sugar. It might participate to a proton relay with Lys 16 to protonate the β-phosphate of the nucleoside diphosphate product of the reaction (1a) making it a better leaving group. This could explain why nucleotides missing a 3' hydroxyl are very poor substrates for NDP kinase.

Table 2. Examples of enzymes phosphorylated on histidine during their catalytic cycle. Bacterial proteins involved in sugar transport and signaling systems are not included although several have a similar mechanism. For a recent review see Swanson et al., 1994

1. Phosphotransferases	
Nucleoside diphosphate kinase	$N_1TP + E \iff N_1DP + E\text{-}P$
	$N_2DP + E\text{-}P \iff N_2TP + E$
Succinyl CoA synthase	$ATP + E \iff ADP + E\text{-}P$
(succinyl thiokinase)	$E\text{-}P + \text{succinate} \iff E.\text{succinyl-P}$
	$E.\text{succinyl-P} + CoA \iff$
	$\qquad\qquad E + \text{succinylCoA} + Pi$
2. Mutases	
Phosphoglycerate mutase	$2,3\text{-BPG} + E \iff 2\text{-PG} + E\text{-}P$
	$E\text{-}P + 3\text{-PG} \iff 2,3\text{-BPG} + E$
2,3-Bisphosphoglycerate synthase	$1,3\text{-BPG} + E \iff E\text{-}P + 3\text{-PG}$
(mammalian erythrocytes)	$E\text{-}P + 3\text{-PG} \iff E + 2,3\text{-BPG}$
3. Phosphatases	
6-phosphofructo-2-kinase/	$F\text{-}2,6\text{-BP} + H_2O \longrightarrow F\text{-}6\text{-P} + Pi$
fructose-2,6-bisphosphatase	
(the phosphatase reaction)	
Acid phosphatase (human prostate)	$R\text{-}OP + H_2O \longrightarrow R\text{-}OH + Pi$

6. Conclusion

Table II show examples of enzymes which are phosphorylated on a histidine residue during the catalytic cycle. The acceptor can be a nucleotide (NDP kinase), succinate (succinyl CoA synthase), phosphoglycerate (mutases), or water (phosphatases). Because of the high reactivity of the phosphohistidine, a secondary phosphatase reaction can be measured in all cases. It is as low as 10^{-7} with respect to the transfer reaction for NDP kinase and 10^{-4}-10^{-2} for the mutase reactions. Three-dimensional structures are available for most of these enzymes, but detailed reaction mechanisms have not yet been presented. We expect that many features of our description of the NDP kinase mechanism will apply to them.

References

Benkovic, S. J. & Sampson, E. J. (1971) *J. Amer. Chem. Soc.* **93**, 4009-4016
Bominaar, A. A., Tepper, A. D. & Véron, M. (1994) *FEBS Letters* **353**, 5-8
Cherfils, J., Moréra, S., Lascu, I., Véron, M. & Janin, J. (1994) *Biochemistry* **33**, 9062-9069
Chiadmi, M., Moréra, S., Lascu, I., Dumas, C., LeBras, G., Véron, M. & Janin, J. (1993) *Structure* **1**, 283-293
De la Rosa, A., Williams, R.L. & Steeg, P.S. (1995) *BioEssays* **17**, 53-62
Dowler, M. J. & Nakada, H. I. (1968) *J. Biol. Chem.* **243**, 1434-1440
Dumas C., Lascu I., Moréra S., Glaser P., Fourme R., Wallet V., Lacombe M. L., Véron M. & Janin J. (1992) *EMBO J.* **11**, 3203-8

Edlund B., Rask L., Olsson P., Walinder O., Zetterqvist O. & Engstrom L. (1969) *Eur. J. Biochem.* 9, 451-5

Frey, P. A. (1992) *The Enzymes* 20, 142-186

Garces E. & Cleland W. W. (1969) *Biochemistry* 8, 633-40

Gilles, A.M., Presecan, E., Vonica, A. & Lascu, I. (1991) *J. Biol. Chem.* 266, 8784-9

Hultquist, D. E., Moyer, R. W., & Boyer, P. D. (1966) *Biochemistry* 5, 322-331

Johnson, L. N. & Barford, D. (1993) *Ann. Rev. Biophys. Biomol. Struct.* 22, 199-232

Knowles, J. R. (1980) *Ann. Rev. Biochem.* 49, 877-919

Lascu I., Deville-Bonne D., Glaser P. & Véron M. (1993) *J. Biol. Chem.* 268, 20268-75

Lascu I., Pop R. D., Porumb H., Presecan E. & Proinov I. (1983) *Eur. J. Biochem.* 135, 497-503

Lascu, I., Duc, M. & Cristea, A. (1981)*Analyt. Biochem.* 113, 207-11

Lecroisey, A., Lascu, I., Bominaar, A. Véron, M. & Delepierre, M. (1995) submitted

Lukat, G. S., McCleary, W. R., Stock, A. M. & Stock, J. B. (1992) *Proc. Natl. Acad. Sci. USA* 89, 718-722

Moréra, S., Chiadmi, M., LeBras, G., Lascu, I. & Janin, J. (1995) *Biochemistry* in press

Moréra, S., Lascu, I., Dumas, C., LeBras, G., Briozzo, P., Véron, M. & Janin, J. (1994a) *Biochemistry* 33, 459-67

Moréra, S., LeBras, G., Lascu, I., Lacombe, M.-L., Véron, M. & Janin, J. (1994b) , *J. Mol. Biol.* 243, 873-890

Parks R. E. Jr. & Agarwal R. P. (1973) *The Enzymes* 8, 307-334

Sheu K. F., Richard J. P. & Frey P. A. (1979) *Biochemistry* 18, 5548-56

Strelkov, S. V., Perisic, O., Webb, P. A. & Williams, R. L. (1995) *J. Mol. Biol.* 249, 665-674

Swanson, R. V., Alex, L. A. & Simon, M. I. (1994) *Trends Biochem. Sci.* 19, 485-490

Tepper, A. D., Dammann, H., Bominaar, A. A. & Véron, M. (1994) *J. Biol. Chem.* 269, 32175-32180

Walinder O., Zetterqvist O. & Engstrom L. (1969) *J. Biol. Chem.* 244, 1060-41.

Wei, Y.-F. & Matthews, H. R. (1991) *Meth. Enzymol.* 200, 388-414

Williams R. L., Oren D. A., Munoz-Dorado J., Inouye S., Inouye M., Arnold E. (1993) *J. Mol. Biol.* 234, 1230-47

Wolodko, W. T., Fraser, M. E., James, M. N. G. & Bridger, W. A. (1994) *J. Biol. Chem.* 269, 10883-10890

SECTION V

Manipulation of Sulfur in Proteins

Selenol-Catalyzed Reduction of Disulfide Bonds in Peptides and Proteins

Rajeeva Singh
ImmunoGen, Inc.
Cambridge MA 02139

I. Introduction

Disulfide-reducing reagents are used in biochemistry to reduce the disulfide bonds in proteins and peptides and to maintain the essential thiol groups in proteins by preventing their oxidation to the disulfide state (1, 2). Typically, dithiothreitol (DTT) and sodium borohydride are used for reducing disulfide bonds in proteins and peptides. Both DTT and sodium borohydride, however, are slow in reducing disulfides and require several hours for complete reduction (1).

Selenols (e.g., selenocysteine, 2-hydroxyethaneselenol) and their precursors (e.g., diselenide, selenocyanate) were reported to catalyze the interchange reactions of organic dithiols and disulfides (3). The rates of activation of sulfhydryl enzymes by cysteine were reported to be accelerated using excess selenocystine (4). These previous studies did not identify the rate-determining step in the selenol-catalyzed reaction.

In this study, we have measured the kinetics of selenol-catalyzed reduction of disulfide bonds in proteins and in small organic disulfides using DTT and sodium borohydride as reductants. Selenocystamine, a commercially available diselenide, is used as the precursor of the catalyst and is reduced to its selenol (selenocysteamine) in situ. We have identified the rate-determining step in the selenol-catalyzed reduction of disulfides as the reaction of selenolate anion with disulfide. A rapid and convenient assay for disulfide group has been developed. In

Selenol-Catalyzed Reduction of Disulfide Bonds in Peptides and Proteins

Rajeeva Singh
ImmunoGen, Inc.
Cambridge, MA 02139

I. Introduction

Disulfide-reducing reagents are used in biochemistry to reduce the disulfide bonds in proteins and peptides and to maintain the essential thiol groups in proteins by preventing their oxidation to the disulfide state (1, 2). Typically, dithiothreitol (DTT) and sodium borohydride are used for reducing disulfide bonds in proteins and peptides. Both DTT and sodium borohydride, however, are slow in reducing disulfides and require several hours for complete reduction (1).

Selenols (e.g., selenocysteamine, 2-hydroxyethaneselenol) and their precursors (e.g., diselenide, selenocyanate) were reported to catalyze the interchange reactions of organic dithiols and disulfides (3). The rates of activation of sulfhydryl enzymes by cysteine were reported to be accelerated using excess selenocystine (4). These previous studies did not identify the rate determining step in the selenol-catalyzed reaction.

In this study, we have measured the kinetics of selenol-catalyzed reduction of disulfide bonds in proteins and in small organic disulfides using DTT and sodium borohydride as reductants. Selenocystamine, a commercially available diselenide, is used as the precursor of the catalyst, and is reduced to its selenol (selenocysteamine) *in situ*. We have identified the rate determining step in the selenol-catalyzed reduction of disulfide as the reaction of selenolate anion with disulfide. A rapid and convenient assay for disulfide group has been developed in

TECHNIQUES IN PROTEIN CHEMISTRY VII

which the disulfide is reduced using sodium borohydride in the presence of a catalytic amount of selenol, followed by acid quench and measurement of the resulting thiol by Ellman's assay. Practical conditions using selenocystamine and DTT are described to achieve rapid reductions of disulfide bonds in proteins.

II. Materials and Methods

Selenocystamine dihydrochloride, dithiothreitol (DTT), sodium borohydride, oxidized glutathione, insulin (bovine pancreas), α-chymotrypsinogen A (bovine pancreas), 5,5'-dithiobis(2-nitrobenzoic acid) (DTNB, Ellman's reagent) were purchased from Sigma Chemical Co. (St. Louis, MO). Cystamine dihydrochloride and bis(2-hydroxyethyl) disulfide were purchased from Fluka (Switzerland). The Fab fragment of immunoglobulin was prepared by the digestion of murine monoclonal anti-B4 antibody (IgG_1) using papain at pH 6.3 in the presence of cysteine (10 mM), followed by purification using Protein A-Sepharose and S-Sepharose. Pre-cast 14% polyacrylamide gels (Tris-Glycine) were purchased from Novex (San Diego, CA).

A. Reduction of Oxidized Glutathione Using Sodium Borohydride in the Presence or Absence of Selenol

The rates of reduction of oxidized glutathione (GSSG) by sodium borohydride in the presence or absence of selenol were compared. Selenocystamine ($^+NH_3CH_2CH_2SeSeCH_2CH_2NH_3{}^+$) was used as the precursor of the selenol selenocysteamine ($^+NH_3CH_2CH_2SeH$). A solution containing GSSG (0.4 mL of a 5 mM solution in water), selenocystamine (20 μL of a 4.1 mM solution in water), water (0.58 mL) and ethanol (0.5 mL) was incubated at 23 °C, and the reduction was initiated by the addition of sodium borohydride (0.5 mL of a 0.4 M solution in ethanol). The final volume was ~1.93 mL. The uncatalyzed reaction mixture contained water in place of the selenocystamine solution. A control solution was prepared using water (0.98 mL), selenocystamine (20 μL of a 4.1 mM solution in water), ethanol (0.5 mL) and sodium borohydride (0.5 mL of a 0.4 M solution in ethanol). A second control solution without selenol was similar to the above control solution but contained water in place of the selenocystamine solution.

At several time intervals, aliquots (50 µL) from the above solutions were quenched using acetic acid (50 µL of a 1.7 M solution), and then were assayed for thiol content by the addition of buffer (0.9 mL of a 0.5 M sodium phosphate buffer, pH 7, 1 mM in EDTA) and Ellman's reagent (10 µL of a 0.1 M solution in DMSO) and by measurement of absorbance (A) at 412 nm. The values of change in absorbance (ΔA) for the reaction mixture containing selenol and that without selenol were calculated by subtracting the absorbance value of the appropriate control solution. As expected, the control solution with selenol showed a slightly higher absorbance than that without selenol, because of the reaction of selenol with Ellman's reagent. The ratio of ΔA at time t (ΔA_t) to the maximum observed ΔA (ΔA_{max}) was calculated, and the values of % reduced disulfide [= ($\Delta A_t / \Delta A_{max}$) x 100] were plotted vs time (Fig. 1A). The values of -ln [(remaining disulfide)/(total disulfide)] (= -ln [($\Delta A_{max} - \Delta A_t$)/($\Delta A_{max}$)]) were then plotted vs time (Fig. 1B); the slope of this plot equals the value of apparent pseudounimolecular rate constant (k') (Table I). A vigorous effervescence was observed on addition of sodium borohydride to GSSG, and therefore the kinetic analysis was performed using the measurement obtained 1 min after sodium borohydride addition as the first time point (Fig. 1B).

For assaying disulfide groups, we used a similar procedure as above using sodium borohydride (0.1 M) and a catalytic amount of selenol (10-15 mole% of the disulfide) for reducing the disulfide (in the concentration range 0.1 mM to 1 mM) in ethanol/water (1:1, v/v) for ~1 h at ambient temperature. The reaction mixture was then quenched using acetic acid and analyzed for thiol content by Ellman's assay as described above. An extinction coefficient of 14150 M^{-1} cm^{-1} was used for Ellman's assay (5). The amount of disulfide was calculated as half of the amount of thiol formed. Special precautions to exclude air (such as degassing) were not required, therefore making the assay convenient.

The stock selenocystamine solution (5 mM, based on the weight of selenocystamine dihydrochloride) was prepared in water, and its concentration was measured as 4.1 mM by its reduction to selenol using sodium borohydride and by Ellman's assay of the selenol.   This stock selenocystamine solution (~5 mM) was stable for more than one year of storage at -20 °C.

B. Reduction of Disulfide Bonds in Insulin Using Sodium Borohydride in the Presence or Absence of Selenol

A stock solution of insulin (2.4 mg/mL) was prepared in 2.5 mM HCl (0.5 mM in EDTA). The concentration of insulin (M_r 5730) was estimated using the $A^{0.1\%}{}_{1\text{ cm}}$ value of 1.06 at 278 nm (6). Samples containing insulin (1.1 mg/mL) were reduced using sodium borohydride (0.1 M) in aqueous ethanol (1:1, v/v) at 23 °C in the presence or absence of a catalytic amount of selenol (0.084 mM), and were quenched at several time intervals using acetic acid, and then were analyzed for thiol content using Ellman's assay. After 90 min of selenol-catalyzed reduction, 2.5 disulfide groups per insulin molecule were reduced, and this value was assumed as the total disulfide for kinetic analysis (Table I).

C. Reduction of Disulfide Bond in α-Chymotrypsinogen A by Dithiothreitol in the Presence or Absence of Selenol

Of the five disulfide bonds in α-chymotrypsinogen A, only one disulfide bond (191-220) is reported to be reduced by dithioerythritol under non-denaturing conditions (7). Samples containing α-chymotrypsinogen A (3.1 mg/mL) were reduced using DTT (5 mM) in 50 mM bis-tris buffer (pH 6, 1 mM in EDTA) at 23 °C in the presence or absence of selenocystamine (0.41 mM; i.e., 0.82 mM selenol), and were quenched at several time intervals using acetic acid (to pH ~4), and immediately filtered through a Sephadex G-25 (fine) column equilibrated with 100 mM sodium acetate buffer (pH 4.5, 1 mM in EDTA). The reduced samples obtained from gel filtration were analyzed for protein content ($A^{0.1\%}{}_{1\text{ cm}}$ = 2.0 at 280 nm; M_r = 25 000) and for thiol content (Ellman's assay).   The values of reduced disulfide per chymotrypsinogen molecule were plotted vs time (Fig. 2).   The maximum value of reduced disulfide per chymotrypsinogen molecule was ~1, and was assumed as total disulfide for the kinetic analysis. The pseudounimolecular rate constant (k') was determined from the plot of -ln [(remaining disulfide)/(total disulfide)] vs time. The rates of reduction of the disulfide bond in α-chymotrypsinogen A were also determined at pH 7 (Table I).

D. *Reduction of Disulfide Bond in Fab Fragment of Immunoglobulin Using Dithiothreitol in the Presence or Absence of Selenol*

A sample of Fab fragment of immunoglobulin (1.7 mg/mL) in phosphate/EDTA buffer (50 mM sodium phosphate, pH 7, 1 mM EDTA) was reduced using DTT (5 mM) at 23 °C in the presence or absence of selenocystamine (0.41 mM; i.e., 0.82 mM selenol). At several time intervals, aliquots (50 µL) were quenched using iodoacetamide (100 µL of a 0.3 M solution in phosphate/EDTA buffer, pH 7) at 0 °C, and were analyzed by 14% SDS-PAGE under non-reducing conditions (Fig. 3).

III. Results

Table I shows a comparison of the rate constants for the reductions of disulfides in the presence or absence of selenol using sodium borohydride or DTT as reductant. The pseudounimolecular rate constants for the selenol-catalyzed reduction (k'_{cat}) and the uncatalyzed reduction (k'_{uncat}) were compared and the values of rate enhancements (k'_{cat}/k'_{uncat}) were calculated (Table I).

The rate of reduction of oxidized glutathione (GSSG) by sodium borohydride is enhanced by a factor of 10 using a catalytic amount of selenol (8 mole% of the initial amount of disulfide). Use of a smaller amount of selenol (4 mole% of the initial amount of disulfide) results in a 5-fold rate enhancement of the reduction of GSSG by sodium borohydride. The rate enhancement of the reduction of GSSG using selenol is therefore directly proportional to the amount of selenol used (Table I).

The reduction of oxidized glutathione by sodium borohydride in the presence of a catalytic amount of selenol was rapid, resulting in the generation of a maximum amount of thiol within 10 min (Fig. 1A). This method of reduction of disulfide by sodium borohydride using a catalytic amount of selenol, followed by the measurement of thiol using Ellman's assay, is therefore a rapid, sensitive and quantitative assay for disulfide groups. Small amounts of cystamine (10 nmol) were accurately measured by this method (96 ± 1 % of that expected based on the weight of disulfide, assuming that each disulfide group generates two thiol groups upon reduction).

Table I. Comparisons of Apparent Rate Constants for Reductions of Disulfides Using Selenol as Catalyst vs Those for Uncatalyzed Reactions[a]

Disulfide (initial conc., mM)	Reductant (initial conc., mM)	Selenol conc.[b] (mM)	k'_{uncat} (min⁻¹)	k'_{cat} (min⁻¹)	$\dfrac{k'_{cat}}{k'_{uncat}}$	k_{uncat} (M⁻¹ min⁻¹)	k_{cat} (M⁻¹ min⁻¹)
HOCH₂CH₂SSCH₂CH₂OH (1.0)	NaBH₄ (100)[c]	0.082	0.020	0.23	12	0.20	2800
	NaBH₄ (100)[c]	0.041	0.022	0.14	6	0.22	3300
Oxidized glutathione (GSSG) (1.0)	NaBH₄ (100)[c]	0.082	0.051	0.52	10	0.51	6300
	NaBH₄ (100)[c]	0.041	0.061	0.33	5	0.61	8000
Insulin (0.55)[d]	NaBH₄ (100)[c]	0.084	0.031	0.36	12	0.31	4300
α-Chymotrypsinogen A (0.12)	DTT (5.0)[e]	0.82	0.023	2.07	90	4.6	2500
	DTT (4.9)[e]	0.41	0.024	1.08	45	4.8	2600
	DTT (2.4)[e]	0.41	0.011	0.99	90	4.6	2400

[a] The values of k' were calculated from the pseudounimolecular rate equation: $-d[\text{disulfide}]/dt = k'[\text{disulfide}]$. For the uncatalyzed reaction, k_{uncat} was calculated from the second order rate equation: $-d[\text{disulfide}]/dt = k_{uncat}[\text{reductant}][\text{disulfide}]$. For the catalyzed reaction, k_{cat} was calculated from the second order rate equation: $-d[\text{disulfide}]/dt = k_{cat}[\text{selenol}][\text{disulfide}]$.

[b] The catalyzed reactions initially contained selenocystamine, which was reduced to the active selenol ($^+\text{NH}_3\text{CH}_2\text{CH}_2\text{SeH}$) in situ. The values of concentration of selenol are listed in the table.

[c] The reductions using sodium borohydride (NaBH₄) were carried out in aqueous ethanol (1:1, v/v) at 23 °C.

[d] The concentration of reactive disulfide in insulin (1.1 mg/mL) was calculated assuming three reactive disulfide groups per insulin molecule.

[e] α-Chymotrypsinogen A (3.1-3.2 mg/mL) was reduced using dithiothreitol (DTT) at pH 7 at 23 °C. The concentration of reactive disulfide in α-chymotrypsinogen A was calculated assuming one reactive disulfide group per protein molecule.

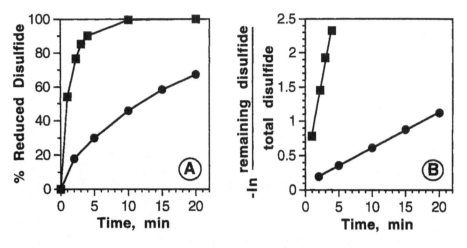

Figure 1. Reduction of oxidized glutathione by sodium borohydride in the presence (■) or absence (●) of selenol. (A) Plot of the percent of reduced disulfide vs time. (B) Plot of -ln [(remaining disulfide)/(total disulfide)] vs time (see Methods).

The accessible native disulfide bonds in proteins were reduced under non-denaturing conditions using DTT and selenol, which were then separated from the reduced protein by gel filtration. The rate of reduction of the accessible disulfide bond in α-chymotrypsinogen A (3 mg/mL) by DTT (5 mM) at pH 7 was enhanced in the presence of selenol (0.82 mM) by a factor of 90. The rate determining step in the selenol-catalyzed reduction of disulfide was identified by varying the concentration of DTT while keeping the concentration of selenol constant, which resulted in a proportional change in the value of k'_{uncat} but no change in the value of k'_{cat} (Table I). The value of k'_{cat} for the selenol-catalyzed reaction is therefore independent of the concentration of DTT but is directly proportional to the concentration of selenol. The rate determining step in the selenol-catalyzed reaction is therefore the reaction of selenolate anion with disulfide. The second-order rate constants for the selenol-catalyzed and the uncatalyzed reactions (k_{cat} and k_{uncat}, respectively) were therefore calculated based on the concentrations of selenol and DTT, respectively (Table I). The reduction of the disulfide bond in α-chymotrypsinogen A using DTT and selenocystamine was rapid even at pH 6 (Fig. 2). The selenol-catalyzed reaction at pH 6 (k'_{cat} = 0.91 min⁻¹) was ~250-fold faster than the uncatalyzed reaction (k'_{uncat} = 0.0037 min⁻¹).

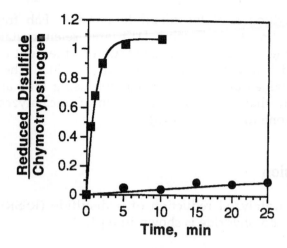

Figure 2. Reduction of disulfide bond in α-chymotrypsinogen A by dithiothreitol at pH 6 in the presence (■) or absence (●) of selenol (see Methods).

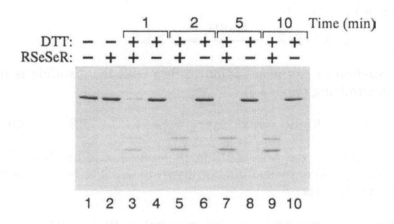

Figure 3. Reduction of Fab fragment of immunoglobulin by dithiothreitol (DTT) at pH 7 in the presence or absence of selenol. Selenol was generated *in situ* from selenocystamine (RSeSeR). At several time intervals, aliquots were quenched using iodoacetamide and analyzed by SDS-PAGE under non-reducing conditions. Lane 1 is Fab alone. Lane 2 is from a mixture of Fab and selenocystamine. Lanes 3, 5, 7, and 9 are from a mixture of Fab with both DTT and selenocystamine, after 1-, 2-, 5- and 10 min of reaction, respectively. Lanes 4, 6, 8, and 10 are from a mixture of Fab and DTT, after 1-, 2-, 5- and 10 min of reaction, respectively (see Methods).

The reduction of the disulfide bond in Fab fragment of immunoglobulin at pH 7 using DTT and selenocystamine was rapid, resulting in complete fragmentation of the light chain from the truncated heavy chain within 5 min (Fig. 3). The extent of reduction obtained after 1 min of reaction using DTT and selenol was similar to that obtained after 90 min of uncatalyzed reaction using DTT alone (data not shown).

IV. Discussion

The uncatalyzed reduction of a disulfide (RSSR) to thiols using sodium borohydride is shown in eq 1.

$$RSSR \xrightarrow{\text{NaBH}_4} 2\ RS^- \qquad (1)$$

The first step in the catalyzed reaction is the rapid generation of selenol (R'SeH) by the reduction of its precursor diselenide (R'SeSeR') (eq 2).

$$R'SeSeR' \xrightarrow{\text{NaBH}_4} 2\ R'Se^- \qquad (2)$$

The reaction of selenolate anion (R'Se$^-$) with the disulfide is the rate determining step (eq 3).

$$R'Se^- + RSSR \longrightarrow R'SeSR + RS^- \qquad (3)$$

The resulting mixed selenosulfide (R'SeSR) is rapidly cleaved by selenolate (eq 4) or by sodium borohydride (eq 5) to regenerate the selenolate catalyst.

$$R'Se^- + R'SeSR \longrightarrow R'SeSeR' + RS^- \qquad (4)$$

$$R'SeSR \xrightarrow{\text{NaBH}_4} R'Se^- + RS^- \qquad (5)$$

The selenol-catalyzed reaction using DTT as reductant has a similar reaction sequence as above. Both sodium borohydride and DTT are strongly reducing and maintain selenol in the reduced state. The pK_a of selenol group of selenocysteamine is 5.3 (8). Selenol catalyzes the reduction of disulfide because (i) selenol is present as the nucleophilic selenolate anion at pH 7, and (ii) selenolate anion is both a better nucleophile and a better leaving group than thiolate anion.

V. Conclusions

Disulfide groups in peptides and small organic molecules are reduced rapidly using sodium borohydride and a catalytic amount of selenol. An assay for disulfide group has been developed based upon Ellman's assay of the thiol formed by selenol-catalyzed reduction of the disulfide. For the complete reduction of disulfides at initial concentrations of ~0.1 to 1 mM, we recommend the use of 0.1 M $NaBH_4$ and a catalytic amount of selenol (10-15 mole% of the initial amount of disulfide) in aqueous ethanol for ~1 h at ambient temperature. Small amounts of a disulfide (10 nmol) were accurately measured using this assay.

The rates of reduction of disulfide bonds in proteins by DTT at pH 7 are enhanced by a factor of ~100 in the presence of selenol. These rates of reduction using both DTT and selenol are also significantly faster than those obtained using a dithiol reagent of low pK_a such as bis(2-mercaptoethyl) sulfone (BMS) (9). We therefore recommend the combined use of DTT (5 mM) and selenol (~1 mM) to achieve rapid reduction of disulfide bonds in proteins at pH 6 or 7.

Acknowledgment
We are thankful to Thomas J. Reidy and Rita M. Steeves for purifying Fab fragment of immunoglobulin.

References
1. Jocelyn, P. C. (1987). *Methods Enzymol.* **143**, 246-256.
2. Cleland, W. W. (1964). *Biochemistry* **3**, 480-482.
3. Singh, R., and Whitesides, G. M. (1991). *J. Org. Chem.* **56**, 6931-6933.
4. Dickson, R. C., and Tappel, A. L. (1969). *Arch. Biochem. Biophys.* **131**, 100-110.
5. Riddles, P. W., Blakeley, R. L., and Zerner, B. (1983). *Methods Enzymol.* **91**, 49-60.
6. Weil, L., Seibles, T. S., and Herskovits, T. T. (1965). *Arch. Biochem. Biophys.* **111**, 308-320.
7. Sondack, D. L., and Light, A. (1971). *J. Biol. Chem.* **246**, 1630-1637.
8. Tan, K.-S., Arnold, A. P., and Rabenstein, D. L. (1988). *Can. J. Chem.* **66**, 54-60.
9. Singh, R., and Whitesides, G. M. (1994). *Bioorg. Chem.* **22**, 109-115.

Cysteine Modification with Acrylonitrile for Amino Acid Analysis and Protein Sequencing

Daniel C. Brune, Donna G. Hartzfeld, and Thomas W. Johnson

Dept. of Chemistry and Biochemistry, Arizona State University, Tempe, AZ 85287-1604

I. Introduction

Unmodified cysteine (Cys) is not detected during analysis of amino acids derivatized with the fluorogenic reagent o-phthaldialdehyde (OPA). For this reason, various procedures have been developed for converting Cys to derivatives that give fluorescent products upon reaction with OPA (1). These methods include performic acid oxidation to cysteic acid, disulfide exchange, and alkylation with 4-vinylpyridine. It is known from the early literature that Cys is easily alkylated with other vinyl compounds, such as acrylonitrile and acrylamide (2), but use of these derivatives was initially rejected because their hydrolysis product, Cys-S-β-propionate (also referred to as S-carboxyethylcysteine), was not well separated from glutamic acid (3). Preliminary experiments with proteins in which cysteine was derivatized with acrylamide for sequencing (4) indicated that, after reaction with OPA, Cys-S-β-propionate was easily resolved from Glu. This result suggested that the use of acrylonitrile as a derivatizing agent should be reconsidered.

In the experiments presented here, insulin, which contains 3 disulfide bonds, was reduced by tributylphosphine (Bu$_3$P) and alkylated by acrylonitrile in the presence of ammonium bicarbonate buffer, pH 8.5. All of these reagents, including the buffer, are volatile and thus in principle can be removed from the protein during the normal vacuum drying prior to hydrolysis. The time course of the reaction was monitored by laser desorption time-of-flight mass spectrometry to obtain conditions resulting in complete Cys alkylation and a minimum of side reactions (e.g. alkylation of amino groups). In addition, acrylonitrile was used in place of acrylamide as a Cys alkylating agent prior to protein sequencing and found to give a well-resolved and easily identified derivative.

TECHNIQUES IN PROTEIN CHEMISTRY VII

II. Materials and Methods

A. *Insulin reduction and alkylation*

The procedure for reducing and alkylating insulin was basically a modification of those of Rüegg and Rudinger (5) and Andrews and Dixon (6). Bovine insulin (Sigma Chemical Co.) was added to water (2.5 mg/2 ml) and the pH adjusted to 10 with a few μl of 1 M NaOH to facilitate dissolution, followed by addition of 1 ml 2-propanol (2-PrOH), giving an approximate insulin concentration of 150 μM. Reduction of disulfide bonds was accomplished by adding 16 μl of 100 mM Bu_3P (Sigma) in 2-PrOH to 80 μl of the insulin solution, followed by addition of 40 μl of 100 mM ammonium bicarbonate, pH 8.5, taking care to avoid aerating the solution during mixing. After 5 min, 40 μl of a 2.9% or 10% acrylonitrile (Aldrich Chemical Co.) in 2-PrOH was added, giving a final acrylonitrile concentration of 100 mM or 340 mM, respectively. All reactions were carried out at room temperature (about 23°C). Aliquots (25 μl) were removed at 2, 5, 10, 20, and 40 min intervals and the reaction stopped by adding 5 μl of 10% trifluoroacetic acid (TFA). These samples were used for amino acid analysis and mass spectrometry. Samples for mass spectrometry were diluted 5-10x with a saturated solution of α-cyano-4-hydroxycinnamic acid (α-CHCA) (Aldrich) in 0.1% aqueous TFA/acetonitrile (2/1). Aliquots (2 μl) of these samples were dried on stainless steel sample pins and analyzed using a Vestec Research mass spectrometer incorporating a nitrogen laser (347 nm pulses of 3 ns duration).

B. *Amino acid analysis*

Samples for amino acid analysis were vacuum dried and subjected to vapor phase hydrolysis (105°C, 24 h) using 6 N HCl containing 1% phenol. The hydrolysate was vacuum dried and redissolved in 0.4 M borate buffer, pH 10.4. Aliquots (1 μl) of this solution were derivatized by OPA and fluorenyl-methylchloroformate (FMOC) using a Hewlett-Packard AminoQuant analyzer according to the manufacturer's instructions. Cys-S-β-propionamide, prepared as described by Brune (4), was also hydrolyzed to give a Cys-S-β-propionate standard.

C. *Protein sequencing*

Alkylation of the protein sequencing standard α-lactalbumin was as described by Brune (4) modified by replacing acrylamide with acrylonitrile. Basically, this involves interrupting sequencing after completion of the coupling reaction, opening the reaction cartridge, and adding Bu_3P and the alkylating agent (0.5 M acrylonitrile in aqueous 1.3% diisopropylethylamine. After 20 min incubation at 45°C, the sample was given three 15-second washes in the sequencer with ethyl acetate and sequencing resumed. Chromatographic conditions for separating PTH amino acids were those recommended by the manufacturer of the sequencer and are given in reference (4).

III. Results

A. Mass spectrometry

The time course for insulin alkylation by 100 mM and 340 mM acrylonitrile is shown in Figure 1. Alkylation of the insulin B chain was complete in 10 min at the lower concentration and in about 2 min at the higher concentration. No unreduced insulin remained, as indicated by the absence of a mass = 5734 peak (data not shown). The insulin A chain and its alkylation products were difficult to observe, but a small peak at 2553 Da corresponding to the tetraalkylated A chain can be seen in some of the spectra. Possibly the lack of basic amino acids in the A chain makes its ionization by protonation inefficient, although low solubility and poor interaction with α-CHCA during sample drying might also be a factor.

At longer alkylation times, overalkylation occurs, and after 40 min of alkylation with 340 mM acrylonitrile, a prominent peak due to trialkylation of the B chain and a shoulder due to tetraalkylation were observed. These are probably due to reaction with Lys ε and N-terminal amino groups, respectively (see below).

For comparison, insulin alkylation with 210 mM 4-vinylpyridine (conditions similar to those in ref. 6) was also examined (Figure 2). The time course for alkylation was similar to that with 100 mM acrylonitrile and there was no evidence for overalkylation, even after a reaction time of 40 min, indicating that 4-vinylpyridine is a weaker alkylating agent than acrylonitrile. As with acrylonitrile, the insulin A chain was much less easily observed than the B chain, even though four basic S-β-(4-pyridylethyl) Cys groups are generated by alkylation.

Spectra in the low mass region were dominated by a strong peak at 256 Da in the case of acrylonitrile and at 308 Da with 4-vinylpyridine (data not shown). These may be due to reaction of Bu_3P with the alkylating agent to form a tetraalkylphosphonium compound, contrary to the general assumption that Bu_3P does not react with vinylic alkylating agents. Problems due to this possible side reaction were avoided by prior addition of the Bu_3P reductant and the use of a large excess of the alkylating agent as in the experiments described here.

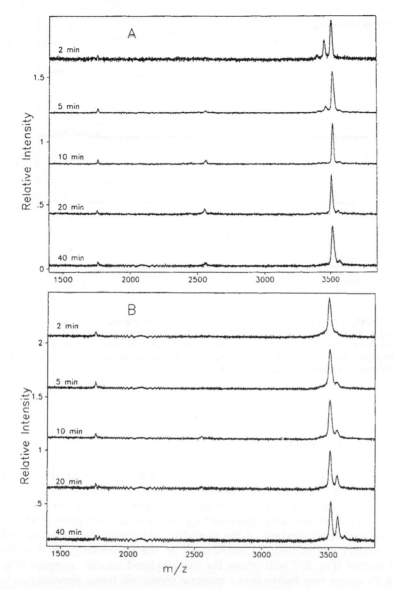

Figure 1. Mass spectra of bovine insulin after reduction and alkylation with acrylonitrile for different periods of time. A 100 mM Acrylonitrile; B 340 mM acrylonitrile. The major peak in each mass spectrum, at about 3507 Da, is due to the doubly alkylated B chain, and a small peak (*ca.* 1754 Da) due to the doubly charged form is also apparent. Some of the traces also show a peak at about 2553 Da due to the tetraalkylated A chain. Calibration was done using horse heart cytochrome *c* as an external standard.

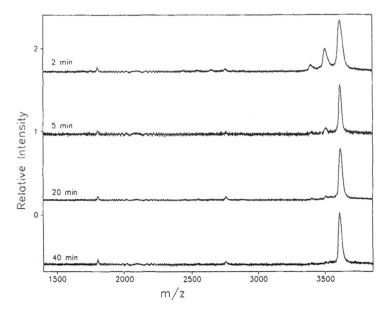

Figure 2. Mass spectra of bovine insulin after reduction and alkylation with 4-vinylpyridine for varying periods of time. The major peak due to the dialkylated B chain is at about 3611 Da. Peaks due to the unalkylated and monoalkylated B chain as well as very small peaks due to the monoalkylated through tetraalkylated A chain can be seen in the 2 min trace. The mass of the tetraalkylated A chain is 2752 Da.

B. Amino acid analysis

Chromatograms obtained from amino acid analyses of samples after reaction with 340 mM acrylonitrile for 2, 5, and 40 min as well as an unalkylated insulin control are shown in Figure 3. Also shown are chromatograms obtained after hydrolyzing synthetic Cys-S-β-propionamide and a vacuum dried mixture of the reducing and alkylating reagents and buffer without any protein. Comparison of the chromatogram for the insulin control (Fig. 3A) with those for the alkylated insulin samples (Fig. 3D, E & F) shows two fluorescence maxima (retention times approximately 1.63 and 2.13 min) that also occurred in the Cys-S-β-propionamide hydrolysate (Fig. 3B) and thus are due to Cys-S-β-propionate. Because only a single hydrolysis product was expected, it is unclear why 2 peaks due to Cys were found. Only the larger, later-eluting peak (indicated by an arrow in the chromatograms) was used for quantitative Cys determinations.

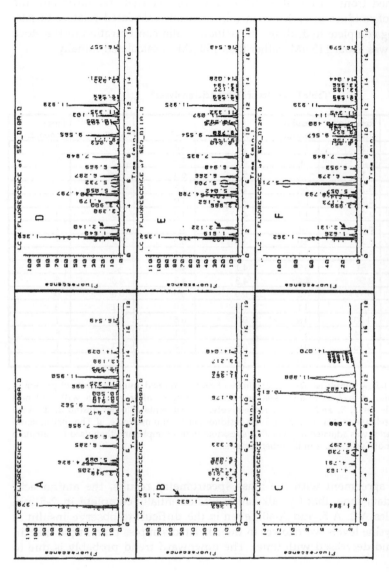

Figure 3. Chromatograms for separation of fluorescent derivatives of amino acids after hydrolysis of insulin alkylated with 340 mM acrylonitrile. **A** Untreated insulin control. **B** Cys-S-β-propionate standard from hydrolysis of the propionamide. **C** Control with Bu3P, acrylonitrile, and buffer only. **D** Insulin after 2 min reaction with acrylonitrile. **E** Insulin after 5 min reaction. **F** Insulin after 40 min reaction.

Quantitative results from the experiments on the insulin-containing samples are summarized in Table 1. Due to the lack of a precisely quantified Cys-S-β-propionate standard, Cys alkylation was assumed to be 100% after 40 min, and the instrumental response factor adjusted to give 6 Cys per 3 Ala in the 40 min hydrolysate. Amounts of all other amino acids were determined from a calibration table based on external standards and the number of residues of each per insulin molecule calculated relative to Ala. Assuming complete hydrolysis, the actual insulin concentration in the stock solution was about 115 µM, rather than 150 µM as calculated initially.

Table I Results of amino acid analyses[a]

Amino acid	Retention time	Insulin control	A-2[b]	A-5	A-40	From Sequence
Asx	1.24	3.1	2.9	3.0	2.7	3
Glx	1.36	7.6	7.7	7.5	7.3	7
Cys	2.13	-	5.7	6.3	6.0[d]	6
Ser	4.17	2.7	2.6	2.6	2.3	3
His	4.72	1.9	2.1	1.9	2.1	2
Gly	4.79	4.1	4.1	4.0	4.0	4
Thr	5.05	1.1	1.1	1.1	1.2	1
Ala[c]	6.27	3.0	3.0	3.0	3.0	3
Arg	6.96	1.1	1.1	1.1	1.2	1
Tyr	7.84	3.9	4.1	4.0	3.9	4
Val	9.56	4.4	4.3	4.3	4.3	5
Met	(9.77)	0.2	0.0	0.0	0.0	0
Phe	11.10	3.1	3.0	3.0	3.1	3
Ile	11.33	0.6	0.5	0.5	0.4	1
Leu	11.94	6.1	9.3	9.1	9.6	6
Lys	12.56	0.9	0.5	0.4	0.3	1
Pro	16.55	1.0	1.0	0.9	1.1	1

a Columns below sample names indicate moles of each amino acid found per mole of protein in that sample. b A-2, A-5, and A-40 are insulin samples after reaction with 340 mM acrylonitrile for 2, 5, and 40 min, respectively. c Amounts of other amino acids were calculated relative to Ala, assuming 3 Ala residues per insulin molecule. d The instrumental response factor was adjusted to give 6.0 Cys residues in this sample, and then used to calculate the number of Cys residues in the other samples.

In agreement with the mass spectrometry results, the amino acid analysis data show that Cys alkylation was essentially complete in 2-5 min (cf. samples A-2, A-5, and A-40), even in the difficult-to-observe insulin A chain. Lysine levels were lower in the alkylated samples than in the insulin control and decreased with time. The downward trend probably is due to alkylation of the ε-amino group after prolonged exposure to acrylamide. In contrast, the relative amounts of the N-terminal amino acids (Gly in the A chain and Phe in the B chain) were not noticeably affected.

A control experiment, in which a portion of the reaction mixture without insulin was vacuum dried and hydrolyzed, gave broad peaks eluting at about 10.6 and 11.9 min (Fig. 3C). The source of these peaks is not yet known, although their size is independent of the amount of acrylonitrile in the reaction mixture (data not shown). The 11.9 min peak coeluted with Leu and caused it to be overestimated in the alkylated samples (Table I). The chromatograms in Fig. 3C, D, E, and F all show a peak near 5.72 min (enclosed by parentheses) that became progressively larger at longer reaction times. This may be due to β-Ala formed in a side reaction between acrylonitrile and ammonia in the buffer. β-Ala is known to elute between Thr and Ala with chromatographic conditions similar to those used here (7).

C. Protein sequencing

Chromatograms for the first 8 cycles of sequencing of α-lactalbumin following alkylation with 500 mM acrylonitrile are shown in Figure 4. Although the acrylonitrile concentration seems excessive, overalkylation apparently is prevented by completing the coupling steps of the first sequencing cycle prior to alkylation. Thus the lysine peak on cycle 5 is not noticeably attenuated. Residue 6, Cys, gives a peak at 14.7 min that is well separated from those due to other PTH amino acids (the closest is PTH-Tyr at 13.8 min) and sequencing side products. Its position is fairly similar to that of PTH N-isopropylcarboxamidomethyl Cys, which was recommended by Krutzsch and Inman (8) for Cys determinations during sequencing because of its unique retention time as well as its high extinction coefficient.

IV. Discussion

The results presented here illustrate the use of mass spectrometry for monitoring the kinetics and extent of reactions involving proteins. In this case, reduction and alkylation of cysteine residues in insulin was shown to proceed rapidly, with further alkylation of amino groups as an unwanted side reaction at high acrylonitrile concentrations. With 100 mM acrylonitrile, Cys alkylation at room temperature was essentially complete in 10 min with insignificant alkylation of amino groups.

Acid hydrolysis of alkylated insulin converted Cys-S-β-propionitrile residues to Cys-S-β-propionate. Reaction with OPA gave 2 fluorescent

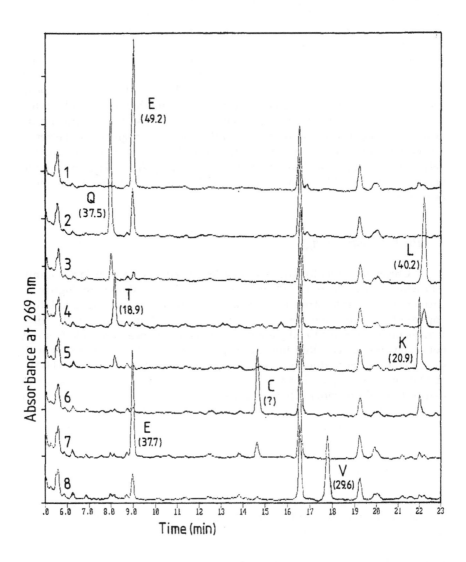

Figure 4. Chromatograms from the first 8 cycles of sequencing after treating α-lactalbumin with 0.5 M acrylonitrile. Letters adjacent to peaks in the chromatograms indicate the amino acid responsible for the peak, and numbers in parentheses indicate the amount of that amino acid in picomoles. Approximately 100 picomoles of α-lactalbumin initially were placed on a Beckman fiberglass support.

derivatives that were separated from those of other amino acids by HPLC. Use of the major fluorescence peak for quantitative determination of Cys in insulin gave reproducible results that were consistent with the data from mass spectrometry. Tributylphosphine, acrylonitrile, and ammonium bicarbonate were selected as the reductant, alkylating agent, and buffer, respectively, in these experiments because they are volatile and, in principle, can be removed from the protein sample by vacuum drying prior to hydrolysis. This strategy was not entirely successful, in that these reagents left behind small amounts of impurities resulting in fluorescence peaks, one of which coeluted with Leu and interfered with its determination. Further work is needed to identify the source of these impurities.

So far this reduction and alkylation procedure has only been applied to insulin for purposes of amino acid analysis. Other proteins should also be tested. In some cases, it may be necessary to reduce and alkylate the protein in the presence of a denaturant such as urea or guanidinium chloride followed by removal of the denaturant by HPLC, which would make the method more complicated. Alternatively, it might be possible to selectively alkylate cysteine in the hydrolysate, although side reactions with amino groups might also occur more readily under those conditions. In this case, a fluorescent derivative of Cys-S-β-propionitrile would be formed. The retention time of this product has not yet been determined.

Acrylonitrile works quite well for cysteine identification during protein sequencing. Although other good derivatizing reagents for Cys determination during sequencing have been reported (1), acrylonitrile is inexpensive and readily available, which may make it useful in some instances.

References

1. Strydom, D.J., Anderson, T.T., Apostol, I., Fox, J.W., Paxton, R.J. and J.W. Crabb (1993) *In* "Techniques in Protein Chemistry IV" (Angeletti, R.H ed.) Academic Press, San Diego, pp. 279-288.
2. Cavins, J.F. and M. Friedman (1968) *J. Biol. Chem.* **243**, 3357-3360.
3. Friedman, M., L.H. Krull and J.F. Cavins (1970) *J. Biol. Chem.* **245**, 3868-3871.
4. Brune, D.C. (1992) *Anal. Biochem.* **207**, 285-290.
5. Rüegg, U.T. and J. Rudinger (1977) *In* "Meth. Enzymol. XLVII" (Hirs, C.H.W, and Timascheff, S.N., eds.) Academic Press, New York, pp. 111-116.
6. Andrews, P.C. and J.E. Dixon (1987) *Anal. Biochem.* **161**, 524-528.
7. Jones, B.N. (1986) *In* "Methods of Protein Microcharacterization" (Shively, J.E., ed.) Humana Press, Clifton, NJ, pp. 121-149.
8. Krutzsch, H.C. and J.K. Inman (1993) *Anal. Biochem.* **209**, 109-116.

SPECIFIC CLEAVAGE OF BLOTTED PROTEINS AT CYSTEINE RESIDUES AFTER CYANYLATION: ANALYSIS OF PRODUCTS BY MALDI-TOF

N.D. Denslow[1,2] and H.P. Nguyen[2]

[1]Dept. Biochem. and Molec. Biol. and [2]Interdisc. Center for Biotech. Res. (ICBR), University of Florida, Gainesville, FL 32610

I. INTRODUCTION

Modification and cleavage at cysteine residues can be used to great advantage in studies examining the structure and function of proteins. Because cysteines are relatively scarce in proteins, cleaving at these residues usually produces large fragments. Equally important is the role that cysteine, with its potential for forming disulfide linkages, plays in the secondary structure of proteins. Pinpointing the location of Cys residues thus becomes an important strategy for determining protein structure.

Cleavage at Cys residues using 2-nitro-5-thiocyanobenzoate (NTCB) has been described by several groups (1-5). The reagent cyanylates reduced cysteine residue thiols followed by cleavage at the N-terminal side under alkaline conditions. While potentially a very useful method, it has seldom been used for sequence determination because the N-terminus of the released peptide fragment becomes blocked by the iminothiazolidine-carboxylyl group (1-2). Though a method to unblock the N-terminus by catalytic reduction is available (6), it has been used rarely. Alternatively, cystine residues can be cleaved with a new reagent N,N-diethylaminopropyl-bis-(3-hydroxypropyl)phosphine (7).

Another reason Cys cleavage by NTCB has been used infrequently is that the reaction conditions cause a number of side reactions that compete with the desired cleavage products. These include the reversibility of the cyanolysis reaction and the elimination of thiocyanate (3). In addition, other side reactions become possible as reactive by-products accumulate during the course of the reaction (4). Efficient cleavage at Cys residues requires careful monitoring of the reaction.

Effective methods for cleaving proteins in solution with NTCB have been described previously (1-5). For many proteins however, it is difficult to have a pure sample in solution. Often the final purification step requires SDS gel electrophoresis. Proteins can then be electro-transfered to PVDF membranes, a surface which is compatible with many chemical cleavages and enzymatic digestions.

We describe efficient procedures for cleaving proteins bound to PVDF membranes. The resulting fragments are analyzed by matrix-assisted laser-desorption ionization time-of-flight (MALDI-TOF) mass spectrometry, an analytical method that is powerful for identifying resulting fragments simply by their mass (8). We find most of the expected fragments for complete cleavage. However, as for most digests performed on membranes, the efficiency of cleavage is somewhat less than that in free solution. In addition to complete cleavage products, several partial cleavage fragments are found, some of which have suffered β-elimination or reacted to form mixed disulfides, and these are evaluated as well.

II. MATERIALS AND METHODS

A. Preparation of samples
Protein samples (50 pmol each of bovine pancreatic ribo-nuclease, bovine ß-lactoglobulin and bovine insulin) were mixed with Laemmli sample buffer (9), heated to 90°C for 5 min and electrophoresed in tris-tricine gels (10) that had been pre-run in gel buffer containing 0.1 mM mercaptoacetic acid to remove all oxidants (11). Gels were electroblotted to Immobilon P membranes (Millipore) in 10 mM MES, pH 6, 5mM DTT, 20% MEOH and stained with Ponceau S. Membranes were then incubated for 1 hr at 50°C in buffer A (200 mM Tris-acetate, pH 8, 1 mM EDTA, 5 mM DTT) under N_2. In the case of insulin, one sample was prepared as above and another was prepared in the absence of reducing agents. Samples in solution were dissolved in Buffer A and incubated for 1 hr at 50°C under N_2.

B. Cyanylation and cleavage reactions
For samples treated with NTCB in the presence of a reducing agent, the DTT level was reduced to 1 mM. Samples were cyanylated by adding NTCB to a final concentration of 4 mM and were incubated for 1 hr at 50°C.

Membrane-blotted samples were rinsed three times with water to remove NTCB and by-products of the cyanylation reaction. Cleavage was achieved by placing membranes in 50 ul Buffer C (25 mM Sodium borate, pH 9) and incubating for 1 hr at 50°C. Membranes were rinsed once with water and fragments were extracted with 60% Acetonitrile, 2.5% TFA by sonication for 30 min.

Liquid samples were treated essentially as membrane-bound samples except that after cyanylation they were precipitated with 9 volumes of ice-cold acidified acetone (pH 3) to remove NTCB. Pellets were rinsed once with acetone before proceeding with the cleavage reaction.

C. Analysis by MALDI-TOF
MALDI mass spectra were acquired on a Voyager RP (PerSeptive Biosystems - Vestec) MALDI-TOF mass spectrometer equipped with a 337 nm nitrogen laser. The spectra were acquired in the linear mode. The matrices used were saturated solutions of α-cyano-4-hydroxycinnamic acid in acetone or 3,5-dimethoxy-4-hydroxycinnamic acid (sinapinic acid) in acetone. External mass calibration was used.

D. Calculations
Theoretical masses were calculated by adding 25 mass units for cyanylation, subtracting 34 mass units for ß-elimination and adding 197 mass units for mixed disulfides. Some of the observed fragments appear 16 (or multiples of 16) larger than the predicted masses presumably due to oxidations.

III. RESULTS AND DISCUSSION

The reaction of NTCB with thiol compounds has been well characterized by others (1-5). The predominant reaction expected is the cyanylation of the protein thiol followed by cleavage (N-terminal to the cys) of the protein chain under alkaline conditions. As others have noted, several side reactions are also possible, including ß-elimination of the thiocyanoalanine residues

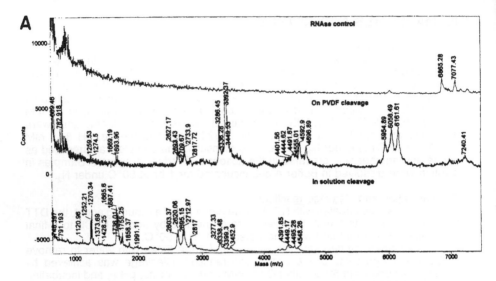

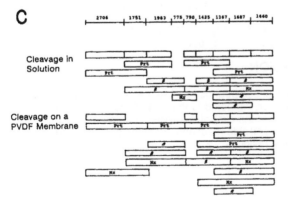

B

Fragments Theor. Mr	Fragments in Solution		Fragments in PVDF	
Complete cleavage				
775				
790	791		782	
1267	1270	1252	1274	1259
1425	1428			
1660	1665		1669	
1687	1687		1693	
1751	1755			
1983	1991			
2706	2712		2709	2693
Partials				
2714	2713		2733	
3393	3399		3392	
4421	4449		4445	
4551	4548		4555	
5907			5954	

Fragments Theor. Mr	Fragments in Solution		Fragments in PVDF	
Partials + Beta Elimination				
2614	2620	2603	2627	
2680	2695		2693	
2833	2817		2817	
3269	3277		3286	
3325	3338		3336	
4379	4391		4401	
4483	4495		4491	
Partials + Mixed disulfide				
1718	1736			
3500	3452		3449	
4610			4592	4696
6104			6058	6161
7280			7240	

C

Figure 1. Cleavage of bovine pancreatic ribonuclease with NTCB. (A)Mass spectra of starting material (top panel), fragments produced by cleavage in solution (middle panel) or on a PVDF mem- brane (bottom panel).(B) Table of theoretically calculated masses for complete cleavage with NTCB and empirically obtained masses (C) Schematic representation of fragments produced by cleavage with NTCB. Prt, partial cleaved fragment, 6, a fragment resulting from 6-elimination and mx, a fragment resulting from addition of a mixed disulfide.

instead of cyclization and cleavage (3) or the formation of mixed disulfides (4). Our analyses show that all of these possibilities occur. As expected, however, the predominant reaction observed under our conditions was cyclization and cleavage. A large number of partially cleaved fragments are also observed. Most of these appear not to have reacted with NTCB at the internal Cys residues, perhaps because of oxidation of the Cys. Some fragments clearly include internal Cys residues which have undergone either δ-elimination or disulfide addition. But only a few of the possible ions of this type are accounted for, suggesting that not all Cys are equally susceptible to the side reactions. The procedures we used minimized the side reactions in favor of complete cleavage.

A. Ribonuclease

NTCB cleavage of bovine pancreatic ribonuclease occurs well both in solution and on PVDF membranes (Figure 1). Of the 8 complete cleavage fragments expected, we were able to detect 7 when the reaction was done in solution and 5 when the reaction was done on a membrane. Some fragments are larger than expected, and best match fragments that overlap two or more of the complete cleavage products, indicating that some cysteines were not cut. In the solution reaction we could identify 13 partially cleaved fragments, 7 of which showed evidence of δ-elimination and 2 of mixed disulfide addition. Similarily, in the reaction done on a PVDF membrane, 15 partially cleaved fragments are observed, 7 of which showed evidence of δ-elimination and three which are mixed disulfides. Presumably the rest of the partial fragments are due to oxidized cysteines which were not targeted by the reagent. These results are summarized in Fig. 1.

B. δ-Lactoglobulin

We obtained similar results for δ-lactoglobulin (Fig. 2). In this case six complete cleavage products are expected ranging in size from 275 to 7189 mass units. After cleaving with NTCB in solution we are able to clearly identify 3 of the expected products. It is not surprising that we do not see the two smallest fragments (275 and 397 mass units) since they would fall in the region reserved for matrix peaks. Thus only one fragment is missing from the spectra illustrated in Fig 2A. However, we were able to detect this ion in other spectra obtained from different areas of the sample spot. This result emphasizes the need to look for ions throughout the sample spot. Cleavage on the PVDF membrane resulted in clear identification of 2 of the expected fragments. Again, as in the case of ribonuclease, several partial fragments were observed in each digest. Of the 7 partials seen in the solution experiment, three showed evidence of δ-elimination and two of mixed disulfide addition. Of the 8 partials seen for the cleavage on the membrane, one appears to be due to δ-elimination, and 4 show evidence of mixed disulfide addition. Again there appear to be more partials in the case of the reaction on a membrane. This may be due to the geometry of the sample on the surface of the membrane, perhaps blocking the reagent from the cutting sites, or it may reflect cys modifications occurring during the gel and transfer processes despite the precautions that were taken to avoid them.

C. Insulin

The third experiment we present involves the use of NTCB on insulin. Insulin is composed of two different chains, A and B, containing 6 cysteines held together by 3 disulfide bonds. Because of its simplicity, we selected this protein as a model for testing NTCB modification using our procedure of a disulfide-linked sample that has

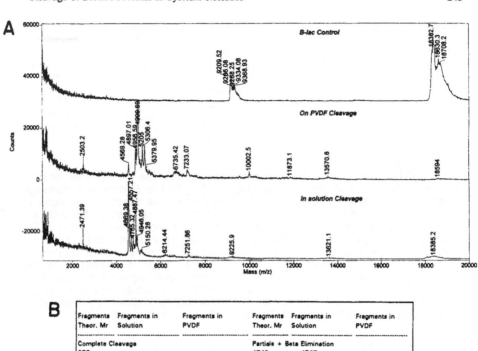

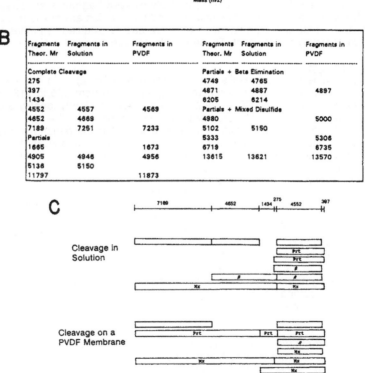

B

Fragments Theor. Mr	Fragments in Solution	Fragments in PVDF	Fragments Theor. Mr	Fragments in Solution	Fragments in PVDF
Complete Cleavage			Partials + Beta Elimination		
275			4749	4765	
397			4871	4887	4897
1434			6205		6214
4552	4557	4569	Partials + Mixed Disulfide		
4652	4669		4980		5000
7189	7251	7233	5102	5150	
Partials			5333		5306
1665		1673	6719		6735
4905	4946	4956	13615	13621	13570
5136	5150				
11797		11873			

Figure 2. Cleavage of bovine 6-lactoglobulin with NTCB. Mass spectra of intact 6-lactoglobulin (top panel) and fragments produced by cleavage on a PVDF membrane (middle panel), or in solution (bottom panel). (B) and (C) as in Figure 1.

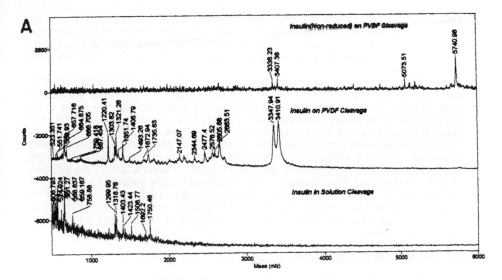

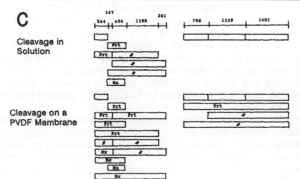

Fig. 3. Cleavage of insulin with NTCB. (A) Mass spectra of non-reduced insulin treated with NTCB on a PVDF membrane (top panel), reduced insulin treated on a PVDF membrane (middle panel) and reduced insulin treated in solution (bottom panel). (B) and (C) as in Figure 1.

been electrophoresed and blotted. As expected, the native molecule bound to PVDF resists cleavage by NTCB (Fig. 3A). This result is in contrast to the cleavage obtained of the reduced protein, either blotted to PVDF membranes or in solution (Fig. 3A). Thus, NTCB is able to discriminate between free Cys residues and Cystine residues.

Cleaving efficiency of insulin appeared to be rather good in both the solution and the membrane experiments. Four of the eight expected fragments are clearly seen in both. Three of the missing fragments are small, 147 to 404 mass units range and thus not clearly resolved from the matrix ions. The remaining missing ion could be seen in some spectrograms and not in others. As with RNase and ß-lactoglobulin, several larger fragments resulting from partial cleavage were also observed. In the reaction done in solution, 6 partially cleaved fragments were observed of which 3 match masses expected for ß-elimination and 1 for the addition of a mixed disulfide. The reaction on the PVDF membrane yielded 13 partials, of which 4 are due to ß-elimination at an internal cysteine and 4 are due to the addition of a mixed disulfide. Some of the larger fragments may also exhibit multiple oxidation states since their masses vary by multiples of 16 from the theoretically calculated masses.

IV. SUMMARY AND CONCLUSIONS

In conclusion, NTCB works almost as well on PVDF membranes as it does in solution, producing most of the expected cleavage products as measured by MALDI-TOF. Cleavage on the membrane, however, appears to be somewhat less efficient than in solution, producing a larger number of partial fragments. This may be due to oxidation of Cys residues during electrophoresis and electro-transfer to membranes, even in the presence of reducing agents. Alternatively, it may be due to the geometry of the reaction where the protein is bound in high concentration to a surface, rather than being free in solution. Degani and Patchornik have shown that the concentration of reactants affects the distribution of products between the primary reaction and the side reactions (3).

The reagent 2-nitro-5-thiocyanobenzoic acid (NTCB) is ideal for cleaving proteins at fully reduced Cys residues under relatively mild conditions. The cleavage is efficient and offers high potential for distinguishing cysteine residues from cystine residues in proteins where both are present and for assigning disulfide bonds, as we demonstrate with insulin in Fig. 3.

We describe a procedure for cleaving proteins with NTCB after they have been separated by SDS gel electrophoresis and electroblotted to PVDF membranes. The best conditions require the protein to be fully denatured and reduced. Cyanylation occurs best at pH 8 with a 4-fold excess of NTCB over SH groups, whereas cleavage occurs best at pH 9. However after cyanylation, there is always an equilibrium between cleavage and ß-elimination. Both products are observed. In addition some fragments are mixed disulfides, another possible alternative reaction of this procedure. Analyzing the products by MALDI-TOF gives a clear idea of the extent of the side reactions.

NTCB has been under-utilized in recent years because it forms an N-terminal iminothiazolidine-4-carboxyl residue which effectively blocks N-terminal sequencing. However, as new methods become developed for analyzing samples by MALDI-TOF, these constraints are no longer important. It is now possible to ladder-sequence proteins using exopeptidases such as aminopeptidase M and the carboxy-peptidases (12). In addition, new methods with MALDI-TOF involving the use of fragmentation by post source decay (PSD) have allowed unequivocal identification of fragments and direct sequence analysis of peptides (13-14). With these new methods of analysis, a blocked N-terminus no longer hinders the use of NTCB in structural studies.

Acknowledgment
        This research was supported by the Interdisciplinary Center for Biotechnology Research (ICBR), University of Florida.

References

1.Catsimpoolas, N. and Wood, J.L. (1966) J. Biol. Chem. 241,1790-1796.

2.Jacobson, G.R., Schaffer, M.H., Stark, G.R., and Vanaman,T.C. (1973) J. Biol. Chem. 248,6583-6591.

3.Degani, Y. and Patchornik, A. (1974) Biochemistry 13,1-11.

4.Price, N.C. (1976) Biochem. J. 159,177-180.

5.Lu, H.S., and Gracy, R.W. (1981) Arch. Biochem. Biophys. 212,347-359.

6.Schaffer, M.H. and Stark, G.R. (1976) Biochem. Biophys. Res. Commun. 71,1040-1047.

7.Mitchell, C., Hinman, L., Miller, L. and Andrews, P.C.(1995) in Techniques in Protein Chemistry VI ed. by J.W. Crabb, Academic Press, San Diego, pp 193-199.

8.Hillenkamp, F., Karas, M., Beavis, R.C. and Chait, B.T. (1991) Anal. Chem. 63,1193A-1203A.

9.Laemmli, U.K. (1970) Nature 227,680-685.

10.Schagger, H. and von Jagow, G. (1987) Anal. Biochem. 166,368-379.

11. Moos Jr., M., Nguyen, N.Y., and Liu, T-Y. (1987) J. Biol. Chem., 263,6005-6008.

12.Whittmann-Liebold, T.B., Bienert, M., Krause, E. (1995) FEBS Lett. 357,65-69.

13.Spengler, B., Kirsch, D., Kaufmann, R., Jaeger, E. (1992) Rapid Commun. Mass Spectrom. 6,105-108.

14. Kellner, R., Talbo, G., Houthaeve, T. and Mann, M. (1995) in Techniques in Protein Chemistry VI ed. by J.W. Crabb, Academic Press, San Diego, pp 47-54.

Disulfide Characterization of CD31 (PECAM)

John H. Robinson, Michael D. McGinley, J. Christopher Leidli, David E. Lyons, Chi-Hwei Lin, Barbara Karan-Tamir, Mark M. Zukowski, and Michael F. Rohde
Amgen Inc., Amgen Center, Thousand Oaks, CA 91320

I.   Introduction

CD31 (PECAM) is a 100kDa integral membrane glycoprotein found mainly in endothelial intercellular junctions, platelets and monocytes (1).   Although its function has not been determined, CD31 is believed to be involved in the regulation of endothelial cell migration; thus it is probably involved in vascular development, wound repair, and angiogenesis (2).   Recently, CD31 has been shown to be required for transendothelial migration of leukocytes (3).   Although binding of CD31 is thought to be mainly homophilic, there is some evidence that it may bind other members of the cellular adhesion molecule family (1).   Common to the adhesion molecules is the immunoglobulin (Ig) homology unit which consists of two ß-sheets stabilized by a disulfide bond (4).   CD31 is thought to have six of these homology units, thus six putative disulfide bonds (4,5).   Since these folds are important for the structural function of CD31, we wish to elucidate the disulfide structure to further our understanding of this molecule.

II.   Materials and Methods

A.   Materials

Recombinant peptide-N^4-(N-acetyl-glucosaminyl) asparagine amidase (N-glycanase) was a product of Genzyme (Cambridge, MA).   Trypsin and Asp-N endoproteinases, sequencing grade, were obtained from Boehringer Mannheim (Indianapolis, IN).   Reverse-phase C_4, 2.1 x 150 mm and 4.6 x 250 mm columns were obtained from Vydac (Hesperia, CA).   Reverse phase HPLC solvents were products of Burdick and Jackson (Muskigon, MI).   Trifluoroacetic acid for use in reverse-phase was from J.T. Baker (Phillipsburg, NJ).   Sequencer reagents and buffers were obtained from Applied

Biosystems (Foster City, CA) or Hewlett Packard (Palo Alto, CA). Dithiothreitol was from Calbiochem (La Jolla, CA). Iodoacetic acid was a product of Sigma (St. Louis, MO). All other routine lab chemicals were of the highest quality commercially available.

B.   CHO expression and purification of r-sol-Hu-CD31
The cloning, expression, and purification of r-sol-hu-CD31 was similar to the techniques described by Muller *et al.* (3).

C.   N-site deglycosylation of CD31
An aliquot of one milligram of r-sol-hu-CD31 in PBS was adjusted to pH 7.5 or pH 8.0 with 0.1M TRIS. 35μg of N-glycanase was added to remove potential N-glycosylation sites (6,7). Digestion was allowed to proceed for three days at 37°C.

D.   Proteolytic Digestion, Reduction and Carboxymethylation
The sample was treated with trypsin at an enzyme to substrate ratio of 1:30. Digestion was then allowed to proceed for 18 hours at 37°C. Twenty percent of the digest was reduced in 5mM dithiothreitol at 45°C for 20 minutes, then immediately carboxymethylated with 10mM iodoacetic acid for 20 minutes in the dark at room temperature.

E.   HPLC and Mass Spectrometric Analysis
Digests were injected on a Hewlett Packard 1090 HPLC with a 4.6 x 250 mm C4 reverse phase column, equilibrated with 95% buffer A (0.1% TFA in water) and 5% buffer B (90% acetonitrile, 0.09% triflouroacetic acid in water). Elution was performed using a linear gradient of 5% to 45% buffer B over 110 minutes. Elution of peptides was monitored at 214 nm and fractions manually collected. Approximately 20% of the flow was diverted to a PE Sciex API III electrospray mass spectrometer.

F.   N-terminal Sequence Analysis
Protein sequences were determined using either the Applied Biosystems 470 and 477 protein sequencer or the Hewlett Packard G1000A sequencer. For samples loaded on Applied Biosystems sequencers, fractions were loaded on a precycled polybrene treated glass fiber filter. Samples loaded on the Hewlett Packard sequencer were loaded on the

hydrophobic portion of a biphasic protein sequencing column. On line phenylthiohydantoin amino acid analysis was performed using the Applied Biosystems 120 or Hewlett Packard 1090 HPLC, respectively.

G.   Subdigestion of selected peptides

Any fraction that required secondary digestion was dried in a Speed Vac (Savant). The sample was then re-dissolved in 0.1M TRIS pH 7.5 to 200μL, and endoproteinase Asp-N added at an enzyme to substrate ratio of approximately 1:30. Digestion was allowed to proceed for 18 hours at 37°C. HPLC and mass spectrometric analysis was then performed as described above, except a shorter gradient of 0% to 50% buffer B over 80 minutes was used.

H.   Matrix Assisted Laser Desorption Ionization Mass Spectrometry.

Selected peptides were subjected to mass spectrometry using the KRATOS MALDI III, with 33mM α-cyano-4 hydroxy-cinnamic acid as the matrix. Laser power was manually varied to maximize the signal to noise ratio. Selected disulfide containing fractions were subjected to prompt fragmentation by increasing the laser power (8).

III.   Results and Discussion

The primary sequence of soluble human CD31 is shown in figure 1 (note the 12 conserved cysteines and the two unconserved cysteines). Our goal was to verify the proposed disulfide structure, and the state of the two unconserved cysteines. To identify disulfide containing peptides, we first deglycosylated the protein with N-glycanase. This greatly improved the simplicity of the tryptic peptide map, as well as mass spectrometric analysis. A small aliquot of this digest was reduced and carboxymethylated. Both the native and reduced aliquots were run on the HPLC and fractions were collected from the native map. Figure 2 shows both native and the reduced peptide maps (corrected for differential amounts injected).

Several peptides readily disappeared upon reduction, as shown in figure 2. Major reducible peaks observed at retention times 18.6, 68.1, 69.7, 71.1, 81.9, and 85.1 minutes were analyzed. Minor peaks which disappeared upon reduction were analyzed and shown to represent similar disulfide

NH2-terminus (1-22): QENSFTINSVKMSGLFMWKN

Ig-like domains:

```
Domain 1 (23-117):  GRNLFLQCFADVSTTSVKPQHQMLFY-KDEVLFYNIESMKSTESYFI-PEVRIDSGTYKCTVIVNN--KEKTTA---EVQLIV-EVVSPRV--TLFKEAIQ
Domain 2 (118-221): GGIVRNCSVPEEKAPIHFTIEKLEANERWKLKREKSRQ2NFVILEFPVEEQRVLSFPCQARIEGIHR&QTSESTKSELVTVTEFSTPKFHISPTCM-IME
Domain 3 (222-312): GA2LHKCTIQVTHLAQEFPEIII--QKDAIVAHNHGNK-AVYSVWAMVE--HSGNTICKVESSKI-SKVSSI----VVNITELFSKPELESS--FFHLEQ
Domain 4 (313-396): GERLNLGCSIPGA-PPANFTI-----QKEFTIV--SQFQD--FYKI--ASKEDSGTYICTAGIDKVVKSNTV----QTVWCEMLSQPRISYDACF-EVIK
Domain 5 (397-488): GQTIEVRCESISKTLPISYQLL--KTS-KV-LENSTNSND-PAVFKENPT---EEVEYCCVALNCSHAKMLSE--VLRVVAPVLEFQISHLSSN-VVFS
Domain 6 (489-574): GEETVLQCAVNEGSGPTTYK----FYREKEGKFFYQMTSNATQ-FVFKQXASKEQEGEYCTAFNRANHA---SSVFRSKILTVFFV ILAFKCK
```

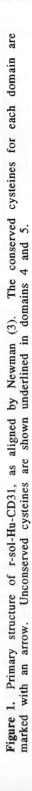

Figure 1. Primary structure of r-sol-Hu-CD31, as aligned by Newman (3). The conserved cysteines for each domain are marked with an arrow. Unconserved cysteines are shown underlined in domains 4 and 5.

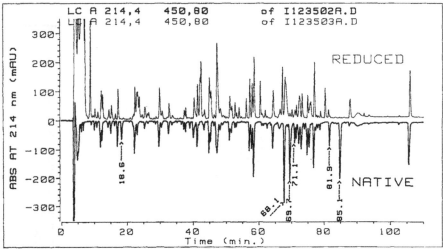

Figure 2. RP-HPLC chromatograph of r-sol-Hu-CD31.   The native tryptic digest is shown on the lower chromatograph.   The upper chomatograph is an aliquot of the main digest which was reduced and carboxymethylated prior to injection.   Note the scale is adjusted to correct for differential amounts adjusted.   Gradient and buffer conditions are described as in methods.

products as seen in the six major peaks above (data not shown).   For each of the major peaks, masses were determined from the corresponding peaks in the total ion current chromatograph.   Fractions representing these peaks were also subjected to N-terminal sequence analysis.   Mass data is shown in figure 3 and summarized with sequence data in table 1.

The mass spectrometry and sequencing data for peaks at 18.6, 68.1, 69.7, and 81.9 minutes retention times gave results that verify corresponding disulfide peptides.   As shown in table 1, each fraction shows two sequences, and one clear mass.   The observed mass for each fraction correlates to the theoretical mass of the corresponding disulfide peptide. Mass corrections were made to account for disulfide bond formation and for deamidations of N-site asparagines. Treatment with N-glycanase deamidates N-sites asparagines to aspartic acid (6,7) which further helps elucidate the location of glycosylated asparagines.   One disulfide peptide was seen to co-elute with a contaminating peptide.   For the peak at 71.1 minutes retention time, a minor sequence was observed.   The predicted mass for this peptide correlates with its predicted sequence and does not appear to be involved in disulfide formation (Table 1).   The main sequence, along with

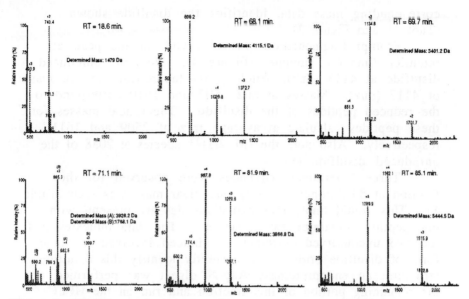

Figure 3. ES-MS spectra corresponding to each of the noted peaks in figure 2. Masses were determined using the software included with the PE Sciex API III electrospray mass spectrometer.

Table 1 Summary of observed sequences, with predicted and observed average masses for each fragment discussed. Deamidated asparagines are noted with an asterisk (*). Predicted masses were corrected for all observed deamidations of asparagines.

RETENTION TIME	Peptide Sequence	Observed Average Mass (Da)	Predicted Average Mass (Da)	NOTES
18.6 MIN	1)V-N*-C-S-V-P-E-E-K	1479	1479.7	DISULFIDE #2
	2)C-Q-A-R			Cys125 to Cys179
68.1 MIN	1)E-Q-E-G-E-Y-Y-C-T-A-F-N-R	4115.1	4115.5	DISULFIDE #6
	2)V-V-E-S-G-E-D-I-V-L-Q-C-A-V-N-E-G-S-G-P-I-T-Y-K			Cys496 to Cys545
69.7 MIN	1)S-D-S-G-T-Y-I-C-T-A-G-I-D-K	3401.2	3401.9	DISULFIDE #4
	2)L-N*-L-S-C-S-I-P-G-A-P-P-A-N*-F-T-I-Q-K			Cys320 to Cys359
71.1 MIN	1)N*-L-T-L-Q-X-F-A-D-V-S-T-T-S-H-V-K-P-Q-H-Q-M-L-F-Y-K	3926.2	3926.6	DISULFIDE #1
	2)C-T-V-I-V-N-N-K			Cys30 to Cys82
	3)F-H-I-S-P-T-G-M-I-M-E-G-A-Q-L-H	1768.1	1769.1	NON DISULFIDE
81.9 MIN	1)C-T-I-Q-V-T-H-L-A-Q-E-F-P-E-I-I-I-K	3866.8	3867.5	DISULFIDE #3
	2)S-V-M-A-M-V-E-H-S-G-N*-D-Y-T-C-K			Cys229 to Cys277
85.1 MIN	1)S-N-T-V-Q-I-V-V-C-E-M-L-S-Q-P-R	5444.5	5445.0	TRI-PEPTIDE
	2)C-E-S-I-S-G-T-L-P-I-S-Y			Cys449 and Cys454 with
	3)D-N-P-T-E-D-V-E-Y-Q-C-V-A-D-N-C-H-S-H-A-K			Cys378 or Cys404

corresponding mass data, identifies the disulfide shown (Table 1 and Figure 3).

Prompt fragmentation was observed for the peak at retention time 68.1 minutes (Figure 4). Note the unreduced disulfide at 4115 (m/z), matching its theoretical average mass of 4117 (m/z). Masses at m/z 2507 and 1610 correspond to the reduced peptides of the disulfide. Theoretical masses for these peptides calculate at average m/z of 2509 and 1611, respectively. Also note the $(M+2H)^{2+}$ species at 2058 of the unreduced disulfide (8).

Three N-terminal sequences were observed for the fraction at 85.1 minutes. Only one clear mass was seen (Table 1). This would imply that two disulfides are present, with one peptide containing two cysteines. The data indicates that the two unconserved cysteines are indeed involved in some form of disulfide bridge. To correctly identify this disulfide structure, an endoproteinase Asp-N digest was performed on the remaining portion of this fraction and run on the HPLC (Figure 5). Sequence analysis and mass spectrometric data are summarized in table 2. It clearly shows that the conserved cysteines at Cys^{378} and Cys^{454} are involved in disulfide formation. The unconserved cysteines at Cys^{404} and Cys^{449} are also covalently linked. This is not surprising, since Cys^{404} is conformationally close to Cys^{449} in the Ig homology domain model (9).

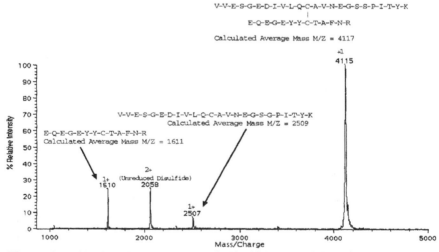

Figure 4. MALDI-TOF spectrum of disulfide containing fraction at 68.1 minutes. Laser power was increased (8) until the reduced fragments are seen at m/z = 2507 and 1610. The unreduced disulfide is seen at 4115 (m/z) along with its 2+ seen at 2058 (m/z).

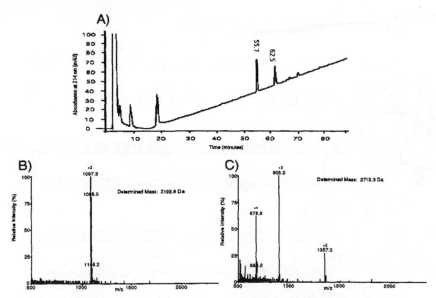

Figure 5. Redigestion with Asp-N of fraction eluting at 85.1 minutes. A)RP-HPLC chromatograph of redigest.   B)ES-MS spectra for peak eluting at 55.7 minutes.   C)ES-MS spectra for peak eluting at 62.5 minutes.

Concerns were raised that this disulfide bond is formed by the oxidation of any free sulfhydryls during deglycosylation or digestion.   The presence of free sulfhydryls was tested using Ellman's reagent under denaturing conditions (10). Results show essentially no free sylfhydryls are present in CD31.   Furthermore, treatment of denatured, non-reduced CD31 with iodoacetic acid does not lead to any carboxymethylation of cysteines (data not shown), thus confirming that the Cys^{404} to Cys^{449} disulfide bond is present in native CD31.

IV.   Conclusion

Theoretical disulfides have been predicted for CD31 based on Ig type homology units.   These six disulfide bonds have been positively identified: Cys^{30} to Cys^{82}, Cys^{125} to Cys^{179}, Cys^{229} to Cys^{277}, Cys^{320} to Cys^{359}, Cys^{404} to Cys^{449} and Cys^{496} to Cys^{545}.   These represent the disulfides stabilizing the six Ig-like domains of the CD31 molecule.   The two unconserved cysteines, Cys^{404} and Cys^{449} are found to be involved in a disulfide bond between domain 4 and domain 5.

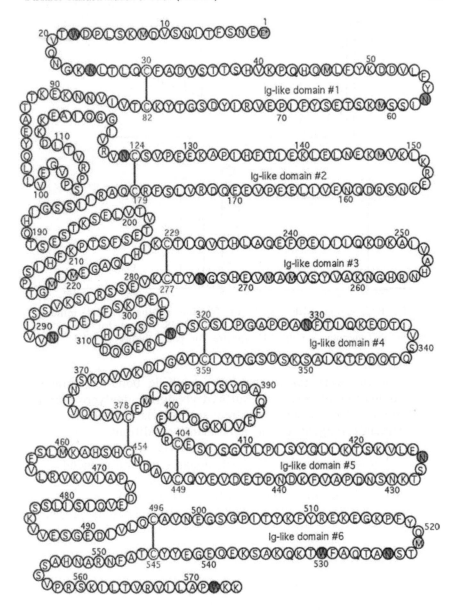

Figure 6. Complete bubble diagram of r-sol-Hu-CD31. Note disulfide linkage between domain 4 and 5. Also, each domain has the predicted disulfide. Potential N-site, methionine, and tryptophan residues are highlighted. A pyroglutamate residue is noted at the NH$_2$-terminus.

258   John H. Robinson *et al.*

Table 2. Summary of observed and predicted masses with observed sequences for Asp-N redigestion

RETENTION TIME	Peptide Sequence	Observed Average Mass (Da)	Predicted Average Mass (Da)	NOTES
55.7 MIN	C-E-S-I-S-G-T-L-P-I-S-Y D-V-E-Y-Q-C-V-A	2193.6	2193.4	Unconserved Cys forming disulfide Cy404 to Cys449
62.5 MIN	S-N-T-V-Q-I-V-V-C-E-M-L-S-Q-P-R D-N-C-H-S-H-A-K	2712.3	2713.1	Predicted Disulfide for Ig domain 5 Cys378 to Cys454

These finding suggest that domain 4 and 5 might be functionally linked. Figure 6 represents the actual disulfide structure of CD31.

Few adhesion molecule disulfide structures have been determined. Most disulfide structures are assumed by structural homology with other proteins in the same family or superfamily. Since most adhesion molecules are relatively large (greater than 100kDa), hydrophobic, and heavily glycosylated, they present a special challenge to the protein biochemist. However, steps can be taken to increase the probability of obtaining disulfide information. Deglycosylation of the native protein (N- and O- sites) reduce the complexity of peptide maps and simplify mass spectral data; allowing for simpler disulfide assignments. Minimization of sample handling is also important. With large hydrophobic proteins, it is not always possible to buffer exchange for optimal digestion conditions. Recoveries after dialysis or diafiltration are usually disappointing. Utilizing buffer conditions that support all manipulations made to the protein avoid sample losses due to buffer exchange and sample handling. These procedures increase the information that can be obtained from small amounts of these large proteins.

Acknowledgments

We wish to thank John Le and Vishwanatham Katta for their expertise in electrospray mass spectrometry, Scott Patterson for has broad knowledge of MALDI-TOF, and all members of the Amgen Protein Structure group whose individual talents have built a tremendous team.

References

1. Albelda, S.M., Muller, W.A., Buck, C.A., and Newman, P.J. (1991) *J. Cell Biol.* **114**, 1059-1068.

2. Schimmenti, L. A., Horng-Chin, Y., Mardi, J. A., and Albelda, S. M. (1992) *J. Cell. Physiol.* **153**, 417-428.

3. Muller, W. A., Weigl, S. A., Deng, X., Phillips, D.M. (1993) *J. Exp. Med.* **178**, 449-460.

4. Newman, P.J., Berndt, M.C., Gorsky, J.,White, G.C., Paddock, L. S., and Muller, W.A. (1990) *Science* **247**, 1219-1222.

5. Williams, A. F. (1987) *Immunol. Today* **8**, 298-303.

6. Elder, J. H., and Alexander, S. (1982) *Proc. Natl. Acad. Sci. USA* **79**, 4540-4544.

7. Tarentino, A. L., Gómez, C. M., and Plummer Jr., T. H. (1984) *Biochemistry* **24**, 4665-4671.

8. Patterson, S. D., and Katta, V. (1994) *Anal. Biochem.* **66**, 3727-3732.

9. Newman, P. J. (1995) personal communication.

10. Ellman, G. L. (1959) *Arch. Biochem. Biophys.* **82**, 70.

Retroaction of a Disulfide Bond in an
Octreotide-Like Peptide: A Multicenter Study

Ruth Hogue Angeletti, Lisa Bibbs, Lynda F. Bonewald, Gregg B. Fields,
John S. McMurray, William T. Moore, & John T. Stults

Dept. Developm. & Molecular Biology, Albert Einstein College Med, Bronx NY 10461
Research Institute of Scripps Clinic, La Jolla CA 92037
Dept. Med. & Biochem., Univ. Texas Health Science Center, San Antonio TX 78284
Dept. Lab Medicine & Pathology, Univ. of Minnesota, Minneapolis MN 55455
Dept. Neuro-Oncology, M.D. Anderson Cancer Center, Univ. of Texas, Houston TX 77030
Dept. Pathology & Lab Medicine, Univ. of Pennsylvania, Philadelphia PA 19104
Genentech, Inc., South San Francisco CA 94080

1. Introduction

Peptide synthesis core facilities produce large numbers of peptide
antigens and biologically active peptides for their academic or industrial research
communities. In recent years, they have been challenged to begin synthesizing
posttranslationally modified peptides, peptides used for structural studies and
physical chemical experiments, as well as combinatorial peptide libraries. The
standards for evaluating the quality of peptide products have also been raised
in the face of recent dramatic improvements in analytical technology, particularly
mass spectrometry. The Peptide Synthesis Research Committee of the
Association of Biomolecular Resource Facilities (ABRF) conducts anonymous
studies to evaluate the ability of ABRF member laboratories to synthesize and
characterize test peptides (1-6). The most recent studies have shown that peptide
assembly and cleavage are no longer significant problems in most core
laboratories. Therefore, the committee has departed from the direction of
previous years, and designed a study not only to test capabilities, but to provide
an opportunity for core member laboratories to efficiently incorporate new
technologies. Since oxidized peptides have been shown to be more effective than
their linear counterparts in many biological systems, the 1995 study had focused
on methods for producing oxidized peptides. ABRF member laboratories were
asked to prepare an octreotide-like peptide with an intact disulfide bond produced
by any of 14 molecules provided by the committee. Coded samples of unpurified
peptides were characterized by AAA, HPLC, electrospray ionization mass
spectrometry (ESI-MS), matrix-assisted laser desorption ionization mass
spectrometry (MALDI-MS), Edman sequencing, and spectrophotometric analysis
following treatment with Ellman's reagent. These methods permitted evaluation
and evaluation of purity, as well as assessment of the amount of oxidized peptide.
Forty one laboratories participated in this study, submitting 92 samples for
analyses.

Formation of a Disulfide Bond in an Octreotide-Like Peptide: A Multicenter Study

Ruth Hogue Angeletti[1], Lisa Bibbs[2], Lynda F. Bonewald[3], Gregg B. Fields[4],
John S. McMurray[5], William T. Moore[6], & John T. Stults[7]

[1]Dept. Develop. & Molecular Biology, Albert Einstein College Med., Bronx NY 10461
[2]Research Institute of Scripps Clinic, La Jolla CA 92037
[3]Depts. Med. & Biochem., Univ. Texas Health Science Center, San Antonio TX 78284
[4]Dept. Lab Medicine & Pathology, Univ. of Minnesota, Minneapolis MN 55455
[5]Dept. Neuro-Oncology, M.D. Anderson Cancer Center, Univ. of Texas, Houston TX 77030
[6]Dept. Pathology & Lab Medicine, Univ. of Pennsylvania, Philadelphia PA 19104
[7]Genentech, Inc., South San Francisco CA 94080

I. Introduction

Peptide synthesis core facilities produce large numbers of peptide antigens and biologically active peptides for their academic or industrial research communities. In recent years, they have been challenged to begin synthesizing posttranslationally modified peptides, peptides used for structural studies and physical chemical experiments, as well as combinatorial peptide libraries. The standards for evaluating the quality of peptide products have also been raised, because of recent dramatic improvements in analytical technology, particularly mass spectrometry. The Peptide Synthesis Research Committee of the Association of Biomolecular Resource Facilities (ABRF) conducts anonymous studies to evaluate the ability of ABRF member laboratories to synthesize and characterize test peptides (1-4). The most recent studies have shown that peptide assembly and cleavage are no longer significant problems in most core laboratories. Therefore, the committee has departed from the direction of previous years, and designed a study not only to test our abilities, but to provide an opportunity for our member laboratories to efficiently incorporate new technologies. Since cyclized peptides have been shown to be more effective than their linear counterparts in many biological systems, the 1995 study has focused on methods for producing cyclized peptides. ABRF member laboratories were asked to prepare an octreotide-like peptide with an intact disulfide bond produced by any of 4 protocols provided by the committee. Coded samples of unpurified peptides were characterized by AAA, HPLC, electrospray ionization mass spectrometry (ESI-MS), matrix-assisted, laser desorption ionization mass spectrometry (MALDI-MS), Edman sequencing, and spectrophotometric analysis following treatment with Ellman's reagent. These methods permitted quantitation and evaluation of purity, as well as a measure of the amount of oxidized peptide. Forty one laboratories participated in this study, submitting 97 samples for analysis.

A. Amino Acid Analysis

Standard analyses were performed according to the method of Spackman, Stein and Moore (6). Samples were hydrolyzed for 24 hr at 110°C in 6 N HCl containing 2% phenol and 1% 2-mercaptoethanol. Cys was analyzed as cysteic acid after performic acid oxidation and hydrolysis in 6 N HCl for 24 hr at 120°C (7). Trp was analyzed after a rapid hydrolysis for 25 min at 166°C in 6 N HCl containing 4% phenol (8). All analyses was performed on a Beckman 6300 amino acid analyzer with a sodium polystyrene sulfonated cation exchange column from Pickering Laboratories.

B. Spectrophotometric Detection of Free Sulfhydryl Groups

The percent of free sulfhydryl groups was detected spectrophometrically at 412 nm after reaction of the peptide samples with Ellman's reagent, 5,5'-dithiobis(nitrobenzoic acid) (9). An extinction coefficient of 13,600 $M^{-1}cm^{-1}$ was used.

C. Analytical Reversed Phase HPLC

Samples were dissolved in 0.1% TFA, 5% acetonitrile, and analyzed on a HP1090 HPLC using a Vydac C18 column (300 Å pore size, 4.6 x 250 mm). The linear gradient extended from 0.1% aqueous TFA to 70% acetonitrile containing 0.09% TFA over 60 minutes. The flow rate was 1 mL/min and the absorbance was monitored at 214 nm using a photodiode array detector. Quantitation was obtained by the Waters 820 software program.

D. Sequence Analysis

Automated Edman degradation of selected samples was performed on an Applied Biosystems 477A Protein Sequencer/120A Analyzer using BioBrene Plus as carrier.

E. Mass Spectrometry

Electrospray ionization mass spectrometry (ESI-MS) (10) was performed with a PE-Sciex API-III triple quadrupole mass spectrometer with an IonSpray source. The samples, approximately 20 pmol/μl, were infused at 2.5 μl/min. Data were acquired in the range of m/z 400-1000, with resolution sufficient to observe the isotope peaks for singly-charged ions at m/z 1000.

Matrix-assisted laser desorption ionization mass spectrometry (MALDI-MS) (11) was performed on a Fisons Instrument (Beverly, MA) VG TOfSpec time-of-flight mass spectrometer (0.6 m flight tube) outfitted with a nitrogen laser (337 nm, 4ns pulse). The accelerating voltage was set to 20 kV and the detector voltage to 1.5 kV. Positive ion data were collected in the linear mode, each spectrum derived from the accumulation of 20 to 50 laser shots. External calibrations were performed using synthetic peptides having masses covering the range of interest. Data was analyzed using Fisons Instruments Opus Software. Samples were dissolved in 50 ml 50% acetonitrile, 0.1% TFA, and diluted 1/1,000 before analysis using α-cyano-4-hydroxycinnamic acid as the matrix.

II. Materials and Methods

ABRF member laboratories were asked to synthesize the following peptide, ABRF95, by the method most frequently used in their facility:

H-Phe-Cys-Phe-Trp-Lys-Thr-Cys-Thr-NH$_2$

This peptide sequence was chosen for several reasons. It is homologous to octreotide, an 8-residue somatostatin analogue, which is readily oxidized to its cyclical form. Its synthesis is straightforward, with no anticipated problems with assembly or cleavage (5). To verify that the assembly, cleavage and oxidation were indeed as straightforward as predicted, members of the committee synthesized the peptide, and oxidized it by each of the four protocols. The presence of a Trp residue provided a site susceptible to modification. The low m/z of the reduced (1034.2) and oxidized (1032.2) peptides would permit the determination of the relative proportions of linear and cyclized forms in the same sample by analysis of the isotopic distribution in ESI-MS.

The participating laboratories were asked to oxidize the peptide by one or all of 4 frequently used protocols, 2 off-resin and 2 on-resin. Laboratories were instructed that the *Acm* group should be used for side chain protection of Cys residues if on-resin protocols were followed. Since Trp would be modified during on-resin oxidation procedures, it was also suggested that the *Boc* group should be used for Trp protection if Fmoc chemistry were employed, or the formyl group for Boc chemistry. Peptide samples containing 5-10 mg of unpurified product were supplied to the committee. When off-resin protocols were used, a sample of the linear product was also provided. The samples were coded, so that the participating laboratories would remain anonymous.

The number of oxidation protocols used was limited to four to restrict the number of samples analyzed by the committee, and to provide enough samples from each protocol to fairly evaluate both the methods and the quality of laboratory handling. The 2 on-resin and 2-off resin procedures were chosen as representative methods, and to some extent, for economy. In protocol I, the cleaved peptide was dissolved in 7g/L ammonium acetate, and stirred for 3 days before lyophilization. Protocol II used direct oxygenation for 24 hr in an ammonium bicarbonate buffer at pH 8.5. In protocol III, the peptide-bound resin was treated with a 1.5 molar excess of 0.4 M thallium trifluoraceetate in dimethylformamide for 1 hr before washing, and cleavage. For protocol IV, the peptide-resin was treated with a 4-fold molar excess of 0.1 M mercuric acetate in dimethylformamide for 1 hr, after which the resin was filtered, treated with a 10-fold molar excess of 2-mercaptoethanol in dimethylformamide for 1 hr, filtered and washed with solvents, and cleaved from the resin with 95% TFA in water. Participants were warned that mercury adducts had been observed in peptides oxidized with this procedure. Three laboratories submitted 5 samples oxidized by a post-cleavage potassium ferricyanide method. Another 3 laboratories used the Ekathiox™ resin. All samples received were analyzed.

The use of external calibrants potentially leads to a 1 to 3 μ error. In order to clearly evaluate cyclization, only a 2 μ mass change, reaction with *p*(hydroxymercuric)benzoic acid (pHMB) was performed (12). PHMB adds to free thiols and results in a 321 u mass shift per thiol. Percent desired product for the peptide samples was estimated from ion signal peak height measurements.

Expected m/z were: linear, 1034.2; cyclized, 1032.2; single sulfhydryl dimer, 2066.5; dual disulfide dimer, 2064.5; linear reacted with pHMB, 1674.3; cyclized with mercury adduct, 1237.

III. Results and Discussion

A total of 97 samples were received for this study, of which 37 were the linear peptide precursors of the peptides cyclized by the post-cleavage procedures. For 33 of the 41 laboratories, >50% of the sample had the correct structure, defined as the desired sequence with no remaining protecting groups. Thus, the assembly and cleavage process for the peptide chosen in the study appears to be routinely achieved in the participating core laboratories.

As noted in the introduction, cyclized peptides are not routinely prepared in core peptide synthesis facilities. Thus, this study tested both the methods as well as the ability of the laboratories to set up these protocols. As seen in Table I, 60 cyclized peptide samples were supplied. Of these, the majority were prepared by protocol I, which did not require the purchase of specialty reagents. Both post-cleavage methods I and II produced the correct cyclized product. However, both of these procedures, particularly method I, yielded dimers in the majority of samples, which reached as high as 70% of the product in one of the samples. The on-resin cyclization with thallium trifluoroacetate (protocol III) produced high yields of cyclized peptide in all samples submitted, with little or no dimer formation. The samples for which either Ekathiox™ resin treatment or potassium ferricyanide oxidation were utilized also appeared to produce the cyclized peptide in good yield, although the sample size was too small to assess. Only a small amount of dimer formation, if any, was observed in these samples.

Table I. Summary of Results from 1995 ABRF Peptide Synthesis Study

Protocol	Total Samples	Samples ≥50% Correct Product	>5% Dimers*
I	29	9	18
II	8	6	4
III	10	10	2
IV	5	0	
ferricyanide	5	5	2
Ekathiox™	3	3	

*as judged by at least 2 of 3 methods.

Table II. Analysis of Selected Peptides

CODE	protocol	nmoles/tube aaa	nmoles/tube abs	% free sulfhydryl spectrophotometric analysis	HPLC % linear	MALDIMS % linear	ESIMS % linear	HPLC %cyclized	MALDIMS % cyclized	ESIMS % cyclized	HPLC % dimer	MALDIMS % dimer	ESIMS %dimer	HPLC % Hg	MALDIMS % Hg	ESIMS % Hg
8808	L	1055			69	59	75	0	6	6	0	6	0	0		
8808	I	1772	1511	3	0	9	0	68	60	60	32	25	20			
8808	II	340	280	0	11	0	0	46	99	50	33	0	40			
8808	III	344	802	0	0	0	0	98	98	95	0	0	0			
8808	IV	1491	1407	33	14	2	30	5	12	0	0	0	0	75	80	65
988	L	178			92	94	90	0	0	0	0	0	0			
988	I	188	127	7	34	7	40	15	66	15	19	7	35			
1609	L	784			0	90	0	0	10	0	0	1	0			
1609	I	757	437	0	0	0	0	0	85	0	0	1	0			
9840	L	721			0	0	0	0	0	0	0	0	0			
9840	I	151	1211	0	0	0	0	0	0	0	0	0	0			
3090	L	75			0	81	0	0	0	0	0	0	0			
3090	I	292	314	0	0	0	0	0	90	0	0	9	0			

Several laboratories prepared cyclized peptide by all 4 protocols requested. Examination of the experimental data for the peptides submitted by laboratory 8808 shows the overall concurrence of the analytical methods used in this study (Table II and Figures 1A-D). The linear peptide sample contained 59-75% of the correct linear product by HPLC, MALDI-MS and ESI-MS analysis, plus a significant amount of incompletely deprotected peptide, also detected by all three methods. A small amount of cyclized product and peptide dimer was observed in this linear sample by MALDI-MS. This analysis points out the utility of performing the MALDI-MS analysis on both the original samples and samples derivatized with pHMB. Reaction with pHMB shifts the mass of the linear product away from that of the cyclized product, effectively enhancing the resolution of the TOF mass spectrometer. The peptide dimer is easily recognized by MALDI-MS with or without derivatization of the peptide. The linear product was used for preparation of cyclized peptides by post-cleavage protocols I and II. The desired product was produced in good yield for either protocol. However, the amount of dimer formation is also very high. The estimate of cyclized peptide from the MALDI-MS data for the protocol II sample is higher than for the other methods. Protocol III produced from 95-98% of cyclized peptide, with no dimeric peptide detected. The sample prepared by protocol IV, the mercuric acetate procedure, showed essentially no correct product. Some linear peptide was still present, but 65-80% was in the form of a mercury adduct. This is easily seen by all analytical methods except one. Analysis of the amount of free sulfhydryl present by Ellman's reagent generally overestimates the amount of cyclized peptide, because both cyclized peptide and peptide dimer are unreactive to this reagent. The mercury adduct is also unreactive to the Ellman's reagent. It should also be noted that the identity of the HPLC peaks was established both by standards prepared for the study and by mass spectrometry. Without this more thorough analysis, the observation of a single HPLC peak could be misleading.

Figure 1. Analysis of Peptide Samples Received from Laboratory 8808 (following pages). The data from the linear sample as well as samples prepared by all 4 oxidation protocols are shown.

A. HPLC analysis
B. MALDI-MS analysis
C. MALDI -MS analysis after reaction with pHMB
D. ESI- MS analysis

For Figures 1A-C, the top panel is the analysis of the linear peptide sample, and in order, peptides prepared by protocols I-IV.

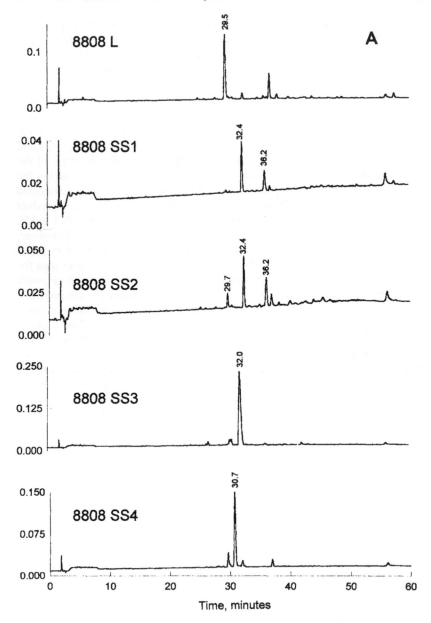

Figure 1A.

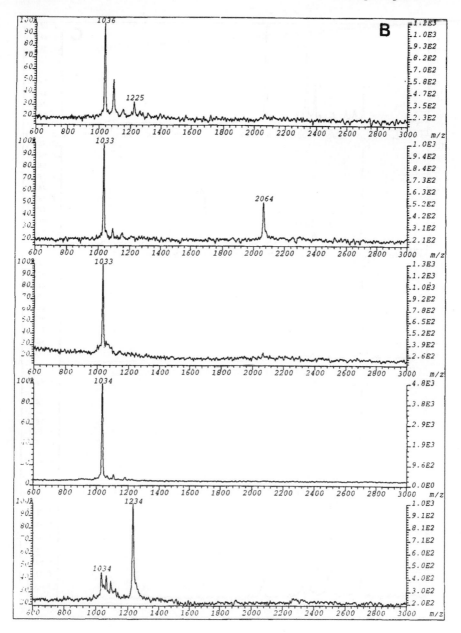

Figure 1B.

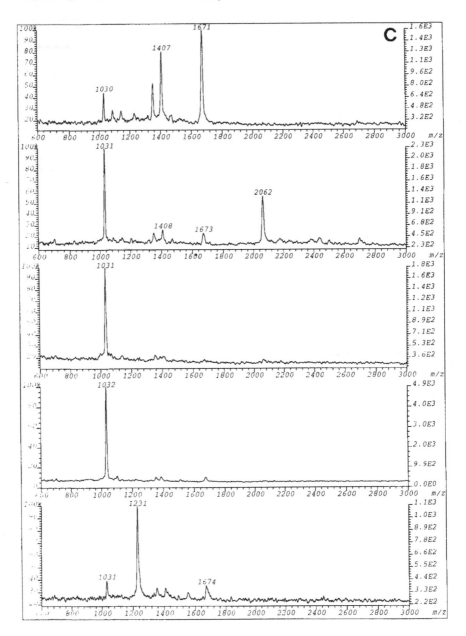

Figure 1C.

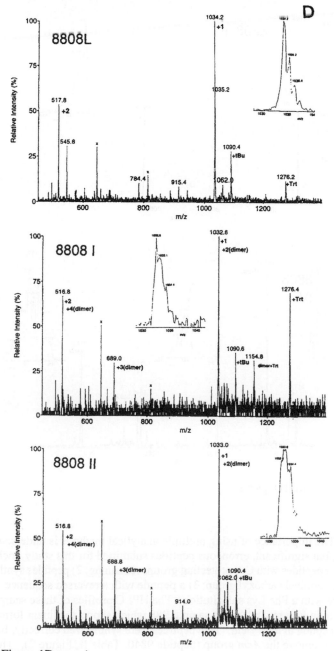

Figure 1D, part 1.

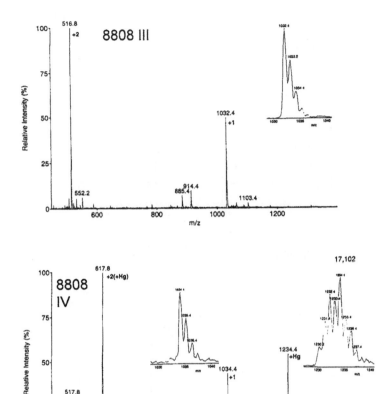

Figure 1D, part 2.

The importance of using multiple analytical methods is underscored by the few, but significant, erroneous peptides submitted for this study, including: 1) correct peptides with Cys protecting groups remaining; 2) peptides synthesized with a C-terminal carboxyl group; 3) a peptide with a reversed sequence; and 4) a peptide with a Phe/Lys substitution. The HPLC profiles of these samples are compared in Figure 2 to that of sample 988 which shows the linear form of the desired peptide in high yield. Two peptides were synthesized correctly, but with failure to remove the *Acm* group (peptide 9840, Table II, Figure 2). Thus, the Ellman's test indicated the presence of cyclized product where none was present.

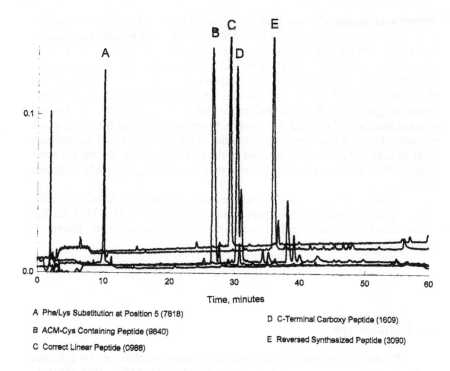

A Phe/Lys Substitution at Position 5 (7818)   D C-Terminal Carboxy Peptide (1609)

B ACM-Cys Containing Peptide (9840)
                                                                      E Reversed Synthesized Peptide (3090)
C Correct Linear Peptide (0988)

Figure 2. HPCL Analysis of Selected Peptides.

There were several peptide samples for which no correct structure had been made. This set included 2 peptides which had been made on a peptide resin that produced the desired sequence with a C-terminal carboxyl group instead of the requested C-terminal amide. The incorrect choice of resin for the synthesis has occurred in previous ABRF studies. This error cannot be detected by AAA, and only with difficulty by linear MALDI-TOF MS. HPLC was able to detect these more acidic peptides (1609), as seen in Figure 2 and Table II, compared to one of the excellent linear peptide products, sample 988. However, it should be noted that this discrepancy was found by HPLC only because this was a controlled study with standard peptides prepared and mass spectrometry of peaks performed. ESI-MS analysis of these peptide samples revealed the difference in 1 amu from the expected mass. One peptide was made in reverse sequence (3090). This discrepancy was particularly difficult to detect (Table II). The amino acid analysis was correct, and the MALDI-MS analysis did not reveal the problem. The retention time of the peptide on reversed phase HPLC was greatly retarded. Edman sequence analysis of the peptide revealed the inverted sequence. ESI-MS showed a difference of 1 amu from the expected mass, indicating only that the peptide had been synthesized on the incorrect resin. The peptide did cyclize. Still another peptide was found to have a low level of Lys by AAA, and a delayed retention time on HPLC (7818). Both MALDI-TOF MS and ESI-MS analysis showed the mass to be incorrect. In the sequence analysis of this peptide, PTH-Phe eluted at cycle 5 instead of PTH-Lys. In addition, the peptide had *t*-butyl protecting groups on the Cys residues instead of *Trt*-protected Cys. On the protocol sheet submitted with this sample, it was indicated that the peptide was synthesized on a multiple peptide synthesizer. It is possible that such instruments must be used with more care because of the greater potential for human error possible than when using an instrument with only a single peptide synthesis capability.

A final interesting sample was the result of a partially failed synthesis. This product contained a deletion peptide missing a residue of Thr. This deletion peptide failed to cyclize, probably because the peptide was too short. However, peptide dimers were formed.

IV. Conclusions

Producing cyclized peptides with the correct structure can be achieved readily by either on-resin or post-cleavage techniques (13). Post-cleavage techniques are less expensive and provide reasonable yields of the desired product. However, on-resin techniques produce greater yields of the final product, but are more expensive to perform. The mercuric acetate method did not yield the correct peptide product, but produces a mercury-peptide adduct which is stable by HPLC and mass spectrometery. This adduct had been noted previously during the synthesis of octreotide (5).

It is best to use a combination of analytical techniques which complement each other in their ability to detect the correct product and unwanted byproducts. The use of Ellman's reagent can record the loss of free sulfhydryl groups, but does not discriminate between sulfhydryls which become oxidized to disulfide bonds in monomers or dimers, mercury adducts, or peptides with incompletely deprotected cysteine residues. HPLC analysis readily separates peptides which are linear, cyclized, dimeric, mercury adducts, incompletely deprotected or with residues deleted. However, unless the peaks are identified by mass spectrometry, one cannot assume that even a single peak is the desired product. Mass spectrometry is very useful in discriminating the peptide products. It is essential to use unit resolution techniques to resolve 1-2 amu differences such as those seen between acids and amides.

Overall, the quality of the peptides produced in this study was excellent. The results of the cyclization tests are encouraging, particularly since many of the participating laboratories may have been performing these protocols for the first time. However, that 5 out of 41 laboratories submitted samples in which the desired peptide had not been made, for reasons which can be attributed only to human error, is a recurring theme in the annual studies performed by the ABRF Peptide Synthesis Research Committee.

Acknowledgments

This study was supported in part by grant from the Department of Energy. The committee also thanks Anthony J. Makusky, Andrew J. Miles, and Edward Nieves for their assistance in carrying out this study.

References

1. A.J. Smith, J..D Young, S.A.. Carr, D.R. Marshak, L.C. Williams & K.R. Williams (1992) Techniques in Protein Chemistry III (ed. R.H. Angeletti): 219-229.

2. G.B. Fields, S.A. Carr, D.R. Marshak, A.J. Smith, J.T. Stults, L.C. Williams, K.R. Williams & J.D. Young (1993) Techniques in Protein Chemistry IV (ed. R.H. Angeletti): 229-238.

3. G.B. Fields, R.H. Angeletti, S.A. Carr, A.J. Smith, J.T. Stults, L.C. Williams & J.D. Young (1994) Techniques in Protein Chemistry V (ed. J.W. Crabb): 501-507.

4. G.B. Fields, R.H. Angeletti, L.F. Bonewald, W.T. Moore, A.J. Smith, J.T. Stults & L.C. Williams (1995)Techniques in Protein Chemistry VI (ed. J.W. Crabb): 539-546

5. B.W. Edwards, C.G. Fields, C.J. Anderson, T.S. Pajeau, M.J. Welch & G.B. Fields (1994) J. Med. Chem. 37, 3749-3757.

6. J. Spackman, W. Stein & S. Moore (1958) Analytical Chemistry 30, 1190-1205.

7. C.H.W. Hirs (1967) Methods in Enzymology 11, 59-62.

8. K. Muramoto & K. Kmiya (1990) Analytical Biochemistry 189, 223-230.

9. J.M. Stewart & J.D. Young (1984) Solid Phase Peptide Synthesis, Second Edition, Pierce Publishing, Rockford IL, p. 116.

10. C.K. Meng, S.F. Wong & C.M. Whitehouse (1989) 246, 64-67.

11. F. Hillenkamp, M. Karas, R.C. Beavis & B.T. Chait (1991) Anal. Chem. 63, 1193A-1203A.

12. E.J. Zaluzec, D.A. Gage & J.T. Watson (1994) JASMS 5, 359-366.

13. D. Andreu, F. Albericio, N.A. Sole, M.C. Munson, M. Ferrer & G. Barany (1994) in Methods in Molecular Biology vol. 35: Peptide Synthesis Protocols (ed., M.W. Pennington & B.M. Dunn) Humana Press, Totowa NJ, pg. 91-169.

The Application of tert-Butylhydroperoxide Oxidation to Study Sites of Potential Methionine Oxidation in a Recombinant Antibody

Felicity J. Shen
May Y. Kwong
Rodney G. Keck
Reed J. Harris

Analytical Chemistry Dept.
Genentech, Inc.
So. San Francisco, CA 94080

I. Introduction

Problems caused by the instability of protein pharmaceuticals have been well-documented, with the instability typically caused by deamidation of Asn residues, isomerization of Asp residues, or by the oxidation of Met or Trp residues [1, 2]. Such instability may limit the product's clinical application or shelf-life if it reduces the specific activity of the molecule or introduces an antigenic determinant.

rhuMAb HER2 is a CDR-grafted (humanized) recombinant IgG$_1$-subclass antibody that binds to the extracellular region of the human epidermal growth factor receptor 2 tyrosine kinase (HER2, also known as *neu* or *C-erbB-2*) [3]. HER2 over-expression correlates with a poor prognosis in a number of cancers [reviewed in reference 4]. *In vitro* administration of the antibody renders HER2-overexpressing cell lines cytostatic and increases their susceptibility to chemo-therapeutic agents [5]. rhuMAb HER2 is now in phase III clinical trials to test its ability to reduce the progression of breast cancer in patients with demonstrated HER2 overexpression.

The only reported instability in rhuMAb HER2 is due to the formation of a relatively stable succinimide at heavy chain Asp-102 [6]. We wanted to test our ability to detect potential Met oxidation sites, and sought to develop a suitable, robust assay. tert-Butylhydroperoxide (tBHP) has been used to selectively oxidize solvent-accessible Met residues (R. G. Keck, submitted). rhuMAb HER2 was incubated with varying levels of tBHP to identify peptides containing the oxidation-susceptible Met residues in the S-carboxymethylated tryptic map. Sensitive detection and reliable quantitation of Met oxidation in rhuMAb HER2 was achieved by hydrophobic interaction chromatography (HIC) after carboxypeptidase B (CpB) and papain digestion.

II. Materials and Methods

Synthetic peptides Met-enkephalin (YGGFMR) and Met(O)-enkephalin were obtained from Bachem (Torrance, CA). Peptide samples (20 μg) were reconstituted with 8M Urea, 0.35 M Tris, 1 mM EDTA at pH 8.3. Dithiothreitol (DTT; Sigma) was added to a final concentration of 10 mM and samples were incubated for 4 hours at 37°C. Iodoacetic acid (Sigma) was added (to 35 mM) and samples were incubated in the dark for 45 minutes at 37°C, then quenched by addition of DTT (to 50 mM). Samples were chromatographed using a Hewlett Packard 1090M HPLC system with a Vydac C18 (4.6 x 250 mm) column operating at 1.0 mL/min. and 35°C. The column was equilibrated with solvent A (0.1%TFA). Upon sample injection, a linear gradient from 0 to 35% solvent B (0.1%TFA in acetonitrile) was generated over 50 minutes.

rhuMAb HER2 samples, purified from transfected Chinese hamster ovary cells, were stored in a buffer containing 5 mM NaOAc, 150 mM NaCl, 0.01% Tween-20, pH 5 at 5 mg/mL. tert-Butylhydroperoxide (tBHP; Sigma, 70%) was added in varying amounts to 0, 0.01, 0.1, 1 and 10% tBHP by volume. Samples stood at room temperature for 20 hours, then were S-carboxymethylated and dialyzed against 10 mM Tris, 100 mM NaOAc, 1 mM $CaCl_2$ at pH 8. TPCK-trypsin (Worthington) was added at a 1:50 (w:w) ratio for a 4 hour digestion at 37°C, after which a 1:100 (w:w) aliquot was added and digestion proceeded for 4 more hours. TFA (Pierce) was then added to bring the samples to 0.1% TFA by volume. Samples were chromatographed using the same instrument and column operating at 1.0 mL/min. and 30°C. The column was equilibrated with solvent A. Upon sample injection, a linear gradient from 0 to 40% solvent B was generated over 80 minutes.

The papain cleavage/HIC method that we employed was a minor modification of the procedures developed by Cacia *et al.* [7]. rhuMAb HER2 samples (250 μg) were diluted from 5 mg/mL to 2 mg/mL by the addition of 0.17M Tris, pH 7.2. DFP-treated carboxypeptidase B (CpB; Boehringer Mannheim) was diluted to 0.5 mg/mL in the same buffer. CpB was added to the samples at a 1:100 weight ratio for a 20 minute digestion at 37°C. The sample was then diluted two-fold with buffer containing 0.1 M Tris, 20 mM cysteine and 4 mM EDTA, pH 7 (the presence of the EDTA inactivates the CpB).

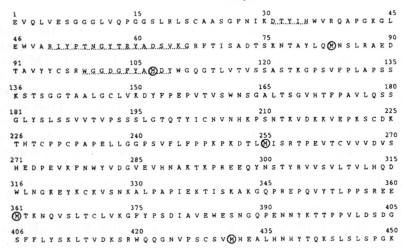

Figure 1. Amino acid sequences of rhuMAb HER2 light and heavy chains. Methionine residues are encircled, and the complementarity-determining regions (CDRs) are underlined [3].

Papain (Boehringer Mannheim) was diluted to 0.25 mg/ml in the same Tris/cysteine/EDTA buffer and then added at a 1:200 weight ratio for a 2 hour incubation at 37°C.

Samples were chromatographed using a HP 1090 system equipped with a TSK Butyl NPR (4.6 x 35 mm) column (TosoHaas) that was operated at 1 mL/min. and 30°C. Solvent A was 2M $(NH_4)_2SO_4$, 20 mM Tris, pH 7, while solvent B contained only 20 mM Tris, pH 7. The column was equilibrated with 10% solvent B. Upon injection of a 5-10 µg sample, a linear gradient from 10 to 100% solvent B was generated over 34 minutes.

III. Results

A potential limitation for detecting methionine sulfoxide [Met(O)] peptides by tryptic mapping is the reported reduction of Met(O) by dithiothreitol (DTT) [8]. Incubation of a Met(O)-containing

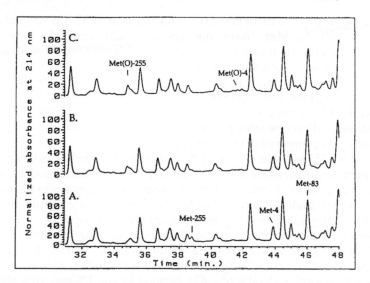

Figure 2. Tryptic map detail, showing peaks containing the Met-4, Met-83 and Met-255 peptides after incubation in A. 0% tBHP (control), B. 1% tBHP and C. 10% tBHP. The sequence of the Met-4 peptide is DIQMTQSPSSLSASVGDR. The sequence of the Met-83 peptide is NTAYLQMNSLR. The sequence of the Met-255 peptide is DTLMISR.

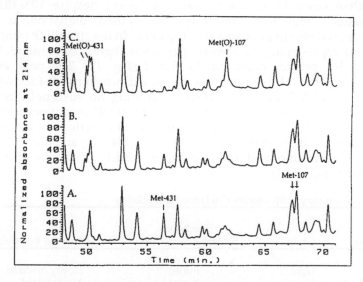

Figure 3. Tryptic map detail, showing peaks containing the Met-107 and Met-431 peptides after incubation in A. 0% tBHP (control), B. 1% tBHP and C. 10% tBHP. The sequence of the Met-107 peptide is WGGDGFYAMDYWGQGTLVTVSSASTK. The sequence of Met-431 peptide is WQQGNVFSCSVMHEALHNHYTQK. Two unrelated peptides coelute with the peptide that contains Met-107; a similar coelution problem exists for the Met(O)-107 peak.

enkephalin peptide in 10 mM DTT for 4 hours at 37°C did not reduce the Met(O) to Met (data not shown), allowing the use of S-carboxymethylation and tryptic mapping to identify potential Met(O) sites. There are six methionines in rhuMAb HER2. Met-255, Met-361 and Met-431 are located in the Fc region, while Met-4, Met-83 and Met-107 are in the Fab region (Fig. 1). Met(O) peptides typically elute by RP-HPLC at earlier retention times than their unoxidized (Met) forms, so methionine sites that are susceptible to oxidation could be identified by comparative tryptic mapping.

Met-255 and Met-431 were the most susceptible to oxidation by tert-butylhydroperoxide (tBHP) treatment, as demonstrated by the reduction in the peak areas of their Met forms and increases in the peak areas of their Met(O) forms at low levels of tBHP. For example, in the 1% tBHP sample the peak representing the Met-255 peptide is reduced in area by 50% and the corresponding peak containing Met(O)-255 peptide has grown by an equal amount (Fig. 2). In the 10% tBHP sample, Met-255 is completely converted to Met(O)-255. However, the peak corresponding to the Met-431 peptide in the 1% and 10% tBHP samples is only reduced in area by 1/3 and 2/3, respectively, and the oxidized form appears as a doublet peak (Fig. 3).

Methionyl peptides containing Met-4 (Fig. 2) and Met-107 (Fig. 3) oxidize more slowly; the Met to Met(O) shift resulted in a loss of less than one-fifth of their total peak areas despite 10% tBHP incubation. Met-361 is unaffected by 1% tBHP oxidation, as no noticeable change in peak area was apparent. In the 10% tBHP sample, the peak representing the Met-361 peptide is slightly diminished; the Met(O)-361 peptide could not be found (Fig. 4). The peak area containing the Met-83 peptide remained unchanged following treatment with 1% and 10% tBHP, indicating no susceptibility to oxidation in this experiment (Fig. 2). The tBHP oxidation/tryptic mapping experiment shows that the location of a methionine strongly influences its ability to be oxidized by tBHP. Met-431 and Met-255 are solvent accessible and oxidize more readily than the buried methionines (Table I).

Table I. Summary of Methionine Oxidation Susceptibility

Residue	Chain	Predicted Solvent Accessibility	Observed Reactivity
Met-4	Light	buried	Slowly oxidized
Met-83	Heavy	buried	Not oxidized
Met-107	Heavy	buried	Slowly oxidized
Met-255	Heavy	exposed	Rapidly oxidized
Met-361	Heavy	buried	Slowly oxidized
Met-431	Heavy	exposed	Rapidly oxidized

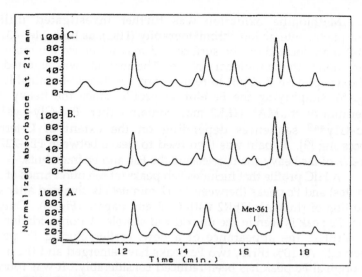

Figure 4. Tryptic map detail, showing the peak containing the Met-361 peptide after incubation in A. 0% tBHP (control), B. 1% tBHP and C. 10% tBHP. The sequence of the Met-361 peptide is EEMTK.

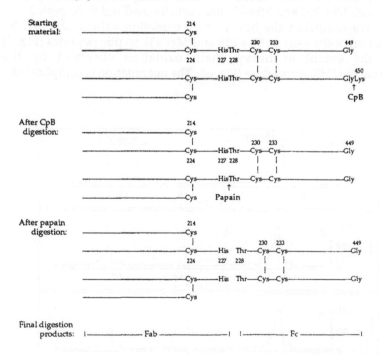

Figure 5. Carboxypeptidase B (CpB) and papain cleavage of rhuMAb HER2. CpB removes the heavy chain C-terminal lysine residue, then papain cleaves the heavy chain between Thr-227 and His-228, releasing the Fab and Fc fragments.

Methionine oxidation was further investigated utilizing hydrophobic interaction chromatography (HIC), as both Met-431 and Met-255 are located on the surface and may influence the molecule's interaction with the stationary phase. The antibody was digested with carboxypeptidase B (CpB) to remove any residual C-terminal lysines (Fig. 5), simplifying the Fc elution profile, since the heavy chain C-termini of rhuMAb HER2 may contain either -Pro-Gly-Lys[450] or -Pro-Gly[449] sequences depending on the extent of C-terminal processing [9]. Papain was then used to cleave between His-227 and Thr-228 of the heavy chain, releasing the Fab and Fc fragments.

A HIC profile that includes Fab peaks (at retention times of 10-13 minutes) and Fc peaks (between 15-17 minutes) is obtained after serial digestion of rhuMAb HER2 with CpB and papain (Fig. 6). Only one major Fc peak is seen in the unoxidized sample. Upon oxidation with 0.01% tBHP, the appearance of a new, earlier eluting Fc peak is observed. At 10% tBHP, two Fc peaks have emerged and the area of the original Fc peak has been reduced considerably. It was found by tryptic mapping that the three Fc peaks shown in Fig. 6C contain fragments with two, one or zero oxidized methionines (data not shown). Met-255 is the site that is predominantly oxidized in Fc peak 1, while both Met-255 and Met-431 are partially oxidized in Fc peak 2.

We compared the heavy chain oxidation values obtained by evaluation of the shift in Met-255 and Met-431 tryptic peptides (Fig. 7) with the extent of heavy chain oxidation observed by the CpB/papain/HIC procedure (Fig. 6). The quantitation is complicated

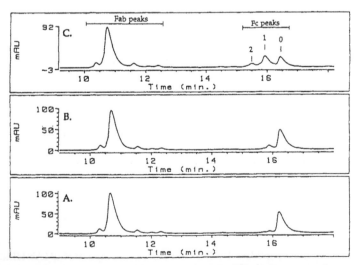

Figure 6. HIC profile, after incubation of rhuMAb HER2 with A. 0% tBHP (control), B. 0.01% tBHP and C. 0.10% tBHP, followed by serial digestion with CpB and papain. The number of Met(O) residues are indicated by the Fc peak labels 2, 1 and 0 as explained in Results.

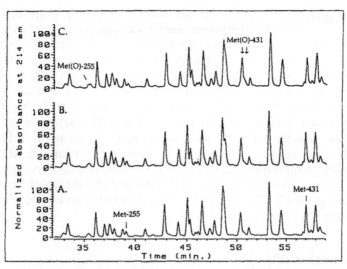

Figure 7. Tryptic map detail, showing peaks containing the Met-255 and Met-431 peptides after incubation in A. 0% tBHP (control), B. 0.01% tBHP and C. 0.10% tBHP. Resolution of the peak pair found at 49 minutes is not consistently achieved; this is not due to oxidation of the sample.

Table IIa. Site Specific Methionine Oxidation After Incubation with tBHP[a]

tBHP	Met-255	Met-431	Unoxidized Heavy Chain Fraction
0 %	0%	0%	$(1.00-0.00)(1.00-0.00) = 1.00$
0.01%	4%	0%	$(1.00-0.04)(1.00-0.00) = 0.96$
0.10%	31%	9%	$(1.00-0.31)(1.00-0.09) = 0.63$

[a]Methionine oxidation levels were assigned by decreases in peak areas of methionyl peptides (Fig. 7). The calculated unoxidized heavy chain fraction is obtained by multiplication of the non-oxidized fractions. Oxidation of Met-431 and Met-255 are assumed to be independent.

Table IIb. Fc Methionine Oxidation Observed After tBHP Oxidation[b]

tBHP	2-Met(O)	1-Met(O)	0-Met(O)	Unoxidized Heavy Chain Fraction
0 %	0%	6.3%	93.7%	$(0.937)^{1/2} = 0.97$
0.01%	trace	13.6%	86.4%	$(0.864)^{1/2} = 0.93$
0.10%	12.4%	46.0%	41.6%	$(0.416)^{1/2} - 0.64$

[b]Methionine oxidation levels were assigned by relative peak areas of Fc fractions observed by HIC analysis after CpB and papain cleavage (Fig. 6) The calculation of the unoxidized heavy chain fraction is explained in Results.

by the fact that Fc fragments are dimers of heavy chain residues 228–449, so complete oxidation at one Fc Met residue will only show as a 50% reduction in peak area in the tryptic map. Quantitation assumes the highest extent of oxidation is not greater than two methionine sites. To obtain a consistent value, we calculated the fraction of heavy chain that is oxidized (or non-oxidized) using the levels of Met(O). Let m represent the fraction of heavy chains with Met(O), then the fraction of 2-Met(O) material is equal to m^2, the 1-Met(O) fraction is equal to $2(m)(1-m)$, and the 0-Met(O) fraction is $(1-m)^2$. As shown in Tables IIa and IIb, close agreement between the extent of oxidation calculated by the tryptic mapping and CpB/papain/HIC methods demonstrates that either can be utilized in measuring the amount of methionine oxidation in rhuMAb HER2. The relative peak areas of the main Fab peak and Fc peaks are not affected by oxidation.

IV. Conclusions

We do not expect that methionine oxidation will become a potency issue for rhuMAb HER2, as only one Met residue is found within a complementarity-determining region, and this residue (Met-107) is not very susceptible to oxidation. No Met residues are found within the Fc$_\gamma$ receptor epitopes, which include heavy chain residues 237–241 (Leu-Leu-Gly-Gly-Pro) and Thr-338 [10–12], nor are any found in the complement C1q epitope, which includes heavy chain residues Glu-321, Lys-323, Lys-325 and Pro-334 [13, 14]. Nonetheless, we wanted to be able to detect Met oxidation if it became a regulatory or stability issue, as it had for the murine monoclonal OKT3 [15].

Several agents have been described that oxidize Met residues, including chloramine T and hydrogen peroxide [2, 16]. We avoided these out of concern that they might also oxidize other residues such as cysteine, tryptophan or histidine, or that they might oxidize methionines to the sulfone. In our experiments using rhuMAb HER2, tBHP was found to selectively oxidize surface-accessible Met residues. New peaks that appeared after oxidation were collected for identification by ESI-MS; only the sulfoxide forms were found. This indicated that tBHP may serve as a model to identify sites likely to oxidize upon extended storage.

We demonstrated that Met(O) residues are not reduced to Met under the conditions employed for *S*-carboxymethylation of rhuMAb HER2. Thus, the sites that are the most susceptible to Met oxidation could be assigned by *S*-carboxymethylation and tryptic mapping after tBHP was added to rhuMAb HER2 at different levels to deliberately oxidize the molecule. The most susceptible Met residues are found in the Fc region of the antibody at heavy chain positions 255 and 431; these are found on the surface of the antibody. Other sites that are slower to oxidize include Met-107 and Met-361, while Met-4 and Met-83 (buried in the variable domains) do not appear to oxidize under the conditions we employed. No other residues (*e.g.*, Trp, Cys, His) were affected by tBHP oxidation.

The tryptic mapping procedure is not trivial; it requires several long incubations and buffer exchanges. Instead, digestion of rhuMAb HER2 with papain (which cleaves the molecule into Fab and Fc fragments) is used to provide rapid, sensitive chromatographic resolution of Fc fragments that have zero, one or two oxidized methionine residues. Some heterogeneity in the Fc region is already present, due to the incomplete processing of C-terminal Lys-450 residues, so we added a carboxypeptidase B digestion step prior to the papain digestion to simplify the Fc profile. Using the combined CpB/papain/HIC method, we can detect Met oxidation in rhuMAb HER2 when only a few percent of the molecule is oxidized. The Fc sites found to be susceptible to oxidation are conserved in most human IgG$_1$-subclass antibodies, so this approach may have general utility.

Acknowledgments

We thank Jerry Cacia and John Frenz for discussions regarding their method of HIC after papain cleavage to resolve antibody Fab and Fc fragments [7]. Use of tBHP to selectively oxidize Met residues was suggested by Robert Garnick. Len Presta and Jeff Cleland provided advice regarding the solvent accessibility of the rhuMAb HER2 Met residues based on the structure determined by Deisenhofer [17].

References

1. "Stability and Characterization of Protein and Peptide Drugs: Case Histories." (Wang, Y.J. and Pearlman, R., eds.) 1993. Plenum Press, New York.
2. Manning, M. C., Patel., K. and Borchardt, R. T. (1989) *Pharm. Res.* **6**, 903–918.
3. Carter, P., Presta, L., Gorman, C. M., Ridgway, J. B. B., Henner, D., Wong, W. L. T., Rowland, A. M., Kotts, C., Carver, M. E. and Shepard, H. M. (1992) *Proc. Natl. Acad. Sci. USA* **89**:4285–4289.
4. Hynes, N. E. and Stern, D. F. (1994) *Biochim. et Biophys. Acta* **1198**:165–184.
5. Fendly, B. M., Winget, M., Hudziak, R. M., Lipari, M. T., Napier, M. A. and Ullrich, A. (1990) *Cancer Res.* **50**:1550–1558.
6. Kwong, M. Y. and Harris, R. J. (1994) *Protein Sci.* **3**:147–149.
7. Cacia, J., Keck, R., Presta, L.G. and Frenz, J. *submitted*.
8. Houghten, R. A. and Li, C. H. (1993) *Methods Enzymol.* **91**, 549–559.
9. Harris, R. J., Murnane, A. A., Utter, S. L., Wagner, K. L., Cox, E. T., Polastri, G. D., Helder, J. C. and Sliwkowski, M. B. (1993) *Bio/Technol.* **11**, 1293–1297.
10. Chappel, M. S., Isenman, D. E., Everett, M., Xu, Y.-Y., Dorrington, K. J. & Klein, M. H. (1991) *Proc. Natl. Acad. Sci. USA* **88**:9036–9040.
11. Chappel, M. S., Isenman, D. E., Oomen, R., Xu, Y. Y. and Klein, M. H. (1993) *J. Biol. Chem.* **268**:25124–25131.
12. Jefferis, R., Lund, J. and Goodall, M. (1995) *Immunol. Lett.* **44**: 111–117.
13. Duncan, A. R. and Winter, G. (1988) *Nature* **332**: 738–740.
14. Xu, Y., Oomen, R. and Klein, M. H. (1994) *J. Biol. Chem.* **269**:3469–3474.
15. Kroon, D. J., Balwin-Ferro, A. and Lalan, P. (1992) *Pharm. Res.* **9**:1386–1393.
16. Brot, N. and Weissbach, H. (1983) *Arch. Biochem. Biophys.* **223**, 271–281.
17. Deisenhofer, J. (1981) *Biochemistry* **20**:2361–2370.

SECTION VI

Methods Used in Primary
Structural Analysis

ENZYMATIC DIGESTION OF PVDF-BOUND PROTEINS: A SURVEY OF SIXTEEN NON-IONIC DETERGENTS

Michele Kirchner, Joseph Fernandez, Quazi Aga Shakey,
Farzin Gharahdaghi, and Sheenah M. Mische

The Rockefeller University Protein/DNA Technology Center
1230 York Ave, New York, NY   10021

INTRODUCTION

The use of multiple approaches for the primary characterization and possible identification of a protein has become the norm rather than the exception in the modern protein chemistry core facility.   With the average sample size for any protein being 1-5 μg, the techniques employed need to be efficient, sample conservative and rapid.   SDS-PAGE or 2D gel electrophoresis followed by electrotransfer to PVDF membrane is the primary method for preparing proteins for direct primary structural characterization (1-3).   PVDF is preferred over nitrocellulose because it can be used for other types of analysis, including amino acid analysis, sequence analysis, and mass spectrometry (4-5), and can be stored for an indefinite period of time.

In the event that a protein has a blocked amino terminus, or noncontiguous stretches of amino acid sequence are required for DNA cloning or confirmation of protein identification, enzymatic digestion of protein contained on PVDF or in gel has become the standard method for obtaining internal protein sequence data (6-14). It has become routine in our laboratory to perform N-terminal sequence analysis (20%), enzymatic digestion for internal peptide sequencing (78-80%), and mass spectrometric analysis of the digestion mixture ($\leq$ 2%) on any one PVDF-bound sample.   In addition, selected HPLC peak fractions are analyzed by Matrix-Assisted-Laser-Desorption-Ionization Time-Of-Flight Mass Spectrometry (MALDI-TOF MS) to assess purity, size, and to assist in determining peptide sequence.

Successful enzymatic digestion of PVDF-bound protein requires the presence of non-ionic detergents in the digestion buffer such as 1% hydrogenated Triton X-100 (RTX-100)/10% acetonitrile/100 mM Tris-HCl, pH 8.0 (6, 8-9). RTX-100 was initially used due to its low chromatographic background, high peptide recovery, and lack of interference to Edman chemistry (9). However, direct mass analysis of peptide/detergent mixtures by MALDI-TOF or ESI MS could not be obtained due to the presence of detergent related adducts signals and peptide signal suppression (7, 11, 15-16). As a result, peptide/detergent mixtures obtained from enzymatic digestion of PVDF-bound proteins could not be used for quick screening and tentative identification of proteins via peptide mass database analysis search (17-18).   In the search for optimizing peptide recoveries and mass spectrometric analysis, we undertook a systematic exploration of the non-ionic detergent types that were commercially available.   Sixteen detergents were chosen for evaluation and the comparative studies employed are presented here.

MATERIALS AND METHODS

Chemicals and Reagents.   Endoproteinase Lys-C (Lys-C) was from Wako. Trypsin (sequencing grade, modified) was obtained from Promega. All the detergents studied were of the highest grade available and were obtained from Calbiochem (see TABLE I) in solid form except for heptyl thioglucopyranoside, Genapol X-080, Pluronic F-127, hydrogenated Triton X-100, Tween 20 and Tween 80 which were purchased as a 10% aqueous solution. Acetonitrile was from Fischer. Alpha-cyano-4-hydroxy-cinnamic acid was from Sigma. All other chemicals and reagents are as previously described (5-9).

Enzymatic Digestion of PVDF-bound protein.   Standard proteins (4μg) were analyzed by SDS-PAGE, transferred to PVDF (Immobilon p^{sq}), stained with amido black and enzymatically digested as previously described (6, 8). Briefly, the excised protein bands were cut into 1x1 mm pieces and 50 μl of either 1% detergent/10%acetonitrile/100mM   Tris-HCl,   pH   8.0   or   1%   detergent/ 10%acetonitrile/25mM NH$_4$HC0$_3$, pH 8.9 was added.   Detergents used are described in **TABLE I**.   Samples were reduced by adding 5 μl of 45mM DTT at 55°C for 30 minutes and carboxyamidomethylated with 5 μl 100mM iodoacetamide at room temperature in the dark for 30 minutes (19). Four microliters of endoproteinase Lys-C (0.1 μg/μl) or 2 μl of trypsin (0.25 μg/ul) was added followed by incubation at 37°C for 24 hours. The supernatant was transferred to an Hewlett-Packard glass vial, and the peptides were extracted with 50 μl of fresh buffer (1% detergent/10% acetonitrile/100 mM Tris-HCl, pH 8.0) followed by 100 μl of 0.1% TFA. The supernatants were pooled and if not immediately analyzed by HPLC, the digestion was stopped with the addition of 1% DFP/ethanol (v/v) and the vial stored at -20°C.

Microbore HPLC isolation of peptides.   The peptides were separated on a VYDAC C$_{18}$ column (2.1 X 250mm) using a Hewlett- Packard 1090 HPLC as described elsewhere (9). The short gradient was 1.6%-100% B (0-60 min) with a flow rate of 300μl/min.   The column was then washed with 100% B for 5 minutes at 300μl/min and equilibrated at 1.6% B for 25 minutes at 300μl/minute.   The long gradient was 1.6-29.6% B (0-63min), 29.6%-60% B (63-95 min), 60%-80% B(95-105 min) with a flow rate of 150μl/minute. The column was then washed at 80% B (12min) at 150μl/min and equilibrated at 1.6% B for 41 minutes at 300μl/minute.   Solvent A was 0.1% TFA and solvent B was 0.08% TFA in acetonitrile. Peptide elution was monitored at 220nm and 280nm. For **Figure 2**, fractions were collected during the long gradient every 0.5 min and stored at -20°C until sequence analysis was performed.

MALDI-TOF Mass Spectrometry.   One microliter of peptide/detergent mixture or purified peptide was mixed with 1μl of bradykinin (50 fmol) and 2 μl of alpha-cyano-4-hydroxy-cinnamic   acid   made   as   a   saturated   solution   in   50% acetonitrile/0.1% TFA as previously described (16). This was then deposited on a sample plate and analyzed on a Perseptive Biosystems Vestec Laser Tec Bench Top II System as described elsewhere (7, 16).

Amino Terminal Sequence Analysis.   Purified peptides were analyzed on a Hewlett-Packard G-1000A as previously described using version 3.0 chemistry (20). An internal sequencing standard test peptide (15 pmol), obtained from   the HHMI Biopolymers Facility at Yale University, was sequenced along with each peptide in **Table III** (21).

TABLE I. Structure and Classification of 16 Non-ionic Detergents

DETERGENT	DETERGENT TYPE	Molecular Weight	STRUCTURE
Decyl glucopyranoside	Alkyl glucoside	322	X - 9
Octyl glucopyranoside (OGP)	Alkyl glucoside	292	X - 7
Decyl maltopyranoside	Alkyl maltoside	483	X - 9
Dodecyl maltopyranoside	Alkyl maltoside	511	X - 11
Heptyl thioglucopyranoside	Alkyl thioglucoside	274	X - 6
Octyl thioglucopyranoside	Alkyl thioglucoside	308	X - 7
Big Chap	Big Chap	878	X - OH
Deoxy Big Chap	Big Chap	862	X - H
MEGA 8	Glucamide	322	X - 6
MEGA 9	Glucamide	336	X - 7
MEGA 10[a]	Glucamide	350	X - 8
GENAPOL X-080	Polyoxyethlene	553	$CH_3(CH_2)_y - O(CH_2CH_2O)_x - H$ X - 8, Y - 12
PLURONIC F-127	Polyoxyethlene	12,600	$HO(CH_2CH_2O)_x - (CH - CH_2O)_y - (CH_2CH_2O)_z - H$ X - 98, Y - 67, Z - 98
Triton X-100, hydrogenated (RTX-100)	Polyoxyethlene	631	$O(CH_2CH_2O)_x - H$ X - 9-10
Tween 20	Polyoxyethlene	1228	W + X + Y + Z = 20 R = $C_{11}H_{23}CO_2$- (laurate)
Tween 80	Polyoxyethlene	1310	W + X + Y + Z = 20 R = $C_{17}H_{33}CO_2$- (oleate)

a. MEGA 10 was not soluble as a 1% solution using either buffer.

Portions of this Table were reproduced from "A Guide to the Properties and Uses of Detergents in Biology and Biochemistry" with permission from Calbiochem.

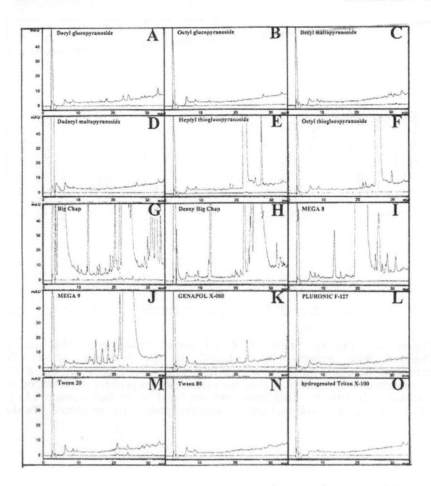

Figure 1. Microbore HPLC analysis of 15 non-ionic detergents. Detergents were prepared as 1% solutions using 10% acetonitrile/100 mM Tris-HCl, pH 8.0. One hundred microliters of the detergent solution was mixed with 100 μl of 0.1% TFA and a total of 200 μl was analyzed by HPLC as described in Materials and Methods using the short gradient. Absorbance was monitored at 220 nm (top line) and 280 nm (bottom line). Detergents used are listed in each panel (A-O). MEGA 10 from **TABLE I** was not analyzed since it was not soluble in the digestion buffer listed above.

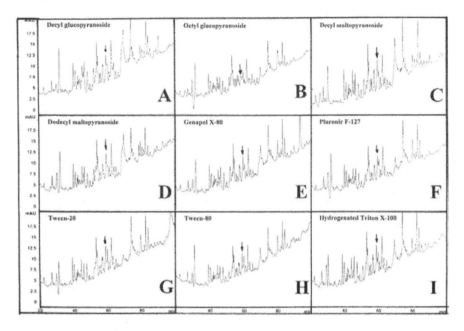

Figure 2. Peptide maps of PVDF-bound transferrin (50 pmol) digested with endoproteinase Lys-C in the presence of a 1% detergent solution in 10% acetonitrile/100 mM Tris-HCl, pH 8.0. Transferrin digestion and HPLC conditions using the long gradient are described in Materials and Methods. Detergents used are listed in the figure panels (A-I). HPLC fractions were collected and the indicated peaks were sequenced (**Table III**).

RESULTS AND DISCUSSION

Non-ionic detergents such as hydrogenated Triton X-100, decyl glucopyranoside and octyl glucopyranoside have been shown to be essential for successful enzymatic digestion and peptide recovery from polyvinylidene difluoride (PVDF) bound proteins (5-9). Other types of non-ionic detergents vary in their hydrophilic head, hydrophobic tail, and molecular weight as summarized in **TABLE I**. To determine which of these detergents (**TABLE I**) could be used to obtain internal sequence data from PVDF-bound proteins the following criteria was established: a) solubility as 1% detergent solutions in digestion buffers, b) chromatographic performance at 220nm and 280nm, c) peptide mapping by HPLC (peak pattern), d) peptide recovery (peak yields), e) MALDI-TOF MS analysis of peptide/detergent mixtures, f) MALDI-TOF MS analysis of HPLC purified peptides, and g) microsequence analysis. A total of sixteen detergents were surveyed and all were soluble as 1% solutions except for MEGA 10. Therefore MEGA 10 was not analyzed further.

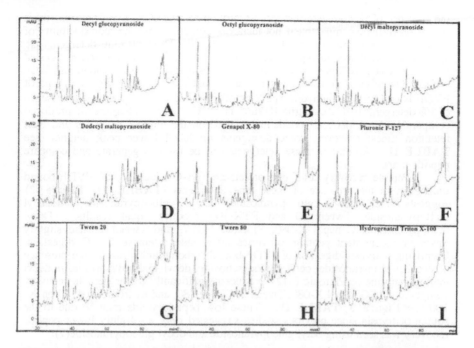

Figure 3. Peptide maps of PVDF-bound carbonic anhydrase (138 pm) digested with trypsin in the presence of 1% detergent/10% acetonitrile/100 mM Tris-HCl, pH 8.0. Trypsin digestion and HPLC conditions using the long gradient are described in Materials and Methods. Detergents used are listed in the figure panels (A-I).

The chromatographic performance of these fifteen detergents at 220nm and 280nm was evaluated by microbore HPLC analysis of 1% detergent solutions in 10% acetonitrile/100mM Tris-HCl, pH 8.0 (**Figure 1**). Clearly, heptyl thioglucopyranoside, octyl thioglucopyranoside, Big Chap, Deoxy Big Chap, MEGA 8, and MEGA 9 produced unacceptable high uv absorbing contaminant peaks at 220nm which would interfere with peptide identification during HPLC. No further studies were done with these detergents. Decyl glucopyranoside, Genapol X-080 and Tween 20 generated some minor artifact peaks at 220nm which would have minimal to no effect on peptide identification. Tween 20 also produced a few artifact peaks at 280nm. Octyl glucopyranoside, decyl maltopyranoside, dodecyl maltopyranoside, Pluronic F-127, Tween 80 and hydrogenated, Triton X-100 gave no extraneous artifact peaks at 220 and 280nm and are therefore ideal for HPLC analysis. Similar results were generated for each detergent as a 1% solution in 10% acetonitrile/25mM NH4HCO3, pH 8.9 (data not shown). Only these nine detergents that contained minimal or no contaminants were deemed useful for further analysis.

HPLC analysis of PVDF-bound transferrin (50 pmol) digested with endoproteinase Lys-C in the presence of 1% detergent/10% acetonitrile/100 mM Tris-HCl, pH 8.0 (**Figure 2**) shows that there is no loss of enzyme activity due to the presence of detergents or enzyme adsorption to PVDF. As can be seen, these HPLC profiles are comparable if not identical, demonstrating the reproducibility of the digestion procedure for the detergents analyzed. Thus, all nine detergents are applicable to peptide mapping. Peptide maps of PVDF-bound transferrin digested with endoproteinase Lys-C in the presence of 1% detergent/10% acetonitrile/25mM NH_4HCO_3, pH8.9 produced similar results (data not shown). Peptide maps of PVDF-bound carbonic anhydrase (138 pmol) digested with trypsin in the presence of 1% detergent/ 10% acetonitrile/100mM Tris-HCl, pH 8.0 (**Figure 3**) produced comparable if not identical peptide maps based on the number of peaks and their retention times; however, some detergents produced lower peak heights (see **TABLE II**). All nine of these detergents can be used to generate and compare peptide maps.

Peptide recovery after endoproteinase Lys-C digestion of PVDF-bound transferrin in the presence of these nine detergents (**TABLE II, Figure 2**) suggests that decyl glucopyranoside, octyl glucopyranoside, dodecyl maltopyranoside, Tween 80 and RTX-100 produce equal results. Decyl maltopyranoside, Genapol X-080, Pluronic F-127, and Tween 20 had slightly lower or inconsistent peptide recoveries after endoproteinase Lys-C digestion. Surprisingly, trypsin digestion of PVDF-bound carbonic anhydrase in the presence of decyl glucopyranoside, octyl glucopyranoside, decyl maltopyranoside, dodecyl maltopyranoside or Pluronic F-127 produced significantly lower peptide recovery compared with Genapol X-080, Tween 20, Tween 80, and hydrogenated Triton X-100 (see **Figure 3, TABLE II**). These low peptide yields may be due to the source of trypsin since previous studies suggested only a slightly lower peptide recovery was seen using octyl glucopyranoside and decyl glucopyranoside(7).

TABLE II. Peptide Recovery after Enzymatic Digestion of PVDF-Bound Protein in the Presence of Nine Non-ionic Detergents[a]

Detergent	Human Transferrin				Carbonic Anhydrase	
	Tris-HCl, pH 8.0		NH4HCO3, pH 8.9		Tris-HCl, pH 8.0	
	Peptide[b] Recovery (%)	n[c]	Peptide[b] Recovery (%)	n[c]	Peptide[b] Recovery (%)	n[c]
Decyl glucopyranoside	48.1 ± 3.5	4	38.9	1	37.9 ± 6.2	1
Octyl glucopyranoside	45.4 ± 3.4	4	48.9	1	28.4 ± 8.0	1
Decyl maltopyranoside	42.6 ± 2.0	4	49.6	1	35.2±13.7	1
Dodecyl maltopyranoside	45.5 ± 3.0	4	38.5	1	33.9 ± 5.2	1
Genapol X-080	43.9 ± 3.5	4	58.7	1	50.2±13.9	1
Pluronic F-127	47.1±10.2	3	46.5	1	33.1 ± 3.3	1
Tween 20	38.3 ± 5.5	4	36.3	1	44.2 ± 4.8	1
Tween 80	48.8 ± 5.4	4	33.7	1	42.3±10.8	1
hydrogenated Triton X-100	46.9 ± 2.5	4	39.6	1	42.0 ± 8.8	1

a. PVDF-bound transferrin (50pmol) was digested with endoproteinase Lys-C and carbonic anhydrase (138 pmol) was digested with trypsin using the appropriate detergent or buffer as described in Materials and Methods.

b. Peptide recovery is based on the total peak height of selected peptides (10-12) compared to the same protein digested in solution (4 μg).

c. Number of protein bands digested.

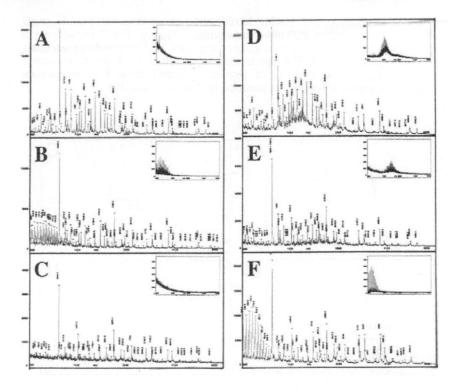

Figure 4. **MALDI-TOF Mass Spectrometric analysis of peptide/detergent mixtures after endoproteinase Lys-C digestion of PVDF-bound transferrin (50 pmol) in the presence of various detergents.** Mass spectrometric analysis and enzymatic digestion were performed as described in Materials and Methods. Approximately 125-250 fmol of sample was analyzed by MALDI-TOF mass spectrometry. Decyl glucopyranoside, octyl glucopyranoside, and dodecyl maltopyranoside produced comparable results to decyl maltopyranoside (A) and are therefore not shown. Other detergents are as follows: B) Genapol X-080, C) Pluronic F-127, D) Tween-20, E) Tween-80, and F) hydrogenated Triton X-100. Bradykinin (1060.1 Da) was used as an internal calibrant. Inset spectra represents mass spectrometric analysis of the blank digestion.

MALDI-TOF MS analysis of peptide/detergent mixtures after endoproteinase Lys-C digestion of PVDF-bound transferrin in the presence of six detergents is shown in **Figure 4**. MALDI-TOF MS was also done on blank digestions and performed exactly as the transferrin band but minus the protein (see **Figure 4** inset spectra). As can be seen from the blanks, detergent clusters are a persistent problem with Genapol X-080 (500-1300 Da),Tween 20 (1200-2000 Da), Tween 80 (1500-2400 Da), and hydrogenated Triton X-100 (500-1200 Da). These clusters interfere with peptide identification and suppress peptide signal resulting in less useful information and making identification of proteins via peptide mass database analysis searches difficult to impossible.   Pluronic F-127 shows only a slight signal suppression.   The two glucopyranosides, and two maltopyranosides produced superior MALDI-TOF spectra due to the absence of detergent related signals and high signal intensity of peptides.   Clearly these four pyranoside detergents are preferred for peptide mass database analysis searches.

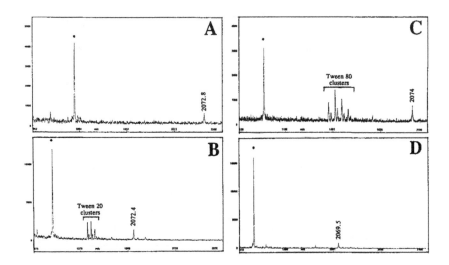

Figure 5.   MALDI-TOF MS analysis of HPLC purified peptides generated from peptide maps of transferrin digestions shown in Figure 2.   Detergents shown are octyl glucopyranoside (A), Tween 20 (B), Tween 80 (C), and hydrogenated Triton X-100 (D), while other detergents produced results similar to A and D (data not shown).   MALDI-TOF MS was performed using an internal calibrant (asterisk labeled signal) as described in Materials and Methods.   Edman sequence analysis (**TABLE III**) of all peptides produced the same sequence.   The Tween 20 clusters indicated in B and C are not additional peptides (see **TABLE III**) but are directly attributable to the detergent.

To perform Edman sequence analysis of HPLC purified peptides, fractions of the peptide maps shown in **Figure 2** were collected.   MALDI-TOF MS analysis of HPLC purified peptides is a sensitive and fast method for screening peptides prior to sequence analysis to obtain valuable information such as peak purity, peptide length, and assistance in identifying peptide sequence.   Three peptides from each HPLC profile were analyzed by MALDI-TOF MS and representative spectra for one peptide is shown in **Figure 5** with the observed masses listed in **TABLE III**.   A single signal was observed with all detergents except Tween 20 (**Figure 5B**) and Tween 80 (**Figure 5C**) which were contaminated with additional detergent clusters.   The two additional HPLC purified peptides that were not sequenced produced the same contamination problem with Tween 20 and Tween 80.   These additional detergent clusters, present even after HPLC purification, compromise the ability to screen peptides by MALDI-TOF prior to sequence analysis.

TABLE III. Sequence Analysis of HPLC Purified Peptides after Endoproteinase Lys-C Digestion of PVDF-Bound Transferrin in the Presence of Nine Non-ionic Detergents

Detergent	Observed Mass[a] (Da)	SER[b] (pm)	ASP (pm)	ASN (pm)	CYS (pm)	GLU (pm)	ASP (pm)	IY[c] nLEU (pm)	RY[c] nLEU (%)
Decyl glucopyranoside	2073.4	3.3	9.5	6.8	2.5	6.9	7.3	10.7	94.8
Octyl glucopyranoside	2072.8	3.9	8.4	6.8	3.6	7.0	7.3	11.9	96.6
Decyl maltopyranoside	2074.0	3.1	7.0	6.3	3.0	5.7	6.6	9.9	97.4
Dodecyl maltopyranoside	2071.5	4.6	9.3	8.0	3.9	5.1	7.8	11.6	94.2
Genapol X-080	2071.2	3.7	9.4	7.7	4.1	7.0	8.6	11.8	96.0
Pluronic F-127	2072.9	3.6	8.7	7.6	3.6	6.8	7.7	10.9	96.2
Tween 20	2072.4	2.8	5.9	4.5	2.1	4.5	4.1	11.4	95.4
Tween 80	2074.0	2.9	5.7	4.8	2.6	3.9	4.2	11.4	96.8
hydrogenated Triton X-100	2069.5	3.5	7.3	6.6	3.5	6.0	6.2	15.2	94.6

a.   One percent of purified peptide was removed and analyzed by MALDI-TOF prior to sequence analysis. The theoretical mass of the peptide is 2072.2 Da.
b.   The sequence is SDNCEDTPEAGYFAVAVVK. Pmol values are background corrected. Cysteine was derivatized to carboxyamidomethylcysteine (CAMC) after transfer to PVDF but prior to digestion as described in Materials and Methods.
c.   Performance of an internal sequencing standard (14 pm) used to evaluate the instrument performance (21). Repetitive yield of nLEU is calculated from positions 1 and 6. Residues 2-5 are succinyl-LYS. The higher nLEU initial yield in the hydrogenated Triton X-100 run was due to a pipetting error.

Mirosequence analysis of the purified peptides shown in **Figure 2** are presented in **TABLE III**. As can be seen, the same peptide sequence was obtained. No additional sequences were observed for the Tween 20 and Tween 80 peptides demonstrating that only one peptide was present in the HPLC fractions and further suggesting that the extra signal in **Figure 5B and 5C** are detergent related. The initial yields (ASP in cycle #2) were all comparable (average 8.5 ± 1.0 pm) except for Tween 20 and Tween 80 (average 5.8 ± 0.1 pmol) which were approximately 30% lower. It should be noted that CYS at residue 4 was identified as CAMC and that these lower yields for all detergents, approximately 50% of ASN in cycle #3, are attributed to the Hewlett-Packard sequencer and not the peptide (19). An internal sequencing standard was loaded at the same time as the peptide to evaluate the instrument's performance and the data is shown in the last two columns (21). The initial and repetitive yields of this internal sequencing standard are extremely consistent (average 11.2 ± 0.6 pmol, and 95.9 ± 1.1 pmol respectively) demonstrating that the lower peptide initial yields for Tween 20 and Tween 80 are significant and not due to instrument variation.

CONCLUSIONS

Non-ionic detergents heptyl thioglucopyranoside, octyl thioglucopyranoside Big Chap, Deoxy Big Chap, MEGA 8, and MEGA 9 produced unacceptably high uv absorbing contaminants that would interfere with indentification of peptides during HPLC purification. The remaining nine detergents produced either a few minor or no artifact peaks and therefore are applicable for peptide mapping by HPLC. These remaining nine detergents were further evaluated by MALDI-TOF, HPLC, and Edman sequence analysis.

When MALDI-TOF MS is not a consideration, all nine detergents produce satisfactory peptide maps by HPLC. Endoproteinase Lys-C digestions resulted in comparable peptide recoveries except for decyl maltopyranoside, Genapol X-080, and Tween 20 which produced only slightly lower yields. Peptide recoveries after trypsin digestion in the presence of decyl glucopyranoside, octyl glucopyranoside, decyl maltopyranoside, dodecyl glucopyranoside, and Pluronic F-127 are significantly lower than that of the other detergents. Edman sequence analysis of HPLC purified peptides indicate that there are approximately 30% lower initial yields when Tween 20 or Tween 80 is used. Consistent performance of the internal sequencing standard suggests that these lower initial yields are significant and are not attributed to inconsistent instrument performance.

When MALDI-TOF MS is available for additional sample characterization, all nine detergents produce satisfactory peptide maps by HPLC but not by mass spectrometric analysis. Octyl glucopyranoside, decyl glucopyranoside, decyl maltopyranoside, and dodecyl maltopyranoside all produce superior mass spectra due to the absence of detergent related signal that could suppress peptide response. While peptide/Pluronic F-127 mixtures produced no detergent related signals, the peptide signal response did appear lower than the four pyranoside detergents. Genapol X-080 and hydrogenated Triton X-100 produced detergent related signals with masses less than 1000 Da while Tween 20 and Tween 80 produced an unacceptably high number of detergent related signals that not only suppress peptide signal, but interfere with peptide identification in the 1000-2500 Da region. In fact, after digestion in the presence of Tween 20 and Tween 80, artifact signals were observed by MALDI-TOF MS even after HPLC purification of peptides.

Analysis of the primary structure of proteins is much more powerful with the inclusion of MALDI-TOF MS data; however, this may not be available to all laboratories due to the high cost of the instrumentation. Therefore, our recommendations are as follows: A) when only HPLC and Edman sequence analysis are to be employed Genapol X-080, Pluronic F-127, and hydrogenated Triton X-100 appear to produce a higher peptide recovery while obtaining good sequencing initial yields; B) when incorporating MALDI-TOF MS into the characterization scheme, the two glucopyranoside and two maltopyranoside detergents appear good choices due to the strong signal response and absence of detergent clusters; C)Tween 20 and Tween 80 are not recommended due to lower sequencing initial yields and the detergent's interference during MALDI-TOF MS analysis.

Acknowledgments

The authors wish to thank Jeffrey C. Mathers for computer programs which assisted in calculating the peptide molecular masses from the theoretical digestion of human transferrin.

References

1. Laemmli, U.K., (1970) *Nature*, **227**, 680-685.
2. O'Farrell, P., (1975) *J. Biol. Chem* **250**, 4007-4021.
3. Matsudaira, P., (1987) *J. Biol. Chem.*, **21**, 10035-10038.
4. Gharahdaghi, F., Atherton, D., DeMott, M., and Mische, S.M. (1992) *in* Techniques in Protein Chemistry III (Angeletti, R.H., Ed.) pp 249-260, Academic Press, New York.
5. Altherton, D., Fernandez, J., DeMott, M., Andrews, L., and Mische, S.M., (1993) *in* Techniques in Protein Chemistry IV (Angeletti, R.H., Ed.) pp 409-418, Academic Press, New York.
6. Fernandez, J., Andrews, L., and Mische S.M., (1994) *Anal Biochem* **218**, 112-117.
7. Fernandez, J., Gharahdaghi, F., and Mische, S.M., (1995) *in* Techniques in Protein Chemistry VI (Crabb,J., Ed) pp 135-142, Academic Press, New York.
8. Fernandez,J., Andrews, L., and Mische, S.M., (1994) *in* Techniques in Protein Chemistry V (Crabb, J., Ed.) pp 215-222, Academic Press, New York.
9. Fernandez, J., DeMott, M., Atherton, D., and Mische,S.M., (1992) *Anal. Biochem.* **201**, 255-264.
10. Erdjument-Bromage, H., Lui, M., Sabatini, D., Snyder, S.H., and Tempst, P., (1994) *Protein Science* **3**:2435-2446.
11. Merewether, L. A., Clogston, C., Patterson, S., Lu, H., (1995) *in* Techniques in Protein Chemistry VI. (Crabb, J. Ed) pp. 153-160, Acedemic Press, New York.
12. Best, S., Reim, D.F., Mozdzanowski, J., and Speicher, D., (1994) *in* Techniques in Protein Chemistry V (Crabb, J., Ed) pp 205-213, Academic Press, New York.
13. Zhou, S., and Admon, A., (1995) *in* Techniques in Protein Chemistry VI (Crabb, J., Ed.) pp 161-167, Academic Press, New York.
14. Rosenfeld J., Capdeville, J., Guillmot, J., and Ferrara, P. (1992) *Anal. Biochem.* **203**, 173-179.
15. Vorm, O., Chait, B.T., and Roepstorff, P., (1993) Proc. 41st ASMS Conf., 654-655.
16. Gharahdaghi, F., Fernandez, J., and Mische, S.M., (1995) Proc. 43rd ASMS conf. *in press.*
17. Henzel, W., Billeci, T.M., Stults, J.T., Wong, S.C., Grimley, C., and Watanabe, C., (1993) *P.N.A.S., U.S.A.* **90**, 5011-5015.
18. Pappin, D.J.C, Hojrup, P., and Bleasby, A.J., (1993) *Current Biology* **3**, 327-332.
19. Fernandez, J., Agashakey, Q., Andrews, L., Gharahdaghi, F., Kirchner, M., Mathers, J.C., and Mische, S.M., (1994) *Protein Science* **3** (**suppl.1**), 77.
20. Hewlett-Packard User's Guide for the HP G1000S Protein Sequencing System.
21. Elliot, J.I., Stone, K.L., and Williams, K.R., (1993) *Anal. Biochem.* **211**, 94-101.

Development and Optimization Of A SE-HPLC Method For Proteins using Organic Mobile Phases

Michael R. Schlittler, Barbara A. Foy, James J. Triska,
Bernard N. Violand, and Gerald L. Bachman
Monsanto Company, St. Louis MO 63198

I. INTRODUCTION

Recombinant proteins expressed in *E. coli* such as bovine somatotropin (bST), porcine somatotropin (pST) and Tissue Factor Pathway Inhibitor (TFPI) form oligomers during their isolation. Size exclusion HPLC (SE-HPLC) is the method of choice to monitor removal of these oligomers during purification of these proteins. Most SE-HPLC methods for proteins have used aqueous mobile phases containing salts or chaotropic agents such as sodium dodecyl sulfate (SDS) or urea (1,2,7). Hydrophobic and ion-exchange interactions occurring between the protein and the chromatography resin with these mobile phases may result in non-ideal size exclusion chromatography (1,3). Limited sample solubility and non-covalent oligomer formation may also occur with these mobile phases resulting in inconsistent performance of the method (6). There are several examples where mobile phases containing acetonitrile and trifluoroacetic acid (TFA) have been shown to overcome non-ideal SE-HPLC of proteins and peptides (1,2,8). This paper describes the development and optimization of an organic mobile phase SE-HPLC method for bST, pST and TFPI.

II. MATERIALS AND METHODS

Burdick and Jackson HPLC grade acetonitrile and Pierce sequanal grade TFA were obtained from Baxter Products. HPLC grade phosphoric acid and sodium phosphate were obtained from Fisher Scientific and SDS was obtained from BioRad. SE-HPLC was performed with either a BioRad BioSil 250 (30 x 0.78 cm), TosoHaas TSK 3000SW$_{XL}$ (30 x 0.75 cm) or a DuPont GF250 column (30 x 0.78 cm) using a Perkin-Elmer Series 4 HPLC and ISS-100 autosampler. The flow rate was 1 ml/min and the column eluate was monitored with a Perkin-Elmer LC-95 detector at 278 nm. Data were collected and analyzed with a PE-Nelson Turbochrom data system. Bovine and porcine somatotropin were purified as previously described (4). TFPI was purified according to reference 5. Proteins were solubilized at 1 mg/ml in 40 % acetonitrile, 80 mM TFA and twenty micrograms was injected for each analysis.

III. RESULTS

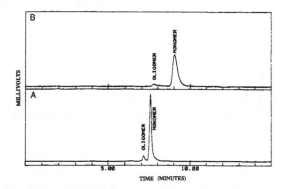

Figure 1: SE-HPLC chromatograms of bovine somatotropin (bST)
           using a DuPont GF250 column.   The chromatographic
           conditions were A. 50 mM sodium phosphate, 0.05 %
           SDS, pH 6.8 and B. 220 mM sodium phosphate, 0.05 %
           SDS, pH 6.8.

Figure 1 shows SE-HPLC chromatograms of bST on a DuPont GF250
column in 50 mM sodium phosphate, 0.05 % SDS pH 6.8 (A) and 220
mM sodium phosphate, 0.05 % SDS, pH 6.8 (B).   As the
concentration of sodium phosphate was increased, the monomer and
oligomer peaks broadened and eluted later off the column.   This
indicates that hydrophobic interactions between the resin and
the protein are occurring.   SE-HPLC of bST using mobile phases
containing 0.5 % SDS and 220 mM sodium phosphate pH 6.8 did not
decrease the hydrophobic interactions significantly.
Hydrophobic interactions were also observed with BioRad BioSil
250 and TosoHaas 3000SW$_{XL}$ columns when a SDS/phosphate mobile
phase was used (unpublished results).

There are several instances where TFA and acetonitrile have
been used to minimize non-ideal SE-HPLC of prtoeins and peptides
(1,2,7,8).   Mobile phases containing various amounts of TFA and
acetonitrile were examined using the BioRad BioSil 250 column
with bST (Figure 2).

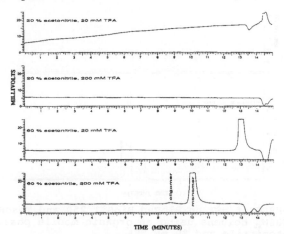

Figure 2: The effect of acetonitrile and trifluoroacetic acid
           on the elution of bST on a BioRad BioSil 250 column.

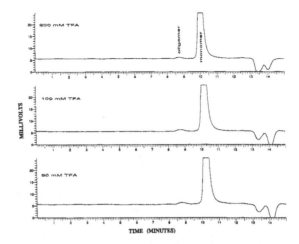

Figure 3:   SE-HPLC of bST using a BioRad BioSil 250 column with
            60 % acetonitrile and various amounts of TFA in the
            mobile phase.

    The 60 % acetonitrile/200 mM TFA showed acceptable resolution
of bST oligomers from monomer. Similar results were obtained at
lower TFA concentrations with 60 % acetonitrile in the mobile
phase (Figure 3). Since there was no difference in resolution
for TFA concentrations of 80-200 mM and TFA can degrade silica
support at these concentrations (2), it was decided to use 60 %
acetonitrile and 80 mM TFA as the SE-HPLC mobile phase to
minimize the degradation. Figure 4 shows SE-HPLC chromatograms
of bST on a BioRad BioSil 250 column and a TosoHaas 3000SW$_{XL}$
column using the optimal mobile phase. Porcine somatotropin
(pST) oligomers and monomer can also be separated using the same
conditions as bST (data not shown). Forty percent acetonitrile
and 100 mM TFA was found to be the optimal mobile phase for SE-
HPLC of Tissue Factor Pathway Inhibitor (TFPI) using the BioRad
BioSil 250 column (Figure 5).

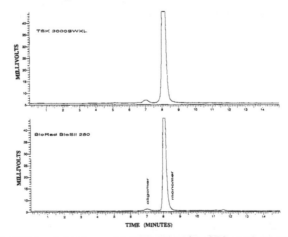

Figure 4: SE-HPLC of bST on BioRad BioSil 250 and TosoHaas
          3000SW$_{XL}$ columns using an organic mobile phase of 60 %
          acetonitrile and 80 mM TFA.

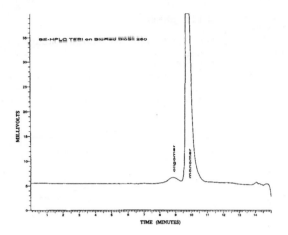

Figure 5:   SE-HPLC of Tissue Factor Pathway Inhibitor (TFPI)
            using a BioRad BioSil 250 column with 40 %
            acetonitrile and 100 mM TFA as the mobile phase.

Several other organic mobile phases containing different acids
were examined. Figure 6 shows SE-HPLC chromatograms of bST
using a Bio-Rad BioSil 250 column and mobile phases containing
60 % acetonitrile with either HCl, H_3PO_4, TFA, or
heptafluorobutyric acid (HFBA).

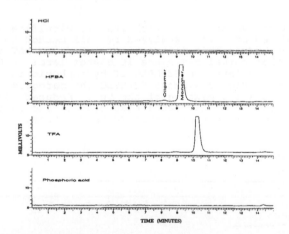

Figure 6:   SE-HPLC chromatograms showing the effect of various
            acids on the elution of bST oligomers and monomers
            from a BioRad BioSil 250 column. All the mobile
            phases were 60 % acetonitrile and pH 1.0.

The pH of all the mobile phases was 1.0. The acetonitrile/TFA
or HFBA mobile phases showed elution of bST within the included
volume. Lack of sample solubility in the mobile phase was not
the reason bST did not elute with the organic mobile phases
containing HCl or H_3PO_4. Since the HCl or H_3PO_4 mobile phases did
not have bST elute off within the included volume, this would
indicate non-ideal SE-HPLC occurs with these mobile phases. The
retention time of the bST oligomer and monomer is about 1 minute
earlier with the acetonitrile/HFBA mobile phase as compared to

the acetonitrile/TFA mobile phase.

Several proteins having different molecular weights were
analyzed by SE-HPLC to check the linearity of the method.
Figure 7 is a plot of log(MW) versus the ratio of elution volume
(V_e) to the void volume (V_0) using the BioRad BioSil 250 column
and 60 % acetonitrile/80 mM TFA. The plot shows a good linear
relationship for proteins in the molecular weight range of 1,350
to 670,000 daltons.

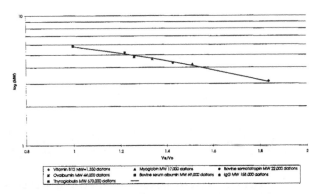

Figure 7: Plot of log(MW) versus V_e/V_0 of various proteins
          using a BioRad BioSil 250 column and 60 %
          acetonitrile, 80 mM TFA as the mobile phase.

BST in its native state and with its two disulfide bonds
reduced (reduced bST) were analyzed by this method to determine
the extent of hydrophobic interactions. Reduced bST is more
hydrophobic than normal bST since it elutes later on RP-HPLC and
should also elute later on SE-HPLC if hydrophobic interactions
were occurring. Figure 8 shows SE-HPLC chromatograms of bST,
reduced bST and a 1:1 mixture of the two. Reduced bST eluted
earlier than bST which indicates that hydrophobic interactions
do not play a major role with this SE-HPLC method.

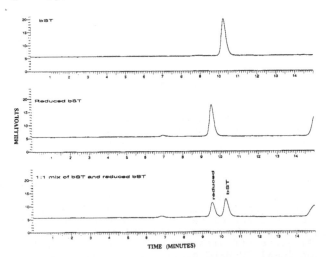

Figure 8: SE-HPLC chromatograms of bST, reduced bST and a 1:1
          mixture of the two. The chromatographic conditions
          were 60 % acetonitrile and 80 mM TFA as the mobile
          phase with a BioRad BioSil 250 column.

IV. DISCUSSION

This report describes a SE-HPLC method using organic mobile phases for several proteins. With an acetonitrile/TFA mobile phase, non-ideal SE-HPLC (ionic, hydrogen bonding, and hydrophobic interactions) is minimized. The resolution factor of the three protein oligomers from their monomers is 1.4, well above the USP recommended factor of 1.0 and near baseline resolution of 1.5. Although the amount of TFA in the mobile phase is high which may degrade the resin, one SE-HPLC column has been used to analyze over 1500 samples with no loss in resolution. To extend column life, the column is stored in 90 % acetonitrile after analysis. Like the aqueous mobile phase SE-HPLC methods, acetonitrile/TFA mobile phases must be optimized for each particular protein of interest.

There are several advantages to the use of an organic mobile phase SE-HPLC system. Electrospray mass spectrometry (ESMS) analysis requires the removal of all salts from the protein. If one desires to analyze by ESMS heterogeneity in the monomers and the oligomers, they can be easily collected by SE-HPLC and then analyzed directly by ESMS after evaporation of the volatile acetonitrile/TFA mobile phase. With an aqueous based solvent system, another step must be used to remove the interfering salts before ESMS analysis.

Another advantage of the organic mobile phase system is little if any adsorption of protein onto the column. In general, proteins have better solubility at low pH. Column lifetime should be better with acetonitrile/TFA mobile phases since silica dissolves at neutral or basic pH (3). HPLC pump seals should also have increased lifetime since no salts are used in the organic mobile phase. An organic mobile phase is less likely to have bacterial growth as compared to aqueous mobile phases. To avoid bacterial growth, it is recommended that SE-HPLC columns be stored in organic solvents such as 10 % isopropanol (3). In conclusion, this report has demonstrated that acetonitrile/TFA mobile phases can be effectively used for analyzing several different proteins.

REFERENCES

1. Irvine, G.B. and Shaw, C. (1986). *Analytical Biochemistry*. **155**, 141-148.
2. Mant, C.T., Parker, J.M. Robert, and Hodges, R.S. (1987). *Journal of Chromatography*. **397**, 99-112.
3. Gooding, K.M. and Regnier, F.E. (1990). *HPLC of Biological Macromolecule*, 47-75.
4. Violand, B.N., Schlittler, M.R., Lawson, C.Q., Kane, J.F., Siegel, N.R., Smith, C.E., and Duffin, K.L. (1994). *Techniques in Protein Chemistry VI*, 99-106.
5. Diaz-Collier, J.A., Palmier, M.O., Kretzmer, K.K., Bishop, B.F., Combs, R.G., Obukowicz, M.G., Frazier, R.B., Bild, G.S., Joy, W.D., Hill, S.R., Duffin, K.L., Gustafson, M.E., Junger, K.D., Grabner, R.W., Galluppi, G.R., and Wun, T.C. (1994). *Thrombosis and Haemostasis*. **71**, 339-346.
6. Scopes, R.K. (1984) *Protein Purification: Principles and Practices*, 39-66.
7. Welinder, B.S. (1984), *CRC Handbook of HPLC for the Separation of Amino Acids, Peptides and Proteins, Volume II*, 413-419.
8. Alpert, Andrew J., *How Big are Amino Acids?*. Poster abstract T157, Protein Society Symposium 1992.

Interfacing Polyacrylamide Gel Electrophoresis with Mass Spectrometry

R. R. Ogorzalek Loo, C. Mitchell, T. Stevenson,[a] J. A. Loo,[a]and P. C. Andrews
Dept. of Biological Chemistry, University of Michigan, Ann Arbor, MI 48109-0674
and
[a]Parke-Davis Pharmaceutical Research, Division of Warner Lambert Company, Ann Arbor, MI 48105

I. Introduction

Protein characterization methodologies providing higher throughput than classical methods are needed to analyze the flood of new gene products identified by recombinant DNA technology in the past decade. One approach is to link the unparalleled resolution of electrophoretic separation methods with the high sensitivity, high throughput, and high information content of mass spectrometry. A number of efforts in recent years have endeavored to achieve this goal using a variety of techniques (1-17). Many of these efforts have been relatively successful in mapping proteins and identifying sites of post-translational modification, but the greatest problems remaining with this approach are encountered at the interface between the gel and the mass spectrometer. Several methods have been developed which rely on post-electrophoretic elution of proteins from gels or membranes sometimes followed by clean-up procedures and enzymatic or chemical cleavages prior to mass analysis (6,7). On-membrane or in-gel digestion followed by elution of the product peptides has also been pursued (8-15). These effective methods do not lend themselves well to the automation required for high throughput analyses. Other methods that electroblot proteins from the electrophoretic gels to a polymeric support followed by desorption directly from the membrane are more amenable to automation,(1,2,11,17) but in their present forms have limitations in sensitivity and resolution or in the availability of the equipment used, although the methodology looks more and more promising (11). Here we demonstrate a method for acquiring mass spectra directly from electrophoretic gels, without electroelution or electroblotting. This method relies upon ultrathin polyacrylamide gels which dry to thicknesses of 10 microns or less and which have the additional advantages of rapid preparation and run times. Spectra have been acquired from isoelectric focusing (IEF), native, and sodium dodecyl sulfate (SDS) gels. The approach also makes it possible to run virtual 2-D gels in which proteins are resolved in the first dimension on the basis of their charge (i.e., IEF gels) while the second dimension is matrix assisted laser desorption ionization-mass spectrometry (MALDI-MS) instead of SDS gel electrophoresis. We present the analysis of simple mixture of proteins employing this approach. A second method, diffusive transfer, is shown to overcome many of the problems associated with electroblotting to membranes.

Mass spectrometric methods provide an efficient way to rapidly characterize proteins via peptide mapping. Mass mapping approaches have found increased usage in the past several years, reflecting improvements in instrumentation and the needs of investigators for more effective biopolymer analysis. Major benefits of peptide mapping of proteins by MS include identification of sites of post-translational modifications, errors in DNA sequencing, and determination of protein identities via database searches. The latter technique has seen considerable interest recently with the development of several programs for database searching based on mass mapping information (12-16). We examine the products of CNBr digestion in-gel and demonstrate its utility for peptide mapping applications.

II. Experimental

A. *Gel Electrophoresis*

Electrophoresis was carried out on a PhastSystem automated gel electrophoresis unit (Pharmacia, Uppsala, Sweden) using Pharmacia's precast thin polyacrylamide gels marketed for use with this equipment. Isoelectric focusing (IEF) employed 5% polyacrylamide gels, 0.35 mm thick when hydrated, precast to generate a pH 3-9 gradient. The procedure detailed in Pharmacia Separation Technique File No. 100 was followed, except that the cooling temperature was reduced to 10°C and the final volt-hours were extended to 600 to ensure that all proteins reached their pI's. Staining, when performed, followed conditions detailed in Pharmacia Development Technique File No. 200. Unstained gels were manually soaked in 20% trichloroacetic acid/H_2O for about 45 minutes and then washed in at least two changes of 10% acetic acid/30% acetonitrile for 30 minutes each and allowed to dry, lightly covered at room temperature. If gels were required for mass spectrometry before they were dry, (despite standing overnight), drying was completed by placing the gel in a vacuum dessicator or by gently blowing a stream of air over the gel for a few hours.

Native gel electrophoresis employed gels with 12.5% polyacrylamide that were 0.45 mm thick when hydrated. Electrophoresis conditions recommended in Separation Technique File No. 121 were followed and gels were stained as described in Pharmacia Technique File No. 200. Unstained gels were washed with at least 2 changes of 30% methanol/10% acetic acid for at least 30 minutes each with gentle rocking. The higher acrylamide content of these gels rendered them more brittle after drying and they tended to pull away from the polyester film backing after a few days, unless specially treated. These problems were obviated by soaking the unstained gels in 0.1 g sinapinic acid in 5 mL acetic acid and 5 mL acetone. Stained native gels incorporated 10% glycerol as a preservative. Gels were dried as described above.

Sodium dodecyl sulfate (SDS) electrophoresis employed gels with the same characteristics as those used for native gel electrophoresis. SDS was only present in the buffer strips. Samples were dissolved in 10 m$\underline{M}$ Tris/HCl, 1 m$\underline{M}$ EDTA pH 8.0, 2.5% SDS, 0.01% bromophenol blue, and 0.55% β mercaptoethanol and heated at 80°C for 5 minutes (except for the rainbow molecular weight markers which were not heated). The samples were loaded onto the gel, and electrophoresis was carried out (Pharmacia Separation Technique File No. 111) with a discontinuous buffer system. Unstained gels were washed as described for native gels.

The long term stability of the gels has not been thoroughly characterized, because gels were usually examined by MALDI within 4 days of preparation. Cracking of gels may be more likely as they age and is affected by the composition of the gel (i.e., higher percent acrylamide gels are more prone to cracking) and the surface treatment of the gel; e.g., whether it is soaked in matrix or whether matrix is spotted onto the surface. We have obtained spectra from IEF gels up to 3 months old. While cracking of the gels has not been a serious problem in our hands, we are aware that it can be a difficulty. Our experience is limited to Phast gels; gels from other manufacturers may be more or less robust. Gels are no older than 6 months from the date of manufacture. We currently recommend that SDS gels be examined a day after preparation and that transfers in and out of the vacuum chamber be limited.

B. Diffusive Membrane Transfers

A blotting sandwich was assembled with a polyethylene (18) or Teflon membrane (prewet in 95% ethanol) placed on top of the gel (with gel bond backing still in place) followed by 2 sheets of blotting paper soaked in 10 m$\underline{M}$ CAPS/10% ethanol. Four more sheets of dry paper were laid on top followed by a 200 g weight. IEF gels were transferred for 1 hour at room temperature. SDS PAGE and native gels were transferred overnight at 4°C.

C. Reactions

In-gel CNBr digests were performed by spotting 5 μL of 10 mg/mL CNBr in 50% trifluoroacetic acid (TFA) on the band of interest. TFA was substituted for the formic acid traditionally employed in CNBr digest protocols to prevent O-formylation of the products.(19) The tube was incubated in the dark at room temperature for 2-4 hours. The sample was then dried on a centrifugal dryer.

D. Mass Spectrometry

Matrix solutions, prepared from saturated sinapinic acid in 1:1 methanol:0.1% trifluoroacetic acid/H_2O or 1:2 CH_3CN:0.1% trifluoroacetic acid/H_2O were spotted (0.3-0.5 μL) onto the bands of interest. Most gels were examined unstained; bands were located by comparing a duplicate, stained gel.

Time-of-flight mass spectra were acquired at 337 nm using a Vestec LaserTec ResearcH reflectron time-of-flight mass spectrometer (PerSeptive Biosystems, Framingham, MA) operated in linear mode (1.3 meter pathlength). It is pumped by two turbomolecular pumps and two rough pumps and could acquire spectra from a dry gel 5 minutes after its introduction into the vacuum chamber. The instrument was fitted with a 100-position sample plate modified by milling a 4 cm x 4 cm square to a depth of 0.5 mm to better accommodate gels. Gels with polyester backing intact were trimmed to fit this size and taped at their edges to the plate. CNBr digested bands were also taped to the sample plate.

III. Results and Discussion

Spectra were obtained from unstained native and SDS gels (12.5% acrylamide) and IEF gels (5% acrylamide) to which matrix had been spotted. Figure 1 illustrates a spectrum acquired from a 4.5 pmol loading of the α and β chains of human hemoglobin on an IEF gel. IEF bands are particularly sharp, yielding low detection limits by MALDI. Spectra have been acquired from gels

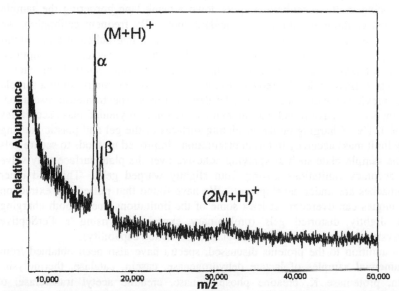

Figure 1. MALDI spectrum directly from an IEF gel on which 4.5 pmol of α and β chain human hemoglobin had been loaded.

loaded with as little as 1 pmol of myoglobin and 0.65 pmol of each hemoglobin subunit.

Protein spectra were also obtained from bands stained with Coomassie blue or silver at sample loadings of 100 pmol or less. Silver-stained bands tended to yield broad and poorly resolved spectra. The degraded resolution probably arose from reaction of the protein with glutaraldehyde employed in the silver staining procedure. Coomassie-blue adducted protein ions were typically observed for Coomassie staining, distributing the ion signal over several species and thereby reducing the overall sensitivity of the method.(1) However, for relatively low molecular weight proteins, such as bovine ubiquitin (Mr 8565), the 800 Da adducts can be easily resolved. The degree of Coomassie adduction was variable; a myoglobin band in a commercial IEF marker solution (approximately 1 μg loading) showed only a small amount of a single Coomassie adduct.

Mass accuracy is extremely important if this method is to have general utility. MALDI has been reported to yield accuracies as high as +/-0.01%. We find that accuracy is compromised when desorbing from gels and considerable attention must be paid to running frequent calibration spots if good masses are to be obtained. For our purposes, external calibration refers to spotting a standard protein *near* a band of interest on the gel, while internal calibration refers to spotting the standard on *top* of the band of interest. When we scanned down an entire lane after calibrating with a spot at the top of the gel, we saw a 1% drift in calibration, with the size of the drift reflecting the distance from the standard spot. This behavior is consistent with a distorted gel, which is not surprising since our gels were only taped down at the edges and did not lie flat on the probe. Clearly such accuracies are unacceptable, but they are easily improved by placing multiple calibration spots down a blank lane bordering the sample lane and calibrating with nearby standard spots. By frequent calibration, we have achieved 0.1-0.2% mass accuracy. In principal, internal calibration would be superior, but it is particularly difficult with gels, because non-optimal standard loadings can overwhelm sample signals, and once spotted, the chosen loading is irrevocable. Moreover, respotting standard and matrix onto a sample band previously examined by MALDI degraded mass spectrometric resolution. In practice, we have found that internal calibration also yielded mass accuracies of +/- 0.1%. Charging on the insulating surfaces of the gel and plastic backing may limit mass accuracy in this configuration. Improved methods to secure gels to the sample plate such as spraying adhesive over the plate surface may solve the accuracy limitations arising from slightly warped gels. This and other approaches are under development. We have found that pulsed ion extraction techniques can overcome at least some of the limitations due to both charging and slightly distorted gels (preliminary experiments using a PerSeptive Biosystems Elite instrument with Delayed Extraction capability).

In addition to the proteins discussed, spectra have also been obtained from commercial samples of lactate dehydrogenase, casein, catalase, thermolysin, actin, proteinase K, creatine phosphokinase, creatine acetyl transferase, α lactalbumin, α amylase, and a broad range of other proteins, demonstrating the wide applicability of the technique. It is also rugged; an IEF gel reexamined 80 days after spotting with matrix yielded good spectra. (We do not recommend reexamining ancient high acrylamide composition gels, e.g., SDS, 20% acrylamide, however, since they may crack under vacuum.)

A mixture of proteins including the β subunit of mouse nerve growth factor 2.5S, bovine hemoglobin, bovine carbonic anhydrase, and bovine serum albumin was loaded onto an IEF gel and the gel was run as described earlier. Matrix was spotted in discrete spots down the lane (approximately 1.5 mm between centers of spots). The laser position was manually scanned down the lane. For each step down the lane, the laser position was "dithered" to obtain the optimum spectrum. Spectra were obtained for each protein. However, some streaking was observed for nerve growth factor and carbonic anhydrase; i.e., protein spectra were obtained over a wider area than the width indicated by the stained bands. Further study is needed to determine if the apparent electrophoretic resolution yielded by laser scanning will be poorer than by staining. Streaking is common in IEF as proteins often precipitate near their isoelectric point. Unfortunately, the exquisite sensitivity of MALDI renders a small amount of streaking problematic. Improvements in the electrophoresis are anticipated to reduce this problem.

If mixtures and low level samples are to be examined by MALDI directly off gels, it is desirable to develop an alternative to spotting for depositing matrix on the surface of the gels. We have soaked the gels in sinapinic acid in various solvent compositions and noted that spectra can be easily obtained under those conditions, also. Other methods for matrix application are also under investigation.

Most protein-protein interactions are disrupted under the denaturing conditions employed in SDS gel electrophoresis (20). However non-denaturing conditions can be employed in native and IEF electrophoresis, enabling non-covalent interactions to be studied. Heterosubunit associations can be identified easily by combining non-denaturing electrophoretic methods with MALDI.

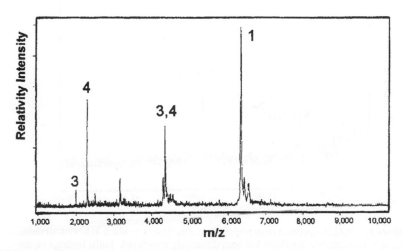

Figure 2. MALDI spectrum directly from an IEF gel on which 3 pmol of bovine carbonic anhydrase had been loaded and CNBr digested.

This approach is powerful, because MALDI alone (without covalent cross-linking methods) has been unsuccessful at elucidating significant non-covalent interactions (21); complexes are either too fragile to survive the desorption event, or non-specific oligomers are obtained. Complexes are amenable to study if we rely on the gel to demonstrate that subunits migrate together, and use MALDI to detect those subunits. One example of a complex which we have studied by this method is lentil lectin. Lentil lectin is heterogeneous, like many lectins, and migrates as numerous bands. We observe that every band on an IEF gel yields components at 5 *and* 20 kDa, consistent with its predicted α/β heterodimer structure.

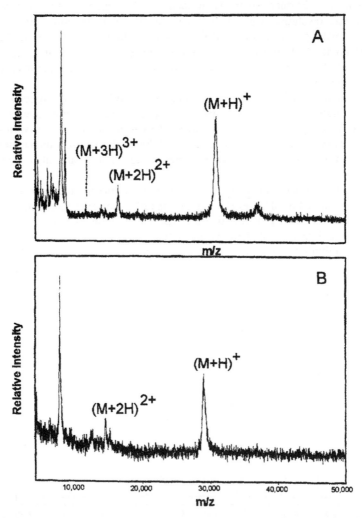

Figure 3. MALDI spectrum from polyethylene membranes to which a) bovine albumin, and b) bovine carbonic anhydrase had been diffusively transferred. Initial loadings on the IEF gel were 11 and 14 pmol, respectively.

Beyond measuring masses for intact proteins from gels, an equally important capability to be demonstrated is digestion with subsequent mass analysis. Excised bands were incubated with CNBr and the resulting products were mass analyzed in-gel. Figure 2 illustrates the spectrum obtained from a digest of bovine carbonic anhydrase, with 3 picomole loaded on the gel. Three of four complete digest products were observed (222-239, 240-259, and 1-58) labeled **3**, **4**, and **1**, respectively, and the incomplete digest product 222-259 labeled **2,3**. The 59-221 fragment, predicted mass 17933 for the homoserine lactone, was not observed. Mass accuracy from excised gel slices was not as good as from the gels. The worst case yielded an accuracy of only 0.5%. Problems appear to arise from warping at the edges of the gel slices. Use of an intact gel will improve accuracy. We have had success with gas phase CNBr digests of entire gels and are investigating the mass accuracy under these conditions.

Previous studies have employed electroblotting to transfer proteins from gels to membranes. This approach has limited the type of membrane employed for the transfer and the analyst's ability to obtain MALDI spectra from the intact protein. Membranes which perform excellently for dot blotting perform less well for electroblotting, probably because the proteins penetrate the membrane deeply under electroblotting conditions. We have employed diffusive transfer to passively elute proteins from gels to membranes. The method is particularly suited to thin gels. Fig. 3 illustrates spectra obtained for bovine albumin (11 pmol) and bovine carbonic anhydrase (14 pmol) transferred to polyethylene membrane.

IV. Conclusions

The *direct* combination of thin layer gel electrophoresis and MALDI mass spectrometry has been demonstrated in IEF, SDS, and native gels with good sensitivity and mass accuracy, offering advantages in speed and reduced complexity. CNBr digests were performed and the products were probed in-gel, yielding sufficient information for identification based on isoelectric point, intact mass, and fragment masses. Non-covalent complexes can also be examined by this approach, because some gel electrophoretic methods can be run under non-denaturing conditions where many macromolecules maintain their native structures, resulting in comigration of subunits.

Acknowledgments

We thank James Gagnon of 3M Corporation for kindly providing the microporous polyethylene membrane used in these experiments and J. E. Hoover for helpful comments regarding cation adducts.

References

1. C. Eckerskorn, K. Strupat, M. Karas, F. Hillenkamp, and F. Lottspeich, *Electrophoresis* **13**, 664-665 (1992).
2. K. Strupat, M. Karas, and F. Hillenkamp *Anal. Chem.* **66**, 464-470 (1994).
3. M. le Maire, S. Deschamps, J. V. MØller, J. P. Le Caer, and J. Rossier, *Anal. Biochem.* **214**, 50-57 (1993).
4. C. Bonaventura, J. Bonaventura, R. Stevens, and D. Millington, *Anal. Biochem.* **222**, 44-48 (1994).
5. H. Yanase, S. Cahill, J. J. M. de LLano, L. R. Manning, K. Schneider, B. T. Chait, K. D. Vandegriff, R. M. Winslow, and J. M. Manning, *Protein Science* **3**, 1213-1223 (1994).
6. S. C. Hall, D. M. Smith, F. R. Masiarz, V. M. Soo, H. M. Tran, L. B. Epstein, and A. L. Burlingame, *Proc. Natl. Acad. Sci USA* **90**, 1927-1931 (1993).
7. K. R. Clauser, S. C. Hall, D. M. Smith, J. W. Webb, L. E. Andrews, H. M. Tran, L. B. Epstein, and A. L. Burlingame, *Proc. Natl. Acad. Sci. USA* **92**, 5072-5076 (1995).
8. W. Zhang., A. J. Czernik, T. Yungwirth, R. Aebersold, and B. T. Chait (1994) *Protein Science* **3**, 677-686, and references therein.
9. J. Bai, M. G. Qian, Y. Lui, X. Liang, and D. M. Lubman *Anal. Chem.* **67**, 1705-1710 (1995).
10. E. Mortz, O. Vorm, M. Mann, and P. Roepstorff, *Biol. Mass Spectrom.* **23**, 249-261 (1994).
11. S. D. Patterson *Electrophoresis*, in press.
12. P. James, M. Quadroni, E. Carafoli, and G. Gonnet, *Biochem. Biophys. Res. Commun.* **195**, 58-64 (1993).
13. J. R. Yates III, S. Speicher, P. R. Griffin, T. Hunkapillar, *Anal. Biochem.* **214**, 397-408 (1993).
14. W. J. Henzel, T. M. Billeci, J. T. Stults, S. C. Wong, C. Grimley, and C. Watanabe, *Proc. Natl. Acad. Sci. USA* **90**, 5011-5015 (1993).
15. D. J. Pappin, P. Hojrup, and and A. J. Bleasby, *Curr. Biol.* **3**, 327-322 (1993).
16. M. Mann, P. Hojrup, and P. Roepstorff, *Biol. Mass Spectrom.* **22**, 338-345 (1993).
17. M. M. Vestling and C. Fenselau *Anal. Chem.* **66**, 471-477 (1994).
18. J. A. Blackledge and A. J. Alexander, *Anal. Chem.* **67**, 843-848 (1995).
19. P. C. Andrews, M. H. Allen, M. L. Vestal, and R. W. Nelson *Techniques in Protein Chemistry III*, ed. R. H. Angeletti, Academic Press, San Diego, 1992, pp. 515-523.
20. K. Weber, J. R. Pringle, and M. Osborne, *Methods in Enzymology* **26**, 3-27 (1971).
21. T. B. Farmer and R. M. Caprioli *Biol. Mass Spectrom.* **20**, 796-800 (1991).

IMMUNOPRECIPITATION AS A MEANS OF PURIFICATION FOR ANALYSIS WITH MASS SPECTROMETRY

A. Grey Craig, Steven W. Sutton, Joan Vaughan and Wolfgang H. Fischer

The Clayton Foundation Laboratories for Peptide Biology, The Salk Institute, La Jolla, CA 92037

I. Introduction

Corticotropin Releasing Factor-Binding Protein (CRF-BP) is a 35 kDa glycoprotein which was originally purified from human plasma (1). Native and recombinant CRF-BP bind rat/human CRF and inhibit CRF-induced ACTH release from pituitary cells *in vitro* in a dose dependent fashion (2). The anatomical distribution of CRF-BP (3,4) suggests CRF-BP might act to limit the action of CRF. For example, the plasma level of CRF-BP in pregnant women is typically 4 to 7 nM but decreases to about 1 nM at term, thus increasing the level of free CRF in plasma and potentially contributing to parturition (5).

Matrix assisted laser desorption mass spectrometry (MALD-MS) (6) is a highly sensitive ionization technique which when coupled with time-of-flight mass analysis allows the detection of sub pmol amounts of proteins up to 1 MDa (7). Indeed MALD-MS has been used to detect lactalbumin, lactoglobulin and casein from milk samples without the need for prior purification (8). Since proteins are commonly isolated using gel electrophoresis, considerable effort is being expended towards the analysis of proteins either from gels directly or after blotting to a membrane using MALD (9-12). These efforts to determine appropriate methods for MS analysis are motivated by the increased mass accuracy available with MS compared to gel electrophoresis. In addition, MALD-MS sensitivity is comparable, if not greater, than that observed with gel electrophoresis using non radioactive detection (e.g., Coomassie Blue or silver staining).

An alternative method of isolating proteins is immunoprecipitation using an antiserum raised against the intact protein, or a portion of the protein. Immunoprecipitation is commonly used as a specific method of trapping ligands for further analysis, such as in immunoassays or immunoblotting which can detect fmol amounts of proteins. MALD-MS can be used in combination with immunoprecipitation procedures to detect proteins without further purification. Investigators have identified lactoferrin from urine (13) and transferrin (14), myotoxin and mojave toxins in blood (15) with immunoprecipitation techniques.

We began developing an immunoprecipitation procedure as a means of improving the analysis of glycoproteins with MALD. The excellent sensitivity we observed in our initial studies stimulated this investigation to determine whether analysis of proteins such as CRF-BP in plasma at physiological levels was feasible. Here, we vary immunoprecipitation conditions at successively lower concentrations of CRF-BP using MALD-MS and gel electrophoresis in order to optimize this procedure.

II. Materials and Methods

CRF-BP was expressed as a recombinant protein in Chinese Hamster Ovary Cells and purified as previously described (16). Briefly, media enriched in CRF-BP through incubation with recombinant CHO cells was centrifuged and sterile filtered through a 0.22 μm filter to remove cells and debris. CRF-BP was first purified from the media by affinity chromatography using r/hCRF coupled to Affi-prep 10 (Bio Rad, Richmond, CA) and an eluent of 80 mM triethylammonium formate (pH 3.0) containing 20% acetonitrile. The resulting fraction was further purified using tandem gel chromatography (2 Pharmacia Superose 12 HR 10/30 columns in series) in a mobile phase of 6M guanidine HCl - 0.1M ammonium acetate (pH 4.75). Final purification and desalting was achieved by C4 HPLC (Vydac, Hesperia, CA) using an eluting gradient of acetonitrile in 0.1% trifluoroacetic acid. The quality and quantity of the resulting material was evaluated by amino acid analysis.

Polyacrylamide gel electrophoresis was carried out according to Laemmli (17) in 15% gels. Proteins were electrotransfered to PVDF membranes (PVDF-Plus, MSI). For Western staining, (18) the same antiserum was employed that was used for the precipitation. Antisera were raised in rabbits by intradermal injection of CRF-BP and antisera were purified using methods previously described (19). Non-specific binding was blocked with 5% non-fat dry milk solution. Visualization was achieved using a goat-anti-rabbit alkaline phosphatase conjugated antiserum (Bio-Rad) and BCIP/NBT substrate.

MALD mass spectra were measured using a Bruker Reflex time-of-flight mass spectrometer fitted with a gridless reflectron energy analyzer and a nitrogen laser. An accelerating voltage of +31 kVwas employed. Targets were precoated with a solution of α-cyano-4-hydroxy cinnamic acid in acetone (20). Samples were applied to the pre-coated target, allowed to dry and rinsed with 5 μl of H_2O.

Immunoprecipitation: Purified recombinant CRF-BP (20- 200 ng) was added to either buffer (20 μl, 50 mM Hepes, pH 7) or normal rabbit serum (NRS) (20 μl) to give a concentration of 30-300 nM. Affinity purified antisera raised against either a synthetic N-terminal CRF-BP peptide or against the recombinant CRF-BP was added to the CRF-BP solution. In addition, either carrier IgG, a sheep-anti-rabbit IgG (SAR) or both the carrier IgG and the SAR were added to the CRF-BP solution. The reaction was incubated at 4 °C overnight. The mixture was then centrifuged (20 min. @ 15000 rpm) and the supernatant removed. The pellet was rinsed with buffer and again centrifuged and the supernatant removed. An aliquot (5.0 μl) of 0.1 % TFA was added to the pellet, aliquots were removed for analysis by SDS gel electrophoresis and for application to the MALD target (either directly or after further dilution with 0.1 % TFA).

III. Results & Discussion

When 200 ng (300 nM) of CRF-BP was immunoprecipitated from either buffer or normal rabbit serum we were able to analyze the precipitate with both gel electrophoresis and MALD-MS. Figure 1 shows the SDS gel of CRF-BP + anti-CRF-BP in either buffer (lane 1), NRS (lane 2), with carrier IgG in buffer (lane 3) or carrier IgG in NRS (lane 4). Figure 2 shows the MALD mass spectrum of the precipitate from the CRF-BP + anti-CRF-BP in buffer (c.f., Figure 1, lane 1). A number of species are observed which were assigned to correspond with the intact $[M+H]^+$, $[M+2H]^{2+}$, $[M+3H]^{3+}$ and $[M+4H]^{4+}$. Although a band corresponding with the CRF-BP in Figure 1, lane 1 was not observed, measurement of the precipitate did produce intact molecule ions in the MALD spectrum (Figure 2). These intact molecule ions were not observed when the immunoprecipitate was diluted (1/10) and measured with MALD. The absence of a CRF-BP band on the gel and the MALD-MS observations suggest that only a small amount (<<10 ng) of CRF-BP is being immunoprecipitated.

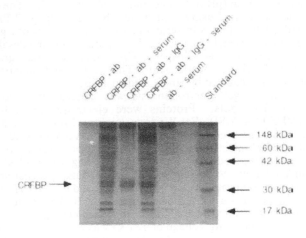

Figure 1. Silver stained polyacrylamide gel of CRF-BP (200 ng) immunoprecipitates (80 % aliquot).

The MALD mass spectrum of the immunoprecipitate of CRF-BP (200 ng) from normal rabbit serum (c.f., Figure 1, lane 2) did not contain intact molecule ions (data not shown). However, by diluting an aliquot of the immunoprecipitate (1/100 or 1/1000) intense intact molecule ions were observed as shown in Figure 3 (1/1000 dilution). In this case, both the intense CRF-BP band observed in the gel, and the MALD-MS observations before and after dilution indicate that considerably more protein is immunoprecipitated compared with the precipitate in the absence of NRS. The MALD mass spectra of the precipitates which were incubated with carrier IgG (and analyzed with gel electrophoresis, see Figure 1, lanes 3 and 4) also gave intact molecule ions after 1/100 and 1/1000 dilution. The dilution prior to MALD analysis appears to be required to lower the overall protein loaded or to dilute inhibitory factors. Based on these results, we propose that the immunoprecipitation was significantly assisted by the presence of either carrier IgG or NRS (the latter containing a number of proteins including IgG). The increased concentration of IgG may allow cross linking between immunoglobulins specific for CRF-BP and carrier or other IgGs, thereby improving the efficiency of CRF-BP immunoprecipitation.

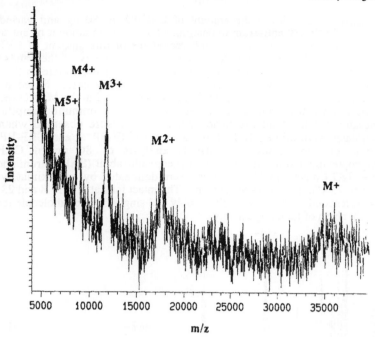

Figure 2.   MALD mass spectrum of a 20 % aliquot of 200 ng of CRF-BP immunoprecipitated with anti-CRF-BP serum from a Hepes buffer.

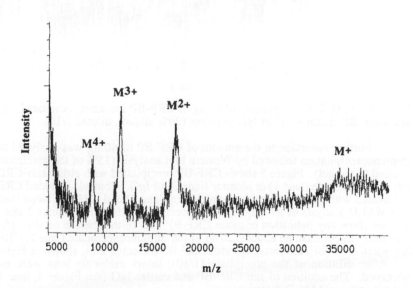

Figure 3.   MALD mass spectrum of 200 ng of CRF-BP immunoprecipitated with anti-CRF-BP serum from NRS (20% aliquot, diluted 1/1000).

Next we reduced the amount of CRF-BP to 80 ng and varied the amounts of primary antiserum to determine the optimum amount of antiserum for precipitation. Although immunoprecipitates of this amount of CRF-BP would produce bands on gel electrophoresis, the intensity of the bands with silver staining was not reliable (applying 80% of the precipitate to the gel). These precipitates were analyzed with MALD in any case. Based on the tolerance to dilution of the intact molecule ions observed in the MALD spectra, we optimized the amount of antiserum used for immunoprecipitation. Generally, the most intense intact molecule ions were observed when the precipitates were diluted 10-fold as shown for CRF-BP + anti CRF-BP and carrier IgG in Figure 4. MALD analysis of 80 ng of CRF-BP immunoprecipitated in the presence of sheep anti-rabbit (SAR) serum and the carrier IgG (intact and 1/10 dilution) were dominated by approximately 8.2, 11.8 and 23 kDa ions (data not shown). The intact ions at 8.2, 11.8 and 23 kDa may correspond to the intact triply, doubly and singly charged molecule ions of the light chain of IgG, respectively.

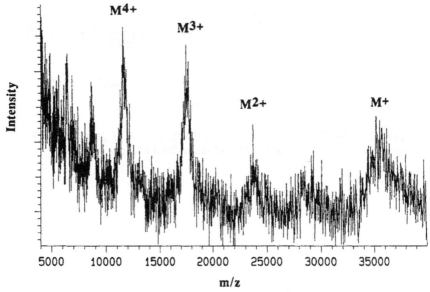

Figure 4. MALD mass spectrum of 80 ng of CRF-BP immunoprecipitated with anti-CRF-BP serum + carrier IgG in Hepes (20% aliquot, diluted 1/10).

Further reduction in the amount of CRF-BP to 40 ng was analyzed by immunoprecipitation followed by Western blot analysis (15% of the precipitate applied to the gel). Figure 5 shows CRF-BP precipitated with either anti-CRF-BP in the presence (lane 1) or absence (lane 2) of IgG. Controls included CRF-BP treated with carrier IgG without antiserum (lane 3) and CRF-BP alone (lane 4). MALD analysis of an aliquot of the precipitate (20%) from lanes 3 and 4 did not show any indication of intact CRF-BP ions (either before or after 1/10 dilution). In contrast, an aliquot (20%) of the immunoprecipitate (lane 2, CRF-BP + anti-CRF-BP) did result in intact CRF-BP molecule ions, shown in Figure 6. After dilution of the precipitate (1/10), intact molecule ions were not observed. The addition of anti-CRF-BP and carrier IgG (see Figure 5, lane 1) did not improve MALD detection.

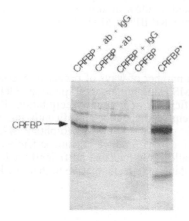

Figure 5. Western blot of CRF-BP (40 ng) precipitates (20 % aliquot);
* = CRF-BP applied directly (8 ng).

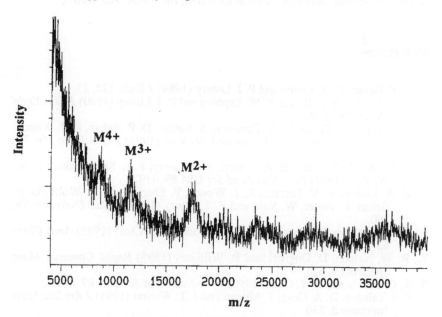

Figure 6. MALD mass spectrum of 40 ng of CRF-BP immunoprecipitated with
anti-CRF-BP serum from Hepes (20% aliquot).

Although an indication of CRF-BP intact molecule ions was observed, the intact
CRF-BP ions were superimposed with a species at approximately 12 and 24
kDa (data not shown). While the MALD spectrum improved after a 1/10
dilution, both intact CRF-BP and the impurity were still observed (data not

shown). Neither species were observed after further dilution. Presumably, the efficiency of the immunoprecipitation and dilution protocol at the 200 ng and 40 ng levels is not the same. With only 40 ng of protein applied, insufficient CRF-BP is immunoprecipitated to allow dilution of IgG related interference species while retaining sufficient CRF-BP for MALD analysis.

IV. Conclusion

We have investigated a number of ways of preparing recombinant CRF-BP for analysis with MALDI-MS using immunoprecipitation. Gel electrophoresis was used to monitor the efficiency of immunoprecipitation. It was found that the amount of CRF-BP precipitated could be significantly increased in the presence of serum or carrier IgG. However, in order to obtain MALD spectra of CRF-BP in the presence of carrier IgG or serum, sufficient CRF-BP must be immunoprecipitated to allow dilutions that sufficiently reduce the concentration of the IgG and its degradation products relative to that of CRF-BP.

Acknowledgments

We would like to thank M. Park, R. Craig, J. Svoboda and A. Goradia for technical assitance. This work was supported by the National Institute of Health (1S10RR-8425, HD-13527, DK-26741, CA-54418, GM-48677) and supported in part by the Foundation for Medical Research, Inc. (AGC and WHF).

References

1. D. P. Behan, E. A. Linton and P. J. Lowry (1989) *J Endo* **122**, 23.
2. E. A. Linton, D. P. Behan, P. W. Saphier and P. J. Lowry (1990) *J Clin Endo Metab* **70**, 1574.
3. F. Petraglia, E. Potter, V. A. Cameron, S. Sutton, D. P. Behan, R. J. Woods, P. E. Sawchenko, P. J. Lowry and W. Vale (1993) *J Clin Endo Met* **77**, 919.
4. E. Potter, D. P. Behan, E. A. Linton, P. J. Lowry, P. E. Sawchenko and W. W. Vale (1992) *Proc Natl Acad Sci USA* **89**, 4192.
5. E. A. Linton, A. V. Perkins, R. J. Woods, F. Eben, C. D. A. Wolfe, D. P. Behan, E. Potter, W. Vale and P. J. Lowry (1993) *J Clin Endo Met* **76**, 260.
6. F. Hillenkamp, M. Karas, R. C. Beavis and B. T. Chait (1993) *Anal Chem* **63**, 1193A.
7. R. W. Nelson, D. Dogruel and P. Williams (1995) *Rapid Commun Mass Spectrom* **9**, 625.
8. R. C. Beavis and B. T. Chait (1990) *Proc Natl Acad Sci USA* **87**, 6873.
9. E. J. Zaluzec, D. A. Gage, J. Allison and J. T. Watson (1994) *J Am Soc Mass Spectrom* **5**, 230.
10. M. M. Vestling and C. Fenselau (1994) *Anal Chem* **66**, 471.
11. K. Strupat, M. Karas, F. Hillenkamp, C. Eckerskorn and F. Lottspeich (1994) *Anal Chem* **66**, 464.
12. C. Eckerskorn, K. Strupat, M. Karas, F. Hillenkamp and F. Lottspeich (1992) *Electrophoresis* **13**, 664.
13. T. W. Hutchens and T. T. Yip (1993) *Rapid Commun Mass Spectrom* **7**, 576.

14. T. Nakanishi, N. Okamoto, K. Tanaka and A. Shimizu (1994) *Biological Mass Spectrometry* **23**, 230.
15. R. W. Nelson, J. R. Krone, A. L. Bieber and P. Williams (1995) *Analytical Chemistry* **67**, 1153.
16. S. W. Sutton, D. P. Behan, S. L. Lahrichi, R. Kaiser, A. Corrigan, P. Lowry, E. Potter, M. H. Perrin, J. Rivier and W. W. Vale (1995) *Endocrinology* **136**, 1097.
17. U. K. Laemmli (1970) *Nature* **227**, 680.
18. H. Towbin, T. Staehelin and J. Gordon (1979) *Proc Natl Acad Sci USA* **76**, 4350.
19. J. M. Vaughan, J. Rivier, A. Z. Corrigan, R. McClintock, C. A. Campen, D. Jolley, J. K. Voglmayr, C. W. Bardin, C. Rivier and W. Vale, (1989). In *"Methods in Enzymology"* (P. M. Conn, ed.) Academic Press, Inc., Orlando, FL, 588.
20. O. Vorm, P. Roepstorff and M. Mann (1994) *Anal Chem* **66**, 3281.

AMINO ACID ANALYSIS - RECOVERY FROM PVDF MEMBRANES: ABRF-95AAA COLLABORATIVE TRIAL

A. M. Mahrenholz[1], N. D. Denslow[2], T. T. Andersen[3], K. M. Schegg[4], K. Mann[5], S. A. Cohen[6], J. W. Fox[7], and K. Ü. Yüksel[8]

1. Dept. Biochemistry, Purdue Univ. West Lafayette, IN 47907
2. Dept. Biochem. and Molec. Biol., Univ. Florida, Gainesville, FL 32610
3. Dept. Biochem. and Molec. Biol., Albany Medical College, Albany, NY 12208
4. Dept. Biochemistry, Univ. Nevada, Reno, NV 89557
5. Max-Planck-Inst. Biochemie, 82152 Martinsried, Germany
6. Waters Corp, Milford, MA 01757
7. Dept. Microbiology, Univ. Virginia Medical School, Charlottesville, VA 22908
8. CryoLife, Inc. Marietta, GA 30067

I. INTRODUCTION

For eight consecutive years, the Amino Acid Analysis Research Committee of the Association of Biomolecular Resource Facilities (ABRF) has conducted collaborative amino acid analysis (AAA) trials among its member laboratories. These efforts provide an opportunity for individual facilities to determine their performance relative to other member facilities as well as to create a data base to judge the merits of alternative methods for AAA. The data from these studies are published [1-8].

This year, the ABRF-95AAA sample revisited the challenge of analyzing protein bound to PVDF membrane, which was last examined with the 1990 ABRF samples [4]. For the present study, ample protein was provided in triplicate so that sensitivity *per se* was not an issue and all laboratories could participate. Pieces of membrane without added protein (blanks) were included in triplicate to assess the utility of this control. Data analysis correlates the results with the various methods employed.

II. MATERIALS AND METHODS

A. Sample Preparation and Distribution

For this study, PVDF membranes (Immobilon-P^{sq}, Millipore), were prepared as follows. (1) A set of 8 membranes (10 cm x 10 cm) were wetted with methanol and rinsed in double distilled H_2O. (2) Each sheet was immersed in 180 mL of a solution containing 0.1 mg/mL horse heart myoglobin (MW 18.8 kDa, Sigma) in 10% methanol and gently agitated for 3 h at room temperature in a glass petri dish. A second set of 8 membranes was treated as above but agitated in 10% methanol without added protein to generate blanks. (3) Membranes were then rinsed twice in distilled water for 15 min and air dried. Samples were cut out of the membranes with a hole-puncher, wrapped in glassine weigh-paper, and distributed to members by regular mail. Along with reporting forms, the sample packet also included a recommended extraction protocol following hydrolysis (30% methanol/ 70% 0.1 N HCl for 30 min with vortexing twice) and general references for further information [9-14]. Due to the way protein was loaded on the membranes, it was not possible to know the exact amount of protein adsorbed. Preliminary studies by the committee indicated that 5-10 µg protein were adsorbed.

B.   Calculations

Raw data were received by an independent collaborator, identifying marks removed, and the anonymous results forwarded to the 1995 ABRF AAA Research Committee as a spreadsheet. Data reduction was as described [4,8]. Briefly, total nmols myoglobin were estimated on the basis of data from individual amino acids by dividing the total nmols of each amino acid on the membrane by the known number of residues of that amino acid per molecule of myoglobin. These data were averaged. A corrected average was then calculated by excluding individual yield values differing ≥ 15% from the average obtained above. Composition (number of each amino acid per molecule) was obtained by dividing the nmols amino acid by the corrected nmols myoglobin for that analysis. Accuracy (internal error) of each residue was calculated as:

% Error = 100 * (| experimental composition value - true value |)/true value

Average % Error per analysis = (Σ % error of 16 amino acids) /16

Average yield and average error were calculated from the triplicate results for each participant. These values were then used to obtain overall averages across participants. The amino acid contents of blanks (µg) were determined by multiplying the total pmol each amino acid by its molecular weight.

III.   RESULTS AND DISCUSSION

A.   Participation

A record number of 77 sites participated in this study, representing 52% of the 147 sites offering amino acid analysis out of 243 ABRF member laboratories. Pre-column (52%; Table I) and post-column techniques (48%) were almost equally represented in the analyses of ABRF-95AAA. Ninhydrin and PITC were overwhelmingly the preferred methods.

B.   Yield and Accuracy

The overall average yield (±S.D.) for the sample was 0.220 ± 0.102 nmol (4.1 ±1.9 µg; see Fig. 1 and Table I). There was no statistically significant difference in yield for the pre- or post-column methods collectively, or between the individual methods. The range of yields extended from 0.02 nmol to 0.56 nmol. Although individual membranes (A - H) were identified in the reporting process so that variations arising from sample preparation could be traced, analysis of the yield data on this basis revealed no statistically significant differences (Table II). It is worth noting that the use of internal standards, performed by 25 of 77 sites, did not appear to have any influence on yield (data not shown). Figure 1 shows that the majority of laboratories (73%) obtained yields within ±1 S.D. of the average.

The accuracy achieved in this study (21.4% overall average error, Table I and Fig. 2) is improved compared with ABRF-90AAA3 [4], which was also a PVDF membrane-bound protein, (26.9% average error, n=30). However, compared with the 11% average error obtained for ABRF-94AAA1, which was

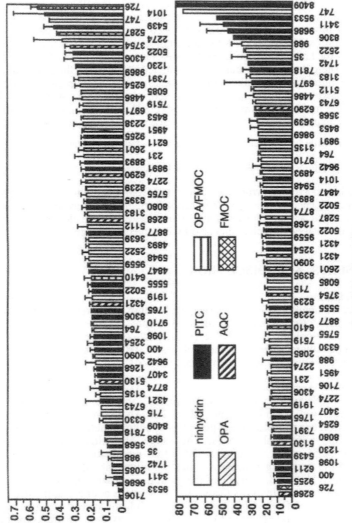

Figure 1. (top) The absolute yields of ABRF-95AAA.

Figure 2. (bottom) The quality of the analyses of ABRF-95AAA.

TABLE I. Correlation of method with yield, error and % of sample analyzed

Method	Yield (nmol) average±S.D. (range)	n	Error (%) average±S.D. (range)	n	(%) Sample Analyzed average±S.D. (range)	n
Overall	0.220 ± 0.102 (0.021 - 0.559)	77	21.4 ± 12.5 (8.7 - 79.4)	77	41.7 ± 32.3 (1 - 100)	†72
Post-column	0.225 ± 0.086 (0.021 - 0.481)	37	21.0 ± 10.8 (12.7 - 74.7)	37	54.7 ± 28.3 (20 - 100)	35
ninhydrin	0.223 ± 0.089 (0.021 - 0.481)	35	21.3 ± 11.0 (12.7 - 74.7)	35	55.8 ± 28.7 (20 - 100)	33
OPA	0.250 ± 0.009 (0.244 - 0.256)	2	15.4 ± 2.6 (13.6 - 17.2)	2	37.5 17.7 (25 - 50)	2
Pre-column	0.220 ± 0.117 (0.032 - 0.559)	40	21.4 ± 13.9 (8.7 - 79.4)	40	29.4 ± 31.4 (1.00 - 100)	37
PITC	0.201 ± 0.106 (0.032 - 0.512)	32	22.9 ± 15.0 (10.0 - 79.4)	32	36.0 ± 32.5 (1 - 100)	29
AQC	0.239 ± 0.070 (0.158 - 0.349)	5	16.9 ± 6.7 (8.7 - 25.1)	5	4.3 ± 1.5 (1.60 - 5.00)	5
OPA/FMOC	0.319 ± 0.15 (0.212 - 0.425)	2	16.9 ± 2.0 (15.4 - 18.3)	2	5.5 ± 6.3 (1 - 10)	2
FMOC	0.559	1	9.0	1	12.5	1

† 5 sites did not report the amount of sample analyzed

TABLE II. Correlation of different membranes with yield, error and % sample analyzed

Membrane	Yield (nmol) average ±S.D. (range)	Error (%) average ±S.D. (range)	(%) Sample Analyzed average ±S.D. (range)	n
A	0.201 ± 0.083 (0.021 - 0.331)	18.0 ± 5.09 (10.0 - 25.4)	49.1 ± 34.1 (10 - 100)	13
B	0.180 ± 0.081 (0.037 - 0.283)	20.1 ± 13.8 (8.7 - 47.5)	47.5 ± 35.5 (2 - 100)	10
C	0.212 ± 0.077 (0.032 - 0.330)	25.1 ± 10.8 (13.9 - 51.7)	47.6 ± 25.2 (10 - 100)	13
D	0.225 ± 0.124 (0.080 - 0.481)	33.7 ± 27.4 (13.4 - 79.4)	46.5 ± 28.4 (13 - 100)	8
E	0.212 ± 0.125 (0.076 - 0.456)	19.4 ± 8.38 (11.8 - 34.9)	31.0 ± 35.2 (1 - 100)	7
F	0.295 ± 0.140 (0.116 - 0.559)	16.4 ± 5.92 (9.0 - 29.2)	47.1 ± 42.5 (5 - 100)	8
G	0.191 ± 0.067 (0.070 - 0.271)	19.9 ± 6.80 (10.4 - 30.5)	38.6 ± 33.3 (5 - 100)	11
H	0.306 ± 0.112 (0.192 - 0.512)	17.5 ± 3.43 (12.7 - 21.7)	11.7 ± 11.4 (1 - 30)	7

Abbreviations used: AQC, 6-aminoquinolyl-N-hydroxysuccinimidyl carbamate; DTPA, dithiopropionic acid; FMOC, N-(9-Fluorenylmethoxycarbonyl); OPA, orthopthaldialdehyde; PITC, phenylisothiocyanate; TGA, thioglycolic acid a free protein mixture hydrolyzed and analyzed by the participants, the error in analysis of PVDF-bound samples is still essentially 2-fold greater.

In general, there is evidence that there has been improvement in accuracy among the participants. For example, comparing results from analysis of free protein samples, the accuracy of ABRF-94AAA1 was better than ABRF-90AAA1 (10.9% and 13.5% average error, respectively). Hence, the improvement seen in this year's results relative to those obtained in 1990 with PVDF-bound protein may be a result of generally improved technique among the participating laboratories. However, examination of the data shown in Table II suggests that the analysis of PVDF-bound samples is troubled by high variability (indicated by the broad range of results) relative to unbound samples [8]. While it is curious that the top ten sites all used pre-column derivatization, when viewed in the context of previous years' studies [5-8], this may be a statistical aberration, as in the case of ABRRF-94AAA2, when 9 of 10 best sites used post-column methods. In all other studies, pre- and post-column technologies were equally represented among the highest quality sites.

Table III correlates the accuracy obtained for individual amino acids with the method of analysis. As observed previously [4], Arg and Met, present in low abundance in myoglobin, are problem residues. The same argument may apply to Tyr, Pro, and Ser, with the latter also subject to hydrolytic losses. The same cannot be said of His and Lys, abundant residues which show higher than average error. Interestingly, the data from the less commonly reported techniques (last 4 columns in Table III), do not show high errors consistently associated with any specific amino acid.

Table IV correlates the various additives used in hydrolysis with yield, error, and proportion of sample analyzed. The large majority of participants used only phenol in their hydrolyses, and it appears that this group achieved slightly better yield and accuracy. However, the low number of sites using other additives and the wide range of the data make definitive conclusions impossible.

TABLE III. Correlation of accuracy (%Error) of determination for individual amino acids with analysis method

Amino Acid	Total (n=77)	Post-column (n=37)	Pre-Column (n=40)	Nin-hydrin (n=35)	PITC (n=32)	AQC (n=5)	FMOC/OPA (n=2)	OPA (n=2)	FMOC (n=1)
Ala	9.3	7.2	11.1	7.4	12.8	6.0	4.3	3.9	0.9
Arg	31.5	28.6	33.8	30.0	34.5	41.5	25.0	4.3	21.6
Asp	22.6	17.4	27.1	17.6	29.6	15.1	23.9	15.3	21.9
Glu	13.0	14.1	12.0	14.5	13.2	7.3	10.1	7.6	4.3
Gly	18.6	17.0	20.1	17.5	20.2	27.4	12.1	9.0	1.3
His	34.7	35.9	32.7	37.3	36.0	21.0	29.7	11.0	15.1
Ile	13.4	12.6	14.6	12.0	15.9	7.1	9.8	23.5	8.3
Leu	9.3	7.0	11.2	7.3	13.1	4.5	6.0	3.0	2.3
Lys	26.7	25.6	26.8	26.8	27.2	15.6	65.7	4.9	16.4
Met	39.9	47.6	35.3	44.7	36.5	28.9	15.0	98.8	8.6
Phe	11.0	9.1	12.6	9.3	14.6	5.7	6.1	5.7	3.3
Pro	34.4	41.0	27.3	42.8	29.8	19.3	18.4	9.0	17.5
Ser	25.5	25.5	25.9	25.5	26.3	30.9	16.4	24.3	1.6
Thr	12.0	10.5	13.7	10.6	14.8	12.9	3.4	8.8	0.8
Tyr	26.5	25.7	27.3	26.9	32.0	9.3	18.6	4.7	11.5
Val	11.2	11.9	10.8	11.9	12.1	3.7	5.4	12.9	8.5
Average	21.4	21.1	21.3	21.4	22.9	16.0	16.9	15.4	9.0

TABLE IV. Correlation of additives used with yield, error and % sample analyzed

Additive	Yield (nmol) average (range)	Error (%) average (range)	(%) Sample Analyzed average (range)	n*
none	0.203 ± 0.094 (0.076 - 0.330)	24.1 ± 19.4 (13.9 - 79.4)	57.1 ± 33.1 (10 - 100)	11
phenol	0.238 ± 0.099 (0.032 - 0.559)	18.3 ± 7.8 (8.6 - 51.7)	32.3 ± 29.3 (1 - 100)	44
phenol +reducer	0.263 ± 0.106 (0.147 - 0.481)	25.7 ± 21.9 (12.7 - 74.7)	51.5 ± 32.1 (20 - 100)	8
2-mercapto-ethanol	0.178 ± 0.097 (0.021 - 0.262)	22.2 ± 7.5 (14.0 - 32.0)	67.5 ± 32.6 (25 - 100)	5
thioglycolate/ dithiopropionate	0.167 ± 0.037 (0.142 - 0.193)	19.3 ± 3.5 (16.8 - 21.8)	80.0 ± 28.3 (60 - 100)	2

* 7 sites did not report the type of additive used.

C. Background

Data from the blank membranes are tabulated in Table V. Columns 1 and 2 relate yield from blanks to chemistry used and show that all methods detected significant amounts of amino acids. Columns 3 and 4 indicate that additives are largely irrelevant in affecting the background, with the possible exception of TGA/DTPA (n=2), which seemed to be associated with lower background levels. Columns 5 and 6 indicate that the amount of background did not depend on the sheet of membrane used, and there was wide variation in background levels obtained from each sheet of membrane. Also, it was not useful to subtract blank values from sample values in determining amino acid composition (data not shown). This amount of background contamination is surprising, and would certainly degrade the quality of these analyses. Whether PVDF-bound samples obtained from electroblotting procedures give similar results is a question not addressed by this study. Obviously, this issue should be considered by laboratories performing such experiments.

Table V. Correlation of total background with method, additives, and membrane batch

Chemistry	µg, ±S.D. (range)	Additive	µg, ±S.D. (range)	Membrane	µg, ±S.D. (range)
Post-column	0.446±0.569 (0.037-2.89)	none	0.215±0.173 (0.037-0.607)	A	0.372±0.189 (0.081-0.76)
ninhydrin	0.454±0.585 (0.037-2.89	phenol only	0.874±1.53 (0.109-9.45)	B	1.32±0.79 (0.186-2.9)
OPA	0.308±0.067 (0.303-0.313)	phenol + reducer	0.426±0.36 (0.135-1.0)	C	0.376±0.711 (0.044-2.71)
Pre-column	0.754±1.53 (0.069-9.45)	2mercapto-ethanol	0.219±0.09 (0.091-0.307)	D	0.318±0.295 (0.044-0.99)
PITC	0.759±1.71 (0.069-9.45)	TGA/ DTPA	0.047±0.053 (0.043-0.051)	E	0.235±1.92 (0.38-0.61)
AQC	0.858±0.581 (0.304-1.53)			F	0.503±0.444 (0.163-1.37)
FMOC/OPA	0.305±0.186 (0.174-0.436)			G	1.16±2.76 (0.76-9.45)
				H	0.334±0.234 (0.125-0.82)

Table VI. Best analysis of ABRF-95AAA by pre- and post-column methods

Site / Technique [*] Amino Acid	Theory	8268 / AQC Rank: 1	7391 /ninhydrin Rank: 11	Overall Avrg.± S. D.
Ala	15	15.72	14.38	14.56 ± 2.33
Arg	2	2.39	2.30	2.29 ± 0.71
Asp	10	10.95	12.54	11.56 ± 4.33
Glu	19	18.10	18.83	19.35 ± 3.88
Gly	15	16.67	16.59	16.69 ± 3.87
His	11	9.73	7.37	7.13 ± 2.08
Ile	9	8.76	8.29	7.91 ± 1.12
Leu	17	17.16	18.08	16.15 ± 2.5
Lys	19	16.29	15.11	14.35 ± 4.49
Met	2	1.97	2.68	1.36 ± 0.81
Phe	7	6.80	6.78	6.88 ± 1.54
Pro	4	4.81	4.68	4.84 ± 2.94
Ser	5	5.93	4.92	5.86 ± 1.21
Thr	7	7.18	6.39	6.66 ± 1.30
Tyr	2	1.81	1.77	1.85 ± 0.73
Val	7	7.28	7.28	6.91 ± 1.17
Amt.analyzed (%)		5	20	
Total Yield (nmol)		0.234	0.295	
Error (%)		8.65	12.74	

[*] These are the best pre- and post-column sites; the overall rank includes the full set of sites.

The data of the best pre- and post-column analyses, shown in Table VI, also reflect this difficulty with PVDF-bound sample, especially when compared to last year's two best results from an unbound protein sample [8], which had, on average, errors of 5.6%. However, with approximately 10% average error, the current best results can be considered reasonably good analyses.

IV.   SUMMARY AND CONCLUSIONS

The goal of this study was to determine the accuracy of amino acid analysis of a protein adsorbed to PVDF membranes. Adequate sample was provided so that analyses could be conducted using the preferred method at any site. Pre- and post-column methods gave equivalent overall accuracy, a result consistent with a previous study where relatively high levels of sample were provided [4]. Analysis of PVDF-bound protein is still problematic; 2-fold higher average error is obtained compared with results obtained from analysis of free protein [8]. The inclusion of control blank PVDF in this year's sample revealed high and variable background levels of amino acids. It is most likely that high background levels generally precluded high quality analyses from PVDF membranes.

In conclusion, there was improvement in analysis of PVDF-bound proteins compared to the earlier study [4]. It appears that this result is due to generally improved performance among the participants, a major objective of the ABRF collaborative trials. Nevertheless, average errors observed from analysis of PVDF-bound protein are still 2-fold higher relative to analysis of free protein [8] and there was little difference between the major methodologies utilized.

Acknowledgments

We would like to thank Dr. Y. Bao (U. Virginia) for tabulating the data and maintaining the anonymity of the participants, Hung Nguyen and Sara Reynolds (U. Florida) for sample preparation, and all the ABRF facilities who have taken part in this project. This work was supported in part by NSF grant DE-FG02-95ER61839, to J.W. Crabb, on behalf of the ABRF.

References

1. Niece, R. L., Williams, K. R., Wadsworth, C. L., Elliot, J., Stone, K. L., McMurray, W. J., Fowler, A., Atherton, D., Kutny, R., and Smith, A. (1989) *in* "Techniques in Protein Chemistry" (T. E. Hugli, ed.), Academic Press, San Diego, pp 89-101.
2. Crabb, J. W., Ericsson, L. H., Atherton, D., Smith, A. J. and Kutny, R. (1990) *in* "Current Research in Protein Chemistry" (J. J. Villafranca, ed.) Academic Press, San Diego, pp 49-61.
3. Ericsson, L. H., Atherton, D., Kutny, R., Smith, A. J. and Crabb, J. W. (1991) *in* "Methods of Protein Sequence Analysis" (H. Jörnvall, J.-O. Höög, and A. M. Gustavsson, eds.) Birkhauser Verlag, Basel pp. 143-150.
4. Tarr, G. E., Paxton, R. J., Pan, Y.-C. E., Ericsson, L. H., and Crabb, J. W. (1991) *in* Techniques in Protein Chemistry II" Academic Press, pp 139-150.
5. Strydom, D. J., Tarr, G. E., Pan, Y.-C. E., and Paxton, R. J. (1992) *in* "Techniques in Protein Chemistry III" (R. H. Angeletti, ed.) Academic Press, San Diego, pp 261-274.
6. Strydom, D. J., Andersen, T. T., Apostol, I., Fox, J. W., Paxton, R. J., and Crabb, J. W. (1993) *in* "Techniques in Protein Chemistry IV" (R. H. Angeletti, ed.) Academic Press, San Diego, pp 279-288.
7. Yüksel, K. Ü., Andersen, T. T., Apostol, I., Fox, J. W., Crabb, J. W., Paxton, R. J., and Strydom, D. J. (1994) *in* "Techniques in Protein Chemistry V" (J. W. Crabb, ed.) Academic Press, San Diego, pp 231-240.
8. Yüksel, K. Ü., Andersen, T. T., Apostol, I., Fox, J. W., Paxton, R. J., and Strydom, D. J. (1995) *in* "Techniques in Protein Chemistry VI" (J. W. Crabb, ed.) Academic Press, San Diego, pp 185-192.
9. Atherton, D. (1989) in "Techniques in Protein Chemistry" (T. Hugli, ed.) pp. 273-294.
10. Chen S.-T., Chiou, S.-H., Chu, Y-H., and Wang, K.-T. (1987) Int. J. Peptide Protein Chem. **3 0**, 572-576.
11. Choli, T., Kapp, U., and Wittman-Liebold, B. (1989) J. Chromatog. **476**, 59-72.
12. Ozols, J. (1990) Meth Enzymol. **182**, (Deutscher, M. P. ed.) Academic Press, San Deigo pp. 587-601.
13. Tous, G. I., Fausnaugh, J. L., Akinyosyye, O., Lackland, H., Winter-Cash P., Vitorica, F. J., and Stein, S. (1989) Anal. Biochem., **179**, 50-55.
14. Gharahdgaghi, F., Atherton, D., DeMott, M., and Mische S. M. (1992) in "Techniques in Protein Chemistry III" (R. H. Angeletti, ed.) Academic Press, San Diego, pp 249-260

Amino Acid Analysis Using Various Carbamate Reagents for Precolumn Derivatization

Daniel J. Strydom

Center for Biochemical and Biophysical Sciences and Medicine,
and Department of Pathology, Harvard Medical School, Boston, MA 02115

I. Introduction

Chromatographic analysis of pre-derivatized amino acids is at present the most popular method of amino acid analysis. The recent introduction of 6-aminoquinolyl-N-hydroxysuccinimidyl carbamate (AQC) as a derivatizing agent (1,2) demonstrated for the first time the practical utility of a novel class of derivatizing agents, which had first been studied as activated carbamate reagents, but had not for the most been developed fully as amino acid analysis reagents (3,4). The fluorescent naphthylaminecarbamate reagent, which was proposed for automated amino acid analyses (3), has not been used widely.

The AQC-derivatives are fluorescent and as a consequence have their greatest utility in sensitivity and selectivity of detection after chromatographic separation. They are very stable, in contrast to most other pre-column derivatives, due to their chemical structures, which are those of N-substituted ureas.

The diversity of samples and sample matrices requiring amino acid analysis makes such a varied demand on resolution, accuracy and sensitivity, that no single method yet suffices. Some applications need different chromatographic conditions not easily provided by current separations of AQC-amino acids, while selectivity of detection and sensitivity may not be an absolute requisite. The properties of amino acid derivatives of two non-fluorescent reagents, based on the same chemistry as that of the AQC-reagent were therefore investigated. These are the simplest aromatic homologue, with an anilino group replacing the aminoquinolyl function (first reported by Nimura et al. (4)), and a reagent with potential near-UV absorbance, based on amino-azobenzene.

II. Materials and Methods

A. Synthesis of Reagents

The reagents were synthesized according to the methodology of Nimura et al. (4) by reaction of aromatic amines with di-(N-succinimidyl) carbonate.

Anilino-N-hydroxysuccinimidyl carbamate (AC): 1 mmol aniline in 3 ml dry acetonitrile was added dropwise to 1.2 mmol disuccinimidyl carbonate (DSC) in 5 ml dry acetonitrile. After 1 h the solution was dried by evaporation under a stream of dry nitrogen, redissolved in 15 ml ethylacetate, and washed successively with 1 N HCl, 4% sodium bicarbonate and water. The organic layer was dried over anhydrous sodium sulfate and evaporated under dry nitrogen to yield the reagent.

4-Phenylazoaniline-N-hydroxysuccinimidyl carbamate (PAPC): 1 mmol phenylazoaniline in 3 ml acetonitrile and 4 ml butylchloride was reacted with 1.2 mmol DSC in 13 ml acetonitrile and 2 ml butylchloride. The reaction mixture was worked up as above.

B. Derivatization of Amino Acids

Both the PAPC and AC derivatives were prepared by reaction of a solution of the reagent in dry acetonitrile (1-5 mg/ml) with the sample in borate (0.2 M, pH 9) buffer, by the same methodology as was used for the AQC reagent (1).

C. Chromatography of Derivatives

Chromatographic separations of the AC- and PAPC-derivatized amino acids were developed as described below. The final chromatographic conditions were as follows:

AC derivatives: The column was a Nova-Pak® C18, 3.9 x 150 mm (Waters), used at 36°. Solvent A was 0.14 M sodium acetate, 0.14% triethylamine, pH set to 5.5 with o-phosphoric acid, while solvent B was 60 % methanol (v/v). The gradient was in linear segments from 9 to 12 %B in 10 min, to 80 %B in 20 min, and to 85%B in 5 min at a flowrate of 1 ml/min. Turnaround time was 56 min. PAPC derivatives: The column was a Symmetry™ C18, 3.9 x 150 mm (Waters), at 45°. Solvent A was 0.14 M sodium acetate, 0.05% triethylamine, pH set to 5.20 with o-phosphoric acid, and solvent B was 60 % acetonitrile (v/v). The gradient was in linear segments from 37 to 40%B in 15 min, to 70%B in 15 min, and to 100%B in 5 min, at a flowrate of 1 ml/min. Turnaround time was 55 min.

III. Results

A. Chromatographic Separation of Phenylazophenylamino- and Anilinocarbamate Derivatives of Common Amino Acids.

The separation of the AC-derivatives of the common amino acids was designed, based on the conditions developed for the AQC-amino acids (1). The use of acetonitrile as organic modifier, under a variety of conditions, did not separate all of the common amino acid derivatives, and a change to methanol provided the foundation for the good separation finally developed, as shown in fig.1.

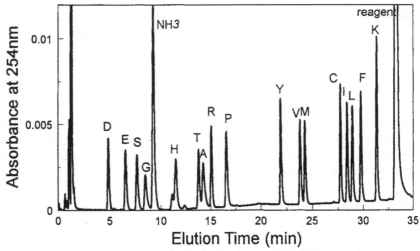

Figure 1. Separation of the AC-derivatives of the common amino acids. A Nova-Pak C18 column at pH 5.5, 36°, was eluted with a gradient in methanol as described in the text.

Likewise the separation of the PAPC-derivatives started from the AQC-separation conditions, but in this instance no satisfactory separation could be attained on the Nova-Pak column, with acetonitrile, methanol, and isopropanol as organic modifiers. The recently developed Symmetry™ column however provided the basis for a very good separation, shown in fig. 2. The high organic modifier content of the initial conditions illustrate the extreme hydrophobicity of the phenylazophenyl derivatives, which is also manifested in difficulties in solubilizing these derivatives and keeping them in solution.

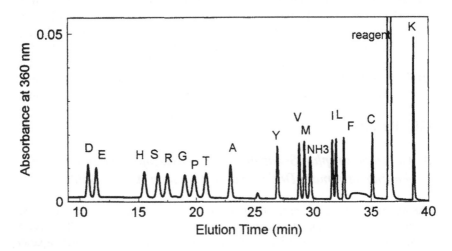

Figure 2. Separation of the PAPC-derivatives of the common amino acids. A Symmetry C18 column at pH 5.2, 45°, was eluted with a gradient in acetonitrile as described in the text.

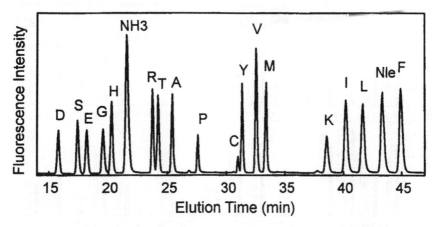

Figure 3. Separation of the AQC-derivatives of the common amino acids. A Nova-Pak C18 column at pH 5.05, 35°, was eluted with a gradient in acetonitrile as described (5).

When comparing these separations mutually and with the standard AQC-separation (fig. 3), it is clear that specific amino acids, such as Arg, His, Lys, Pro and Cys, ammonia, and the hydrolyzed reagent show marked differences in elution times relative to the other amino acids. This provides opportunities to select specific derivatives for use in specific applications.

B. Applications

The good chromatographic separations now available for these derivatives, allows the question to be raised whether they can be used for amino acid analysis. Replicate hydrolysates were therefore prepared of different proteins, by conventional vapor phase hydrolysis with 6 N HCl at 110° for 18 h. Duplicate samples from this set of hydrolysates were derivatized with each of the various reagents (AQC, AC and PAPC) and analyzed, using the established chromatographic conditions. Calibration was based on parallel analysis of derivatized amino acid standards (non-hydrolyzed).

These analyses are compared with the theoretical compositions in fig.4. The derived compositions are quite reasonable, with the AC and PAPC analyses deviating only slightly more than the AQC method, as seen from their average errors in Table I.

Table I. Quality of Amino Acid Analysis, reported as average percentage absolute error in composition

	AQC	AC	PAPC
Chymotrypsin	6.7	9.9	6.3
BSA	4.5	7.8	8.7
Lysozyme	5.6	9.0	11.3

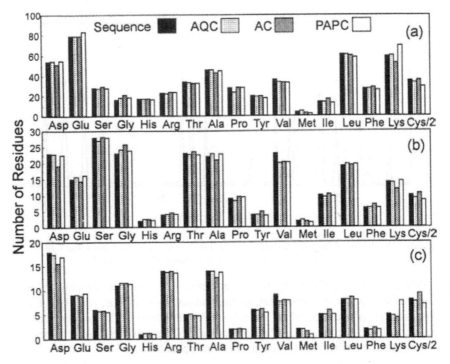

Figure 4. Comparison of amino acid compositions determined by three carbamate reagents. Duplicate analyses of (a) bovine serum albumin, (b) chymotrypsinogen, and (c) human lysozyme were averaged and are compared with the theoretical values.

The success of all these methods in amino acid analysis allows the rational selection of a specific derivatizing agent for specific applications. Two such applications are readily visible.

The separations achievable for the very hydrophilic amino acid derivatives, such as phospho amino acids and hydroxyproline, are most convenient with the PAPC-

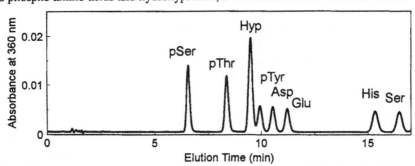

Figure 5. Separation of hydrophilic amino acid derivatives. A mixture of phosphoamino acids, hydroxyproline, and the common amino acids were derivatized with PAPC and chromatographed on a Symmetry™ column as described in the text.

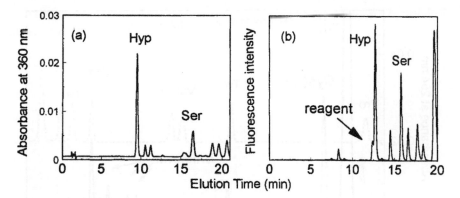

Figure 6. Comparison of hydroxyproline analysis in a "realworld" sample by AQC and PAPC. A plant-protein extract was hydrolyzed with 6N HCl and derivatized either with AQC, or with PAPC. The chromatographic analyses were done for (a) PAPC-derivatives or (b) AQC-derivatives, as in figs 2 and 3, respectively.

derivatives, as shown in fig.5. There is ample chromatographic space for even further manipulation of the separation, and it is noteworthy that, contrary to the other two separations, larger injection volumes can be tolerated, since adsorption to the column is very strong. Since the initial conditions include a very high amount of organic modifier, one can through the simple expedient of starting with a much lower organic modifier content, load a large volume, rapidly change to the appropriate condition for starting the established separation gradient (in < 1 min), and get a successful separation. Fig. 6 demonstrates the easier analysis of hydroxyproline by PAPC, as opposed to AQC where the reagent peak poses some interference.

One example of an application where the AC-derivative is useful, stems from the enhanced separation of cystine and its derivatives. Fig. 7 demonstrates the separation of presumed derivatives and isomers of cystine, formed during disulfide-exchange analyses of a unique protein with dithiodipropionic acid. The unknowns are spread over a considerably wider area than when using AQC as derivatizing agent. In addition *meso*-cystine is separated well from cystine, whereas their AQC-derivatives coelute.

IV. Discussion

The variety of samples and sample matrices requiring amino acid analysis places such a varied demand on resolution, accuracy and sensitivity, that no single method yet suffices, from the arsenal of techniques available for the bioanalytical laboratory. The two most popular techniques are based on postcolumn detection with ninhydrin, and pre-column derivatization with phenylisothiocyanate, both however beset with specific technical problems. Thus ninhydrin while still the most accurate method, suffers from a lack of sensitivity and is a technically difficult technique, while the PTC derivatives although more sensitive, are

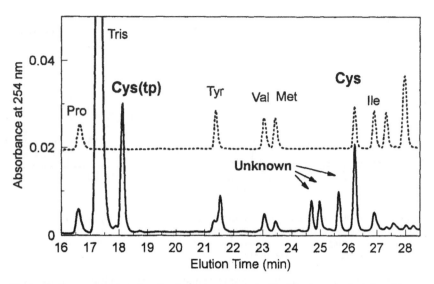

Figure 7. Chromatography of cystine derivatives, reacted with AC. A cystine/cysteine rich protein (10 μg) was hydrolyzed in the vapor phase with 6N HCl for 20 h, in the presence of 0.1 mg dithiodipropionic acid (8). The dried hydrolysate was reacted with AC reagent and chromatographed as in fig. 1. A standard analysis is shown for comparison (......). Cys(tp), thiopropionic acid derivative of Cys.

unstable, leading to loss of accuracy. The introduction of the AQC-reagent for pre-column derivatization (1) has however provided a pronounced improvement in accuracy and ease of use of pre-column derivatization for amino acid analysis (5).

The properties of AQC derivatives that lead to their decided advantage over, e.g., those with PITC, include stability, sensitivity and selectivity due to the fluorescent mode of detection. It can be used very accurately (5). When ultra high sensitivity is not needed for specific applications, the technical difficulties inherent in fluorescent HPLC detection argue for using UV absorbance detection, and more robust quantitation. This should in principle yield even more accurate analyses than possible with the fluorescent AQC method, and if adjustments in chromatography are done (6), the early-eluting reagent peak (aminoquinoline) does not interfere with determination of Asp, and such accuracy is indeed possible for standards. The lack of selectivity due to interference by common contaminants and matrix components of realworld samples will of course act against such improvement.

During the past decade a number of amine-reacting succinimidocarbamate reagents have been made, containing aryl functionalities such as phenyl (3), bromophenyl (1), 2-fluorenyl (7), and α-naphthyl (4). Only the phenyl and α-naphthyl reagents were shown to derivatize amino acids, and for the α-naphthyl derivatives a separation of common amino acids was demonstrated, although the separation was not optimal. The variety of groups one could attach to the succinimido carbamate functionality is endless. When the need for a variant on the AQC-reagent was perceived, the choice fell on the simplest aromatic analog, phenyl, to provide UV absorption for relatively sensitive detection, and on phenylazophenyl, to provide a more selective detection wavelength.

These reagents could be synthesized easily, and both appeared to react well with free amino acids, as borne out by the color yields which were equivalent for all the amino acids, except lysine, which presumably reacted at both α- and ϵ-amino groups, and cystine which reacted at both α-amino groups. Only singly modified His and Tyr was seen with either of the reagents. Stability of the AC-derivatives seems good, but after a week at room temperature, some changes in chromatography were apparent. The PAPC-derivatives likewise appeared stable over periods of 2 days at room temperature, except for di-PAPC-Lys, which seems to crystallize out of solution. No attempts have yet been made to gain solubility by, e.g., increasing the organic solvent content of the derivatization mixture. The reagents provide accuracy in amino acid composition, as shown in fig.4 and Table I. The largest variations are found for the PAPC-derivatives, and specifically in analysis of Lys. The other amino acids are determined with good accuracy.

Comparing the properties of the AQC-, AC- and PAPC-derivatives of amino acids, as to their analysis in different parts of the chromatograms, the strengths and weaknesses of each type of derivative become obvious. Thus even though the hydrophilic amino acids with all three derivatives are very well resolved, the AC-derivatives have limited chromatographic space, which means that amino acid variants such as phospho-amino acids, hydroxyproline and cysteic acid, and amino sugars, do not have adequate space to separate. In contrast, both AQC and PAPC hydrophilic derivatives have ample chromatographic space, which allows the very hydrophilic compounds ample separation space. The AQC-reagent peak, which elutes a few minutes before Asp in the standard separation interferes with estimation of hydroxyproline, and as shown above, this is one aspect of amino acid analysis that is most efficiently handled by the PAPC derivatives. Hydroxyproline is incompletely separated from Asp using AC-derivatives. The phospho amino acids, pSer, pThr, and pTyr, are also separated best as PAPC-derivatives. The extreme hydrophobicity of the PAPC labeling group, although problematic for stability of the more hydrophobic amino acids, provides for disturbance-free injection of these hydrophilic amino acids.

Cystine, which conveniently elutes as one AQC-derivative peak, even after hydrolysis, which partially isomerases cystine to *meso*-cystine, elutes as partially separated PAPC- peaks and as well separated AC-peaks. The preliminary studies on a cystine/cysteine rich protein shown above, showed that better resolution of a variety of cystine-related peaks could be achieved as AC-derivatives. It is interesting that for the three proteins examined here (fig 4), all containing cystine, all three reagents directly provide close to quantitative estimation of cystine.

Overall, therefore, the AQC derivatives provide optimal analyses of amino acids, with PAPC and AC providing solutions to at least the two problem areas discussed above. Complementary use of these reagents should therefore provide the tools to handle most types of samples and purposes of amino acid analysis.

Acknowledgments

I would like to thank Wynford V. Brome and Jerry R. Brito for their excellent technical assistance during various stages of these studies. This work was supported in part by the Endowment for Research in Human Biology, Boston, MA.

References

1. Cohen, S. A., DeAntonis, K., and Michaud, D. P. (1993) *in* Techniques in Protein Chemistry IV (Angeletti, R. H., Ed.), pp. 289-298, Academic Press, San Diego.
2. Cohen, S. A., and Michaud, D. P. (1993) *Anal. Biochem.* **211**, 279-287.
3. Iwaki, K., Nimura, N., Hiraga, Y., Kinoshita, T., Takeda, K., and Ogura, H. (1987) *J. Chromatogr.* **407**, 273-279.
4. Nimura, N., Iwaki, K., Kinoshita, T., Takeda, K., and Ogura, H. (1986) *Anal. Chem.* **58**, 2372-2375.
5. Strydom, D. J., and Cohen, S. A. (1994) *Anal. Biochem.* **222**, 19-28.
6. Liu, H. J. (1994) *J. Chromatogr. A* **670**, 59-66.
7. Hirai, T., Kitamura, M., and Inoue, Y. (1991) *Bunseki Kagaku* **40**, 233-238.
8. Barkholdt, V., and Jensen, A. L. (1989) *Anal. Biochem.* **177**, 318-322.

Analysis of Mixture Sequences
Derived from Edman Degradation Data

William J. Henzel, John T. Stults, Susan C. Wong,
Angela Namenuk, James Yashio and Colin Watanabe

Genentech, Inc., South San Francisco, CA 94080

Introduction

The adaptation of PVDF for electroblotting proteins from gels has greatly facilitated the use of Edman degradation for protein analysis (1). Sequence analysis of a single band on PVDF can provide rapid, direct confirmation of the identity of a protein. For proteins that contain signal sequences N-terminal sequence analysis is often the only method that can directly determine the start of the mature sequence. A limitation to the use of this technique is the presence of multiple components. A sequence mixture can result from the co-migration of several proteins that appear as a single band on a gel. When a mixture sequence is subjected to Edman degradation, it is often difficult to determine what proteins or peptides are present even with known sequences. Often only a very small amount of a protein is available and rather than risk the possibility of obtaining a indecipherable mixture sequence, the decision is made to fragment the protein and isolate and sequence the resulting peptides. A major drawback to this approach is that if a gel band contains several proteins, some of which are at lower levels, the peptides from the lower level proteins may not be detected during peptide isolation.

A number of reports (2-4) have been published that utilize mass spectrometry data to sort mixtures derived from a protein that has a known sequence. More recently Johnson and Walsh (5) described an algorithm for sorting peptide mixtures by combining Edman data with sequence data obtained from tandem mass spectrometry. We have developed two algorithms that are designed to sort mixture sequences. The first is an algorithm (SEQSORT) that allows mixtures that contain known sequences to be easily screened against a protein sequence database. It is capable of identifying known proteins by using all possible combinations of adjacent residues found in a mixture to search a protein sequence database. The second algorithm (MOLWTFIT) was designed for sorting peptide mixtures where the sequence is unknown. This algorithm utilizes Edman degradation data and peptide molecular masses to sort all the amino acids in each cycle to fit a mass within a specified mass tolerance. The MOLWTFIT algorithm utilizes knowledge gained from quantitative sequence analysis to limit the number of possibilities in a search.

SEQSORT

The SEQSORT algorithm finds patterns specified as regular-expression syntax. It is similar in implementation to the UNIX regular-expression matching program grep. SEQSORT has three additional features: (1) a number of allowed mismatches can be specified, (2) the search can be restricted to a region around the N-terminal, and (3) the search can be limited to proteins with a specific molecular weight range. The algorithm begins by compiling the ambiguous sequencer data using a finite automaton to find regular expressions. The finite automaton described by Miller (6) is augmented by allowing transitions on mismatches, so long as the number of mismatches is below a user-specified threshold. Next each sequence of the database is examined. Sequences having a molecular weight outside the specified limits are rejected. If the search is restricted to a region near the N-terminal, the sequence is truncated to the region of interest. The resulting sequence is checked for the existence of the specified pattern. If the pattern is found, the sequence is added to a list which can later be sorted by molecular weight or by the number of mismatches. The algorithm is capable of sorting a mixture of over 100 sequences. However, as the number of sequences increases, the number of random matches also increases, requiring longer sequences to prevent random matches.

A common problem encountered during protein purification is proteolysis. This can result in a ragged N-terminal sequence from aminopeptidase activity or internal cleavage resulting from the presence of endoproteases. For recombinant proteins that have undergone extensive proteolysis, the identity of the protein may be difficult to determine even if the sequence is known. Figure 1 shows an example of purified NT-3 that was cleaved by proteases during purification. The SEQSORT algorithm can easily sort this mixture by searching the mixture against the sequence of NT-3 rather than against the entire database. The output of the SEQSORT algorithm shows the presence of three cleavage sites in the NT-3 sequence. A search of this same mixture against the entire database also finds a match with NT-3 and the same 3 cleavage sites; however, 16 matches are found with other proteins.

At present no commercial protein sequence database is capable of allowing a search that specifies a match at the N-terminus of a protein. Restricting the

Sequence Analysis of NT-3. All units are in pmol.

1				5			
A	A	E	E	K	K	S	K
0.28	0.66	0.88	0.87	0.60	0.81	0.29	0.51
Y	Y	A	H	H	S	H	H
0.11	0.57	0.70	0.21	0.41	0.20	0.25	0.19
R	L	T	S	E	N	N	R
1.27	0.26	0.16	0.10	0.59	0.18	0.19	0.68
	G	R					
	0.21	0.46					
	I						
	0.09						

SEQSORT Input   Sequence file: p1.NT-3
          [AYR][AYLGI][EATR][EHS][KHE][KSN][SHN][KHR]

SEQSORT Output   Flanking regions appear in lower case (number at left gives start of match)
          62 gafresagapanrsr RYAEHKSH rgeysvcdseslwvt
          63 afresagapanrsrr YAEHKSHR geysvcdseslwvtd
          150 khwnsqcktsqtyvr ALTSENNK lvgwrwiridtscvc

Figure 1. SEQSORT Analysis of a Mixture of Several Proteolytic Fragments of NT-3.

mixture sequence matches to the N-terminus of a protein can increase the specificity of the search enabling identification with fewer residues. However, many proteins contain signal sequences. The length and cleavage site of the signal sequence is present as an annotation in a reference in some protein sequence databases, but at present this information can not be searched by current sequence analysis programs. We have devised a simple strategy for screening sequence against the N-terminal region. This is accomplished by utilizing a parameter (N) where N is the length of the sequence to screen plus the maximum length of any expected signal sequences. A value of 60, is usually adequate for a search containing up to 20 residues. When the NT-3 sequence mixture was analyzed by the SEQSORT algorithm with N = 60, only one protein other than NT-3 was found. When the same analysis was carried out with N = 0, allowing the mixture sequence to fit any region of the proteins in the database, false matches with 16 other proteins occurred as noted above.

In another example, sequence analysis of a PVDF electroblotted band containing the rat growth hormone receptor indicated the presence of 3 components. The mixture of 3 sequences was searched against a protein sequence database containing more than 170,000 sequences using the SEQSORT algorithm. With N = 60, the SEQSORT algorithm was able to identify all 3 sequences as fragments of rat growth hormone receptor (Figure 2). With N = 0 only one false match occurred. When residues 1-8 were used by the algorithm, over 100 false matches occurred. Using the same 8 residues with N = 60, the algorithm found only one false match, demonstrating the increased specificity of forcing a match to the N-terminal region of a protein when only a short region of sequence is available for searching.

Figure 3 shows an example of the use of SEQSORT to sort a complex mixture containing several different proteins. Sequence analysis of a 12 kDa band isolated from cystic fibrosis patient sputum showed a mixture of several sequences at equal concentrations. When the SEQSORT program was used with 0 mismatches, only one sequence matched. However, when the number of allowed mismatches was increased to 4, three proteins were found. All three proteins match the data at their N-terminus. The match with neutrophil defensin starts at residue 65, however residues 1-64 contain a signal sequence and

Protein Sequence Analysis of Rat Growth Hormone Receptor. All units are in pmol.

1				5					10	
L	Q	N	P	L	L	E	L	R	S	S
0.8	0.4	0.29	0.25	0.22	0.21	0.20	0.17	0.15	0.16	0.12
I	N	E	I	N	P	R	E	S	E	G
0.15	0.25	0.20	0.19	0.19	0.18	0.15	0.17	0.14	0.12	0.05
R	I	P	S	S	R	S	S			
0.09	0.21	0.15	0.18	0.15	0.15	0.11	0.11			
	P	G								
	0.15	0.06								
	R	R								
	0.05	0.09								

SEQSORT Input   Searching 60 N-terminal residues only
[LIR][QNIPR][NEPGR][PIS][LNS][LPR][ERS][LES][RS][SE][SG]

SEQSORT Output GHR_RAT growth hormone receptor precursor - rattus norvegicus, 71237 Da
36 gsgatpatlgkaspv LQRINPSLRES ssgkprftkcrspel
38 gatpatlgkaspvlq RINPSLRESSS gkprftkcrspelet
39 atpatlgkaspvlqr INPSLRESSSG kprftkcrspeletf

Figure 2. SEQSORT Analysis of Rat Growth Hormone Receptor.

Sequence Analysis of a 12 kDa Band from Human Sputum. All units are in pmol.

1				5					10					15					20	
A	L	Y	E	R	E	P	R	L	R	A	G	E	R	R	Y	G	T	Y	I	R
4.04	4.31	3.55	3.26	8.81	2.56	2.38	4.54	2.32	3.68	2.62	2.20	2.10	7.61	9.22	2.14	1.11	1.21	0.69	0.55	4.20
M	Y	T	R	L	I	K	A	K	N	T	I	R	D	V	K	H	K	Q	S	L
1.90	1.03	2.38	2.09	2.19	1.93	1.74	4.30	0.20	1.77	0.84	1.27	3.98	1.26	1.07	0.21	0.85	0.96	0.11	0.46	0.91
D	R	G	Q	Q	T	A	L		I	S	L	I	L	A	P		D		L	Y
1.20	0.89	0.66	0.46	0.96	1.04	1.18	0.76		1.69	0.80	0.60	1.12	0.37	0.42	0.19		0.26		0.12	0.32
T	V	P	K	I	D	S	Q		H	G			K							A
0.54	0.79	0.58	0.35	0.11	0.38	0.57	0.21		0.21	0.31			0.32							0.06
E	F				H	V	G		S											
0.36	0.39				0.23	0.24	0.09		0.19											
									T											
									0.15											

SEQSORT Input

    [AMDT][LYRVE][YTGPF][ERQK][RLQI][EITDH][PKASV][RALQG]
    [LK][RNIHST][ATSG][GIL][ERI][RDLK][RVA][YKP][GH][TKD][YQ][ISL][RLYA]
    Molecular weight range: 1000-15000
    Mismatches: 4 (mismatches appear in lower case)

SEQSORT Output

    Flanking regions appear in lower case (number at left gives start of match)
    CAGA_HUMAN Calgranulin a - homo sapiens, 10835 Da
      1 MLTELEKALNSIIDVYHKYSL ikgnfhavyrddlkk

    C33178 histone H3 - human (fragment), 2555 Da
      1 ARTKQTARKSTGgKAPrKQLA tka [2 misses]

    DEFN_HUMAN Neutrophil defensins 1, 2 and 3 precursor - homo sapiens, 10201 Da
      65 deslapkhpgsrknm AcYcRIPAcIAGERRYGTcIY qgrlwafcc [4 misses]

Figure 3. SEQSORT Analysis of the Components of a 12 kDa Band from Human Sputum.

a propeptide. The output also contained a number of histone sequences which were identical except for a single amino acid substitution. This program also identified both bovine and porcine calgranulin, which had a difference of 4 residues each when compared to the human sequence. A useful strategy is to increase the number of allowed mismatches until a match occurs or until a number of extraneous matches occur. The confidence of a match can be based on variety of criteria including molecular weight, biological source, and isoelectric point. If residues that are sometimes difficult to determine by Edman degradation (serine, threonine, cysteine and tryptophan) occur as the allowed mismatches, the confidence of the match is increased.

MOLWTFIT

The algorithm calculates all possible sequences from the residues obtained from the Edman degradation data that fit the measured masses within a mass tolerance. Input for the MOLWTFIT algorithm consists of peptide sequences and measured masses for a mixture. The algorithm uses a variety of information obtained from quantitative Edman degradation data to limit the number of possible sequences that result from analysis of a mixture. Residues that may be at the C-terminus are designated by using a $ following the residue. For example, a lysine or arginine would be considered a C-terminal residue if the peptides were derived by tryptic digestion.

Prior to sequence analysis, a small aliquot (10-100 fmol) of the peptide fraction can be analyzed by MALDI mass spectrometry to determine the mass and the number of components present. This information can be entered in the MOLWTFIT algorithm, allowing it to predict a residue not identified by the sequence data. Figure 4 shows the results of analysis by the MOLWTFIT algorithm for a short peptide sequence with an unidentified residue. The peptide

Sequence Analysis of a Lysine-C Peptide from IL-4 Stat. All units are in pmol.

1				5	
R	I	Q	I	X	K
2.0	1.07	0.48	0.47		0.16

MOLWTFIT Input
844.0 (tol=1.0 Dalton)
RIQIXK

MOLWTFIT Output
RIQIWK 844.047 Da

Figure 4. MOLWTFIT Analysis of a Peptide from IL-4 Stat.

was derived by lysine-C cleavage of 100 kDa IL-4 Stat (7). Sequence analysis was performed on a 6 mm micro cartridge using a Applied Biosystem 494 sequencer. Using a dalton mass accuracy of ±1 dalton, the algorithm found a match only if tryptophan was placed at residue 4. The MALDI spectrum (Figure 5) provided additional evidence of a tryptophan since a second mass was observed in the spectrum that was 32 daltons higher then the parent mass, an indicator of oxidation. Oxidation of methionine and tryptophan is frequently observed during MALDI analysis and results in addition of 16 and 32 daltons, respectively. The sequence of this peptide was used to clone IL-4 Stat (7).

Sequence analysis of a fraction obtained from a capillary HPLC separation of a lysine-C digested 98 kDa protein that was induced with high levels of IL-2 (8) resulted in a mixture sequence. MALDI mass analysis of a aliquot of this peptide mixture showed only a single mass of 1025.9 (Figure 5). Using the MOLWTFIT algorithm, it was possible to sort the mixture into two sequences that would fit the measured mass (Figure 6). The sequences obtained by the algorithm were identical except for residue 5, which could be isoleucine or a leucine. Since these two amino acids have identical masses the program was unable to distinguish them. Using the entire mixture sequence as input into the SEQSORT algorithm, a single sequence was obtained (mouse Stat1). This sequence was identical to the second sequence obtained by the MOLWTFIT algorithm.

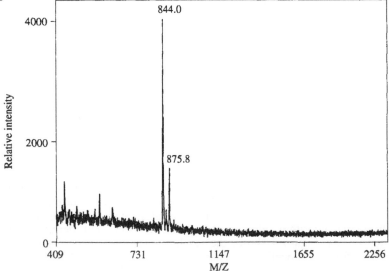

Figure 5. MALDI Mass Spectrum of a Lys-C Peptide of IL-4 Stat.

Sequence Analysis of a Lysine-C Peptide from a 96 kDa Protein Induced by IL-2. All units are in pmol.

1				5			
Y	L	Y	P	N	I	C*	K
0.6	0.3	0.32	0.41	0.31	0.23	0.07	0.22
S	E	T	L	V	L	G	E
0.6	0.05	0.19	0.09	0.18	0.08	0.04	0.04
Q						D	
1.3						0.04	

*Cysteine was alkylated with isopropyliodoacetamide.

MOLWTFIT Input
  Mol wt: 1025.900 (tol=1.0)
  Y L Y P N I C K$
  S E T L V L D E
  Q   G
  C-terminal residue ($)

MOLWTFIT Output
  YLYPNIDK: 1026.177 Da
  YLYPNLDK: 1026.177 Da

SEQSORTInput
  [YSQ][LE][YT][PL][NV][IL][CDG][KE]

SEQSORT Output
  Flanking regions appear in lower case (number at left gives start of match)
  MMU06924_1 Stat1 - Mus musculus, 87197 Da
    666 ykvmaaenipenplk YLYPNIDK dhafgkyysrpkeap

Figure 6. MOLWTFIT Mixture Analysis of a Lysine-C Peptide from a 96 kDa Protein Induced by IL-2

Conclusion

The SEQSORT algorithm allows rapid identification of known proteins that coelute as a single band on a 1 or 2-D gel. The SEQSORT algorithm is capable of rapidly sorting both peptide and protein mixtures that have known sequence. Proteins at lower levels in a mixture can easily be sorted using this algorithm. A typical search of over 170,000 proteins is performed in 1-2 minutes.

The MOLWTFIT algorithm is a useful tool to sort peptide mixtures when the protein sequence is unknown. Proteins that are resistant to proteolysis or that are at the low picomole to subpicomole levels may yield only a few peptides after proteolysis. Sequence analysis of peptides at these low level may result in some ambiguous residues. The MOLWTFIT algorithm can be used to decipher this data which may result in a sequence that can be used to design a cDNA probe.

References

1. Petrilli, P., Sepe, C. and Pucci, P (1991) *Biol. Mass. Spectrom.*, **20**,115-120
2. Kitagishy, T., Hong, Y., Takao, T., Aimoto, S. and Shimonishi, Y. (1982) *Bull. Chem. Soc. Jpn.*, **55**, 575-580
3. Matsuo, T., Matsuda, H., and Katakuse,I. (1981) *Biol. Mass. Spectrom.*, **8**, 137-143
4. Johnson, R.S., and Walsh, K.A. (1992) *Protein Science* 1, 1083-1091
5. Miller, Webb (1987) *A Software Tools Sampler*. Prentice-Hall, Englewood Cliffs, NJ
6. Hou, J., Schindler, U., Henzel, W.J., Ho, T. C., Brasseur, M. and McKnight, S. L. (1994) *Science*, **265**, 1701-1706
7. Hou, J., Schindler, U., Henzel, W. J., Wong, S. C., and McKnight, S. L. (1995) *Immunity*, **2**, 321-329

Design and Analysis of ABRF-95SEQ, a Recombinant Protein with Sequence Heterogeneity

Karen S. De Jongh[1], Joseph Fernandez[2], Jay E. Gambee[3], Gregory A. Grant[4], Barbara Merrill[5], Kathryn L. Stone[6] and John Rush[7]

[1]Department of Protein Chemistry, Cell Therapeutics Inc., Seattle, WA 98119
[2]Protein/DNA Technology Center, The Rockefeller University, New York, NY 10021
[3]Department of Research, Shriners Hospital for Crippled Children, Portland, OR 97201
[4]Department of Molecular Biology and Pharmacology, Washington University School of Medicine, St. Louis, MO 63110
[5]Bioanalytical Sciences, Burroughs Wellcome Co., Research Triangle Park, NC, 27709
[6]W.M. Keck Foundation Biotechnology Resource Laboratory, Yale University School of Medicine, New Haven, CT 06510
[7]Department of Genetics, HHMI/Harvard Medical School, Boston, MA 02115

I. Introduction

One of the current challenges in protein sequencing is obtaining sequence data from proteins of limited availability. This challenge continues to be addressed through improvements in methodology and instrumentation, but it is also crucial that protein sequence instruments are operated under optimal conditions and that instrument operators are well educated in sample preparation and data interpretation. To help achieve this, the Association of Biomolecular Resource Facilities (ABRF) Protein Sequence Research Committee distributes test samples to members for analysis. These samples provide individual facilities with a means of self-evaluation. In addition, the emergence of biopolymer core facilities over the past decade has provided a mechanism for bringing state-of-the-art protein microchemistry to investigators at most institutions and these test samples also define realistic expectations for end-users of these services.

This report summarizes the sequencing results for the current test sample, ABRF-95SEQ, which represents the eighth in a series of unknown samples that have been distributed to ABRF facilities that perform protein sequencing. Previous studies have addressed issues of sequencing sensitivity (1), sample heterogeneity (2), sample support (3, 4), post-translational modifications (5) and identification of difficult residues (6, 7). The current study was primarily designed to assess three important aspects of protein sequence analysis that would routinely be encountered in a typical core facility: the length of sequence assignment possible with a low-level sample; the reliability of cysteine and tryptophan identification; and the ability of members to recognize sample microheterogeneity. This report describes the results

submitted to the Protein Sequence Research Committee by 71 facilities that
analyzed ABRF-95SEQ.

II. Materials and Methods

A. *Design of ABRF-95SEQ*

ABRF-95SEQ is a mixture of two recombinant proteins (variant 1 and variant 2)
of approximately 20,000 daltons each. These proteins are derived from guinea
pig Ocp, a protein from the organ of Corti in the inner ear involved in the
hearing process, which was chosen because it lacks significant homology to
other proteins in sequence databases. Variants of ocp were obtained by
genetically modifying a clone (pGEX/ocp II) encoding ocp fused to the
carboxyl terminus of glutathione S-transferase (GST). The encoded protein can
be separated from GST by thrombin cleavage. To generate variant 1 the clone
was modified to produce the following changes; one Cys residue and one Trp
doublet were introduced between residues 10 and 20; residues 21-23 were
altered to Pro-Arg-Pro to promote lag; residue 5 was changed to Gln and residue
9 was changed to Glu. Variant 2 was modified to generate a protein identical to
variant 1 except for residues 6-9, as shown in figure 1. The proteins were mixed
in a 2:1 ratio (variant 1: variant 2).

The sample allowed a number of questions regarding sequencing
performance to be addressed. First, the use of a protein allowed an assessment
of the number of amino acids that respondents could correctly assign. In
addition, the presence of two Trp residues and one Cys within the first twenty
amino acids allowed evaluation of the ability of respondents to assign these
difficult residues correctly. Cys and Trp assignment was the focus of a previous
study in this series (7), and thus ABRF-95SEQ provided an opportunity to
evaluate improvements in this area. Finally, the mixture of variant 1 and variant
2 generated a region of sample heterogeneity at cycles 6-9, providing an
opportunity to assess the ability of sequencer operators to recognize minor
sequences. The presence of both Glu and Gln at residue 9 was included to
assess whether respondents would recognize major and minor sequences, or
deamidation of Gln, at this position. The presence of both residues would
generate higher levels of Glu than normally seen during deamidation of Gln,
providing a contrast with position 5.

B. *Preparation of the Recombinant Proteins*

A recombinant plasmid allowing overexpression of wildtype ocp as a GST
fusion protein was generously provided by Dr. Hong Chen and Dr. Geoffrey
Duyk (HHMI/Harvard Medical School). Each Ocp variant was constructed by
PCR amplification of the wildtype plasmid, using primers that would change the
amino-terminus as shown in figure 1 and following a protocol similar to that
described previously (8). In the first round of amplification, mutagenic primers

```
Wildtype ocp: G S M P S I K L Q S   S D G E I F E V D V   E I A K Q S V T I K
              T M L E D L G M D D   E G D D D P V P L P   N V N A A I L K K V   I Q W

  Variant 1: G S M P Q I K L E S   S D G E C F E V W W   P R P . . .
  Variant 2: G S M P Q V L G Q S   S D G E C F E V W W   P R P . . .
```

Figure 1. Amino acid sequence of ABRF-95SEQ. The first 63 amino acids of Ocp and the region of variants 1 and 2 containing altered amino acids are shown. ABRF-95SEQ is 45 pmol of a 2:1 mixture of variant 1:variant 2. Residues that differ between the variants and the wildtype sequence are underlined.

annealing to the 5' end of the ocp gene were 178 bases long and were used without purification. These PCR products could not be directly cloned into pGEX expression vectors, presumably because of size heterogeneity at the 5' end of the unpurified primers. To specifically amplify full-length PCR products, the primary PCR products were amplified with a 20-base primer that could anneal to the 5' end of mutagenized ocp only, and these secondary PCR products were readily cloned in pGEX vectors. The variant clone constructions were confirmed by restriction mapping and sequencing the entire ocp gene.

Ocp variant proteins were expressed and purified using Pharmacia's bulk GST purification module and the protocols provided by the manufacturer. After purifying the GST-ocp fusion proteins on glutathione Sepharose 4B (G-S4B) and cleaving the fusion protein with thrombin, the Ocp variants were purified by anion exchange FPLC on Mono Q columns and affinity chromatography on G-S4B.

C. Preparation and Distribution of ABRF-95SEQ

Amino acid analysis was carried out to accurately quantitate the level of each recombinant protein. Variants 1 and 2 were then mixed in a ratio of 2:1, aliquoted from a single stock into pre-washed 1.5 ml polypropylene microfuge tubes such that each tube contained 30 pmol of variant 1 and 15 pmol of variant 2 (45 pmol total), and vacuum dried. Several aliquots were sequenced by the Committee to ensure sample quality. The sample was then distributed to 252 members of the ABRF that carry out protein sequence analysis. Instructions for sample solubilization were provided, and recipients were asked to proceed with derivatization and other handling following their normal procedures before applying the entire sample to the sequencer. Recipients were asked to carry out as many cycles of sequence analysis as possible and to return their results along with a survey that was sent with the sample. They were also informed that the sample contains some heterogeneity within the first 10 residues. Results were returned to a third party who removed identifying marks before forwarding the data to the authors for analysis.

III. Results

A. Survey Results

All the respondents that analyzed ABRF-95SEQ returned the survey, which requested information about the instrumentation and operating conditions employed. The average age of the protein sequencers used was 5.4 ± 3.3 years. These instruments were manufactured by Applied Biosystems Division of Perkin Elmer (ABD; 56/71), Beckman (1/71), Hewlett Packard (HP; 7/71), Milligen (1/71), Porton (5/71) and Shimadzu (1/71). The supports used for sample immobilization were polybrene-treated glass fiber GFC filters (38/71; used with ABD and Porton instruments), PVDF membranes and Prospin cartridges (17/71; used with ABD and Shimadzu instruments), biphasic columns (7/71; used exclusively for HP sequencers), Porton disks (7/71; used on ABD, Porton and Beckman sequencers) and membranes for covalent attachment of sample (2/71; ABD and Milligen instruments). Although 9 of the respondents did not indicate the amount of PTH-amino acid that they analyzed at each cycle, 26 facilities analyze more than 70% of their samples, another 26 analyze 50-70% and 10 facilities analyze less than 50% of their samples. Twenty facilities indicated they conducted most of their sequencing at levels of 1-10 pmol, while 35, 10 and 5 facilities routinely sequence at levels of 10-75, 75-250 and 250-1000 pmol, respectively (one facility did not respond to this question).

B. Sequence Assignments

Although a total of 2076 sequencing cycles were returned for this study, 275 of these were cycles in which either no residue was observed or a residue was not identified (table I). There were 1647 positive assignments and 154 tentative assignments. There was an average of 4 unassigned residues per study result,

Table I. Summary of sequence assignments for ABRF-95SEQ[a]

Total Cycles Assigned	PC+TC+PI+TI	1801
Unassigned Cycles	UR	275
Average Cycles Assigned	Total cycles/No. Responses	25.4
Average Correct Assignments	Correct Assignments/No. Responses	19.2
Average Tentative Assignments	Tentative Assignments/No. Responses	2.2
Average Incorrect Assignments	Incorrect Assignments/No. Responses	6.2
Correct Assignments	PC+TC	1361
Incorrect Assignments	PI+TI	440
Positive Assignments	PC+PI	1647
Tentative Assignments	TC+TI	154
Accuracy of Positive Assignments	PC/(PC+PI)	0.78
Accuracy of Tentative Assignments	TC/(TC+TI)	0.45

[a] Sequence assignments were categorized as positive correct (PC), tentative correct (TC), positive incorrect (PI), tentative incorrect (TI) or unassigned (UR).

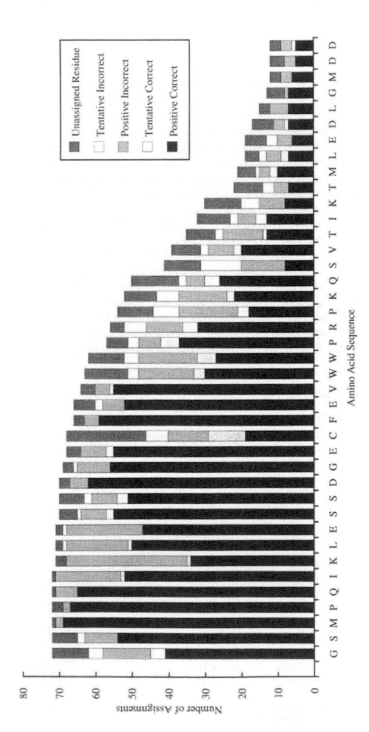

Figure 2. Sequence assignments at each position of ABRF-95SEQ.

The sequence assignments for the first 40 amino acids of ABRF-95 SEQ are indicated. For positions 6-9, the major residues are included. Several responses assigned sequence beyond 40 residues, and one respondent made positive assignments to position 62.

with 1 unassigned residue per 7 attempted assignments. The average response assigned 25 cycles, 19 of them correctly, 6 incorrectly and 2 tentatively. This is similar to the two previous studies in which the average number of cycles assigned were 28 and 21 for ABRF-93SEQ (6) and ABRF-94SEQ (7), respectively.

Of the total of 1801 assigned cycles, 76% were correct and 24% were incorrect. 78% of the positive assignments were correct and 45% of the tentatively called residues were accurately identified. This level of accuracy is lower than observed in previous studies (table II). Presumably this was not due to the amount of sample that was distributed for each study (table II) but rather was sequence related.

Figure 2 outlines the assignments that were made for the first 40 residues of ABRF-95SEQ. The Met at cycle 3 was the residue most frequently assigned correctly, with only 2 incorrect assignments. The reported levels of this residue ranged from 1 to 78 pmol, with an average of 20 pmol. Thus, the average yield observed for this residue was 47%. There also appeared to be few difficulties with the correct assignment of Pro at cycle 4 and Gln at cycle 5.

The correct assignment of Gly at the initial sequencing cycle was poor. 54 of the 71 respondents positively assigned this residue, and of those only 41 were correct. This degree of accuracy for identification of an N-terminal Gly is similar to that observed for previous studies (table II). These results confirm that the presence of a Gly residue at the amino terminus presents major problems for sequence assignment, presumably because this residue is often seen as a contaminant in cycle 1 and hence dismissed.

Within the first 18 residues the Cys at position 15 was the least frequently identified. Position 7 of ABRF-95SEQ is heterogeneous, with Lys and Leu as the major and minor residues respectively, and correct assignment at this position was also poor. 34 respondents positively assigned Lys at this position and 25 respondents assigned Leu as the major amino acid. The ability to make a correct assignment at this position was not dependent on the instrumentation employed. Overall the number of correct assignments dropped dramatically beyond residue 18, probably due to the presence of two Trp residues at positions 19 and 20, Pro residues at positions 21 and 23, and Arg at position 22.

Table II. Comparison of ABRF-95SEQ results with previous studies

	ABRF-91SEQ (4)	ABRF-92SEQ (5)	ABRF-93SEQ (6)	ABRF-94SEQ (7)	ABRF-95SEQ
Amount of sample distributed (pmol)	80	500	50	50	45
Accuracy of positive assignments (%)[a]	83	94	91	96	78
PC assignments of Gly-1 (%)[b]	39	N/A[c]	60	53	58

[a]Sequence assignments were defined as outlined in table I. Accuracy is defined as PC/(PC+PI).

[b]The number of respondents that positively and correctly identified residue 1 as Gly is expressed as a percentage of assignments made.

[c]Not applicable because ABRF-92SEQ did not contain an N-terminal Gly.

C. Cysteine Assignment

ABRF-95SEQ contains a single Cys residue within the first 50 amino acids at position 15. Only 25 of the 71 respondents carried out reduction and alkylation of ABRF-95SEQ prior to sequence analysis (table III). This low number is not surprising, because only 22 of the 78 facilities that sequenced ABRF-94SEQ (7) indicated they routinely alkylate samples before sequencing. Interestingly, 6 of the 7 respondents to ABRF-95SEQ that used HP sequencers derivatized their samples (5 of these correctly identified Cys-15), compared to 20 of the 64 respondents employing other types of sequencers. Ten respondents that derivatized Cys residues in solution employed Prospin cartridges for sample clean-up; in all these cases ABD sequencers were used, and 6 correct assignments of Cys-15 were made. Of the 25 respondents that alkylated ABRF-95SEQ, 17 reduced and alkylated in solution and 8 used an *in situ* procedure. The percentage of correct assignments obtained with the *in situ* method (63%) was similar to that for the solution phase procedure (59%). The most popular alkylating reagent employed was 4-vinyl pyridine, and respondents that used this reagent correctly assigned Cys-15 more often than was observed with acrylamide, iodoacetate or iodoacetamide (table III).

Overall, 56% of the respondents that reduced and alkylated their samples and assigned a residue correctly identified Cys at position 15. This is lower than the correct assignments observed for identification of Cys-10 in ABRF-94SEQ (7), where 82% of the respondents that derivatized correctly identified this residue, but similar to the accuracy reported for residue 20 in ABRF-94SEQ (59% of assignments by those who derivatized were correct). Thus, it seems that the ability to correctly identify derivatized Cys residues increases when they occur earlier in the sequence.

Table III. Identification of Cys-15 in ABRF-95SEQ

Alkylating Reagent	Number (#) of Users[a]	Assignments[b] PC	TC	% Correct[c] (PC+TC)/(# - X)
Acrylamide	4	2	0	50%
Iodoacetate	4	2	0	67%
Iodoacetamide	5	2	0	50%
4-Vinyl Pyridine	12	8	1	75%
None	43	5	9	56%

[a] The total of 68 users reflects the fact that 3 respondents did not make assignments at residue 15 because sequencing was terminated before this.

[b] Assignments at position 15 are categorized as positive correct (PC) or tentative correct (TC) for each reagent employed. Other respondents did not correctly identify Cys-15.

[c] The number of respondents that correctly assigned cysteine 15 is expressed as a percentage of assignments made. X is defined as the number of users in each category that did not attempt to assign a residue at position 15.

D. Tryptophan Assignment

ABRF-95SEQ contains Trp residues at positions 19 and 20, and approximately 60% of the respondents were able to correctly identify these residues (table IV). Higher assignment accuracies were observed for neighbouring positions (88% of assignments for Val-18 and 76% of assignments for Pro-21 were correct), emphasizing the difficulty in correct assignment of Trp. The accuracy of Trp assignment was lower than reported previously; 86% of respondents correctly identified Trp-9 in ABRF-94SEQ, and 58% correctly identified the Trp at position 23 of that sample (7). Users of HP sequencers stood out in their ability to correctly assign Trp in ABRF-95SEQ, and correct assignment by users of ABD instruments was not improved by the addition of isopropanol to solvent B (table IV), in agreement with the conclusions of the previous study (7).

E. Assignment in Region of Heterogeneity

ABRF-95SEQ contains a minor sequence at residues 6-9 derived from Ocp variant 2, and table V shows the ability of respondents to detect this sample heterogeneity. Overall, the ability of respondents to correctly identify the major sequence at these positions (75, 50, 74 and 68 % of major assignments were correct for positions 6, 7, 8 and 9 respectively) was lower than for the surrounding residues (92% and 88% of calls were correct for cycles 5 and 10 respectively). Typically those respondents that correctly identified the major amino acid were also correct in their assignment of the minor residue. At each cycle in this region a number of respondents listed the minor amino acid as the

Table IV. Identification of Trp residues at positions 19 and 20 in ABRF-95SEQ

Instrument and Conditions[a]	Tryptophan 19				Tryptophan 20			
	#[b]	PC[c]	TC[c]	% Correct[d]	#[b]	PC[c]	TC[c]	% Correct[d]
ABD + IPA	24	11	1	60%	24	9	2	55%
ABD - IPA	26	11	1	57%	26	8	2	53%
HP	7	5	0	100%	7	6	1	100%
Other	6	3	1	80%	6	3	0	60%
Total	63	30	3	65%	63	26	5	61%

[a]Responses were categorized into users of ABD, HP or other instruments. ABD instrument users were further divided based on whether isopropanol was (+ IPA) or was not (- IPA) added to solvent B.

[b]Refers to the number of respondents in each category that attempted to assign a residue at that position. Only 63 of the 71 respondents are listed because 8 respondents did not assign amino acids past residue 18.

[c]Assignments are categorized as positive correct (PC) or tentative correct (TC). Other respondents in each category did not correctly identify the amino acid.

[d]The number of respondents that correctly assigned tryptophan at positions 19 and 20 is expressed as a percentage of assignments made. Percentage correct was calculated according to the formula (PC+TC)/(# - X), where X is defined as the number of users in each category that did not attempt to assign a residue.

Table V. Assignments at positions 6-9, the region of heterogeneity in ABRF-95SEQ

Position	Residues[a]	M/m[b]	m/M[c]	(M/m)[d]	M[e]	m[f]
6	Ile/Val	46	7	3	7	1
7	Lys/Leu	31	18	2	4	3
8	Leu/Gly	36	9	2	15	0
9	Glu/Gln	38	8	1	9	1

[a]The major (M) and minor (m) residues at each position are indicated.

[b]The number of respondents that correctly assigned both the major and minor residues.

[c]Number of respondents that assigned the major residue as minor and the minor residue as major.

[d]Number of respondents that identified both residues correctly but did not specify major and minor residues.

[e]Number of respondents that assigned the major residue correctly and did not assign a minor residue.

[f]Number of respondents that assigned the minor residue as the major residue and did not assign a minor residue.

major sequence, and this was particularly evident for the Lys/Leu residues at position 7. A number of respondents did not detect the sample heterogeneity, and in these cases some assigned the major residue while others assigned the minor residue.

F. Best responses

The best responses were tabulated according to those that correctly identified the most residues with the highest degree of accuracy. There were six respondents that correctly identified the first 25 residues or more (table VI), including correct assignment of the major and minor sequences at positions 6-9. Another respondent (response 14) was unable to distinguish between the major and minor sequences at cycles 6 and 8, but otherwise correctly assigned 25 amino acid residues. Two additional study participants (responses 5 and 24) correctly identified at least 24 consecutive residues with the exception of Gly at cycle 1.

Only 2 of the 6 best responses defined in table V reduced and alkylated their samples, and they both positively identified the Cys at position 15. Three of the remaining 4 tentatively identified this residue, presumably by the absence of any other signal. The other (response 59) positively assigned Cys at position 15 without reduction and alkylation, based on the presence of dehydroalanine in this cycle. All the respondents listed in table VI positively assigned the Trp residues at cycles 19 and 20, with the exception of respondent 33 who tentatively assigned Trp-20.

Two of the best responses were generated using HP sequencers, and three employed ABD 494 instruments. Overall these instruments out-performed other sequencers, indicating there have been significant improvements in instrumentation in recent years.

Table VI. Summary of the best responses for ABRF-95SEQ

Response	Consecutive C^a P, T	Assignments[b] #, C	Cys-15[c] Mod'n, ID	Trp-19, 20[d] ID	Instrument
35	49, 1	62, 59	IAA, PC	PC, PC	HPG1000A
29	47, 0	47, 47	VP, PC	PC, PC	HPG1005
59	42, 0	50, 45	None, PC	PC, PC	ABD 494
33	38, 3	41, 41	None, TC	PC, TC	ABD 494
26	32, 1	42, 40	None, TC	PC, PC	ABD 494
12	24, 1	25, 25	None, TC	PC, PC	ABD 473

[a]The number of consecutive, correctly identified residues (C) is indicated as positive (P) and tentative (T) assignments.

[b]The total number (#) of assignments made and the number that were correct (C) is indicated.

[c]The method employed for Cys modification (Mod'n; IAA = iodoacetate; VP = 4-vinylpyridine) and the assignment of Cys at cycle 15 are listed.

[d]The assignments of the Trp residues at cycles 19 and 20 are indicated.

IV. Discussion

ABRF-95SEQ was designed to address a number of problems in sequence analysis as outlined in Methods. The results clearly indicate that the sequencing performance of the study participants is variable (2, 11, 31, 18, 5, and 4 respondents correctly assigned 0, 1-10, 11-20, 21-30, 31-40 and more than 40 residues respectively). While the 6 best responses correctly assigned over 25 consecutive residues, 11 respondents assigned less than 10 residues correctly.

The accuracy of positive calls was lower than observed in previous samples, although this is not simply due to the heterogeneity of ABRF-95SEQ because this figure was not improved significantly by omitting the region of heterogeneity from the calculation (accuracy of positive assignments increased from 78% to 81% when cycles 6 - 9 were omitted). The number of correct calls dropped significantly at the Trp doublet, but the correct identification of Trp was similar to that observed for the previous sample in this series (7). The occurrence of two consecutive Trp residues, coupled with the Pro-Arg-Pro sequence immediately after, may have contributed to the decreased accuracy of ABRF-95SEQ compared to previous samples.

The percentage of assignments that were correct in the region of heterogeneity was lower than for surrounding residues, indicating that sample heterogeneity does present problems for sequence analysis. The Lys/Leu pair at position 7 presented the greatest problem in this region, with many respondents assigning the minor residue as the major. There did not appear to be problems assigning Leu at cycle 8, even though there had been a minor Leu at cycle 7. Just as many respondents assigned the minor Gln at cycle 9 as assigned the minor cycles at 6-8, indicating they did not think that the Glu signal was just due to deamination of the major Gln.

Overall the identification of Cys was similar to that observed for ABRF-94SEQ , the previous study in this series (7). One of the major goals of ABRF-94SEQ was to assess methods for cysteine derivatization. Clearly detailed reduction and alkylation methods were provided with that sample, and the study summarized the success of cysteine identification with each method used. In view of the fact that most respondents to the current study likely also participated in ABRF-94SEQ, it is of concern that so few facilities still do not routinely derivatize Cys residues. Nine respondents that did not reduce and alkylate tentatively assigned residue 15 as cysteine. Some of these respondents had clear evidence of dehydroalanine at this position but that was not the case for the remainder, who presumably assigned Cys because of a lack of any signal in the PTH-chromatogram. It is of concern that a further 5 respondents positively identified this residue as cysteine in the absence of derivatization, because there are a number of unusual amino acids in addition to cysteine that could give rise to blank chromatograms in sequence analysis.

In conclusion, this year's study has provided information on many aspects of protein sequencing. There is still wide variation in the ability of facilities to make correct sequence assignments, shown by the range in the number of correct residues assigned (0-62 residues). Previous studies indicated that operator skill was the main determinant in successful sequencing. However, the somewhat better performance of newer instruments (table VI) for sequence analysis of ABRF-95SEQ suggests that recent advances in instrument technology may improve sequencing capabilities. Identification of Cys and Trp remains problematic. Results from ABRF-94SEQ and ABRF-95SEQ suggest that the accuracy of Trp assignments may depend in part on the position of the residue in a sequence, with earlier assignments being more accurate than later ones. Even though there are a number of simple procedures for modifying Cys, most facilities do not routinely use them, resulting in many inaccurate assignments for this residue. Sequence heterogeneity also reduces assignment accuracy but to a lesser extent: about half the assignments in the heterogeneous region of ABRF-95SEQ were correct. Thus, multiple factors play a role in the accuracy of sequence assignments, but the results of this study indicate that an average of 19 residues will be assigned correctly when sequence analysis of 45 pmol of an unknown protein is performed.

Acknowledgments

This work was partially supported by NSF grant DIR9003100 to John Crabb (W. Alton Jones Cell Science Center) on behalf of the ABRF. The contribution of all respondents that made this study possible by analyzing ABRF-95SEQ is gratefully acknowledged. The assistance of Gary P. Gryan (HHMI/Harvard Medical School Computer Facility) in coordinating the data returns and ensuring the anonymity of the participating laboratories is appreciated. The assistance of the ABRF Business Office for postcard printing and mailings is acknowledged. We thank Dr Hong Chen and Dr Geoffrey Duyk for generously providing the pGEX/ocp II expression plasmid. Several members of the authors' laboratories contributed to this project including Dawn Fitzpatrick and Cara Ruble (HHMI/Harvard Medical School) for construction of ocp variants; Ivar Jensen (HHMI/Harvard Medical School) for protein sequence analysis of ocp variants; Jeffry Bondor (Shriners Hospital for Crippled Children) for sample distribution and Bill Chestnut (Burroughs Wellcome Co.), Quazi Agashakey, Michelle Kirchner and Sheenah Mische (The Rockefeller University Protein/DNA Technology Center) for sequence analysis of ABRF-95SEQ.

References

1. Niece, R.L., Williams, K.R., Wadsworth, C.L., Elliott, J., Stone, K.L., McMurray, W.J., Fowler, A., Atherton, A.D., Kutny, R. and Smith, A.J. (1989) *in* Techniques in Protein Chemistry, T.E. Hugli, ed., pp 89-101, Academic Press, San Diego, CA.
2. Speicher, D.W., Grant, G.A., Niece, R.L., Blacher, R.W., Fowler, A.V. and Williams, K.R. (1990) *in* Current Research in Protein Chemistry, J.J. Villanfranca, ed., pp 159-166, Academic Press, San Diego, CA.
3. Yüksel, K.Ü., Grant, G.A., Mende-Mueller, L.M., Niece, R.L., Williams, K.R. and Speicher, D.W. (1991) *in* Techniques in Protein Chemistry II, J.J. Villanfranca, ed., pp 151-162, Academic Press, San Diego, CA.
4. Crimmins, D.L., Grant, G.A., Mende-Mueller, L.M., Niece, R.L., Slaughter, C., Speicher, D.W. and Yüksel, K.Ü. (1992) *in* Techniques in Protein Chemistry III, R.H. Angeletti, ed., pp 35-45, Academic Press, San Diego, CA.
5. Mische, S.M., Yüksel, K.Ü., Mende-Mueller, L.M., Matsudaira, P., Crimmins, D.L., and Andrews, P.C. (1993) *in* Techniques in Protein Chemistry IV, R.H. Angeletti, ed., pp 453-461, Academic Press, San Diego, CA.
6. Rush, J., Andrews, P.C., Crimmins, D.L., Gambee, J.E., Grant, G.A., Mische, S.M. and Speicher, D.W. (1994) *in* Techniques in Protein Chemistry V, J.W. Crabb, ed., pp 133-141, Academic Press, San Diego, CA.
7. Gambee, J., Andrews, P.C., De Jongh, K., Grant, G., Merrill, B., Mische, S. and Rush, J. (1995) *in* Techniques in Protein Chemistry VI, J.W.Crabb, ed., pp 209-217, Academic Press, San Diego, CA.
8. Cohen, D.M. (1991) *in* Current Protocols in Molecular Biology, F.M. Ausubel et al., ed., pp 15.1.1-15.1.7, John Wiley & Sons, New York, NY.

SECTION VII

Three Dimensional Protein Structure

A Technique of Protein Addition for Repeated Enlargement of Protein Crystals in Solution

Qing Han and Sheng-Xiang Lin

MRC Group in Molecular Endocrinology,
CHUL Research Center, Laval University, Quebec, G1V 4G2, Canada

1. Introduction

The routine preparation of single crystals of suitable size for X-ray diffraction analyses remains one of the major challenges of protein crystallography, though the development of many techniques have been successful in improving this situation.

Vapor diffusion method with hanging and sitting drop is widely used to carry out protein crystallization experiment. The volume of protein droplet needed in this technique is usually between 2-10 µl (Ducruix & Giegé, 1992). The growth of protein crystals will cease after the protein concentration decreases to a certain level. The size of crystals is thus limited particularly when there are many nuclei in solution. For enlargement of crystals, the technique of adding new crystallization solution is often used in small molecular crystal growth, but it has been replaced in biomacromolecule crystallization. More recently, a gel-mediated feeding technique for protein crystal regrowth has been reported (Bernard et al., 1994), while we have developed a technique in our laboratory for repeated enlargement of protein crystals in solution by adding new protein sample. The volume of crystals obtained by this method can be two times greater than before. In many successful experiments, the crystals show good morphology without new nucleation.

A Technique of Protein Addition for Repeated Enlargement of Protein Crystals in Solution

Qing Han and Sheng-Xiang Lin

MRC Group in Molecular Endocrinology,
CHUL Research Center, Laval University, Québec, G1V 4G2, Canada

I. Introduction

The routine preparation of single crystals of suitable size for X-ray diffraction analyses remains one of the major challenges of protein crystallography, though the development of many techniques have been successful in improving this situation.

The vapor diffusion method with hanging and sitting drop is widely used to carry out protein crystallization experiments. The volume of protein droplet needed in this technique is usually between 2-10 µl (Ducruix & Giegé, 1992). The growth of protein crystals will cease after the protein concentration decreases to a certain level. The size of crystals is thus limited particularly when there are many nuclei in solution. For enlargement of crystals, the technique of adding new crystallization solution is often used in small molecular crystal growth, but it has been neglected in biomacromolecule crystallization. More recently, a gel-mediated feeding technique for protein crystal regrowth has been reported (Bernard et al., 1994), while we have developed a technique in our laboratory for repeated enlargement of protein crystals in solution by adding new protein sample. The volume of crystals obtained by this method can be two times greater than before. In many successful experiments, the crystals show good morphology without new nucleation.

TECHNIQUES IN PROTEIN CHEMISTRY VII

The key step of this technique is to choose the proper conditions under which crystals can re-grow but new nucleation will not appear. The determination of a phase diagram should be a prerequisite for such crystal growth, which is routinely used in small molecule crystallization. From the phase diagram, it is possible to get a reasonable estimate of the concentration needed for crystal growth. Even some rough information related to the above diagram helps in setting up the experiments. In order to apply this technique easily , we also introduce a method for determining the simple crystallization diagram for the hanging drop vapor diffusion. The effects of protein concentration, precipitant concentration and crystallization time for regrowing crystals will be described using lysozyme as a model protein which is often a standard material for testing novel methods in protein crystallization (Pusey, 1991). Similar results were obtained for human placental 17ß-hydroxysteroid dehydrogenase (17ß-HSD).

II. Materials and Methods

2.1 Materials and experiment

Hen egg white lysozyme was obtained from Sigma and used without further treatment. 17ß-Hydroxysteroid dehydrogenase(17ß-HSD) (EC 1.1.1.62) from the soluble subcellular fraction of human placenta is purified as previously described (Lin, et al. 1992; Zhu, et al. 1993). All other chemicals were reagent grade. Water was purified by a super-Q system (Millipore Corporation) and all solutions were filtered by a 0.22 μm Millex-GV membrane filter (Millipore).

The hanging drop crystallization experiments were carried out in Linbro boxes at 23 ±1 °C environment. The initial batch experiments of lysozyme were performed with the same protein concentration and three different NaCl concentrations, 0.5%, 1.0% and 1.5%. The volume of drops was 10 μl, containing lysozyme (25 mg/ml), NaN_3 (2 mM), and buffer concentration (0.05 M sodium acetate) at pH 4.55. The reservoir solution contained 6.0% (w/V) NaCl, 0.05 M sodium acetate buffer.In order to obtain phase data, the protein and precipitant concentrations were determined during the procedure of crystallization. 5 μl of protein solution was added to each drop five days after the initiation of the vapor diffusion experiment. The experiments were performed in Linbro boxes for visualization of the phase data: in different columns, the protein concentration was varied from 10 to 60 mg/ml; in different rows, the NaCl concentration was varied from 0 to 5.0 %.

2.2 Measurement of the concentrations of protein and precipitant

The hanging drop experiments were conducted in small volumes. To determine the protein concentration, the Bio-Rad Protein Microassay was used. This method is sensitive to concentrations as low as 1.25-25 µg/ml (Bower-Komro, 1989). The optical density was determined with a Beckman Du-70 spectrophotometer at 595 nm. Disposable micropipets (Fisher Scientific Corporation) were used to measure accurately small volumes. After crystals appear, the operation must be carried out under a microscope to avoid drawing of microcrystals into the pipette. Based on the ratio of the drop volume during the process to the initial volume, we can calculate the precipitant concentration at any time.

III. Results and Discussion

3.1 The simple phase diagram of crystal growth.

The crystallization of biological macromolecules, like other crystallization, includes three steps: nucleation, crystal growth and cessation of growth. Supersaturation can be achieved by vapor diffusion between the drop and reservoir solution. The supersaturation needed for nucleation is significantly higher than that needed for the subsequent crystal growth. The nucleation and crystal growth can be separated into two stages. The Fig. 1 shows the simple phase diagrams of lysozyme crystallization. The nucleation concentration of lysozyme is high, but the lysozyme crystal can grow in concentration ranges from 80 to 100 mg/ml after nucleation. This provided an excellent opportunity to design new crystal growth. For crystal regrowth, the final droplet can be driven into the growth phase again by suitable addition of protein with precipitant. This method can result in appreciable enlargement of crystals, while avoiding new nucleation.

3.2 The schematic plot of the crystal re-growth.

The experiments with lysozyme helped to establish the correlation of regrowth with precipitant and protein concentration. The results of these experiments are presented as schematic plots (Fig.2). The whole region of a phase diagram could be divided into several domains.

crystal dissolve domain

The primary crystals dissolved rapidly after adding protein solution which only contained protein or protein with low concentration salt. This case only occurred when the initial drop contained 0.5% NaCl in our experiments. The concentration of NaCl became low after addition of the new protein solution and which results in the solubility of lysozyme increasing rapidly.

crystal damage domain

The crystals were damaged under the conditions in this domain. The damaged crystals often showed fissures on their surfaces. This is because the droplet after protein addition was no longer supersaturated allowing a partial dissolution of the crystal. After the protein concentration increased through new vapor diffusion, the damaged crystal could grow again, however, the fissures on the surface did not disappear completely.

crystal re-growth domain

Crystals grew into obvious larger size with good optical quality and no new nucleation was apparent in the drop in this domain (Fig. 3b). In this domain, the salt concentration of the new protein solution was close to that of the equilibrated droplet. When the salt concentration was lower than that of the equilibrated droplet, higher protein concentration was required for crystal regrowth and the crystals often regrew first from sides. In the beginning, the surfaces of regrowing crystals could be wrinkled. They gradually became satisfactory after further growth. When the salt concentration of the new drop was a little higher than that

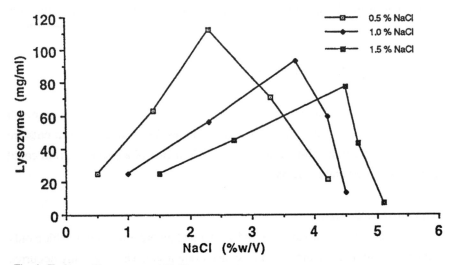

Fig. 1. Tracing of lysozyme and NaCl concentration during the crystal growth. The crystal grew in a broad protein concentration range after nucleation, which is demonstrated by the right side of these triangles. The initial NaCl concentration was: 0.5%, 1.0%, 1.5%. The final concentration was determined after five days of vapor diffusion.

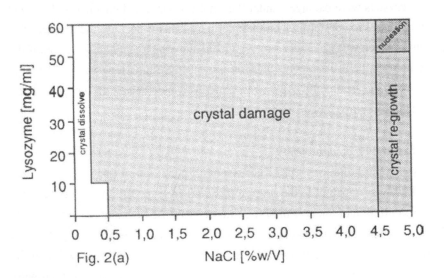

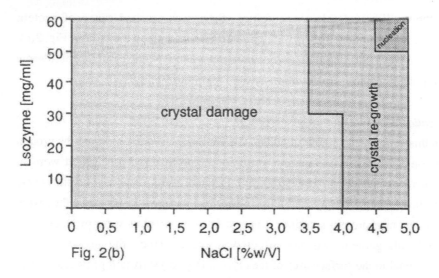

Fig. 2(b)   NaCl [%w/V]

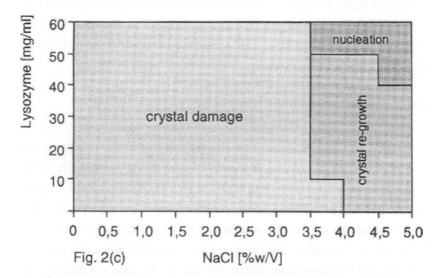

Fig. 2(c)   NaCl [%w/V]

Fig. 2. Schematic representation of crystal regrowth after the addition of 5 μl of new protein solution in different conditions. The volume of initial drops is 10 μl contained 25 mg/ml lysozyme, 2 mM NaN3 and: (a). 0.5% NaCl; (b). 1.0% NaCl; (c). 1.5%. 5 μl of protein solution was added to each drop after five days of the vapor diffusion. The experiments were carried out in Linbro boxes for visualization of the phase data: in different columns, the protein concentration was varied from 10 to 60 mg/ml; in different rows, the NaCl concentration was varied from 0 to 5.0 %.

of the equilibrated salt concentration, there was a broad range of protein concentrations suitable for crystal regrowth. For example in the case of Fig. 2(c), when the NaCl concentration in the new drop was 5.0%, the crystals could grow well in the lysozyme concentration from 10 mg/ml to 50 mg/ml.

nucleation domain

In this domain, some new small crystals were produced after adding the protein solution. This is because the concentration of protein and precipitant were so high that new nucleation was introduced. In some cases, a few new crystals came from the microcrystals present in the initial drop and grew larger when they were fed by the new protein solution. If only a few new crystals were produced, the crystal enlargement was not affected seriously; since the growth rate is proportional to the surface area of the crystal, large crystals will grow faster. The case can be seen in the example of 17ß-HSD. Several new crystals appeared after adding new 17ß-HSD solution, but the primary crystal was obviously enlarged (Fig.4). If many crystals were introduced, the primary crystals could be damaged by overlapping growth with new ones.

3.3 Discussion

In summary, comparing the three phase diagrams, it is clear that the precipitant played a key role in the crystal regrowth. When the concentration of the precipitant in the new drop is lower than that needed for maintaining the crystal, the crystal lattice will be damaged. If the concentration of the precipitant is high enough the crystal will not dissolve and the primary crystal can regrow in a broad range of protein concentrations. We prepared the protein solutions containing all components at the concentrations determined above. As expected, the regrown crystals were excellent and showed no degradation in their appearance (Fig. 5). From these experiments, we can clearly see the major impact of the initial NaCl concentration on the crystal regrowth area. There is a higher probability of enlarging crystals by protein addition with higher a starting NaCl concentration, indicating the important role of the NaCl in the stabilization of the crystal. This may be due to the higher salt concentration in the crystal, which plays an important role in maintaining the latter in the crystal phase.

When preparing the new protein solution, the protein concentration should be lower than that needed for nucleation, while the component concentrations should be very close to the equilibrated drop. If the crystallization phase can not be determined, the enlargement experiment can be designed on the basis of the final precipitant concentration of the drop, which can be evaluated by determining the volume of the drop. Because the final buffer concentration could be several times higher than that of the initial drop, the buffer of the new protein solution should be prepared according to this concentration. A series of protein concentrations with similar precipitant concentration could be used to find the best conditions of crystal enlargement.

The phase diagram is basic information for understanding the crystallogenesis. Up to now, few crystallization phase diagrams for the hanging drop method have been reported because the droplets usually contain only 2-10 µl protein solution and the component concentrations vary during the period of vapor diffusion. Many parameters influence nucleation and crystal growth of biological macromolecules. So does the crystallization method. In the hanging drop method, we could not directly use the phase data obtained through other crystallization methods. The present results show that the simple phase diagram determined by the above method can provide useful data for crystal re-growth.

When there are too many crystals in the equilibrated drop, the number of crystals should be reduced by the "microcrystal selection technique" (Han & Lin) before adding new protein solution. The quality of crystal can be also controlled by eliminating poor quality crystals (Fig. 4).

Based on the above results, we propose a general protocol for crystal regrowth:

(i) Determine the simple phase diagram. Several drops should be prepared under the ordinary crystallization conditions. The volume of the drops can be 2-10 µl. The component concentrations of the drops are determined by the above method. The two states that are necessary to determine are the nucleation state, and the regrowth state. (ii) The new protein solution is prepared according to the results of (i). The protein concentration of the new drop should be lower than that of nucleation. The other component concentrations should be as close as possible to the known concentration of equilibrated drop. (iii) After this regrowth trial, a new trial can be repeated to enlarge crystals again. Fig. 3 shows the lysozyme crystals through two trials of regrowth.

The traditional method of protein crystallization, which proceeds in a sealed chamber, is still widely used. As a result, the crystallization process becomes an isolated procedure after the chamber is sealed and the crystal growth is fixed by the initial conditions. In order to surpass this difficulty, in addition to the above method, we have designed a series of small devices, which can allow us to adjust the reservoir solution easily (Han & Bi, unpublished results), and have developed techniques to control crystal number and quality after nucleation. From our experiments, these micromanipulations can improve crystal growth by assisting in the trial-and-error process of macromolecular crystallogenesis.

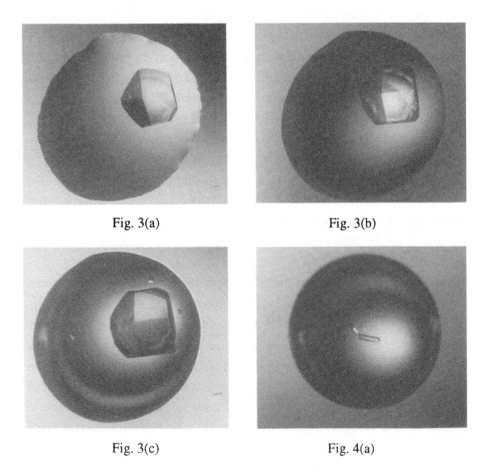

Fig. 3(a)   Fig. 3(b)

Fig. 3(c)   Fig. 4(a)

Legends for these figures are on the following page.

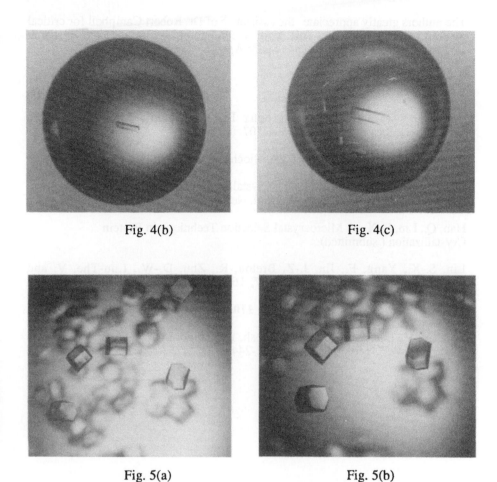

Fig. 4(b)   Fig. 4(c)

Fig. 5(a)   Fig. 5(b)

Fig. 3. The micrographs of growing lysozyme crystals. (a). at the end of the initial 10 ul drop containing lysozyme 25 mg/ml, NaCl 1.5%, NaN$_3$ 2 mM and 0.05 M sodium acetate buffer. (b). the regrown crystals after the first addition of 5 μl new protein solution containing lysozyme 25 mg/ml, NaCl 4.5% and 0.05 M buffer. (c). the regrown crystals after the second addition of 5 μl new protein solution containing lysozyme 25 mg/ml, NaCl 6.5% and 0.05 M buffer.   Fig. 4. (a). The final 17ß-HSD crystals in the initial drop. (b). The 17ß-HSD crystals after the microcrystal selection operation. (c). The regrown 17ß-HSD crystals after the addition of new protein solution.   Fig. 5. (a). The lysozyme crystals grown in the primary drop. (b). The regrown lysozyme crystals after adding new protein solution containing the same component concentrations with primary drop.

Acknowledgments

The authors greatly appreciate the assistance of Dr. Robert Campbell for critical reading of the manuscript. This work was supported by the medical Research Council of Canada and the Canada Space Agency.

References

Bernard, Y., Degoy, S., Lefaucheux F., and Robert, M., (1994), Acta Crystallographica Section **D 50**, 504-507

Bower-Komro, D.M., et al., (1989), Biochem., **28**, 8439

Ducruix, A., and Giegé, R., (1992), Crystallization of Nucleic Acids and Proteins A Practical Approach, Oxford, New York.

Han, Q., Lin, S.-X., A Microcrystal Selection Technique in Protein Crystallization (submitted).

Lin, S.-X., Yang, F., Jin, J.-Z, Breton, R., Zhu, D.-W., Luu-The, V. and Labrie, F. (1992), J. Biol. Chem. **267**, 16182-16187.

Pusey, M., (1991), J. Crystal Growth **110**, 60-64.

Zhu, D.-W., Lee, X., Breton, R., Ghosh, D., Pangborn, W., Duax, W., & Lin, S.-X., (1994), J. Mol. Biol. **234**, 242-244.

A Comparison of the Crystal Structures of Bacterial L-Asparaginases[1]

J. K. Mohana Rao, Michael Gribskov[2], Jacek Lubkowski, Maria Miller[3], Amy L. Swain, and Alexander Wlodawer

Macromolecular Structure Laboratory, NCI - Frederick Cancer Research and Development Center, ABL - Basic Research Program, P. O. Box B, Frederick, MD 21702

I. INTRODUCTION

Type I or type II bacterial L-asparaginase (E.C. 3.5.1.1; L-asparagine amidohydrolase) catalyzes the hydrolysis of L-asparagine to L-aspartate, with the release of ammonia. Enzymes isolated from different sources possess various degrees of asparaginase and glutaminase activities and are broadly classified as asparaginase. Type II L-asparaginase from Escherichia coli and Erwinia chrysanthemi is used extensively in the treatment of acute lymphoblastic leukemia, leukemia lymphosarcoma and lymphosarcoma (1), together with other cytotoxic drugs. Recently, several type II bacterial L-asparaginase structures have been elucidated in our laboratory using X-ray diffraction techniques. They are L-asparaginase from E. coli with bound aspartate (EcA, 2), L-asparaginase from Erwinia chrysanthemi with bound sulphate or with bound L-aspartate (ErA, 3), L-glutaminase-L-asparaginase from Acinetobacter glutaminasificans (AGA, 4), L-glutaminase-L-aspart-ginase from Pseudomonas 7A (PGA, 5), and a mutant of E. coli L-asparaginase (Jubin et al., to be published). Type II L-asparaginase from bacillus licheniformis (6) and the yeast enzyme from Saccharomyces

[1] Supported by the National Cancer Institute, DHHS, under contract No. NO1-CO-46000 with ABL.

[2] Present address: San Diego Supercomputer Center, P. O. Box 85608, San Diego, CA 92186, USA.

[3] Present address: Department of Physical Chemistry, Hoffmann-LaRoche, Inc., Nutley, NJ 07110, USA.

A Comparison of the Crystal Structures of Bacterial L-Asparaginases[1]

J. K. Mohana Rao, Michael Gribskov[2], Jacek Lubkowski, Maria Miller,
Amy L. Swain[3] and Alexander Wlodawer

Macromolecular Structure Laboratory, NCI - Frederick Cancer Research and
Development Center, ABL - Basic Research Program, P. O. Box B
Frederick, MD 21702

I. INTRODUCTION

Type I or type II bacterial L-asparaginase (E.C. 3.5.1.1; L-asparagine amidohydrolase) catalyzes the hydrolysis of L-asparagine to L-aspartate, with the release of ammonia. Enzymes isolated from different sources possess various degrees of asparaginase and glutaminase activities and are broadly classified as asparaginases, functioning biologically as homotetramers with approximate 222 symmetry. Type II L-asparaginase from *Escherichia coli* and *Erwinia chrysanthemi* is used extensively in the treatment of acute lymphoblastic leukemia, leukemic lymphosarcoma and lymphosarcoma (1), together with other cytotoxic drugs. Recently, several type II bacterial L-asparaginase structures have been elucidated in our laboratory using X-ray diffraction techniques. They are L-asparaginase from *E. coli* with bound aspartate (EcA; 2), L-asparaginase from *Erwinia chrysanthemi* with bound sulphate or with bound L-aspartate (ErA; 3), L-glutaminase-L-asparaginase from *Acinetobacter glutaminasificans* (AGA; 4), L-glutaminase-L-aspara-ginase from *Pseudomonas 7A* (PGA; 5), and a mutant of *E. coli* L-asparaginase (Palm *et al.*, to be published). Type II L-asparaginases from *bacillus licheniformis* (6) and the yeast enzyme from *Saccharomyces*

---

[1] Supported by the National Cancer Institute, DHHS, under contract No. NO1-CO-46000 with ABL.

[2] Present address: San Diego Supercomputer Center, P. O. Box 85608, San Diego, CA 92186, USA.

[3] Present address: Department of Physical Chemistry, Hoffman-LaRoche Inc., Nutley, NJ 07110, USA.

cerevisiae (7) as well as the type I asparaginases from *E. coli* (8), *bacillus subtilis* (9), and cytosolic *Saccharomyces cerevisiae* (10) bear considerable sequence resemblance with EcA, ErA, AGA and PGA enzymes. In contrast, plant asparaginases do not seem to have sequence similarity (11) with the bacterial enzymes, as evidenced by the sequence of L-asparaginase from *lupinus angustifolius*. The purpose of this study is to make a systematic comparison of EcA, ErA, AGA and PGA in terms of their sequence, structure as well as the geometry of the active site.

II. MATERIALS AND METHODS

Structural superpositions of the above-mentioned enzymes and analysis of the results were carried out using the program package *UPAMA* developed by one of the authors (Rao). The methodology is described in earlier papers, e.g. in (12). Initial comparisons were made with the C_α coordinates of all L-asparaginase structures with the cut-off distance of 3.0 Å for obtaining the orientation matrix, while final comparisons were made with the cut-off distance of 1.0 Å for superimposed atoms. Using this core, comparisons were made later for a larger value of the deviation between equivalent superimposed atoms. Comparisons were also made using all equivalent N, C, C_α, C_β and O atoms to confirm that the alignment and superposition are not restricted just to the C_α atoms alone. The differences in the conformational angles, ϕ and ψ, for the aligned regions were calculated and an analysis in the deviations of the C_α atoms for the aligned structures was made. The final temperature factors from the refined structures were also analyzed for EcA, ErA and PGA.

III. DISCUSSION

The superposition of the structures of four L-asparaginases is shown in Fig. 1. This figure describes not only the topology of an L-asparaginase monomer, but also represents the deviations in C_α positions in different structures in terms of the thickness of the tube connecting the C_α atoms, the thickest ones corresponding to the largest deviations. Even though the accuracy of the structures vary considerably because of different resolutions (1.8 to 2.9 Å) at which the X-ray data were collected, the overall superposition is indeed very good. Therefore, it is apparent that the type II bacterial L-asparaginases have a unique and characteristic topological fold. The structurally aligned L-asparaginase sequences are shown in Fig. 2.

The L-asparaginase tetramer is quite globular with an approximate diameter of slightly less than 75 Å. In the crystalline state, the 222 symmetry is exact only in the case of AGA and is approximate for the other three enzymes. Even when the angle for the two-fold rotation is not exactly 180°, it is quite close to it. Tetrameric molecules like the dehydrogenases (13) have more or less compact monomers and one monomer is generally sufficient for describing the structure and the activity as well as the binding to substrates and inhibitors. On the other hand, in the case of the

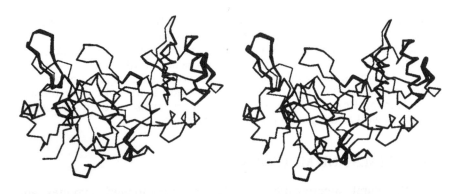

Fig. 1. Stereo view of the structure of a monomer of L-asparaginase from *E. coli*. The virtual bonds between C_α atoms are represented by tubes using a program developed by J. Lubkowski and J.K.M. Rao. Where the deviation between the superimposed structures is large, the virtual bond thickness is more and *vice versa*. The N-terminal domain is on the left side of the picture and the C-terminal domain is on the right side. The left-handed crossover includes the long helix at the left side of the diagram.

L-asparaginase, a dimer is the compact unit as the active sites with the aspartate moiety lie in cavities between structural domains belonging to two *different* monomers. Such a binding wherein the active site pocket lies between two α/β domains of the same protein molecule is quite common as in the case of the arabinose-binding protein (14). A situation similar to that of L-asparaginase wherein the active site lies between two α/β domains belonging to *different* monomers is less common and is present only in a very small number of structures, e.g., in transketolase, a thiamine diphosphate dependent enzyme (15). In each monomer of L-asparaginase, there are two α/β domains (see Fig. 1). The N-terminal α/β domain has a mixed β sheet. The first five strands form a parallel β sheet that has strand connectivities *(-1x, +2x, +1x, +1x)*, reminiscent of flavodoxin (16). The C-terminal domain consists of a four-stranded parallel β sheet with *(+1x, +1x, +1x)* connectivity. A left-handed crossover is present between the fourth and the fifth β strands in the N-terminal domain. The occurrence of left-handed crossovers in protein structures is rare and similar left-handed crossovers were observed in the following structural folds: subtilisin (17), acetylcholine esterase (18), and dihydropteridine reductase from rat liver (19).

The structure of L-asparaginase may best be described as a dimer of dimers. From purely structural considerations, a dimer provides all the residues that interact with the active site moiety. However, so far all the biochemical studies dealt with only the tetrameric form of the enzyme. The inter-subunit interactions do play a role towards this end. Besides, the need for the tetramer may also be due to the fact that globular proteins tend to fold

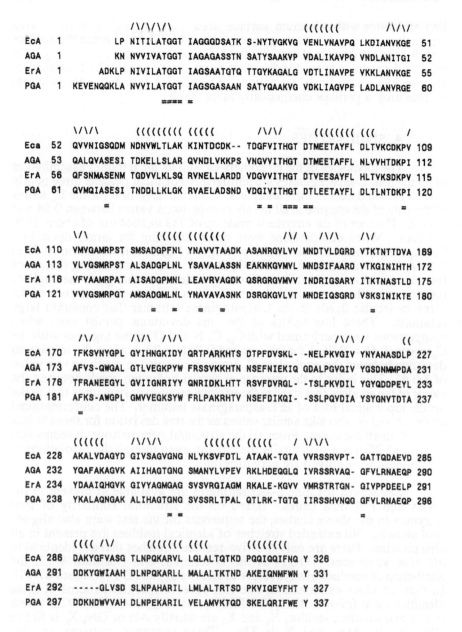

```
             /\/\/\/\                                      ((((((((                /\/\/
EcA   1            LP NITILATGGT IAGGGDSATK S-NYTVGKVG VENLVNAVPQ LKDIANVKGE  51
AGA   1            KN NVVIVATGGT IAGAGASSTN SATYSAAKVP VDALIKAVPQ VNDLANITGI  52
ErA   1         ADKLP NIVILATGGT IAGSAATGTQ TTGYKAGALG VDTLINAVPE VKKLANVKGE  55
PGA   1 KEVENQQKLA NVVILATGGT IAGSGASAAN SATYQAAKVG VDKLIAGVPE LADLANVRGE  60
                  ==== =


             \/\/\          (((((((((( (((((            /\/\/         (((((((((( (((          /
Eca  52 QVVNIGSQDM NDNVWLTLAK KINTDCDK-- TDGFVITHGT DTMEETAYFL DLTVKCDKPV 109
AGA  53 QALQVASESI TDKELLSLAR QVNDLVKKPS VNGVVITHGT DTMEETAFFL NLVVHTDKPI 112
ErA  56 QFSNMASENM TGDVVLKLSQ RVNELLARDD VDGVVITHGT DTVEESAYFL HLTVKSDKPV 115
PGA  61 QVMQIASESI TNDDLLKLGK RVAELADSND VDGIVITHGT DTLEETAYFL DLTLNTDKPI 120
                                   = = ==== ==                             =


             \/\              (((((( ((((((((          /\/ \ /\/\        /\/
EcA 110 VMVGAMRPST SMSADGPFNL YNAVVTAADK ASANRGVLVV MNDTVLDGRD VTKTNTTDVA 169
AGA 113 VLVGSMRPST ALSADGPLNL YSAVALASSN EAKNKGVMVL MNDSIFAARD VTKGINIHTH 172
ErA 116 VFVAAMRPAT AISADGPMNL LEAVRVAGDK QSRGRGVMVV INDRIGSARY ITKTNASTLD 175
PGA 121 VVVGSMRPGT AMSADGMLNL YNAVAVASNK DSRGKGVLVT MNDEIQSGRD VSKSINIKTE 180
                    =         =    =       =                        =


             /\/          /\/\ /\/\              /\/ /       ((
EcA 170 TFKSVNYGPL GYIHNGKIDY QRTPARKHTS DTPFDVSKL- -NELPKVGIV YNYANASDLP 227
AGA 173 AFVS-QWGAL GTLVEGKPYW FRSSVKKHTN NSEFNIEKIQ GDALPGVQIV YGSDNMMPDA 231
ErA 176 TFRANEEGYL GVIIGNRIYY QNRIDKLHTT RSVFDVRGL- -TSLPKVDIL YGYQDDPEYL 233
PGA 181 AFKS-AWGPL GMVVEGKSYW FRLPAKRHTV NSEFDIKQI- -SSLPQVDIA YSYGNVTDTA 237
             =


             (((((( /\/\/\              (((((((( ((((( / \/\/\
EcA 228 AKALVDAGYD GIVSAGVGNG NLYKSVFDTL ATAAK-TGTA VVRSSRVPT- GATTQDAEVD 285
AGA 232 YQAFAKAGVK AIIHAGTGNG SMANYLVPEV RKLHDEQGLQ IVRSSRVAQ- GFVLRNAEQP 290
ErA 234 YDAAIQHGVK GIVYAGMGAG SVSVRGIAGM RKALE-KGVV VMRSTRTGN- GIVPPDEELP 291
PGA 238 YKALAQNGAK ALIHAGTGNG SVSSRLTPAL QTLRK-TGTQ IIRSSHVNQG GFVLRNAEQP 296
                          ==                      =


             (((( /\/       ((((((( ((((       (((((((((
EcA 286 DAKYGFVASG TLNPQKARVL LQLALTQTKD PQQIQQIFNQ Y 326
AGA 291 DDKYGWIAAH DLNPQKARLL MALALTKTND AKEIQNMFWN Y 331
ErA 292 -----GLVSD SLNPAHARIL LMLALTRTSD PKVIQEYFHT Y 327
PGA 297 DDKNDWVVAH DLNPEKARIL VELAMVKTQD SKELQRIFWE Y 337
```

Fig. 2. A structure-based sequence alignment for four bacterial
L-amidohydrolases. The symbols ((and /\ represent the secondary structures
of helices and sheets respectively. The residues that are identical in all nine
L-asparaginases for which the sequences are known are represented by the
symbol =.

into structures with minimum surface area. The accessible surface area for each monomer of L-asparaginase is about 14000 Å² whereas the surface area for the tetramer is about 38500 Å². Thus there is a loss of nearly one-third of the surface area. On the other hand, only one-eighth of the surface area is lost upon dimer formation. Therefore a tetramer with minimal surface area is perhaps energetically more stable.

A. SIMILARITIES

The topologies of the monomers as well as the tetramer assemblages of all the four L-asparaginases are similar. All of them possess three nearly perfect two-fold axes in their tetramers. The rms deviation for the C_α atoms of the core of the enzyme in all the six comparisons varies between 0.58 and 0.71 Å. The core of the enzyme is made up of 753 to 1044 out of about 1320 residues. When comparisons were made allowing for larger deviations between equivalent atoms, not surprisingly a higher rms deviation (0.73 to 1.16 Å) resulted for 1168 to 1272 C_α atoms. A more interesting feature of this superposition was that there was very little or no change in the angle of rotation from the ones with 1 Å cut-off. This is due to the existence of the three molecular dyads in the L-asparaginase tetramer that constrain large rotations. These low values of the rms deviations persist even when comparisons were performed with C_β, C, N and O atoms together with the C_α atoms (0.65 to 0.76 Å for the core and 0.78 to 1.23 Å when the maximum deviation between the equivalent C_α atoms is allowed to reach up to a value of 3.0 Å). This is due to the presence of similar hydrogen bonds between the atoms of the residues that stablize the various secondary structures that make up the topological fold of an L-asparaginase tetramer. The conformational angles, ϕ and ψ, also take similar values as the rms deviation for these is less than 20° in all the cases. From all the structural comparisons, it seems that AGA and PGA are structurally the closest, even though the resolutions at which these were solved are very different.

Sequences for nine L-asparaginases are available at present in the Swiss and PIR data banks. Based on the structural similarity of four enzymes in the above studies, the sequences for the rest were also aligned (not shown). No extended stretches of identical residues are present in all nine proteins. There are only *two* five-residue stretches that are identical in *all* nine sequences: TGGTI at the N-terminal end (residues 9-13; the numbering of residues always refers to EcA) and HGTDT (residues 87-91). In fact, on close examination, one can observe that the patterns at these identities are as follows: $BBX_1TGGTIX_2X_3$ and $BX_4GBVX_5HGTDTB$ where B is a hydrophobic residue, X_1 and X_2 are usually Ala or Gly, X_3 is Ser or Gly, X_4 is Asp, and X_5 is Thr. These sequence patterns in the L-asparaginases occur close to the active site in the strand-turn regions of the N-terminal α/β domain. We therefore conclude that the above two sequence motifs that include identical residues are characteristic *signatures* of all L-asparaginases that possess the unique topological fold found in the above structure determinations. On the other hand, plant asparaginases do *not* possess these signature sequences. Therefore, it is likely that they may have a different topological fold. Such characteristic signature sequences are present also for several other folds, e.g., phosphate-binding sequences in mono- and di- nucleotide-binding proteins (20). Besides the above mentioned residues, Ser58, Pro108, Asp125, Asn128, Ala132, Ala136, Lys162, Phe 171, Gly245, Gly247 and Gly276 are the only other identical residues in all the L-asparaginases for which the amino acid sequences are available.

Another distinguishing feature common to all the four type II L-asparaginase structures is the rarely observed left-handed crossover. Unless there is a reason for these types of connectivities, such as an important role played by a residue therein (e.g., the catalytic His64 in subtilisin), they are not present in nature, as the left-handed twist when viewed along the plane of the pleated β sheet makes these crossovers structurally not preferable. In the L-asparagainase structure, the carbonyl oxygen of Ala114 (the first residue in the left-handed crossover) forms a hydrogen bond with the substrate (e.g., aspartate). This interaction is essential for positioning the substrate and releasing the product of the enzymatic hydrolysis and is present in L-aspartate-bound EcA and ErA (3). Since only the carbonyl oxygen is involved in this interaction, this residue need not have to be identical in all structures and this is indeed the case (see Fig. 2). Four residues in this crossover (Asp125 just before the beginning of the α helix and Asn128, Ala132 and Ala136 within the α helix) are identical in all the L-asparaginase sequences and they are stacked on the interior of the α helix, away from the solvent.

The mean deviation for the four aligned structures (six pairwise deviations if a residue is present in all four structures) yielded a few interesting results. The deviation profiles were similar for all four subunits. The variable C_α-C_α virtual bond thickness in Fig. 1 describes the mean value of this deviation averaged over all the four subunits. At the N-terminal end, the loop region (residues 13-28) after the first β strand seems to be the most flexible region in the entire structure. This region occurs immediately after the first stretch of identical residues (TGGTI). It may be noted that this loop has an alternate conformation in one of the monomers of PGA and was not at all observed in the other three (5). The second α helix of the C-terminal domain is the next region of maximum deviation among the four crystal structures. The core of the structure in terms of its β sheets is quite rigid. The minimum deviation among all the residues is for Asp125 that is identical in all known L-asparaginase sequences.

One of the parameters in the crystallographic refinement is the atomic temperature factor (B-factor). B-factor is indicative of the thermal vibrations of atoms or groups of atoms. Even though the absolute values of B-factors depend primarily on the resolution at which the crystal structures are solved, their relative values are nevertheless good qualitative indicators of structural rigidity (low B-factors) or flexibility (high B-factors). The B-factors for residues in all four subunits in each structure have more or less similar values. More importantly, for EcA, ErA and PGA, the profile of B-factors as a function of the residue number is the same. These profiles are very much in agreement with the pattern of mean deviation at each residue site obtained from the structural comparisons. Those regions that have sizable deviations *between* the structures also have large thermal motions in *each* structure.

B. DIFFERENCES

Usually, active sites in enzymes that have similar functions are highly conserved, with good examples provided in subtilisin and chymotrypsin (21), lactate dehydrogenase and glyceraldehyde-3-phosphate dehydrogenase (22), etc. It is surprising to see *major* structural differences arise in the active site region. Fig. 3 represents the active site cleft for EcA and ErA. One side of the active site is highly conserved in terms of the positions of the residues in the identical stretches TGGTI, HGTDT, Ser58, Lys162 and a few water molecules. On the other side, a helical turn that is present in EcA is absent in ErA. From sequence alignment (not shown), this deletion is also present in the type II L-asparaginase from *S. cerevisiae*. This deletion does not depend on the glutaminase/asparaginase specificity as it is present in AGA and PGA which have more glutaminase activity among all known asparaginase-glutaminases. To what extent the presence or absence of this five-residue stretch influences the enzymatic mechanism or specificity remains unknown.

The second major difference in the structures solved so far concerns the flexible N-terminal loop (residues 10-38). Whereas this loop wraps around the active site in the structures of EcA, ErA and AGA, it has an open conformation in one of the monomers of PGA and was not at all observed in the other three PGA monomers (5). The movement of the tip of this loop in the open conformation amounts to about 20 Å. No aspartate or glutamate was identified in the active site for PGA. Perhaps the enzyme is constrained in the open conformation after the product of the enzymatic reaction has left the active site cleft.

C. ENZYMATIC MECHANISM

Even though we have two structures with the L-aspartate moiety in the active site, the enzymatic mechanism of L-asparaginase is still unclear. However, we are aware of the fact that certain residues like Thr12, Ser58, Thr89, Asp90, Lys162 which are identical in all the aligned L-asparaginases, may play a crucial role. Some mutation studies by Röhm and his colleagues also corroborate this fact (23). Kinetic studies (24) show that the amidation step occurs in two stages: (a) a nucleophilic attack on the C_γ of the substrate leading to an acyl-enzyme intermediate and release of the amine group, and, (b) further attack by another nucleophile to release the product of the enzymatic reaction. The enzymatic mechanism may not be as simple as that of the well-understood serine proteinases where a catalytic triad Ser-His-Asp plays a major role. Though there is an apparently similar triad Thr-Lys-Asp in L-asparaginases, the enzymatic mechanism may be different as there exist several other residues and water molecules in the cleft that may add their contributions (see Fig. 3). It is still unclear whether Thr12 or Thr89 is the nucleophile in the first enzymatic step. At present, structures of mutant enzymes are being elucidated to assign proper roles to these residues. However, it is likely that in L-asparaginase, a threonine residue may play a role normally reserved for Ser in enzymes such as the serine proteinases and the beta lactamases. Only recently a structure of threonine proteinase has been solved, providing another example of such a catalytic residue (25).

ACKNOWLEDGMENTS

We wish to thank all our colleagues who were involved in the crystal structure determinations. We are grateful to Dr Gottfried Palm for many useful discussions.

REFERENCES

(1) Hill, J.M., Roberts, J., Loeb, E., Khan, A., MacLellan A. and Hill, R.W. (1967) *J. Am. Med. Assoc.* 202: 116-122.
(2) Swain, A.L., Jaskólski, M., Housset, D., Rao, J.K.M. and Wlodawer, A. (1993) *Proc. Natl. Acad. Sci. USA* 90: 1474-1478.
(3) Miller, M., Rao, J.K.M., Wlodawer, A. and Gribskov, M.R. (1993) *FEBS Lett.* 328: 275-279.
(4) Lubkowski, J., Wlodawer, A., Housset, D., Weber, I.T., Ammon, H.L., Murphy, K.C. and Swain, A.L. (1994) *Acta Cryst.* D50: 826-832.
(5) Lubkowski, J., Wlodawer, A., Ammon, H.L., Copeland, T.D. and Swain, A.L. (1994) *Biochemistry* 33: 10257-10265.
(6) van Dijl, J.M., de Jong, A., Smith. H., Bron, S. and Venema, G. (1991) *FEMS Microbiol. Lett.* 81: 345-352.
(7) Kim, K.W., Kamerud, J.Q., Livingston, D.M. and Roon, R.J. (1988) *J. Biol. Chem.* 263: 11948-11953.
(8) Jerlstroem, P.G., Bezjak, D.A., Jennings, M.P. and Beacham, I.R. (1989) *Gene* 78: 37-46.
(9) Sun D. and Setlow, P. (1993) *J. Bacteriol.* 175: 2501-2506.
(10) Sinclair, K., Warner, J.P. and Bonthron, D.T. (1994) *Gene* 144: 37-43.
(11) Dickson, J.M.J.J., Vincze, E., Grant, M.R., Smith, L.A., Rodber, K.A., Farnden, K.J.F. and Reynolds, P.H.S. (1992) *Plant Mol. Biol.* 20: 333-336.
(12) Rao, J.K.M., Erickson, J.W. and Wlodawer, A. (1991) *Biochemistry* 30: 4663-4671.
(13) Holbrook, J.J., Liljas, A., Steindel, S.J. and Rossmann, M.G. (1975) in *The Enzymes*, ed. Boyer, P.D. (Academic Press, New York), Vol. XI, pp 191-292.
(14) Gilliland, G.L. and Quiocho, F.A. (1981) *J. Mol. Biol.* 146: 341-362.
(15) Lindqvist, Y., Schneider, G., Ermler, U. and Sundström, M. (1992) *The EMBO J.* 11: 2373-2379.
(16) Ludwig, M.L., Andersen, R.D., Mayhew, S.G. and Massey, V. (1969) *J. Biol. Chem.* 244: 6047-6048.
(17) Wright, C.S, Alden, R.A. and Kraut, J. (1969) *Nature* 221: 235-242.
(18) Sussman, J.L., Harel, M., Frolow, F., Oefner, C., Goldman, A., Toker, L. and Silman, I. (1991) *Science* 253: 872-879.

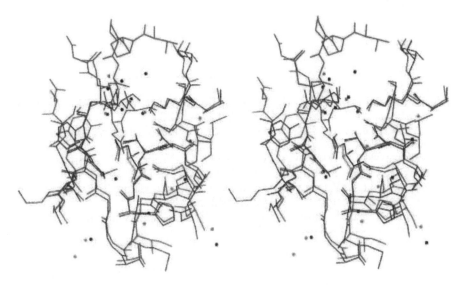

Fig. 3. Environments of the active site in EcA and ErA. The residues from the N-terminal domain of EcA and ErA are shown in red and purple respectively. The residues from the C-terminal domain from a different monomer of EcA and ErA are shown respectively in green and light blue colors. At bottom left is the TGGTI moeity that is identical in *all* bacterial L-asparaginases. At middle right is the second identical region HGTDT. At top right is the lysine residue. At top left and center is the chief difference between the two enzymes. There is a deletion of five residues (see Fig. 2) in ErA (light blue) with respect to EcA (green).

(19) Varughese, K.I., Skinner, M.W., Whitley, J.M., Matthews, D.A. and Xuong, N.H. (1992) *Proc. Natl. Acad. Sci. USA* 89: 6080-6084.

(20) Moller, W. and Amons, R. (1985) *FEBS Lett.* 186: 1-7.

(21) Bode, W., Papamokos, E., Musil, D., Seemueller, U. and Fritz, H. (1986) *EMBO J.* 5: 813-818.

(22) Rossmann, M.G. (1983) in *Colloquim - Mosbach 1983, Biological Oxidations* (Springer-Verlag, Berlin) pp 33-54.

(23) Wehner, A., Derst, C., Specht, V., Aung, H.P. and Röhm, K.H. (1994) *Biol. Chem. (Hoppe-Seyler)*, 375: S108.

(24) Röhm, K.H. and van Etten, R.L. (1986) *Arch. Biochem. Biophys.* 244: 128-136.

(25) Seemüller, E., Lupas, A., Stock, D., Löwe, J., Huber, R. and Baumeister, W. (1995) *Science* 268: 579-582.

Crystal structure of the catalytic domain
of avian sarcoma virus integrase

**Grzegorz Bujacz[1], Mariusz Jaskólski[1,2], Jerry Alexandratos[1] and
Alexander Wlodawer[1]**
[1]Macromolecular Structure Laboratory, NCI-Frederick Cancer Research and Development
Center, ABL-Basic Research Program, Frederick, MD 21702
[2]Center for Biocrystallographic Research, Institute of Bioorganic Chemistry,
Polish Academy of Sciences, Poznań, Poland

George Merkel, Richard A. Katz and Anna Marie Skalka
Institute for Cancer Research, Fox Chase Cancer Center, Philadelphia, PA 19110

Introduction

Retroviruses contain an RNA genome which, upon infection of the host cell, is first reverse transcribed into DNA and then integrated into the host genome. DNA integration is a key target for any consideration of therapeutic intervention for pathogenic human retroviruses, such as HIV-1, which is the causative agent of AIDS. Extensive genetic and biochemical studies have demonstrated that the integration step is catalyzed by a retroviral protein, the integrase (IN) (for review, see 1, 2, 3). IN as well as reverse transcriptase (RT) are encoded in the retroviral *pol* gene. The integration reaction proceeds through two biochemically and temporally separable steps (4, 5). For the first step, the "processing" reaction, IN introduces nicks, usually two base pairs from the 3'-ends of both viral strands, downstream of the highly conserved CA dinucleotides (6, 7). The second, "joining" reaction (4, 5) involves direct attack by the newly created 3'-OH groups of the viral DNA on host DNA phosphates (8) which are separated by 4 to 6 base pairs. The current model for IN function states that i) IN acts as a multimer (9, 10, 11), synchronizing the insertion of the two viral ends at the host target DNA site. ii) IN recognizes specific sequence and structure characteristics in the viral DNA ends, as well as nearly random target sequences in host DNA. iii) IN contains a single active site that carries out both the processing and joining reactions.

IN is composed of three domains that have been defined by amino acid alignments, amino acid replacement studies, and deletion mutagenesis. The prototype avian sarcoma virus (ASV) IN is 286 amino acids in length. The catalytic domain (residues ca. 50 to 200) is defined by a constellation of carboxylate residues, the D,D(35)E motif (Asp-64, Asp-121, Glu-157 in ASV). The conservation of the D,D(35)E array in retroviral and retrotransposon integrases, as well as in some bacterial transposases, suggested that these residues might be involved in the DNA processing and joining activities (12, and references therein). Mutagenesis studies showed that the D,D(35)E carboxylate residues are all essential (12, 13). Here we present the structure of the catalytic domain of ASV IN.

Methods

The expression strategy, purification from *E. coli* and activities of the purified ASV IN 52-207 fragment have been described previously (14). For multiwavelength anomalous diffraction (MAD) experiments, selenomethionine substitution for the core construct was carried out as described previously (15, 16) using the *E. coli* methionine auxotrophic strain DL41 (kindly provided by G. D. Markham). Electrospray ionization mass spectrometry experiments indicated that the four Met residues in the sequence were fully substituted by Se-Met.

The first crystals were produced using the hanging drop vapor diffusion method and two types of precipitants. Optimum conditions consisted of protein concentration between 6 and 7.5 mg/ml, pH 7.5, (HEPES buffer), and 10% isopropanol, 20% PEG 4,000, or alternatively 2M ammonium sulfate (AmS), 2% PEG 400. Both conditions gave tetragonal, $P4_12_12/P4_32_12$, crystals with cell dimensions of a,b = 66.4 Å, c = 81.4 Å, and with one molecule in the asymmetric unit. Crystals of Se-Met derivative were grown in an analogous manner and were handled without any attempt to maintain oxygen-free conditions.

Cu Kα diffraction data (Table 1) were collected at room temperature on a MAR 300 mm image plate detector, using a Rigaku Ru200 rotating anode operated at 50 kV and 100 mA. The diffraction data were processed with Denzo and scaled with Scalepack (17). The native and Se-Met crystals from PEG/isopropanol were highly isomorphous as illustrated by their scaling $R(I)$ of 0.10 (20-3.0 Å). The corresponding diffraction data were converted to normalized structure factors for all reflections with *d*-spacings between 20.0 and 3.7 Å (18). The two *E*-value sets were used to calculate the normalized structure factors corresponding to the difference structure consisting of four 18-electron (Se - S) "atoms". The difference structure was phased by means of the minimal function (19, 20, 21) using a program by Langs et al. (22). All four Se-Met positions were located using the minimal function. Although the minimal function procedure has been demonstrated earlier to successfully determine heavy atoms (Pt, U) from macromolecular SIR data, it had never been used in a case as marginal as Se-Met derivatization and Cu Kα data.

Table 1. Summary of data collection and refinement

Protein	Resol.	R(I)	# Refl.	Rfactor	R-free	r.m.s	dev.	from	ideality
			$F \geq 2\sigma F$			Bond (Å)	Angle (Å)	Planes (Å)	Chiral volume
PEG	2.00	0.077	11100						
SeRT	2.10	0.093	8479	0.148	0.227	0.013	0.045	0.016	0.161
AmS	1.70	0.052	15858	0.154	0.200	0.013	0.034	0.011	0.136
λr^a	1.95	0.065	10188	0.139	0.206	0.013	0.047	0.016	0.168
λp^b	2.20	0.054	14952						
λe	2.20	0.061	8563						

a- Only data to 2.2Å were used (R(I)=0.063) in the phase determination process (MAD).
b- Friedel pairs kept separate.

Multiwavelength anomalous diffraction (MAD) data were collected at line F2 of the CHESS synchrotron in Cornell. The crystal had been transferred from the mother liquor through a layer of a cryoprotectant solution (mother liquor with isopropanol concentration increased to 15%), suspended in a fiber loop (23) affixed to the tip of a syringe needle, shock-frozen in liquid nitrogen and immediately transferred to the goniostat, which was in a stream of cold nitrogen gas (-165°C) flowing coaxially with the spindle axis. An X-ray fluorescence spectrum was used to select the three wavelengths for the multiwavelength diffraction experiments: λp (peak) = 0.9790Å, λr (remote) = 0.9464Å, λe(edge) = 0.9792Å. The MAD data were collected on a CCD detector using the oscillation method (oscillation range 1.75°, oscillation time 10 sec). Images with Friedel pairs were recorded close together in time with inverse beam geometry (24). During scaling, which included partially recorded reflections, the Friedel pairs measured at λp were kept separate, but were merged together at λe and λr for use in isomorphous replacement (Table 1) (17). The integration and scaling procedures did not indicate any signs of crystal deterioration during the data collection.

An anomalous difference Patterson map calculated for the synchrotron λp data as well as a difference Patterson map calculated for λr vs. λe confirmed the four Se sites obtained from direct methods. The parameters of the four Se atoms were refined in a phase refinement process based on the λr and λe data (isomorphous replacement, Se K absorption edge versus remote point scattering) and λp (anomalous scattering) data. Phase refinement was followed by solvent flattening, which assumed 0.42 solvent content in both space group enantiomorphs. Solvent flattening calculations unambiguously indicated the correct space group, $P4_32_12$, with the following statistics (values in parentheses for $P4_12_12$): map-inversion R-factor 0.189 (0.360), correlation coefficient 0.976 (0.912), mean figure of merit 0.943 (0.909). All Fourier calculations, the phase refinement and solvent flattening were done using the PHASES package (25).

Electron density maps, calculated with the programs X-plor (26) and PROTEIN (27), were interpreted using FRODO (28) and O (29). The model underwent one round of refinement by simulated annealing in X-plor (26) followed by multiple cycles of restrained structure-factor least-squares refinement in PROLSQ (30). In addition to protein atoms and 187 water molecules, the final model also includes a well ordered HEPES molecule and an isopropanol molecule which cocrystallized with the protein. The structure of the Se-Met protein at low temperature (λr), with a HEPES molecule but without water molecules, was used as the initial model for the ammonium sulfate structure. During the refinement, all missing C and N terminal residues were modeled and 181 water molecules were added.

Results and Discussion

The catalytic domain of ASV IN is a five-stranded mixed β-sheet flanked by five α-helices. The main part of the β-sheet, formed by three antiparallel β-strands, β1, β2 and β3, is extended by two short parallel β-strands, β4 and β5 (Fig. 1). On one side of the β-sheet there are three short helices, α1, α2, and α3. Helix α2, only 4 residues long, is very close to α3, separated only by one residue (Ser-128). The α3 helix is also separated from the β5 strand by a single residue (Gly-139). The opposite side of the β-sheet is covered by two long antiparallel helices α4 and α5. There is a long 10-residue loop between β5 and α4 which has different conformations depending upon whether the crystal is grown in PEG or AmS. There are no differences between the PEG and AmS structures

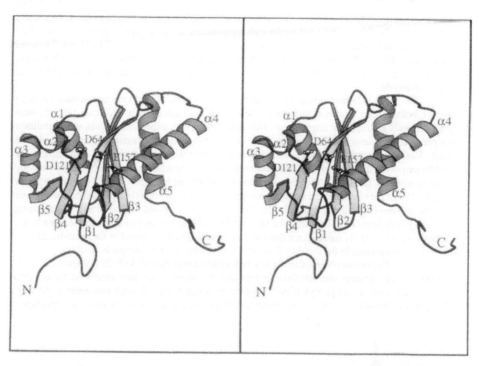

Figure 1. Stereoviews of the monomer, and the active site of ASV IN generated using MOLSCRIPT (33), showing the secondary structure elements, as well as the location of the active site.

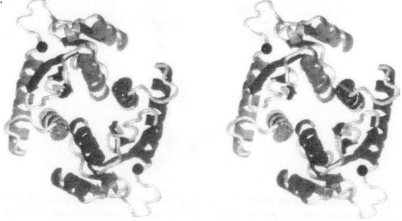

Figure 2. Stereoviews of the dimers of the catalytic domain of IN generated using O (29). A dimer of the ASV IN core domain viewed along its two-fold axis, with Wat-450 positions marked by spheres.

outside of the tip of the β5-α4 loop. This loop extends out of the compact shape of the molecule and must be quite flexible, as indicated by the high temperature factors at its tip. In the PEG structure this loop makes two turns, while in the AmS structure it has one smooth turn. The N and C termini are disordered in the PEG structures. The AmS structure seems to have a more stable conformation of the termini. Supported by the quality and resolution (1.7 Å) of this data set, we modeled both ends completely.

The dimer is created around the crystallographic two-fold axis and has an interface between the α1 helix in one monomer and the α5 helix in the complementary molecule (Fig. 2) with predominantly hydrophobic interactions. This is in accord with experimental data, which indicate the presence of a dimer in solution in the concentration range used for crystallization (unpublished data). Residues from the β3 strand, which is located deeper within the monomer, contribute to the middle part of the dimer interface. This location forms a cavity with several hydrophilic, positively charged residues, filled by a few water molecules. An ion pair between Glu-187 of one monomer and His-103 of the other is found at the outer edges of the helices. Another intermolecular ion pair is made between Glu-133 and Arg-179. The dimer is further stabilized by several polar interactions, including hydrogen bonds between Glu-187 and Trp-134 as well as His-198 and the main chain carbonyl of Ala-110. It seems likely that the dimer organization observed in the catalytic domain crystals is also present in the multimeric structure of the full length proteins in solution.

The active site as observed in the low-temperature structure of ASV IN is shown in Fig. 3. All three acidic residues characteristic for this class of enzymes are seen clearly in the electron density. The carboxylate groups of the active site residues interact through two water molecules, Wat-396 and Wat-450. Another water molecule, Wat-324, appears to be of particular importance,

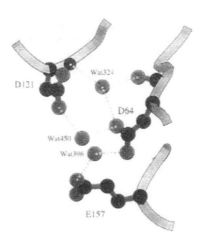

Figure 3. Part of the active site of ASV IN, showing the the side chains of Asp-121, Asp-64, and Glu-157. The water molecules present in the active site, and their hydrogen bonds to the residues above, are also shown.

since it has a relatively low temperature factor, roughly tetrahedral coordination, and plays a role in bridging the two β sheets contributing the active site aspartates. The environment of the active site consists of a long, shallow depression on the surface, rather than a more distinct cavity observed in many other enzymes. This is, however, not unexpected, in view of the very large size of integrase substrates, which need access to the catalytic center.

Comparison with HIV-1 IN

A comparison of the sequences of the catalytic domains of ASV IN (residues 52-207) and HIV-1 IN (residues 50-212) reveals a moderate 32% similarity. Structural alignment of the crystallo-graphic models of the monomers carried out in ALIGN (31) superimposed them with an r.m.s. deviation of 1.4 Å (107 Cα pairs). However, an important segment of as many as 13 residues (containing the third active site amino acid, Glu-152) is missing from the HIV-1 IN model (32). Tentatively, the authors assigned an internal loop to this region, which is positioned between β5 and α4. In the ASV IN structure, however, the corresponding sequence (residues 146-158) covers not only the β5-α4 loop (where for residues 149-150 the temperature factors exceed 70 Å^2 and are the highest in the non-terminal part of the λr low temperature model) but also a very well defined helical segment (154-158) at the N-terminus of the long α4 helix which contains the catalytic Glu-157 as well as Met-155 (perfectly detectable, through its Se-Met analog, even by direct methods based on Cu Kα data). The general architecture of the ASV IN and HIV-1 IN dimers is essentially the same.

The monomer-monomer alignment of the HIV-1 and ASV IN catalytic domains reveals a puzzling picture in the active site. The β strand containing Asp-64 has nearly identical main chain trace in both models. In spite of this, the critical aspartates point in opposite directions in the two structures. On the other hand, the main chains embedding ASV IN Asp-121 and its HIV-1 counterpart, Asp-116, follow significantly different paths. These deviations are larger still at the respective aspartate sites as they again have different orientations. It is unclear why these two proteins, so closely related in sequence, structure, and function should have their absolutely conserved catalytic residues in entirely different conformations. The organization of the active site observed in ASV IN is not an artifact as it was confirmed in several well-refined high resolution structures.

Acknowledgments
Research sponsored in part by the National Cancer Institute, DHHS, under contract NO1-CO-46000 with ABL. Other support includes National Institutes of Health grants CA47486, CA06927, a grant for infectious disease research from Bristol-Myers Squibb Foundation, and an appropriation from the Commonwealth of Pennsylvania. The contents of this publication do not necessarily reflect the views or policies of the Department of Health and Human Services, nor does mention of trade names, commercial products, or organizations imply endorsement by the U.S. Government.

References
1. Katz, R.A. & Skalka, A.M. (1994). The retroviral enzymes. *Annu. Rev. Biochem.* **63**, 133-173.
2. Goff, S.P. (1992). Genetics of retroviral integration. *Annual Review of Genetics* **26**, 527-544.
3. Whitcomb, J.M. & Hughes, S.H. (1992). Retroviral reverse transcription and integration: progress and problems. *Annu. Rev. Cell Biol.* **8**, 275-306.

4. Craigie, R., Fujiwara, T., & Bushman, F. (1990). The IN protein of Moloney murine leukemia virus processes the viral DNA ends and accomplishes their integration in vitro. *Cell* **62**, 829-837.

5. Katz, R.A., Merkel, G., Kulkosky, J., Leis, J., & Skalka, A.M. (1990). The avian retroviral IN protein is both necessary and sufficient for integrative recombination in vitro. *Cell* **63**, 87-95.

6. Roth, M.J., Schwartzberg, P.L., & Goff, S.P. (1989). Structure of the termini of DNA intermediates in the integration of retroviral DNA: dependence on IN function and terminal DNA sequence. *Cell* **58**, 47-54.

7. Katzman, M., Katz, R.A., Skalka, A.M., & Leis, J. (1989). The avian retroviral integration protein cleaves the terminal sequences of linear viral DNA at the in vivo sites of integration. *J. Virol.* **63**, 5319-5327.

8. Engelman, A., Mizuuchi, K., & Craigie, R. (1991). HIV-1 DNA integration: mechanism of viral DNA cleavage and DNA strand transfer. *Cell* **67**, 1211-1221.

9. Jones,K.S., Coleman, J., Merkel, G.W., Laue, T.M., & Skalka, A.M. (1992). Retroviral integrase functions as a multimer and can turn over catalytically. *J. Biol. Chem.* **267**, 16037-16040.

10. Engelman, A., Bushman, F.D., & Craigie, R. (1993). Identification of discrete functional domains of HIV-1 integrase and their organization within an active multimeric complex. *EMBO J.* **12**, 3269-3275.

11. van Gent, D.C., Vink, C., Groeneger, A.A.M.O., & Plasterk, R.H. (1993). Complementation between HIV integrase proteins mutated in different domains. *EMBO J.* **12**, 3261-3267.

12. Kulkosky, J., Jones, K.S., Katz, R.A., Mack, J.P., & Skalka, A.M. (1992). Residues critical for retroviral integrative recombination in a region that is highly conserved among retroviral/ retrotransposon integrases and bacterial insertion sequence transposases. *Molecular & Cellular Biology* **12**, 2331-2338.

13. Engelman, A. & Craigie, R. (1992). Identification of conserved amino acid residues critical for human immunodeficiency virus type 1 integrase function in vitro. *J. Virol.* **66**, 6361-6369.

14. Kulkosky, J., Katz, R.A., Merkel, G., & Skalka, A.M. (1995). Activities and substrate specificity of the evolutionarily conserved central domain of retroviral integrase. *Virology* **206**, 448-456.

15. Hendrickson, W.A., Horton, J.R. & LeMaster, D.M. (1990). Selenomethionyl proteins produced for analysis by multiwavelength anomalous diffraction (MAD): a vehicle for direct determination of three dimensional structure. *EMBO J.* **9**, 1665-1672.

16. Yang, W., Hendrickson, W.A., Kalman, E.T., & Crouch, R.J. (1990). Expression, purification, and crystallization of natural and selenomethionyl recombinant ribonuclease H from *Escherichia coli*. *J. Biol. Chem.* **265**, 13553-13559.

17. Otwinowski, Z. (1992). An Oscillation Data Processing Suite for Macromolecular Crystallography (New Haven: Yale University).

18. Blessing, R.H., and Langs, D.A. (1988). *A priori* estimation of scale and anisotropic temperature factors from the Patterson origin peak. *Acta Crystallogr. A* **44**, 729-735.

19. Hauptman, H.A. & Han, F. (1993). Phasing macromolecular structures *via* structure-invariant algebra. *Acta Crystallogr. D* **49**, 3-8.

20. DeTitta, G.T., Weeks, C.M., Thuman, P., Miller, R., & Hauptman, H.A. (1994). Structure solution by minimal-function phase refinement and Fourier filtering. I. Theoretical basis. *Acta Crystallogr. A* **50**, 203-210.

21. Weeks, C.M., DeTitta, G.T., Hauptman, H.A., Thuman, P., & Miller, R. (1994). Structure solution by minimal-function phase refinement and Fourier filtering.II. Implementation and applications. *Acta Crystallogr. A* **50**, 210-220.

22. Langs, D.A., Guo, D., & Hauptman, H.A. (1995). Use of "Random-Atom" phasing models to determine macromolecular heavy atom positions. *Acta Crystallogr.* (in press)

23. Mitchell, E.P. & Garman, E.F. (1994). Flash freezing of protein crystals: investigation of mosaic spread and diffraction limit with variation of cryoprotectant concentration. *J. Appl. Cryst.* **27**, 1070-1074.

24. Hendrickson, W.A. (1991). Determination of macromolecular structures from anomalous diffraction of synchrotron radiation. *Science* **254**, 51-58.

25. Furey, W. and Swaminathan, S. (1990). Phases - A program package for the processing and analysis of diffraction data for macromolecules. American Crystallographic Association Meeting.

26. Brünger, A. (1992). X-PLOR: A System for X-Ray Crystallography and NMR (New Haven: Yale University Press).

27. Steigemann, W. (1974). Die Entwicklung und Anwendung von Rechenverfahren und Rechenprogrammen zur Strukturanalyse von Proteinen am Beispiel des Trypsin- Trypsininhibitor Komplexes, des freien Inhibitors und der L-Asparaginase. Ph.D. Thesis, Technische Universität, München, Germany.

28. Jones, T.A. (1985). Interactive computer graphics: FRODO. *Methods Enzymol.* **115**, 157-171.

29. Jones, T.A. & Kjeldgaard, M. (1994). O - The Manual (Uppsala: Uppsala University).

30. Hendrickson, W.A. (1985). Stereochemically restrained refinement of macromolecular structures. *Methods Enzymol.* **115**, 252-270.

31. Satow, Y., Cohen, G.H., Padlan, E.A., & Davies, D.R. (1986). Phosphocholine binding immunoglobulin Fab McPC603: An X-ray diffraction study at 2.7 Å. *J. Mol. Biol.* **190**, 593-604.

32. Dyda, F., Hickman, A.B., Jenkins, T.M., Engelman, A., Craigie, R., & Davies, D.R. (1994). Crystal structure of the catalytic domain of HIV-1 integrase: similarity to other polynucleotidyl transferases. *Science* **266**, 1981-1986.

33. Kraulis, P.J. (1991). MOLSCRIPT: A program to produce both detailed and schematic plots of protein structures. *J. Appl. Cryst.* **24**, 946-950.

Structure/Function of the Fourth and Fifth EGF Domains of Thrombomodulin

Elizabeth A. Komives, Michael J. Hunter, David P. Meininger
Lisa R. White & Christopher E. White
Department of Chemistry & Biochemistry, University of California, San Diego,
La Jolla, CA 92093-0601

The structure and function of the two essential EGF-like domains of human thrombomodulin (TM) have been determined. Both the fourth and fifth domains are necessary and sufficient for TM cofactor activity (White et al., 1995). The fragment of TM containing only the fifth and sixth EGF-like domains binds to thrombin, but has no cofactor activity (Stearns et al., 1989; Ye et al., 1992; Kurosawa et al., 1988). The fifth domain alone inhibits thrombin-induced clotting of fibrinogen and activation of protein C by the thrombin-TM complex, both indicators of thrombin-binding. NMR evidence suggests that the fifth domain is flexible. Furthermore, at least three different disulfide bonding isomers of the fifth domain (1-3,2-5,4-6), (1-3,2-4,5-6) and (1-2,3-4,5-6) bind to thrombin and the latter non-EGF-like isomer binds an order of magnitude more tightly than the EGF-like isomer. C-terminal loop peptides from the fifth domain are unstructured in solution but form tri-stranded antiparallel β-sheets upon binding to thrombin. These results suggest that the fifth domain may bind to thrombin by an induced fit mechanism. The fourth domain shows no thrombin binding activity or TM cofactor activity. The structure of the fourth domain, determined by two-dimensional NMR methods, appears rigidly structured. The N-terminal half of the domain resembles other known EGF-like domains, but the structure of the C-terminal loop is novel in that the amino acids C-terminal to the last cysteine form the central strand of an irregular parallel-antiparallel sheet. Thus the fourth and fifth domains function together, the fourth is rigid and unable to bind without the fifth, and the fifth has non-EGF-like disulfide bonds that give it the flexibility to bind thrombin by an induced fit mechanism.

Introduction

Over 300 sequences which are similar to epidermal growth factor (EGF) have been identified and more are continuing to be discovered. These sequences have in common six cysteine residues within a total sequence length of 40 - 50 amino acids and have been termed EGF-like domains. Aside from the cysteines, the rest of the sequences are quite diverse, although certain families of similar sequences are beginning to emerge (Campbell & Bork, 1993).

Although many EGF-like domain sequences are known, very few have well-defined functions. EGF protein and transforming growth factor α bind to the EGF receptor and induce cell growth and proliferation (Carpenter & Cohen, 1979; Moy et al., 1993). The EGF domain from urokinase also binds to a receptor and induces extracellular proteolysis (Apella et al., 1987). EGF-like domains located near the N-terminus of coagulation proteases factors IX, X and VII have been shown to contain a post-translationally hydroxylated aspartic acid residue and to bind calcium ions (Selander-Sunnerhagen et al., 1992). The EGF-like domains numbers 11 and 12 out of 36 in the extracellular domain of the Notch protein are necessary and sufficient for interaction with the Delta and Serrate proteins, interactions which determine cell fate in Drosophila development (Rebay et al., 1991). Finally, the EGF-like domains numbers 4 and 5 from thrombomodulin (TM) interact with thrombin to alter its substrate specificity away from fibrinogen cleavage and toward protein C activation, thus shutting down further production of thrombin (Esmon, 1989).

Besides diversity of function, EGF-like domains show diversity in the region of the domain that is responsible for the function. In the case of EGF and TGFα, Leucine-47, which is C-terminal to the sixth cysteine has been shown to be important for receptor binding, and several other residues have been implicated but these residues are located in several parts of the molecule (Moy et al., 1989). In the case of the urokinase EGF-like domain, it is the sequence between the third and fourth cysteines that is responsible for receptor binding and peptides corresponding to this sequence are potent inhibitors of urokinase binding to its receptor (Apella et al., 1987). In the case of the calcium-binding EGF-like domains, residues within the sequence between the third and fourth cysteine as well as residues N-terminal to the first cysteine bind the calcium (Selander-Sunnerhagen et al., 1992). For the TM EGF-like domains, residues within the sequence between the fifth and sixth cysteines, as well as residues C-terminal to the sixth cysteine of the fifth domain, are involved in thrombin binding (Hayashi et al., 1990; Tsiang et al., 1992; Lougheed et al., 1995). The peptide corresponding to the C5-C6 disulfide bonded loop and the residues C-terminal to this loop has been shown to bind to thrombin by an induced-fit mechanism (Srinivasan et al., 1994).

The one aspect of EGF-like domains which appears to be conserved throughout this family is the structure. EGF has three disulfide bonds connecting the six cysteines in a (1-3, 2-4,5-6) disulfide bonding pattern (Savage & Cohen, 1973). The disulfide bonding pattern of TGFα, and of the factor IX, factor X EGF-like domains are the same as that of EGF. All of these domains also have similar three-dimensional folds as EGF (see above references). Thus, despite the diversity of sequence and function, the definition of an EGF-like domain rests primarily upon its having a structure similar to that found in EGF. It is not clear whether the observation that the structures are the same is merely a function of the small number of structures that have been determined. Furthermore, it is not clear how the structure is related to the function since the location of the functional sequences within each domain appears to vary. We present here structure and function data on two EGF-like domains from TM which challenge the conservation of both disulfide bonding pattern and three-dimensional fold in EGF-like domains.

Materials and Methods

A. Peptide Synthesis

TM4, the 44 amino acid peptide corresponding to the sequence: EPVDPCFRANCYQCQPLNQTSYLCVCAEGFAPIPHEPHRCQMF, was prepared by standard solid phase peptide synthesis on a Milligen/Biosearch 9050 peptide synthesizer. The amino acids used were obtained from Perceptive Biosystems (Burlington, MA), were protected on the amine with fluorenylmethoxycarbonyl and were preactivated as the pentafluorophenyl esters on the carboxylic acid. The solid support was PEG-PS-Phenylalanine with a substitution level of 0.160 meq/gm. High yields of TM4 were obtained by activation of the OPfp esters with hydroxy-7-azabenzotriazole (HOAt, Perceptive Biosystems). Arg(Pmc)-OH was coupled using [0-(7-azabenzotriazol-1-yl)-1,1,3,3-tetramethyluronium hexafluorophosphate] (HATU, Perceptive Biosystems) as the activator and diisopropylethylamine as the base. Dimethylformamide (glass distilled) from EM Sciences (distributed by VWR, Los Angeles, CA) was used as the solvent for all syntheses.

TM5, the 40 amino acid peptide corresponding to the sequence: QMFCNQTACPADCDPNTQASCECPEGYILDDGFICTDIDE, was prepared by the same chemistry as for TM4 except that to direct the formation of specific disulfide bonds, Acm protection of the cysteines was occasionally used. Initially, the peptide was synthesized with all cysteines protected by trityl groups. To direct the synthesis of the EGF-like isomer, Acm protection was used on the second and fourth cysteine and to direct the synthesis of the un-crossed isomer, Acm protection was used on the third and fourth cysteines (Yang et al., 1994). The peptide was cleaved from the resin and the side chain protecting groups were removed by treatment of the resin with a mixture of 90% trifluoroacetic acid, 5% thioanisole, 3% ethanedithiol, and 2% anisole for 4 hours and the peptide was isolated by ether precipitation at -20 °C overnight. The precipitated peptide was collected on a cintered glass funnel, washed with cold ether, dissolved in 5% ammonium acetate buffer, pH 7.5, and lyophilized.

Fully reduced peptide was prepared by dissolving 100 mg portions of crude peptide in 20 mL of 100 mM Tris-HCl pH 8.0 buffer containing 20 mM dithiothreitol under nitrogen for 8-12 hours. Purification of 100 mg portions of crude peptide by reverse phase HPLC on a Waters DeltaPak C18 column (19 x 300 mm) using a gradient of argon-sparged 0.1% trifluoroacetic acid (TFA) to 50% acetonitrile over 1 hour produced a yield of 25-30 mg pure, fully reduced peptide. The purified, reduced peptide was then refolded by slow air oxidation in a redox buffer. The peptide (25 mg) was dissolved in 1L of 0.5% ammonium acetate buffer, pH 5.0, containing 1 mM oxidized glutathione and 9 mM reduced glutathione, the pH was adjusted to 8.3 with 5% NH4OH. The pH of the oxidation solution was maintained at 8 to 8.3 by daily addition of 5% NH4OH for 4 to 5 days. The oxidized, folded product was isolated by loading the solution onto an HPLC column directly through the solvent pump. The column, gradient, and buffers were the same as those above and the gradient was held at 30% acetonitrile for better product separation. After purification to > 95% homogeneity, the overall yield was approximately 10% of the crude.

The TM5 peptides containing Acm protection on two of the six cysteines were prepared in the same manner described above. The Acm groups were then removed by treatment with TFA, anisole, and AgOTf as described in Hunter & Komives, 1995. After ether precipitation, the peptide with S-Ag cysteine residues was then isolated by centrifugation and subjected to mild oxidation in the presence of 50% DMSO / 50% 1 M HCl and purified as already described.

B. Determination of disulfide bonds

EGF-like domains contain six cysteine residues which form 3 disulfide bonds. The disulfide bonding patterns of the peptides were determined by a partial reduction method using tris-carboxyethylphosphine (TCEP) as a partial reducing agent (Gray, 1993). Approximately 500 μg of purified, oxidized peptide was dissolved in 700 μL 0.1% TFA. The TCEP solution was prepared by dissolving 32.4 mg TCEP and 250 mg citric acid in 5 mL MilliQ H_2O and adjusting the pH to 2.5 by the dropwise addition of 1 N NaOH. TCEP solution (700 μL) was added to the peptide solution and the mixture was vortexed. After 30 min - 1 hour, the partially reduced peptides were separated by HPLC on a Vydak C18 HPLC column (4.6 x 250 mm) at a flow rate of 1 ml/min, using a gradient of 0.1% TFA to 10% acetonitrile over 10 minutes and then to 40% acetonitrile over 1 hour. Detection was at 280 nm. . The partially reduced products were collected separately and the free thiols were alkylated by forcibly injecting 500 μL of each HPLC peak into a Falcon tube containing 400 μL of a 0.5 M Tris-acetate buffer pH 8.0, 2 mM EDTA, and 2.2 M iodoacetamide with continuous vortexing during the injection. After 1 minute, the reaction was quenched by acidification with 800 μL of 0.5 M citric acid and the acetylated products were purified separately using the above HPLC conditions. Finally, the purified, alkylated products were characterized by N-terminal sequencing. The PTH derivative of reduced, carboxamidomethylcysteine elutes at the same retention time as the PTH derivative of glutamic acid and was easily distinguished from cystine. For the TM5 peptides, this was problematic for determination of the disulfide bonding of the fourth and fifth cysteines because these cysteines are separated by a single glutamic acid. To circumvent this problem, we utilized N-methyliodoacetamide instead. This compound was prepared by the method of Krutzsch and Inman, 1993 and purified by flash chromatography on silica gel in 50% dichloromethane / 50% ethylacetate. The PTH derivative of N-methyl-S-carboxamidomethylcysteine elutes approximately one min before the PTH derivative of histidine in a standard N-terminal sequencer.

C. TM activity assays

Peptides were dissolved in H_2O at a concentration of 1 - 2 mM and the pH was adjusted if necessary. The peptide solutions were divided into small portions and stored at -20°C. The concentration of each peptide solution was determined by quantitative amino acid analysis using norleucine as the standard. The direct inhibition of thrombin-induced clot formation by the TM EGF-like domains was measured by a method that has been described previously (Lougheed et al., 1995). TM cofactor activity, which results in the thrombin-dependent production of activated protein C, was measured using an assay in which thrombin and TM were incubated with protein C and the resulting activated protein C was then assayed with a chromogenic substrate (Stearns et al., 1989). Cofactor activity of TM4 was measured by incubating various concentrations of each peptide with human thrombin (0.4 µg/mL, 0.875U/mL, 11 nM) in TBS containing BSA (1mg/mL) and $CaCl_2$ (5 mM). After 10 minutes, human protein C (7.25 µg/mL, 120 nM, Hematologic Technologies) was added and the mixture was incubated for 20 minutes. The thrombin activity was quenched with heparin-antithrombin III (80 ng/mL and 220 ng/mL respectively) and activated protein C activity was assayed by monitoring the release of p-nitroaniline at 405 nm from a chromogenic substrate, S-2366 (Chromogenix), at a concentration of 0.74 mg/mL. TM4 and each TM5 isomer were also tested for thrombin-binding as indicated by inhibition of protein C activation in the presence of several concentrations of rabbit TM (10, 15, 20, 25 ng/mL, 0.15, 0.22, 0.29, 0.375 nM) using the same assay, and the reciprocal velocity was plotted against the concentration of the EGF-like domain in a Dixon plot.

D. NMR Spectroscopy

Spectra were acquired on a Bruker AMX spectrometer operating at 500.13 MHz. All 2D data sets were processed on a Silicon Graphics workstation using FELIX 2.30 software (Biosym Technologies, San Diego, CA). Sample preparation, assignments and determination of the structure of TM4 was as previously described (Meininger et al., 1995).

Results

A. Disulfide bonding patterns of TM4 and TM5

Refolding of purified, reduced TM4 resulted in only one major disulfide bonded isomer which was shown to have a (1-3,2-4,5-6) disulfide bonding pattern. This isomer therefore contained the same disulfide bonds found in EGF protein.

Refolding of purified, reduced TM5 resulted in two isolable products, neither of which was EGF-like. The earliest eluting isomer had a (1-2,3-4,5-6) disulfide bonding pattern so that none of the disulfide bonds were crossed, whereas the two N-terminal disulfide bonds are crossed in EGF protein. The later eluting isomer had a (1-3,2-5,4-6) disulfide bonding pattern so that all of the disulfide bonds were crossed, including the C-terminal disulfide bond which is not crossed in EGF protein. In order to obtain the EGF-like disulfide bonding pattern, the differential cysteine protection method of Yang et al., 1994 was employed. This method produced the (1-3,2-4,5-6) disulfide bonded isomer in high yields. Since the disulfide bonding pattern of the fifth EGF-like domain within native TM is not known, all three isomers were tested for TM-like activity.

Functions of TM4 and TM5. TM has two functions, a direct inhibition of thrombin clotting of fibrinogen, which results from competitive binding of

TM to thrombin in the fibrinogen binding site, and stimulation of thrombin cleavage of protein C to form activated protein C. Both domains were tested for both functions. TM4 did not inhibit clot formation, nor did it stimulate or inhibit thrombin activation of protein C at concentrations up to 350 μM.

All of the disulfide bonded isomers of TM5 were active as inhibitors of thrombin clotting of fibrinogen. All were also inhibitors of thrombin-TM complex formation resulting in cleavage of protein C to form activated protein C. Surprisingly, thrombin binding as assessed both by direct inhibition of clot formation as well as by inhibition of protein C activation correlated with lack of crossing of the disulfide bonds (Table 1). The isomer with the (1-2,3-4,5-6) disulfide bonding pattern was a 10 fold more potent inhibitor of protein C activation and almost 50 fold more potent as an inhibitor of clotting than the EGF-like isomer. The EGF-like isomer (1-3,2-4,5-6) showed intermediate potency and the fully-crossed isomer (1-3,2-5,4-6) was the least potent.

Table 1. Summary of Thrombin Inhibition by the Various TM Fifth Domains

Peptide	Peptide concentration required to double the clotting time (μM)[a]	K_i for protein C activation (μM)[b]
TM5(1-2,3-4,5-6)	0.2 +/- 0.02	1.9 +/- 0.2
TM5(1-3,2-4,5-6)	9 +/- 1	13 +/- 1
TM5(1-3,2-5,4-6)[c]	24 +/- 2	35 +/- 3
C-terminal loop[d]	35 +/- 5	65 +/- 10

[a] The concentration of peptide required to double the clotting time was determined as described in the Materials and Methods section. [b] The K_i's were determined from Dixon plots as described in the Materials and Methods section. [c] Currently, the disulfide bonding pattern within the fifth EGF-like domain of native TM is not known. [d] Data for the C-terminal loop peptide is from Blackmar et al, 1995.

B. NMR of TM4 and TM5

Although the TM4 and the TM5 EGF-like domains are of similar size (44 vs. 40 residues) the fingerprint region of the NOESY spectrum of TM4 shows well-dispersed, narrow lines and the expected number of cross peaks, while the corresponding fingerprint regions of the TM5 isomers shows little dispersion and broadened lines (Figure 1). These spectra suggest that TM4 is rigidly structured, while the TM5 isomers are flexible or not well structured. Although all the isomers of the TM5 domain show line broadening, the (1-2,3-4,5-6) isomer, which was the most potent thrombin inhibitor, shows the most resonance dispersion indicative of secondary structure. The differences between the spectra of the fourth and fifth domains are striking considering they are both EGF-like domains of similar size. Such large differences were unexpected simply from looking at the sequences of the domains.

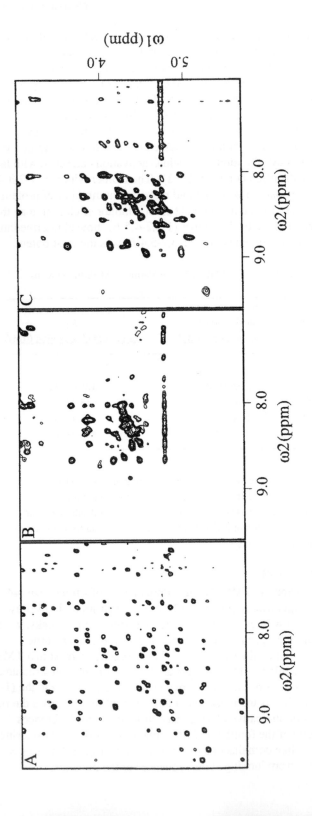

Figure 1. Fingerprint regions of the NOESY spectra of (A) TM4; (B) TM5(1-3.2-4.5-6); and (C) TM5(1-2.3-4.5-6).

The structure of the fourth domain showed an interesting difference from other EGF-like domains. This domain has five more amino acids in the sequence between the fifth and sixth cysteines than EGF protein. The sequence contains three prolines and these distort the tri-stranded β-sheet structure of the C-terminal portion of the domain. Interestingly, this distortion, as well as a few key hydrophobic contacts function to restructure the C-terminal loop so that the amino acids C-terminal to the last cysteine lie between two strands formed by the rest of the loop instead of along side the rest of the loop as is the case for EGF protein (Figure 2). This structure has never before been observed in an EGF-like domain.

Conclusions

A. Functional significance of the structure of TM4

The structure of TM4 can be rationalized in terms of its function. Unlike other EGF-like domains which have been studied to date, the fourth EGF-like domain of TM functions only when it is connected to the fifth domain (White et al., 1995). The fragment of TM containing only the fifth and sixth EGF-like domains binds to thrombin, but has no cofactor activity (Stearns et al., 1989; Ye et al., 1992; Kurosawa et al., 1988). Thus, although the fourth domain is required for TM cofactor activity, it had no observable activity up to concentrations of 350 μM. One possible explanation is that the fourth and fifth domains form a composite surface which interacts with thrombin, and that the region of the fourth domain that interacts with thrombin only when it is attached to the fifth domain. Thus, the connection between the two domains would be expected to be functionally important. It therefore makes sense that the amino acids C-terminal to the last cysteine of the fourth domain , which are

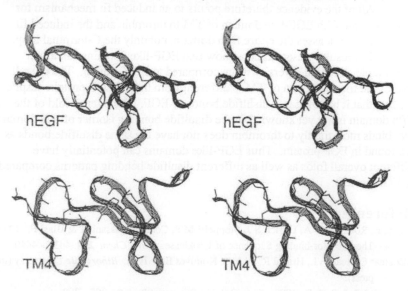

Figure 2. Stereo view of a ribbon diagram of a low energy structure of TM4 compared to a low energy structure of human EGF (3EGF from the Brookhaven protein data bank).

those that connect the fourth domain to the fifth domain, are well-structured and rigidly held in place. Since there are very few other EGF-like domains for which functions have been assigned, it remains to be seen if other EGF-like domains function in tandem and if this is a common structural feature of those EGF-like domains that function only when connected to another domain.

B. Functional significance of the structure of TM5

The fifth EGF-like domain of TM contains the major thrombin-binding region. Peptides corresponding to the sequence between the fifth and sixth EGF-like domains and the five amino acids C-terminal to the sixth cysteine bind to thrombin with affinities near 50 μM (Hayashi et al., 1990; Tsiang et al., 1992; Lougheed et al., 1995). The binding affinities of the two isomers of TM5 with crossed disulfide bonds are only two and four-fold better than the binding affinity of the C-terminal loop peptide. On the other hand, the binding affinity of the uncrossed isomer is 25 fold better as assessed by inhibition of protein C activation. This domain is a 100-fold more potent inhibitor of fibrinogen clotting than the C-terminal loop peptide, and is almost equipotent to full-length TM. Thus, tight thrombin-binding requires more than the C-terminal loop and amino acids C-terminal to the last cysteine. The C-terminal loop region probably interacts with the N-terminal subdomain in the bound structure. Interestingly, the thrombin-binding potency of the TM5 disulfide bonding isomers correlates with lack of crossing of disulfide bonds, which would be expected to correlate with flexibility. Previous studies on C-terminal loop peptides from the fifth domain of TM indicated that the peptides were unstructured in solution but upon thrombin binding, they adopted a tri-stranded β-sheet structure typical of EGF-like domains (Srinivasan et al., 1994). All of the evidence therefore points to an induced fit mechanism for binding of the fifth EGF-like domain of TM to thrombin and the induced fit probably encompasses the entire fifth domain, not only the C-terminal loop.

The results presented here show that EGF-like domains can attain remarkable structural diversity when compared to EGF protein. The overall fold of the fourth domain of TM is different from that of EGF protein despite the fact that it has the same disulfide bonds as EGF. The overall fold of the fifth domain is not yet known, but the disulfide bonding isomer of this domain that binds most tightly to thrombin does not have the same disulfide bonds as are found in EGF protein. Thus EGF-like domains can potentially have different overall folds as well as different disulfide bonding patterns compared to EGF protein.

References

Apella E, Robinson E A, Ullrich S J, Stoppelli M P, Corti A, Cassani G & Blasi F. 1987. The Receptor-binding Sequence of Urokinase. *J Biol Chem* **262**: 4437 - 4440.

Blackmar C, Healy VL, Hrabal R, Ni F, & Komives EA. 1995. *Bioorganic Chemistry* (in press).

Campbell I D & Bork P. 1993. *Curr Opioion Structural Biol* **3**: 385 - 392.

Carpenter G & Cohen S. 1979. *Ann Rev Biochem* **48**: 193 - 216.

Esmon CT. 1989. *J.Biol Chem* **264**: 4743 - 4746.

Hunter MJ & Komives E A. 1995. *Anal Biochem* **228**:173 - 177.

Krutzsch, H. C. & Inman, J. K. 1993. *Anal. Biochem.* **209**: 109 - 116.

Kurosawa S, Stearns DJ, Jackson KW & Esmon CT. 1988. *J Biol Chem* **263**: 5993 - 5996.

Lougheed JL. Bowman CA, Meininger DP, Komives EA. 1995. *Protein Science* **4**:773 - 780.

Meininger, D. P., Hunter, M. J. & Komives, E. A. 1995. *Protein Science* (in press).

Moy FJ, Sheraga HA, Liu J-F, Wu R & Montelione GT 1989. *Proc Nat Acad Sci U. S. A.* **86**: 9836-9840.

Moy FJ, Li Y-C, Rauenbuehler P, Winkler ME, Scheraga HA & Montelione GT. 1993. *Biochemistry* **32**: 7334 - 7353.

Rebay I, Fleming R J , Fehon R G, Cherbas L, Cherbas P & Artavanis-Tsakonas S. 1991. *Cell* **67**: 687 - 699.

Savage CR, Hash JH & Cohen S. 1973. *J Biol Chem.***248**: 7669 - 7672.

Selander-Sunnerhagen M, Ullner M, Persson E, Teleman O, Stenflo J & Drakenberg T. 1992. *J Biol Chem* **267**: 19642 - 19649.

Srinivasan J, Hu S, Hrabal R, Zhu Y, Komives E A & Ni F. 1994. *Biochemistry* **33**: 13553 - 13561.

Stearns D J, Kurosawa S, Esmon C T. 1989. *J Biol Chem* **264**: 3352 - 3356.

White C W, Hunter M J, Meininger D P & Komives E A. 1995. *Protein Engineering.* (in press).

Yang Y, Sweeney WV, Schneider K, Chait BT & Tam JP. 1994. *Protein Science* **3**: 1267 - 1275.

Ye J, Liu L-W, Esmon CT & Johnson AE. 1992. *J Biol Chem* **267**: 11023 - 11028.

The structure of a C-type lectin isolated from bovine cartilage

Peter J. Neame[*]
Raymond E. Boynton
Shriners Hospital for Crippled Children,
and
*Department of Biochemistry and Molecular Biology,
University of South Florida School of Medicine,
Tampa, Florida

I. Introduction

C-type lectins are a class of mammalian, calcium-dependent, carbohydrate-binding proteins which have a broad distribution. Members of this family include the cartilage proteoglycan, aggrecan (1) and related proteoglycans versican (2) and neurocan (3) and also the hepatic lectins which are involved in serum glycoprotein turnover. A 2.5Å structure of a rat mannose-binding protein has been described (4).

We have previously isolated a C-type lectin from shark cartilage (5). The structure of this protein is similar to tetranectin, a 68 kDa tetrameric serum protein which was originally isolated as a result of its affinity for the kringle-4 domain of plasminogen (6). The structure of tetranectin was first described by Fuhlendorff et al.(7). Tetranectin was found to enhance plasminogen activation catalyzed by t-PA in the presence of poly(D-lysine) (6). Tetranectin has a broad distribution in endocrine tissues; it has been suggested that tetranectin might have a dual function, serving both as a regulator in the secretion of certain hormones and as a participant in the regulation of pro-hormone activation by proteolysis(8). It has also been suggested that tetranectin might have a role in osteogenesis (9).

Tetranectin, in conjunction with proteoglycans, may have a function in the packaging of the contents of exocytotic granules or it may be a participant in exocytotic processes(10). A tetranectin-related protein is produced and deposited in extracellular matrix by human embryonic fibroblasts(11).

The gene for tetranectin is about 12 kbp and contains two introns. Tetranectin has a signal peptide of 21 amino acids followed by the mature sequence of 181 amino acids. Southern blot analysis showed hybridization to two genes (12).

We report here the structure of a mammalian cartilage equivalent of the shark cartilage lectin-like protein which we have previously characterized. It is found in both articular cartilage and spinal nucleus pulposus. The cartilage lectins, both shark and bovine, are distinct from but related to tetranectin.

II. Materials and Methods

The methods used for isolation of proteins from cartilage are essentially as described previously (13, 14). Briefly, they are as follows.

Calf spinal nucleus pulposus, or fetal femoral condyle epiphyseal or femoral condyle articular cartilage from 2 year-old steers was extracted with guanidine hydrochloride (4M) in the presence of protease inhibitors. Proteoglycan aggregates were re-formed by dialysis and removed by isopycnic cesium chloride density gradient ultracentrifugation (starting density, 1.5, spun for 40 hours at 40,000 rpm on a Beckman ultracentrifuge using a Ti 50.2 rotor at 10°C). The upper fraction, with a density less than 1.3, was dialyzed against water, concentrated, and adjusted to 4M by the addition of solid guanidine hydrochloride.

Proteins were purified by gel filtration on a Superose 12 column, equilibrated in 4M guanidine HCl. Individual fractions were applied to a reversed-phase column; either a Brownlee Aquapore (Applied

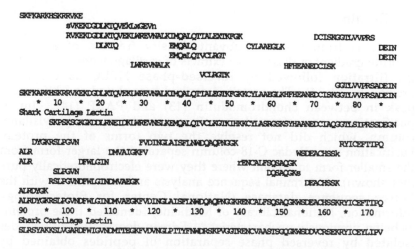

Figure 1. Alignment of peptides against the shark cartilage derived protein.

The bovine cartilage protein was digested with Endoprotease Lys-C and Endoprotease Asp-N to obtain sets of peptides which could be used to define the sequence when compared with the shark cartilage-derived lectin. Lower case letters indicate doubtful calls in the protein sequence analysis.

Biosystems) or a Vydac C18 (4.6 x 250 mm) and proteins were eluted with a gradient of acetonitrile / water in 0.1% trifluoroacetic acid.

Proteolytic digestion with endoprotease Lys-C or endoprotease Asp-N was performed at an enzyme substrate ratio of 1:50 in either 0.1 M Tris-HCl, 10% acetonitrile, 0.1% RTX-100, pH 8 (Lys-C) or 0.1 M sodium phosphate, pH 8 (Asp-N). Reversed-phase HPLC was performed on an Applied Biosystems 130A microbore LC. Separation of peptides was achieved using a Vydac C18 column (2.1 x 30mm) with a gradient of acetonitrile in water (0.1% trifluoroacetic acid - 70% acetonitrile, 0.085% trifluoroacetic acid over 45 minutes). Flow rate was 200 µl/min, eluant was monitored at 220 nm and UV-absorbing peaks were collected by hand for sequence analysis. Protein sequence analysis was performed on an Applied Biosystems 477A protein sequencer with on-line detection of phenylthiohydantoin amino acid derivatives on an Applied Biosystems 120A microbore HPLC.

mRNA was isolated from fetal or calf cartilage by the method of Smale and Sasse (15). cDNA was synthesized by oligo dT priming of crude RNA. 5' RACE (Rapid Amplification of cDNA Ends) analysis of cDNA was performed using a kit supplied by Clontech and following the manufacturer's recommendations. Sequence analysis of PCR products was performed by the ICBR core facilities at the University of Florida, Gainesville.

Analysis of the crystal structure of the carbohydrate-recognition domain of the mannose-binding protein A described by Weis et al (4) was performed using the program RasMol (Roger Sayle, Biomolecular Structure Department, Glaxo, Greenford, UK).

III. Results

Proteins from the low-buoyant density fraction of a cesium chloride gradient of dissociative extracts of cartilage were separated by gel filtration followed by reversed-phase HPLC as described previously (13, 14). The protein described here eluted as a bifurcated peak in between chondromodulin (13) and PARP (14). Initial purification was performed on a Brownlee Aquapore (octylsilane) column, which did not resolve the two forms of the protein. Purification on a Vydac C-18 column separated the larger form from the smaller form to a point where they were electrophoretically pure (not shown). N-terminal sequence analysis and comparison with the Entrez sequence database (National Center for Biotechnology Information, National Library of Medicine) identified these proteins as being previously uncharacterized. Edman degradation of peptides isolated by reversed phase separation of peptides obtained by digestion of the reduced and carboxymethylated protein with endoprotease Lys-C and endoprotease. Asp-N identified it as being similar to the C-type lectin family member that we have previously isolated from shark cartilage. Due to the considerable similarity to

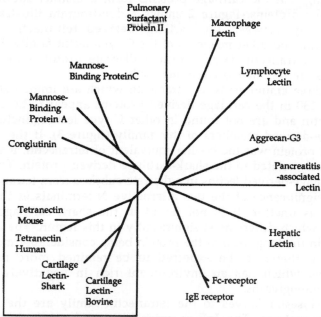

Figure 2. A dendrogram of the relationships between various C-type lectins. An unrooted dendrogram was generated using the program Clustal(16). The box defines the tetranectin/cartilage lectin family.

the shark cartilage homolog of this mammalian protein, it was possible to align the peptides onto the framework of the shark protein. Alignment of the peptides is shown in figure 1. Degenerate PCR primers were designed from the protein sequence and used to amplify a specific product from cartilage-derived cDNA. The product was sequenced and confirmed the protein sequence data. In particular, PCR product-derived sequence data enabled us to confirm ambiguous calls (for example Lys[53], Arg[140]). The resulting cDNA sequence enabled the design of additional 3' primers suitable for RACE analysis. Sequence analysis of the RACE product, using the 3' primer, showed that the protein sequence immediately preceding the larger form of the protein was typical of a signal peptide sequence (not shown).

IV. Discussion

We have characterized a new protein in mammalian cartilage which falls into the C-type lectin family. This protein is strikingly similar to tetranectin, a C-type lectin tetramer found in serum. However, it has a greater similarity to the shark cartilage-derived tetranectin analog which we have isolated from shark neural arch cartilage(5). The protein described here is 61% identical to the shark cartilage protein, and 49% identical to mammalian tetranectin, indicating that the cartilage proteins form a distinct sub-family of lectin-like proteins. Figure 2 shows a dendrogram illustrating the relationship between the cartilage-derived tetranectin analogs, tetranectin, and other members of the C-type lectin family. It is clear that the tetranectin family form a discrete group whose closest relatives are the mannose-binding lectins.

The three amino acids in tetranectin which are immediately after residue 130 in the cartilage derived proteins appear to be unique to tetranectin and are not found in other C-type lectins, including the cartilage-derived members of this family (figure 3). If the cartilage-derived protein was the bovine equivalent of tetranectin, it would be less closely related to the shark-cartilage derived protein. This places the cartilage-derived lectins in a discrete subclass from tetranectin.

The significance of the two alternative N-terminals in the bovine protein is unclear. It is not found in the shark cartilage-derived protein which differs most significantly in this region. The extension results in the long form of the protein being considerably more basic. It might, therefore, be expected to be retained more readily in cartilage, which has an environment rich in negatively charged glycosaminoglycans.

The closest relatives to the tetranectin family are the mannose binding proteins. The IgE receptor has an N-terminal membrane insertion site; this is clearly missing in the tetranectin family. The mannose-binding protein has a collagenous "tail" which also has no corresponding domain in the tetranectin family, although a highly basic region of 20 - 30 amino acids is found at the N-terminal. A striking feature of the tetranectin family is that these proteins are not part of larger precursors and form free proteins; in this respect they resemble the mannose-binding proteins. Whether the cartilage-derived members of the tetranectin family are related to the fibroblast-derived protein reported by Clemmensen et al. (11) is unclear.

The crystal structure of the carbohydrate-recognition domain of the rat mannose-binding protein A has been determined by Weis et al. (4) and forms a basis for structural analysis of other members of this family of calcium-binding lectins. The protein described here, together with tetranectin, contains many of the consensus features of this family. In particular, the spacing of the cysteines is well preserved, as are features such as the "WIGL" sequence at residue 105 and the "NW" sequence at residue 129. However, few of the consensus residues are involved with ligand binding, but appear to be involved primarily in calcium binding or forming some of the features of the protein's secondary structure. Figure 3 shows an alignment of the tetranectin family with the IgE receptor and with the mannose-binding protein A The residues associated with ligand binding are primarily in the C-terminal half of the protein. Some of

```
          10        20        30        40        50        60
SKFKARKHSKRRVKEKDGDLKTQVEKLWREVNALKEMQALQTVCLRGTKFG-KKCYLAAEG
   *S P-SKSG G -D  RNEID        S            K   IH-        SR
P KIVNAKKDVVNTKMFEE  SRLDT AQ  AL   Q        K   VH-M  F  FTQ
  P A KAANAKKDLVSSKMFEELKNRMDVLAQEVALLKEKQA Q   VN-L  L  FTQ
  *GDRGDS AIEV LANMEAEINT KSKLE-LTN-K- HAFSM- K S -  FFVTNHE
NIKSLGLNE  TASDSLEK  QEE A     I I-- ISKGTACNI PKNWLHFQQ   YFGK
KGEAGERGPPGLPAHL EE  QATLHDFRHQILQTRGALS  G-SIM- -V -E VFSSNGQ
                                                        *  *

          70        80        90        100       110       120
LKHFHEANEDCISKGGTLVVPRSADEINALRDYGKRSLPGVNDFWLGINDMVAEGKFVDI
S SY A        AQ    SI   S GS S AK V AR   I V  TT        V
T T     S        R    GT QTGS ND  YE LRQ VGNEAEI   L   A  TW  M
P T     S        Q    GT Q EL NE  FE ARH VGNDANI   L   A  AW  M
RMP SKVKAL SELR   VAI  N E NK IQEVA -T-SA--- -    T E T   Q MYV
S QWIQ RFA SDLQ R - SIHSQK-E--Q FLMQH-INKK S I LQ LNM   E  WS
SIT DAIQ A ARA   RIAV  NPE NE IASFV K-Y-NTYA-YV LTEGPSP D RYS
   *      *    *** * * *   * *           * *      * *** *  ** ** *
                                         ^

          130       140       150       160       170
NGLAISFLNWD---QAQPN-GGKRENCALFSQSAQGKWSDEACHSSKRYICEFTIPQ   BLec
  P TYF     ---RSK V-  T     VAA T G       DV R E        YL  V   SLec
T AR AYK   ETEIT   D-   T     VL GA N   F KR RDQLP      Q G VL   HTet
T GLLAYK   ETEITT  D-   T     VL GA N   F KR RDQLP      Q GL     MTet
T GRLTYS   K---KDE  DH SG D VTIVD--N L N IS QA HTAV     *        ManA
D SPVGYS   N---PGE N-   QG D VMMRG G -- N AF R LDAWV    QLATC>    IgER
D TPVNYT   Y---RGE A-  RGK Q VEMYTDG -- N RN LY RLT     *        PulSur
   *  *  *  **       *   *  ****     ** *  *   *        ****
                      ^  ^ ^  ^ ^   ^^ ^   ^  ^        ^
```

Figure 3. Alignment of the Bovine Cartilage Lectin (BLec) with Shark Cartilage Lectin (SLec), human Tetranectin (HTet) and murine Tetranectin (MTet), the mouse low affinity IgE receptor (IgER, residues 141 - 291)(17),the pulmonary surfactant protein (PulSur) and the carbohydrate-recognition domain of the rat mannose-binding protein A (ManA) that has recently been crystallized (4) (residues 109 - 221). Gaps (-) have been inserted to optimize the alignment. Asterisks (*) indicate the residues that are conserved, or have similar properies, between members of the tetranectin family and the mannose binding protein. Carets (^) indicate residues that are associated with ligand binding in the mannose-binding protein.

the residues associated with ligand binding in the mannose-binding protein are identified (mannose-binding protein residues Ser[154], Lys[182], Glu[185], His[189], Thr[197], Val[199], Asp[200], Asn[205], Ile[207], and His[213]); these are conspicuously different in the tetranectin family (they correspond to bovine cartilage protein residues Asp[103], Asp[131], Gln[134], Gly[138], Leu[197], Ser[147], Gln[148], Ser[155], Glu[157], and Lys[163]). One amino acid, Glu[141] in the bovine cartilage protein, appears to be both conserved and associated with ligand binding. It is also associated with the calcium found adjacent to the ligand binding site. Another potential ligand-associated residue in the bovine cartilage protein, Asn[135], is conserved between the bovine lectin and mannose-binding protein, but is Val in the shark protein and Asp in tetranectin. A region on the opposite side from the ligand binding site is highly conserved (for example; Gly[165], Glu[166], Phe[168], Leu[176], Tyr[178], Asn[180], Trp[181]; these correspond to Gly[114], Glu[115], Phe[117], Ile[125], Phe[127],

Asn[129], Trp[130] in the bovine cartilage-derived lectin) and has several hydrophobic side chains exposed to solvent. Tetranectin forms aggregates, as does the mannose-binding protein. It is possible that this region is involved in association of the protein monomers. The cartilage-derived lectin has not been shown to form aggregates. However it is isolated under rather harsh conditions (4M guanidine hydrochloride) and so may lose some of its structural features.

There is little similarity between the binding site of the mannose-binding protein and the tetranectin family, so mannose is unlikely to be the ligand for the latter. We have been unable to adsorb the cartilage-derived lectin to Sepharose-immobilized sugars (N-acetyl galactosamine, N-acetyl glucosamine, D-galactose, D-mannose). This may be due to denaturation during isolation, or may indicate that this protein is not a true lectin. We are currently attempting to use less extreme conditions for cartilage extraction to answer this question. At present, the function of the tetranectin family is unclear and the carbohydrate ligand, if any, has not been identified.

Acknowledgments

Funded by the Shriners of North America and NIH grant AR 35322. The sequence described here has been submitted to Genbank under the accession number U22298.

References

1. Halberg, D.F., Proulx, G., Doege, K., Yamada, Y. and Drickamer, K. (1988) *J. Biol. Chem.* **263**, 9486-9490
2. Zimmermann, D.R. and Ruoslahti, E. (1989) *EMBO J* **8**, 2975-2981
3. Rauch, U., Karthikeyan, L., Maurel, P., Margolis, R.U. and Margolis, R.K. (1992) *J. Biol. Chem.* **267**, 19536-19547
4. Weis, W.I., Kahn, R., Fourme, R., Drickamer, K. and Hendrickson, W.A. (1991) *Science* **254**, 1608-1615
5. Neame, P.J., Young, C.N. and Treep, J.T. (1992) *Protein Science* **1**, 161-168
6. Clemmensen, I., Petersen, L.C. and Kluft, C. (1986) *Eur J Biochem* **156**, 327-333
7. Fuhlendorff, J., Clemmensen, I. and Magnusson, S. (1987) *Biochemistry* **26**, 6757-6764
8. Christensen, L., Johansen, N., Jensen, B.A. and Clemmensen, I. (1987) *Histochemistry* **87**, 195-199
9. Wewer, U.M., Ibaraki, K., Schjørring, P., Durkin, M.E., Young, M.F. and Albrechtsen, R. (1994) *J. Cell. Biol.* **127**, 1767-1775
10. Clemmensen, I. (1989) *Scand J Clin Lab Invest* **49**, 719-725
11. Clemmensen, I., Lund, L.R., Christensen, L. and Andreasen, P.A. (1991) *Eur. J. Biochem.* **195**, 735-741
12. Berglund, L. and Petersen, T.E. (1992) *FEBS Lett* **309**, 15-19
13. Neame, P.J., Treep, J.T. and Young, C.N. (1990) *J. Biol. Chem.* **265**, 9628-9633
14. Neame, P.J., Young, C.N. and Treep, J.T. (1990) *J. Biol. Chem.* **265**, 20401-20408
15. Smale, G. and Sasse, J. (1992) *Anal. Biochem.* **203**, 352-356
16. Higgins, D.G., Bleasby, A.J. and Fuchs, R. (1991) *CABIOS* **8**, 189-191
17. Bettler, B., Hofstetter, H., Rao, M., Yokoyama, W.M., Kilchherr, F. and Conrad, D.H. (1989) *Proc. Natl. Acad. Sci.* **86**, 7566-7570

Sequence Determinants of Oligomer Selection in Coiled Coils

Derek N. Woolfson and Tom Alber

Department of Molecular and Cell Biology, Stanley Hall #3206, University of
California, Berkeley, CA 94720-3206

I. Introduction

The coiled coil mediates protein-protein interactions within
numerous biological contexts, including muscle proteins,
transcription factors, cell and viral surface proteins, tumor
suppressor molecules, motors and cytoskeletal components. The
sequences of coiled coils contain a characteristic seven-residue repeat,
abcdefg, in which the a and d positions are occupied generally by
hydrophobic amino acids. The motif codes for amphipathic α-helices
that oligomerize through their hydrophobic faces (Figures 1a, b & c).
Despite the presence of a common sequence pattern, coiled coils form
helical ropes with different oligomerization states and alternative
topologies. Two-, three- and four-stranded coiled coils with parallel
or antiparallel chains have been described, with the most common
forms being dimers and trimers [1–8]. We sought to identify
features superimposed on the basic heptad repeat that distinguish
parallel dimer and trimer sequences.

II. Results

A. Identification of heptad repeats

As a first step toward identifying the determinants of oligomer
specification, we developed a statistical method for detecting coiled-
coil sequence motifs in protein sequences. Two related reports were
identified on the basis of the following considerations: (1) Consistent
with coiled-coil structure [9], sequences were required to have at least
four successive heptad repeats with at least five of their a and d sites

---

Present Address: Department of Biochemistry, School of Biological Sciences,
University of Sussex, Falmer, Brighton BN1 9QG, UK

PEPTIDES IN SYSTEMS BIOLOGY
Copyright © 1996 by Academic Press, Inc.
All rights of reproduction in any form reserved.

Sequence Determinants of Oligomer Selection in Coiled Coils

Derek N. Woolfson[1] and Tom Alber

Department of Molecular and Cell Biology, Stanley Hall #3206, University of California, Berkeley, CA 94720-3206

I. Introduction

The coiled coil mediates protein-protein interactions within numerous biological contexts, including muscle proteins, transcription factors, cell and viral surface proteins, tumor suppressors, molecular motors and cytoskeletal components. The sequences of coiled coils contain a characteristic seven-residue repeat, **abcdefg**, in which the **a** and **d** positions are occupied generally by hydrophobic amino acids. This motif codes for amphipatic α-helices that oligomerize through their hydrophobic faces (Figures 1a, b & c). Despite the presence of a common sequence pattern, coiled coils form helical ropes with different oligomerization states and alternative topologies. Two-, three- and four-stranded coiled coils with parallel and antiparallel helices have been described, with the most common classes being parallel dimers and trimers [1-8]. We sought to identify the features superimposed on the basic heptad repeat that distinguish these parallel dimer and trimer sequences.

II. Results

A. *Identification of heptad repeats*

As a first step toward identifying the determinants of oligomer specification, we developed a restrictive method for detecting coiled-coil sequence motifs in protein sequences. Heptad repeats were identified on the basis of the following considerations: (1) Consistent with coiled-coil structures [9], sequences were required to have at least four consecutive heptad repeats with at least 6/8 of their **a** and **d** sites

---

[1]Present Address: Department of Biochemistry, School of Biological Sciences, University of Sussex, Falmer, Brighton BN1 9QG, UK.

occupied by hydrophobic side chains. (2) As expected for helical structures [10-12], the content of the helix-destabilizing residues P[2] and G was limited. (3) To exclude membrane-spanning helices, the total hydrophobic content of the sequences was limited to 65%. This 65% cut-off corresponds to the average hydrophobic content of the transmembrane helices of bacteriorhodopsin [13].

Initially, we determined which combinations of the seven hydrophobic side chains, A, F, I, L, M, V and Y, predominated at the buried sites of coiled-coil heptad repeats. W was omitted because it was considered too large to be compatible with the hydrophobic core of a parallel coiled coil [3]. An automated search of a combined data base containing 21 protein sequences known to form parallel, dimeric and trimeric coiled coils retrieved 721 different heptad repeats (see the legend to Table I for a list of these sequences). The occupancies of the seven hydrophobic side chains at sites **a** and **d** were compared with the frequencies expected by chance using the statistical method of standard error of proportion (data not shown). This analysis revealed that of the seven selected amino acids, only I and L were enriched at the **a** sites and only L was enriched at the **d** sites. In addition, V and A occured at the **a** and **d** sites, respectively, with frequencies that were high, but comparable with those expected by chance. A second search was performed on the separate data bases of dimer and trimer sequences limiting the hydrophobic residues allowed at three quarters of the **a** and **d** sites to combinations of A, I, L and V. After the removing redundant sequences, the searches returned 298 and 256 heptad repeats, respectively. Thus, 77% of the heptads from the less stringent search were found despite using only four of the seven hydrophobic residues.

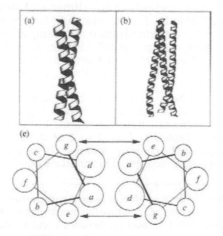

Figure 1.   (a) & (b) Ribbon diagrams [22] of parallel dimeric [6] and trimeric [7] coiled coils, respectively. (c) The sequence repeat **abcdefg** arrayed on a helical wheel representation of a dimeric coiled coil.

---

[2] Abbreviations: A, alanine; C, cysteine; D, aspartic acid; E, glutamic acid; F, phenylalanine; G, glycine; H, histidine; I, isoleucine; K, lysine; L, leucine; M, methionine; N, asparagine; P, proline; Q, glutamine; R, arginine; S, serine; T, threonine; V, valine; W, tryptophan; Y, tyrosine.

B. Amino acid profiles for dimeric and trimeric coiled coils

The amino acid compositions of the heptads retrieved in restrictive searches of separate data bases of dimer and trimer sequences[3] are given in Table I. After normalization for the different sizes of the two data sets, these tables constituted *position-dependent scoring matrices*, or *amino-acid profiles* for dimeric and trimeric coiled coils. Similar profiles for other protein structures have formed the basis for some of the more successful protein-fold-recognition methods [14-16]. In these studies, test sequences are given a score based on their alignment with the target profile. The score returned is compared with the distribution of scores from of a set of random sequences or those for sequences of proteins unrelated to the target structure. The resulting difference (quantified in terms of a z-score) gives a measure of the similarity of the test sequence and the target structure. Our implementation of the profile method was slightly different: Each test sequence was given two scores, one on the basis of alignment with the dimer profile and the other from alignment with the trimer profile. The difference between these scores, the oligomer-score, was taken as a measure of similarity to either a dimeric coiled coil or a trimeric sructure:

$$\text{Oligomer - score} = \sum (\log(\text{Profile}_{AA,y,O}) - \log(\text{Profile}_{AA,y,\Delta})) + c$$

Which simplifies to:

$$\text{Oligomer - score} = \sum \log P + c$$

Where P is the ratio of elements in the dimer and trimer composition tables (Table I) after normalizing for the different sizes of the two data bases. c is a normalisation constant, which was equal to 0.17 in the final scoring system described here. In this scheme, dimer predictions have positive scores and trimer predictons negatives scores. Values for P are referred to as dimer/trimer profiles and are listed in Table II.

Oligomer-scores for the sequences in the data bases of known dimeric and trimeric coiled coils are plotted in Figure 2a. In this case, the oligomer-scores were calculated using all 140 elements of the dimer/trimer profile (Table II). It is clear that the scores for the two groups are well separated: dimeric coiled coils give positive scores and trimers give negative scores. This method, however, can be

---

[3] Sequences used: Dimers: the ß-isoform of human, cardiac, heavy chain myosin; human tropomyosin, α and γ chains; paramyosin; human desmoplakins I and II; the transcriptional activator, GCN4; human cytokeratins 1 and 10; the N-terminal 80 amino acids of the human adenomatous polyposis coli (APC) protein. Trimers: human laminins A, B1 and B2; the B chain of influenza hemagglutinin; the trimerization region of the human heat shock transcription factor, hsf2; human fibrinogen, ß and γ chains; the human macrophage scavenger receptor; the DNA packaging protein, GP17; human tenascin; the whisker antigen control protein, fibritin. Test sequences: two proteins from *Giardia*, a microtubule bundling protein and ß-giardin; TlpA from *Salmonella typhimurium*; human and *Drosophila* kinesin light chains; proteins of the Myc oncogene system, Myc, Mad, Max and Mxi1; heat shock transcription factors from yeast; the *E. coli* seryl-tRNA synthetase; human spectrin, A and B chains; human dystrophin; human α-actinin.

Table I Amino-acid Composition[1]

	a		b		c		d		e		f		g	
	O	Δ	O	Δ	O	Δ	O	Δ	O	Δ	O	Δ	O	Δ
A	29	51	26	19	25	21	68	48	9	13	28	24	23	16
C	4	0	0	2	2	2	1	0	2	0	1	0	0	2
D	0	4	42	20	37	26	3	4	12	10	21	25	25	23
E	2	2	62	45	54	48	14	4	72	45	45	25	73	34
F	3	0	0	2	2	4	3	4	0	7	1	1	0	4
G	2	2	3	7	2	8	0	0	1	4	4	4	3	2
H	2	2	10	3	8	7	3	5	5	4	8	6	3	7
I	55	41	4	4	3	5	11	3	5	4	9	4	4	10
K	16	2	34	25	26	18	4	32	7	3	32	27	39	32
L	109	64	10	8	9	2	132	77	33	32	15	13	13	33
M	2	6	3	3	2	5	6	10	14	24	7	7	2	2
N	11	4	21	14	27	31	3	4	15	20	4	7	7	18
P	0	1	0	2	0	1	0	1	0	0	0	0	0	0
Q	1	10	27	28	32	17	8	4	52	20	20	22	40	12
R	15	1	16	16	40	17	3	1	28	20	41	26	24	10
S	3	13	20	22	11	24	5	6	14	21	21	21	13	12
T	3	10	9	21	7	10	7	12	23	13	11	15	18	19
V	35	38	8	8	4	8	22	35	6	7	8	11	9	13
W	0	1	0	0	0	0	0	0	0	2	0	0	0	2
Y	6	4	3	6	3	1	8	3	1	4	2	2	0	5
h	298	256	298	256	298	256	298	256	296	256	296	256	296	256

Table II Dimer/Trimer Profile[2]

	a	b	c	d	e	f	g
A	0.49	1.18	1.02	1.22	0.59	1.00	1.23
C	n.o.	n.o.	0.86	n.o.	n.o.	n.o.	n.o.
D	n.o.	1.80	1.22	0.64	1.03	0.72	0.93
E	0.86	1.18	0.97	3.01	1.37	1.55	1.84
F	n.o.	n.o.	0.43	0.64	n.o.	0.86	n.o.
G	0.86	0.37	0.98	n.o.	0.21	2.36	1.29
H	0.86	2.86	0.52	n.o.	1.07	1.15	0.37
I	1.15	0.86	2.29	0.30	2.00	1.93	0.34
K	6.87	1.17	1.24	0.43	0.89	1.02	1.05
L	1.46	1.07	1.29	1.47	0.50	0.99	0.34
M	0.29	0.86	0.34	0.52	0.25	0.49	0.86
N	2.36	1.29	0.75	0.64	0.64	0.67	0.33
P	n.o.	n.o.	n.o.	n.o.	n.o.	n.o.	n.o.
Q	0.09	0.83	1.62	1.72	2.23	0.78	2.86
R	12.89	0.86	2.02	2.58	1.20	1.35	2.06
S	0.20	0.78	0.39	0.72	0.57	0.86	0.93
T	0.26	0.37	0.60	0.50	1.52	0.63	0.81
V	0.79	0.86	0.43	0.54	0.74	0.62	0.59
W	n.o.	n.o.	n.o.	n.o.	n.o.	n.o.	n.o.
Y	1.29	0.43	0.29	2.29	0.21	0.86	n.o.

[1] Key: O, head columns containing count data from dimeric sequences; Δ, head columns containing count data from trimeric sequences; h, is the total of each column and, thus, corresponds to the number of heptad repeats analyzed.

[2] n.o. indicates that there were no observations in either or both of the dimer and trimer composition tables (see Table I).

biased by amino acids that are rare at a particular position of the heptad repeat (e.g. C and W at the **a** sites, Table I). The profile values for such amino acids may dominate the oligomer-score for certain sequences. Consequently, sequences can be classified on the basis of poorly represented amino acids, rather than the sequence features that determine oligomerization state.

To assess the contributions of rare amino acids, oligomer-scores were calculated using profile tables derived from two hybrid data bases containing mixtures of dimer and trimer sequences (Figure 2). Sequences were assigned to the hybrid data bases to distribute the heptads from dimer and trimer sequences approximately equally. The process of mixing dimer and trimer sequences in the two data bases was intended to scramble any information concerning oligomer specification, and, as a result, the oligomer-scores calculated using the resulting hybrid profiles should coincide. The finding that the two sets of oligomer-scores calculated using the hydrid profiles were separated (Figure 2a) suggests that the profile values of rare amino acids were dominating the scores.

This problem was overcome by excluding all but three types of profile entries. First, only entries residues at the **a**, **d**, **e** and **g** sites were used. These sites make the majority of interhelical contacts in coiled-coil structures (Figure 1) [6, 17]. Second, profile values of amino acids that were rare in either dimers or trimers were excluded. Finally, because residues with similar values in both profiles cannot aid discrimination in the absence of correlated sequence effects, only values that differed significantly between the dimer and trimer profiles were included. This last point was judged statistically using a standard error of proportion. All three criteria were satisfied by only seventeen values in Table I: A, K, L, R and S at **a** positions; I, L and V at **d** positions; E, L, Q and S at **e** positions; and E, L, N, Q and R at **g** positions. The oligomer-scores calculated using only the seventeen corresponding values of Table II are shown in Figure 2b. In this case the oligomer-scores derived from the two hybrid profiles overlap, indicating that spurious effects of underrepresented amino acids were suppressed. In addition, with the exception of the GCN4 dimer, the dimer and trimer profiles correctly separated the dimer and trimer sequences.

The statistical analysis included profile values only for those amino acids that made up at least 5% of the residues at a given site in both the dimer and trimer data bases. As a consequence, amino acids that were poorly represented at a particular heptad position in one data set but reasonably well represented in the other were ignored, and some amino acids that could contribute to oligomer discrimination may have been overlooked. For example, there are thirty times more

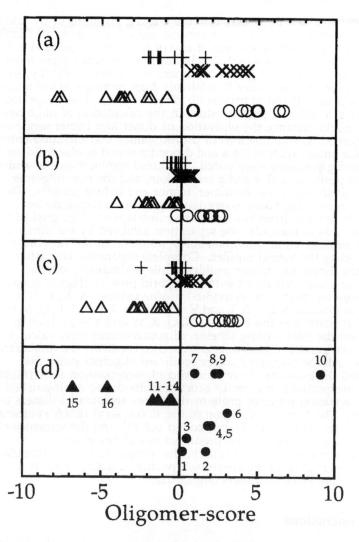

Figure 2.
of both dimers and trimers. (a) Scores calculated using all 140 individual profile
values of Table II. (b) Scores calculated using the set of statistically significant
individual profile values. (c) Scores calculated using the final scoring system. (d)
Oligomer-scores for coiled-coil protein sequences excluded from the derivation of the
dimer and trimer profiles. **Key:** proteins from *Giardia*, 1, 2; TlpA, 3; kinesin light
chains, 4, 5; seryl tRNA synthetase, 6; the Myc oncoproteins, 7, 8, 9, 10; spectrin
family, 11, 12, 13, 14; yeast heat shock transcription factors, 15, 16. As predicted by
our profile method, sequences 1-10 form dimers (filled circles) and 11-16 form trimers
(filled triangles).

aromatic residues, F , W and Y, at the **e** and **g** positions of the trimers than the corresponding sites of dimeric coiled coil sequences (Table I). Even so, the largest proportion of any of these amino acids at either sites **e** or **g** in trimers was 0.027. In addition, leucines occur more frequently at the **e** and **g** sites of trimers than dimers. Finally, some hydrophilic residues occur in relatively high numbers at the **a** sites of dimers (K and R) and of trimers (S) (Tables I & II). This suggested that including additional profile values in the calculation of oligomer-scores might enhance the separation of dimer and trimer sequences.

In particular, combinations of profile values that included entries for polar amino acids at the **a** and **d** positions and apolar residues at the **e** and **g** positions were tested. Different combinations of amino acids at positions **a**, **d**, **e** and **g** were taken, and the new oligomer-scores calculated using the dimer, trimer and hybrid profiles. Two sets of scores were taken as overlapping if the 99% confidence intervals derived from Student's t-test overlapped. The goal of this analysis was to maximize the separation achieved by the dimer and trimer profiles while simultaneously minimizing the separation achieved by the hybrid profiles. Complete separation was attained with the dimer and trimer profiles while maintaining overlap of the oligomer-scores calculated with the hybrid profiles (Figure 2c) using the following thirty-one individual profile values: A, K, L, N, Q, R, S and T at position **a**; I, L, S, T and V at position **d**; E, F, I, L, Q, S, V, W and Y at position *e*; and E, F, I, L, N, Q, R, W and Y at position **g**.

To test the final scoring scheme, oligomer-scores were calculated for eleven parallel coiled coils not used to construct the dimer/trimer profile. All the calculated scores predicted oligomer states that matched the experimentally observed oligomerization state (Figure 2d). Unexpectedly, the profile scoring method also distinguished the oligomerization states of proteins that form antiparallel dimers and trimers. The N-terminal region of the *E. coli* seryl tRNA synthetase contains an antiparallel dimeric coiled coil [2], and the sequence of this motif was retrieved and predicted as a dimer in our computational method. Similarly, the antiparallel, three-stranded, coiled-coil regions of the spectrins (spectrin, α-actinin and dystrophin [8]) were predicted as trimers (Figure 2d).

III. Conclusions

A. *Sequence patterns that distinguish dimeric and trimeric coiled coils*

The analyses presented above indicate that differences exist in the amino-acid compositions of heptads of dimeric an trimeric coiled coils, and that these direct oligomer choice. We have determined the elements of Tables I & II that appear to dominate oligomer selection. Our results agree with the descriptions of Conway and Parry [18], and can be summarized as follows:

1.   Hydrophobic amino acids at a and d positions

The hydrophobic residues A, I, L and V predominate at the **a** and **d** sites of both dimers and trimers. Major roles in oligomer specification are played by L and I (Table II). Dimers are favored by sequences enriched for I at the **a** sites and L at **d**. At the same time, I is strongly depleted at the **d** positions of dimers.

The distinct preferences for I and L at the **a** and **d** sites of dimers and trimers coincide with recent experimental observations [17, 19]. This disposition has been explained in terms of differences in packing geometry in dimeric and trimeric structures. The packing at the **a** and **d** sites of dimeric coiled coils differ and best accommodate the side chains I and L, respectively [17, 19, 20]. The packing at the **a** and **d** sites in trimers, on the other hand, is more similar [7, 17, 20]. Consistent with these structural patterns, the oligomer interface in trimers contains a more even distribution of hydrophobic side chains (Tables I & II). In this sense, the trimer can be thought of as the "default" oligomerization state favored by a "random" distribution of hydrophobic amino acids at the core positions.

Interestingly, the packing spaces of the core positions in antiparallel coiled coils are similar to those in parallel coiled coils. In both classes of dimers, for example, the **a** side chains point out of the interface (parallel packing) and the **d** side chains point into the interface (perpendicular packing) [2, 6]. Trimers, both parallel and antiparallel, show distinct arrangements of core residues in which the side chains make an acute angle with the adjacent helices (acute packing) [3, 7, 17, 20]. These similarities between parallel and antiparallel coiled coils may account for the ability of our method to correctly assign the oligomerization states of antiparallel sequences.

2.   Hydrophilic amino acids at a positions

In accord with earlier observations by Parry [21], a number of polar side chains occurred at the **a** positions more often than was expected by chance (data not shown). Our analysis (Table II) supports the following generalizations: K and R are favored at the **a** positions of dimers but depleted from trimers. In addition, N is found three times more often at the **a** sites of dimers than in the corresponding sites in trimers. In contrast, trimers are enriched Q, S and T residues at the **a** sites. Enrichment of polar side chains at position **d** is limited to preferences for S and T in trimeric coiled coils.

Encouraged by the uneven distributions of polar residues in dimers and trimers, we recently characterized mutants of the GCN4 leucine zipper in which the **a** residue N-16 was replaced by Q and K (L. Gonzalez, Jr., D.N.W. and T.A., unpublished results). Consistent with the sequence analysis presented here, the K mutant formed only dimers, but the Q substitution formed dimers and trimers in solution and in crystals. The different distributions of polar residues (Tables I & II) indicate that polarity *per se* is not the sole determinant of oligomer specification. As with the hydrophobic amino acids, the complementarity of the side-chain shape with the distinct structures of dimers and trimers also must be important.

3.   Residues at the e and g positions

The **e** and **g** sites of the heptad repeat flank the **a** and **d** residues in coiled-coil interfaces (Figure 1c). Consequently, amino acids at the **e** and **g** sites may influence the oligomerization state [6, 17]. Compared to the corresponding sites of dimers, the **e** and **g** positions of trimers are enriched for hydrophobic residues (I, L, V, F, Y and W) and depleted of specific hydrophilic residues (E, Q, S and R) (Tables I & II). These patterns are consistent with the extension of the hydrophobic interface of trimers, relative to that in dimers, to partially include the **e** and **g** sites [17]. In the crystal structure of a trimeric mutant of the GCN4 leucine zipper, for example, the **e** and **g** sites are approximately 50% more buried than they are in the wild-type dimer [17]. Consistent with this difference in surface exposure, we found more oppositely-charged pairs of residues at neighboring **g** and **e** positions (g_n/e_{n+1}) in the sequences of dimeric coiled coils ($\approx$23%) than in trimers ($\approx$12%).

To summarize, we have determined differences in the amino-acid compositions of dimeric and trimeric coiled-coil structures. We show that these can be exploited to predict accurately the oligomerization states of coiled-coil sequences. The method incorporates two ideas that may be generally useful in calculations of profile scores — rare amino acids were excluded to reduce bias introduced by the limited sequence data bases, and exposed residues were excluded to focus the analysis on key positions. Although the current method is relatively accurate, potential improvements can be anticipated. For example, better profiles might be derived from larger data bases of coiled coil sequences. It is hoped that our analyses will increase confidence in the identification of coiled-coil sequences and aid the development of a thorough understanding of this important class of proteins.

Acknowledgments

We thank J.J. Plecs, P.B. Harbury, P.S. Kim and S. Nautiyal for many useful discussions on coiled coils, and H. Ballard and I. Badcoe for advice on the statistical analyses. This work was supported by a grant from the National Institutes of Health (GM48958 to T.A.).

References

1.   Banner, DW, M Kokkinidis & D Tsernoglou. J. Mol. Biol., 1987. **196**, 657-675.
2.   Cusack, S, *et al.* Nature (London), 1990. **347**, 249-255.
3.   Lovejoy, B, *et al.* Science, 1993. **259**, 1288-1293.
4.   O'Shea, EK, R Rutkowski & PS Kim. Science, 1989. **243**, 538-542.
5.   O'Shea, EK, *et al.* Science, 1989. **245**, 646-8.
6.   O'Shea, EK, *et al.* Science, 1991. **254**, 539-44.
7.   Wilson, IA, JJ Skehel & DC Wiley. Nature (London), 1981. **289**, 366.
8.   Yan, Y, *et al.* Science, 1993. **262**, 2027-2030.
9.   Lau, SYM, AK Taneja & RS Hodges. J. Biol. Chem., 1984. **259**, 13253-13261.
10.   Barlow, DJ & JM Thornton. J. Mol. Biol., 1988. **201**, 601-619.
11.   Woolfson, DN & DH Williams. FEBS Lett., 1990. **277**, 185-188.
12.   MacArthur, MW & JM Thornton. J. Mol. Biol., 1991.
13.   Henderson, R, *et al.* J. Mol. Biol., 1990. **213**, 899-929.
14.   Gribskov, M, AD McLachlan & D Eisenberg. Proc. Natl. Acad. Sci. (USA), 1987. **84**, 4355-4358.

15. Gribskov, M, R Lüthy & D Eisenberg. Methods in Enzymology, 1990. **183**, 146-159.
16. Bowie, JU, R Lüthy & D Eisenberg. Science, 1991. **253**, 164-170.
17. Harbury, PB, PS Kim & T Alber. Nature, 1994. **371**, 80-83.
18. Conway, JF & DAD Parry. Int. J. Biol. Macromol., 1991. **13**, 14-16.
19. Harbury, PB, *et al.* Science, 1993. **262**, 1401-1407.
20. DeLano, WL & AT Brünger. Proteins, 1994. **20**, 105-123.
21. Parry, DAD. Bioscience Reports, 1982. **2**, 1017-1024.
22. Kraulis, PJ. J. Appl. Crystallogr., 1991. **24**, 946-950.

Techniques for Searching for Structural Similarities Between Protein Cores, Protein Surfaces and Between Protein-Protein Interfaces

Chung-Jung Tsai[1], Shuo L. Lin[1], Haim J. Wolfson[2]
and Ruth Nussinov[3,4*]

[1] Laboratory of Mathematical Biology
NCI-FCRF
Bldg 469, Rm 151
Frederick, MD 21702

[2] Computer Science Department
School of Mathematical Sciences
Tel Aviv University, Tel Aviv 69978, Israel

[3] Laboratory of Mathematical Biology
SAIC
NCI-FCRF
Bldg 469, Rm 151
Frederick, MD 21702

[4] Sackler Institute of Molecular Medicine
Sackler Faculty of Medicine
Tel Aviv University, Tel Aviv 69978, Israel

August 22, 1995

---

*correspondence should be addressed to R. Nussinov at NCI-FCRF Bldg 469, rm 151,
Frederick, MD 21702. FAX: 301-846-5598

I. Introduction

With the rapid growth in the number of elucidated protein structures, comparison techniques are critically needed. To constitute routinely used tools, employed in a variety of studies, such techniques should preferably possess a number of attributes. (i) As the number of structures deposited in the Protein Data Bank (PDB) (1) increases at a high rate, such techniques should be highly efficient, comparing the three dimensional structures of a pair of proteins in short times (preferably on the order of seconds). (ii) As the common, recurring substructural motif in the two proteins is a-priori unknown, it should not be pre-defined. (iii) High quality results should be sought, matching as large a number of residue-pairs between the two structures in as low as possible root-mean-squared deviations (RMSD). (iv) The technique should preferably be general, i.e., be "real" 3D, amino acid sequence-order independent. While for overall similar protein structures the chain topologies are quite similar, as the proteins diverge, the similarity between them might not necessarily be composed of fragments of consecutive, superimposed, amino-acids. That is particularly the case for proteins that have converged during evolution. A classical example in this regard is that of trypsin and subtilisin (2,3). While the sequences of the two proteins are different, they share 3D structural similarity, unrelated to the order of the amino acids in their respective chains. Optimal superpositioning of two proteins might involve matching of fragments of amino-acids in different order, having different directionalities, and of isolated amino acids. (v) Being able to simultaneously compare one protein structure with many others is highly advantageous, as it circumvents the task of an analysis of comparing numerous pairs.

Having a protein structure comparison technique that is geared toward such tasks and performs well is important in a number of investigations. First, it enables filtering the highly redundant protein databank (1), retaining only structurally dissimilar protein. Such a dataset of proteins (e.g., 4–7) serves as an input for statistical analysis of protein structures. It is needed for derivation of preferences of neighboring amino acid pairs (e.g., 8), for atom-atom potentials, and for threading an amino acid sequence through numerous, different, protein structures in the quest for an optimal fitting of a chain unto a fold (e.g., 9,10). Second, it enables matching the active sites of proteins, which conceivably may be composed of isolated amino-acid point-pairs between the two proteins (3). Third, it facilitates investigations of some evolutionary questions. Thus, for example, if the optimal 3D match between two proteins conserves their amino acid sequence-order, one may argue that the structures have diverged evolutionarily. If, on the other hand, the sequential order differs, convergent evolution may be argued. An efficient structural comparison technique thus constitutes an invaluable tool for studies of protein folding.

A number of protein structure comparison techniques have been developed during the last decade (e.g., 11–18). In general, they fall into three categories. The first is based on dynamic programming techniques, and applies an adaptation of the Needleman-Wunsch type algorithm to the matching of 3D structures (for example, 12). The second matches the 3D structures belonging to fragments of contiguous amino acids (11,14). Fragments whose superpositioning is achieved with similar transformations are clustered to yield a complete 3D structural match. The third approach is derived from Computer Vision (19). It views protein structures as collections of points (atoms) in 3D space, and seeks to optimally match the largest subset of these points (3,17,18). Each of the above methods has advantages as well as drawbacks. The dynamic programming technique is fast. As it matches the inter-C_α-atoms distances between any two amino-acids in the two proteins, it can tolerate insertions and deletions. However, as it is rooted in the dynamic programming algorithm, the order of the amino acids is strictly conserved. The second approach can match fragments whose order in the chains differs; however, it cannot match single, isolated amino acids, nor can it handle insertions or deletions within the fragments. The third approach is immune to both of these deficiencies: namely, the order of the amino acids in the chain is immaterial, and deletions and insertions are tolerated as well. Both of these advantages are the outcome of the disregard for the connectivity of the amino acids in the chain. In Computer Vision unordered pixels are matched, with connectivity being entirely meaningless (19). While it appears that taking connectivity into account reduces the complexity of the problem, leading to faster, more efficient techniques, this is not the case. The speed of the Geometric Hashing, computer vision based technique is impressive. Comparison of the structures of two proteins of average size (say, about 220 amino acids each) takes about 3 seconds on a Silicon Graphics workstation (3).

The Geometric Hashing algorithm is uniquely applicable to carrying out additional tasks that cannot be handled by other, order-dependent approaches. Two such applications will be cited here. The first is the generation of a dataset of non-redundant protein-protein interfaces. The generation of such a dataset requires carrying out structural comparisons between two interfaces, where each interface is composed of at least two protein chains. Furthermore, only pieces of each of the chains actually constitute an interface. Depending on the definition of an interface, the amino acids of the first protein that belong to an interface are roughly those whose distances from the amino acids of the second protein are under pre-defined thresholds. Clearly, an interface may be composed of isolated amino acids, and of fragments of amino acids in different order. An order-dependent structural comparison technique cannot carry out comparisons of interfaces, or comparisons of interfaces with single chains. Yet, such comparisons are of interest, as the principles underlying protein folding and protein-protein associations are not dissimilar. For this reason, motifs found in protein cores are likely to recur at protein-protein interfaces. The second application of the Geometric Hashing is for docking. Docking a ligand onto a receptor surface requires matching of unordered surface descriptors. In our docking application (20), the input to the docking routine is the "critical points", sparsely and accurately placed on peaks of hills (convex molecular faces) and bottoms of valleys (concave faces), and their associated normal vectors (21). Here, we focus on the protein structure comparison.

The Geometric Hashing technique also suffers from some drawbacks. By far the most prominent one is its sensitivity to random matches. Because single, isolated points (C_α atoms in the protein structure comparison) are matched, it is prone to chance matches of these point-atoms when they happen to fall at the same location in 3D space. Furthermore, as the prime determinant in the choice of the matching C_α atoms pairs is the distance from each other, the choice that is being made is not always the biologically sound one. The incorporation of connectivity considerations into the final stages of the matching in the Geometric Hashing and into the scoring improves the quality of the obtained substructural motifs. Although isolated C_α atom matches are still allowed, enabling matchings of active sites, preference is given to architectural considerations, alleviating the problem of the random point C_α atom matches. Furthermore, it makes equivalent many of the commonly encountered topological matches, circumventing the difficulty outlined above. The results of the matches are depicted graphically by a graphic routine. This routine, originally written to display water clusters (22), has been adapted to such a purpose. The resulting Geometric Hashing, with its graphic capabilities, is a useful tool. To demonstrate the capabilities of this approach, we compare it with a number of other methodologies. To facilitate such a task, we make use of a recently compiled comparison survey by May and Johnson (23).

Recently, a combination-approach to the structure comparison problem has been presented (23). This approach includes a genetic algorithm, a dynamic algorithm and a least squares optimization. Inspection of the results obtained by this multi-ingredient methodology illustrates that it achieves at least as many matching C_α pairs as the other methodologies that the authors have tested (15,16), with a lower RMSD. Like the May and Johnson approach, in our current implementation (3), we also match the coordinates of the C_α atoms. However, here we show that at least as many matching C_α atom pairs with roughly similar RMSD's are obtained at a fraction of the matching times. For example, for the 18 cases presented in the above matches, the run time obtained by May and Johnson varied between 1.55 min (for the comparison of two proteins with about 25 residues each) to 110.19 minutes (for the comparison of the two largest proteins tested, with sizes around 225 residues). While taking account of the topological (connectivity) considerations increased the run time of our method from the 3 seconds cited above, it was consistently under 10 seconds.

II. Connectivity Considerations in the Geometric Hashing

The Geometric Hashing, Computer Vision based algorithm has already been described before (3,17,18). It consists of three main stages: *detection of seed matches, clustering* and *extension*. A flow chart of the Geometric Hashing as implemented for the protein structure comparison is given in Figure 1. The basic idea behind the approach is that there is no need to scan the entire conformational space in the quest for matching substructural motifs (19). It suffices to adequately represent the conformation of one of the proteins, and to compare the second protein to it. To this end, the coordinates of all C_α atoms in each of the proteins are redundantly represented and matched as transformation invariants. The least squares between the matching pairs is calculated. Seed matches having similar transformations are subsequently

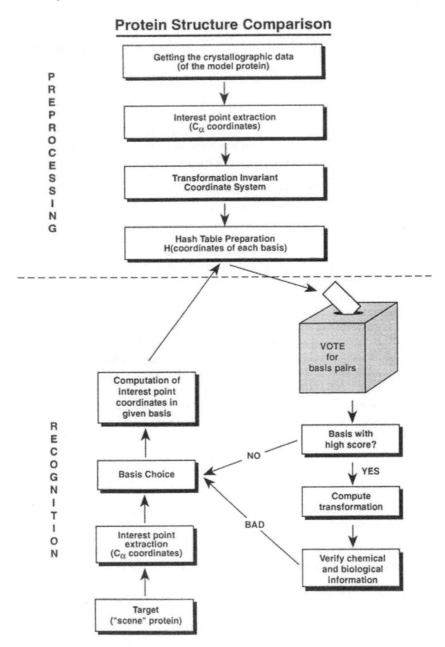

Figure 1. Flowchart of algorithm.

clustered. The clustered match is next extended. For further details of the algorithm see Refs. 3, 17, 18.

The representation and matching of the C_α atoms take no account of their connectivity. Connectivity is used only to reduce the combinatorial complexity in the construction of the reference frames (3). Three non-collinear atoms are needed to construct a reference frame: three points define a plane. One of these is chosen to be the origin. The X-axis is the vector pointing from the origin in the direction of a second point. The Y-axis is counter- clockwise, perpendicular to the X-axis in that plane, and the Z-axis is orthogonal to the plane at the origin. For a protein having n residues, the number of reference frames is n^3. By building reference frames on every three consecutive C_α atoms, the complexity of the reference frame construction is reduced to n.

In the Geometric Hashing, the output of the seed matches is in the form of a listing of matching C_α pairs from the two proteins. The next stage involves using this listing for the calculation of the least squares fit, obtaining the transformation that optimally superimposes the pairs. Seed matches having similar transformations are clustered. To possess an optimal transformation, the least squares is recalculated, using all recurring pairs in all clustered seed matches. This listing, along with its transformation, is the input to the last, extension stage.

The first step in the extension, is applying the transformation to all C_α atoms of the second protein, to superimpose it on the first. The clustered seed matches are used only for this initial superposition. In the extension stage, matching C_α pairs are sought without taking into account those already obtained previously. After all pairs fulfilling the specified criteria are picked, the transformation is recomputed, and additional C_α pairs are sought, to be added to the matching substructural motifs. Several iterations are executed to achieve the final listing of matching C_α pairs.

The strategy for optimizing the matches is non-trivial. The goal is to obtain the largest number of potential matching C_α pairs with the lowest RMSD. While it appears that for the extension stage, the most straightforward strategy is to match pairs of C_α's that have the shortest distances (so long as these are under a maximal allowed threshold), this simple route does not work (see Table I). For the Geometric Hashing, which matches single isolated C_α residue-pairs, incorporating an architectural preference results in a larger number of matching pairs. While isolated pairs are still allowed, a connectivity score is implemented. Residues whose neighbors in the chain are matched as well, though not necessarily to each other, are given a larger weight. This yields not only a larger number of pairs in the match with a roughly equivalent RMSD, but even more importantly, it makes sense from the biological standpoint. There is less random noise of meaningless matchings of amino acid pairs. An added advantage is that it allows a higher distance threshold. Previously, a larger threshold resulted in too large a number of such random matching C_α pairs. Now, the higher threshold enables matching of further diverged proteins. Moreover, these added benefits are indicative that for the Geometric Hashing, the strategy adopted in the extension is biologically reasonable, as these advantages are not unrelated. Figure 2 displays a match between amicyanin (apo form) (PDB file name 1aaj, in white) and pseudoazurin (cupredoxin) (1pmy, in gray). The sequence similarity in the match is under 25 percent.

Table I. Investigation of the effect of connectivity considerations on the quality of the match

Protein pair	GH - no connectivity dist. threshold 3.0 A			GH - with connectivity dist. threshold 3.0 A			GH - with connectivity dist. threshold 3.5 A		
	Nmp	RMSD	Score	Nmp	RMSD	Score	Nmp	RMSD	Score
2hhbA: 2hhbB	134	1.27	124.5	136	1.31	128.5	137	1.36	128.5
2hhbA:1mbd	133	1.34	120.0	135	1.37	124.0	139	1.46	132.5
2hhbA:1ecd	105	1.62	74.5	114	1.87	87.5	123	2.08	104.5
2hhbB:1mbd	138	1.31	123.5	140	1.35	129.5	143	1.48	133.0
2hhbB:1ecd	117	1.75	89.0	126	1.88	103.3	128	1.96	106.0
1mbd :1ecd	133	1.53	122.5	133	1.54	121.8	135	1.58	127.5
3cytI:1ccr	103	0.59	103.0	103	0.59	103.0	103	0.59	103.0
3cytI:3c2c	98	0.98	91.0	98	0.98	91.0	100	1.16	93.0
3cytI:351c	67	1.52	48.3	71	1.67	52.8	71	1.80	52.5
1ccr :351c	67	1.68	47.0	67	1.68	48.5	71	1.91	53.0
1ccr :3c2c	100	1.13	94.0	100	1.13	94.0	101	1.25	95.0
3c2c :351c	66	1.47	47.5	67	1.58	48.0	69	1.78	52.3
2gch :3ptn	210	0.91	193.3	211	0.96	197.5	216	1.08	202.3
2gch :2sga	138	1.53	87.8	141	1.64	93.5	152	1.88	106.0
2gch :3sgbE	136	1.53	87.5	137	1.63	89.3	147	1.90	101.8
3ptn :2sga	134	1.51	88.0	138	1.62	92.3	148	1.96	105.2
3ptn :3sgbE	127	1.44	83.0	132	1.61	87.3	145	1.86	97.5
2sga :3sgbE	171	0.67	161.0	171	0.69	160.0	171	0.78	160.0
2aza :1pcy	77	1.72	51.3	73	1.71	49.5	83	2.00	60.5
9pap :2act	210	0.75	203.0	210	0.75	203.0	210	0.75	203.0

Below we describe the implementation of the connectivity in the Geometric Hashing. The connectivity is used both in the extension stage, in the choice of which amino acid pairs should be matched, as well as in the scoring of the obtained matches. The scores are subsequently used to rank the matches. Matches having higher connectivity scores are ranked at the top of the list.

When matching amino acid a_i from protein a with amino acid b_j from protein b, the status of residues a_{i-1}, a_{i+1} and b_{j-1}, b_{j+1} is investigated. If the distance between a_i and b_j is under the threshold, but neither of the their flanking residues is matched to any other amino acid, a score (S) of

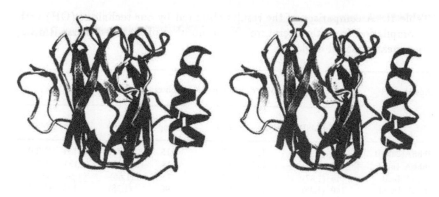

Figure 2. A stereo view of the match obtained for 1aaj and 1pmy. 76 residues are matched, with an RMSD of 1.19 Å.

0 is assigned. If a_{i-1} is matched to b_{j-1}, and the amino acids following a_i and b_j in the corresponding chains are not matched, $S = 1.5$. If both a_{i-1} is matched to to b_{j-1} and a_{i+1} is matched to b_{j+1}, $S = 2.0$. If a_{i-1} is matched to b_{j-1}, and a_{i+1} is matched, though not to b_{j+1}, $S = 1.0$. If a_{i-1} is unmatched under otherwise similar circumstances, $S = 0$. Clearly, symmetry considerations apply to the inverted matching. In the extension stage, several iterations of these connectivity investigations are carried out.

For the ranking of the matches, the *total similarity score* is computed. To compare the goodness of matches between proteins a and b vs a and c, or vs c and d (as, for example, in database comparisons), the *relative score* is calculated. The *relative score* is the (*total similarity score*)/(2* *size of the smaller of the two protein chains*). The range of the *relative similarity score* is between 0 and 1.0. These straightforward connectivity assignments, improve the results considerably. Table I demonstrates that with the exception of one case (2sga - 3sgb) connectivity often results in considerably higher numbers of matched amino acid pairs from the two proteins, while the RMSD's are still highly desirable.

III. Comparison of the Performance of the Geometric Hashing With a Recent Survey

Recently, May and Johnson (23) have compiled a survey, comparing the performance of their multi-ingredient program with other approaches (15,16). Table II compares our results with those of May and Johnson, as well as with others, for all cases presented in their report. It is immediately apparent that not only do we obtain a larger number of matched C_α atom pairs between the two proteins, but the RMSD's are comparable as well. Furthermore, Table III shows the time difference between the approaches. The execution

Table II. A comparison of the results obtained by our technique (GH) with a compilation from the literature. Nmp, number of matched pairs; RMSD, root mean squared deviation

Protein pair	GH (this study)		May & Johnson (23)		Subbarao & Haneef (15)		Luo et al. (16)	
	Nmp	RMSD	Nmp	RMSD	Nmp	RMSD	Nmp	RMSD
2hhhA:2hhhB	136	(1.31)	134	(1.27)	115	(1.34)	131	(1.29)
2hhbA:1mbd	135	(1.37)	135	(1.37)	114	(1.31)	131	(1.29)
2hhhA:1ecd	114	(1.87)	106	(1.61)	75	(1.66)	99	(1.93)
2hhbB:1mbd	140	(1.35)	139	(1.33)	98	(1.28)	133	(1.34)
2hhhB:1ecd	126	(1.88)	116	(1.77)	91	(1.86)	109	(1.91)
1mbd :1ecd	133	(1.54)	131	(1.50)	98	(1.55)	123	(1.52)
3cytI:1ccr	103	(0.59)	101	(0.56)	101	(0.52)	101	(0.55)
3cytI:3c2c	98	(0.98)	99	(1.04)	93	(1.03)	95	(1.00)
3cytI:351c	71	(1.67)	59	(1.52)	29	(1.40)	47	(1.58)
1ccr :351c	67	(1.68)	61	(1.67)	37	(1.41)	43	(1.73)
1ccr :3c2c	100	(1.13)	99	(1.10)	95	(1.13)	93	(1.07)
3c2c :351c	67	(1.58)	60	(1.42)	35	(1.30)	39	(1.62)
2gch :3ptn	211	(0.96)	209	(0.92)	189	(0.80)	98	(0.86)
2gch :2sga	141	(1.64)	121	(1.43)	74	(1.33)	82	(1.30)
2gch :3sgbE	137	(1.63)	118	(1.40)	90	(1.42)	87	(1.30)
3ptn :2sga	138	(1.62)	120	(1.41)	65	(1.40)	92	(1.32)
3ptn :3sgbE	132	(1.61)	117	(1.36)	49	(1.17)	87	(1.24)
2sga :3sgbE	171	(0.69)	170	(0.65)	169	(0.65)	157	(0.64)
2aza :1pcy	73	(1.71)	70	(1.68)	47	(1.69)	50	(1.63)
9pap :2act	210	(0.75)	209	(0.75)	203	(0.73)	199	(0.66)

times of May and Johnson are taken from their paper. Both programs have been executed on a Silicon Graphics Iris workstations. Our results have been obtained in seconds, rather than on a time scale of minutes or hours as has been obtained with the combination of the genetic algorithms, dynamic programming and least squares optimization.

IV. Discussion and Conclusion

The Geometric Hashing is a highly efficient tool for structural comparisons of proteins, and for docking. As it is based in Computer Vision, it matches unconnected points in space. This enables matching protein structures in a manner that is entirely independent of their amino acid sequence order, and carrying out docking of a ligand onto a receptor surface. At the same time, it avoids the very time consuming search of entire conformational space by representing and matching the points in a transformation invariant manner, regardless of whether they are C_α atoms or critical points describing the molecular surfaces. Consequently, high quality matches are obtained in short times.

Here we show that for comparisons of protein structures, implementing considerations of connectivity into the matching procedure, further improves the Geometric Hashing technique. This Computer Vision point-based approach has numerous advantages, such as not being hampered by deletions

Table III.   A comparison of the CPU times required for the matching between our technique and a recently published one

Protein pair	Number of amino acid	Run time (May et al.)	Run time (This study)
2aza :1pcy	129:99	1730 (seconds)	3.8 (seconds)
3ptn :2sga	223:181	5416	9.7
2sga :3sgbE	181:185	4469	9.0
3ptn :2sga	223:185	5700	9.4
2gch :2sga	236:181	5336	10.1
2gch :3ptn	236:223	6619	11.4
2gch :3sgbE	236:185	5757	10.3

and insertions (a serious constraint in many matching algorithms) or by chain directionality, as well as allowing single point matches. As such, it enables detection of motifs between evolutionarily distant proteins. A drawback is the occurrence of biologically meaningless matches of isolated C_α pairs that happen to superimpose well in space; however, the implemented connectivity preferences, while still allowing single point matches, result in biologically more meaningful structural matches.

Acknowledgments

We thank Drs. Nickolai Alexandrov, Daniel Fischer, Robert Jernigan and Jacob Maizel, for helpful discussions, encouragement and interest. We thank the personnel at the Frederick Cancer Research and Development Center for their assistance. The research of R. Nussinov has been sponsored by the National Cancer Institute, DHHS, under Contract No. 1-CO-74102 with SAIC, and in part by by grant No. 91-00219 from the BSF, Israel, and by a grant from the *Israel Science Foundation* administered by the *Israel Academy of Sciences*. The research of H. Wolfson has been supported by a grant from the *Israel Science Foundation* administered by the *Israel Academy of Sciences*. The contents of this publication do not necessarily reflect the views or policies of the DHHS, nor does mention of trade names, commercial products, or organization imply endorsement by the U.S. Government.

References

(1) Bernstein, F. C., Koetzle, T. F., Williams, G. J. B., Meyer, E. F. Jr., Brice, M. D., Rodgers, J. R., Kennard, O., Shimanouchi, T. and Tasumi, M. (1977) *J. Mol. Biol.* **112**, 535–542.

(2) Branden, C. and Tooze, J. (1991) Introduction to protein structure, Garland Publishing, Inc. New York and London.

(3) Fischer, D., Wolfson, H., Lin, S. L. and Nussinov, R. (1994) *Protein Science* **3**, 769–778.

(4) Boberg, J., Salakoski, T. and Vihinen, M (1992) *Proteins* **14**, 265–276.

(5) Hobohm, U., Scharf, M, Schneider, R. and Sander, C. (1992) *Protein Science* **1**, 409–417.

(6) Orengo, C. (1994) *Curr. Opinion in Struct. Biol.* **4**, 429–440.

(7) Fischer, D., Tsai, C. -J., Nussinov, R. and Wolfson, H. (1995) *Prot. Engng.* in press.

(8) Miyazawa, S. and Jernigan, R. L. (1985) *Macromolecules* **18**, 534–552.

(9) Bowie, J. V., Luthy, R. and Eisenberg, D. (1991) *Science* **253**,164–170.

(10) Jones, D. T., Taylor, W. R. and Thornton, J. M. (1992) *Nature* **358**, 86–89.

(11) Alexandrov, N. N., Takahashi, K. and Go, N (1992) *J. Mol. Biol.* **225**, 5–9.

(12) Orengo, C. A. and Taylor, W. R. (1990) *J. Theor. Biol.* **147**, 517–551.

(13) Holm, L. and Sander, C. (1993) *J. Mol. Biol.* **233**, 123–138.

(14) Vriend, G. and Sander, C. (1991) *Proteins* **11**, 52–58.

(15) Subbarao, N. and Haneef, I. (1991) *Prot. Engng.* **4**, 877–884.

(16) Luo, Y., Lai, L., Xu, X. and Tang, Y. (1993) *Prot. Engng.* **6**, 373–376.

(17) Nussinov, R. and Wolfson, H. J. (1991) *Proc. Natl. Acad. Sci (USA)* **88**, 10495–10499.

(18) Bachar, O., Fischer, D., Nussinov, R. and Wolfson, H. (1993) *Prot. Engng.* **6**, 279–289.

(19) Lamdan, Y., Schwartz, J. T. and Wolfson, H. J. (1988) In *Proceedings of IEEE International Conference on Robotics and Automation.* Philadelphia, Pennsylvania, April 1988. pp 1407-1413.

(20) Fischer, D., Lin, S. L., Wolfson, H. and Nussinov, R. (1995) *J. Mol. Biol.* **248**, 459–477.

(21) Lin, S. L., Nussinov, R., Fischer, D. and Wolfson, H. (1994) *Proteins* **18**, 94–101.

(22) Tsai, C. -J., Ph.D. dissertation, Univ. of Pittsburgh, 1992.

(23) May, A. C. W. and Johnson, M. S. (1994) *Prot. Engng.* **7**, 475–485.

SECTION VIII

Folding and Stability of Proteins

Modeling Volume Changes in Proteins Using Partial Molar Volumes of Model Compounds

Kenneth E. Prehoda, Stewart N. Loh[a], and John L. Markley

Department of Biochemistry, University of Wisconsin, Madison, WI 53706
and
[a]Department of Biochemistry, Stanford University Medical Center, Stanford, CA 94305-5307

I. Introduction

Pressure and volume change logically represent variables of as much thermodynamic importance as temperature and enthalpy change. However, compared to thermal investigations, relatively little is known about the pressure behavior of biological molecules. Furthermore, the physical basis of the volume change for important biological reactions, such as protein folding, are poorly understood. This work attempts to rationalize volume changes for protein folding reactions in terms of the volume behavior of model systems, aqueous hydrocarbons, and amides.

Experimental pressure denaturation has been demonstrated with a number of proteins, including ribonuclease A (Gill & Glogovsky, 1965; Brandts et al., 1970), chymotrypsinogen (Hawley, 1971), metmyoglobin (Zipp & Kauzmann, 1973), and, more recently, lysozyme (Samarasinghe et al., 1992) and staphylococcal nuclease (Royer et al., 1993). The observed unfolding volumes found in these studies typically are between -50 and -150 mL mol^{-1} at room temperature and atmospheric pressure. These changes are small considering that the partial molar volumes of these proteins are greater than 10,000 mL mol^{-1}. The magnitude and sign of the experimental ΔV°_{d} values indicate that the partial molar volumes of the folded and unfolded states are similar, with that of the unfolded state smaller than that of the folded state by only about 0.5% of the partial molar volume of either state.

The use of partial molar volumes to model protein systems gives direct insight into the volume effect of solvation of these compounds. By contrast, the use of transfer quantities (e.g. the transfer of pure liquid hydrocarbon into aqueous solution) is complicated by the properties of the reference state (solid, liquid, or gas).

II. Theory

The total volume of a solution, V, is modeled in terms of the molar solvent-excluded, hard-sphere volumes of the solvent ($\overline{V_1^*}$) and solute ($\overline{V_2^*}$), plus the volume interstitial to these solvent–excluded volumes (V_{int})

$$V = V_{int} + n_1 \overline{V_1^*} + n_2 \overline{V_2^*} \tag{1}$$

The partial molar volume of the solute is the rate of change in total volume of solution with respect to number of moles of solute

$$\overline{V_2} \equiv \left(\frac{\partial V}{\partial n_2} \right)_{n_1, T, P} = \left(\frac{\partial V_{int}}{\partial n_2} \right) + \overline{V_2^*} \tag{2}$$

assuming that the excluded, hard-sphere volume of the solvent is independent of n_2. The derivative of the interstitial volume with respect to n_2 is the partial molar interstitial volume, $\overline{V}_{int}$. Equation 2 can be extended to protein unfolding, because ΔV°_d is the difference in partial molar volumes of the unfolded and folded states

$$\Delta V^\circ_d = \Delta \overline{V_2^*} + \Delta \overline{V}_{int} \tag{3}$$

where $\Delta \overline{V_2^*}$ is the difference in solvent excluded volumes between the unfolded and folded states. If $\Delta \overline{V}_{int}$ is broken down into the contributions from hydration of nonpolar and polar surfaces, $\Delta \overline{V}_{int,np}$ and $\Delta \overline{V}_{int,p}$, respectively

$$\Delta V^\circ_d = \Delta \overline{V_2^*} + \Delta \overline{V}_{int,np} + \Delta \overline{V}_{int,p} \tag{4}$$

The difference in excluded volume between the two states, $\Delta \overline{V_2^*}$, is estimated from comparison of the solvent excluded volume of the high-resolution structure of the native with an extended chain representation of the unfolded state. The hydration contributions, $\Delta \overline{V}_{int,np}$ and $\Delta \overline{V}_{int,p}$, are evaluated from the difference in solvent accessible surface areas of the two states and the respective $\overline{V}_{int}$ from the model compounds.

This model neglects the volume effects of charged groups. Whereas charged groups can have large volume effects in aqueous solution (known as electrostriction), proteins contain very few buried charged groups. Charged groups that are exposed in the native state are expected to have little net effect on the volume change. The small pH dependence of ΔV°_d supports this assumption (Royer et al., 1993).

Table I. Volume Parameters for Hydrocarbons and Amides in Water at 25°C and Atmospheric Pressure

	$\overline{V}_2$ (ml)	$\overline{V}^{\bullet}_2$ (ml)	$\overline{V}_{int}$ (ml)	$(\overline{V}_{int}/A)$ (ml / Å^2)
water	18.0	8.51	9.49	0.069
methane	37.3[a]	13.6	23.7	0.167
ethane	51.2[a]	23.0	28.2	0.156
propane	67.1[a]	32.8	34.3	0.162
benzene	81.3[b]	42.7	38.6	0.160
toluene	97.0[b]	52.4	44.6	0.164
formamide	39.2[c]	28.6	10.6	0.042
acetamide	56.1[c]	39.3	16.8	0.055

[a]Taken from Masterton (1954). [b]Taken from Shahidi (1981). [c]Taken from Shahidi, et al. (1977).

The many approximations made in this model prevents its use fore the quantitative determinations of volume changes. However, the model may be useful for rationalizing the different components that determine volume changes.

III. Results

The volume parameters for the aqueous polar and apolar compounds used in this study are shown in Table I. The partial molar interstitial volume, $\overline{V}_{int}$, is a measure of the solute-solvent interaction in volume terms for these solutes. In other words, $\overline{V}_{int}$ is the amount of void introduced into solution per mole of solute. Figure 1 shows $\overline{V}_{int}$ as a function of solvent accessible

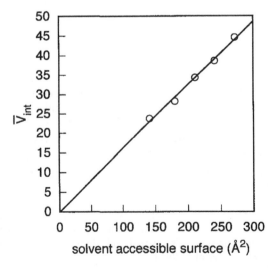

Figure 1. Partial molar interstitial volume as a function of solvent accessible surface area for the series of hydrocarbons in Table I (methane - toluene).

Table II. Contribution of Free Volume and Hydration of Nonpolar and Polar Surfaces To ΔV°_d[a]

	$10^{-3} \, \Delta \overline{V}_2^*$ (ml/mol)	$10^{-3} \, \Delta \overline{V}_{int,np}$ (ml/mol)	$10^{-3} \, \Delta \overline{V}_{int,p}$ (ml/mol)	$\Delta V^{\circ}_{d,obs}$ (ml/mol)
ribonuclease A	-1.39	0.948	0.217	-60[b]
lysozyme	-1.79	1.13	0.423	-35[c]
staphylococcal nuclease	-1.85	1.28	0.284	-30[d]
chymotrypsyinogen	-3.52	2.37	0.705	-20[e]

[a]Protein solvent accessible surface areas (used to calculate interstitial volume changes) taken from Spolar et al. (1992) [b]Gill & Glogovsky (1965). Value of ΔV°_d reported at pH 2.8. [c,e]Li et al. (1976). Value of ΔV°_d reported at pH 7.6. [d]Royer et al. (1993).

surface area for the series of hydrocarbons in Table I. The relationship between $\overline{V}_{int}$ and the solvent acessible surface area of the model compounds allows the computation of the contribution of solvation to the unfolding volume change ($\Delta \overline{V}_{int,np}$ and $\Delta \overline{V}_{int,p}$) for proteins, provided that the proteins surface area can be determined (i.e. if a high-resolution structure is known).

Table II shows the application of the results from the model systems to several proteins that have been studied with high-pressure and have high-resolution crystal structures. Since protein unfolding is accompanied by a large increase in the amount of solvent exposed polar and apolar surface area, and both of these types of surface increase the amount of void in solution, solvation is a large, positive contribution to ΔV°_d. The folded state contains a larger amount of void (from imperfect packing) than the unfolded state. This results in a much smaller excluded volume for the unfolded state (excluded volumes are calculated from the high-resolution native structure and an extended chain to approximate the unfolded chain). These results indicate that volume changes in protein unfolding reactions result from two opposing effects: 1) a large negative term from the difference in excluded volume of the two states and 2) a large positive term from the differential solvation of the two states.

Apomyoglobin (apoMb) is a protein known to exhibit an equilibrium molten globule folding intermediate (I_1) that is highly populated at mildly acidic pH (Barrick & Baldwin, 1993). Molten globules are typically characterized by a high secondary structure content with little fixed side chain structure. Of the eight α-helices in native apoMb, three are structured in I_1 while the rest appear to be unfolded (Loh et al, 1995; Hughson et al, 1990). The fact that molten globules have been postulated to be universal folding intermediates emphasizes the importance of determining their structures and the forces that underlie their stability (Ptitsyn, 1987).

Of the three equilibrium states of apoMb–native (N), unfolded (U), and I_1–I_1 exhibits the highest fluorescence emission while N exhibits the lowest (Kirby & Steiner, 1970). Application of pressure to native apomyoglobin under conditions described in Figure 2 causes a cooperative transition to a state with increased fluorescence emission. It is unlikely that this state is U, because a similar increase in pressure does not unfold I_1; rather, a slight

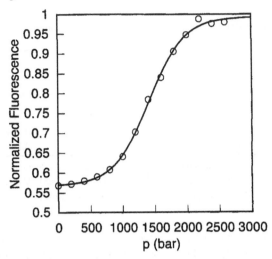

Figure 2. Pressure effect on apoMb. The fluorescence emission (excitation 295nm) of apoMb at pH 5.8, 7°C was monitored as a function of pressure. The data were fitted to a two-state nonlinear equation to yield $\Delta V° = -84$ ml/mole and $\Delta G° = 2.9$ kcal/mole.

(~5%) increase in fluorescence is observed (data not shown). It therefore appears that Figure 3 depicts a transition from N to a state similar to I_1. Fitting the data assuming a two-state transition gives a free energy difference ($\Delta G°_d = 2.9$ kcal/mol) somewhat smaller than the value previously determined for N $\rightarrow$ I_1 ($\Delta G°_d = 4.6$ kcal/mol; Barrick & Baldwin, 1993), further suggesting that U is not appreciably populated at high pressures. The discrepancy in $\Delta G°_d$ values may be due to the fact that in the earlier study, $\Delta G°_d$ was calculated at high pH, where native apoMb is significantly more stable (Barrick & Baldwin, 1993). Interestingly, the volume change for the transition in Figure 3 is negative (-84 ml/mole) indicating that I_1 has a *smaller* partial molar volume than the native state.

Our model suggests that volume effects are dictated by excluded volume and differential solvation terms. The volume change for the N $\rightarrow$ I_1 transition is similar to that measured for the unfolding transition of many other proteins. The fact that pressure induces the the N $\rightarrow$ I_1 transition, but not the N $\rightarrow$ D transition, in apomyoglobin is notable and suggests that the I_1 state occupies a lower volume than either the N or D states. The similarity of the magnitude of the N $\rightarrow$ I_1 volume change for apomyoglobin to those of N $\rightarrow$ D transitions in other proteins reveals that the I_1 state is characterized by extensive solvent penetration and solvation.

Acknowledgment

The authors thank Prof. Catherine A. Royer for access to the high-pressure fluorescence equipment used in this study.

References

Barrick D, Baldwin, RL. (1993). *Biochemistry* **32**, 3790-3796.
Brandts JF, Oliveira RJ, and Westort C. (1970). *Biochemistry* **9**, 1038-1047.
Gill SJ, and Glogovsky RL. (1965). *J. Phys. Chem.* **69**, 1515-1519.
Li TM, Hook JW, Drickamer HG, and Weber HG (1976) *Biochemistry* **15**, 5571-5580.
Kirby EP, and Steiner RF. (1970). *J. Biol. Chem.* **245**, 6300-6306.
Hawley SA. (1971). *Biochemistry* **10**, 2436-2442.
Hughson FM, Barrick D, and Baldwin RL. (1990). *Science* **249**, 1544-1548.
Loh SN, Kay MS, and Baldwin RL. (1995). *Proc. Natl. Acad. Sci. USA* **92**, 5446-5450.
Masterton WL. (1954). *J. Chem. Phys.* **22**, 1830-1835.
Ptitsyn, OB. (1987). *J. Protein Chem.* **6**, 273-293.
Royer CA, Hinck AP, Loh SN, Prehoda KE, Peng X, Jonas J, and Markley JL. (1993). *Biochemistry* **32**, 5222-5232.
Samarasinghe SD, Campbell DM, Jonas A, and Jonas J. (1992). *Biochemistry* **31**, 7773-7778.
Shahidi F, Farrel PG, and Edward JT. (1977). *J. Chem. Soc., Faraday Trans* **73**, 715.
Shahidi. (1981). *J. Chem. Soc., Faraday Trans* **77**, 1511-1514.
Spolar RS, Livingstone JR, and Record MT. (1992). *Biochemistry* **31**, 3947-3955.
Zipp A, and Kauzmann W. (1973). *Biochemistry* **12**, 4217-4228.

A General Approach for the Design and Isolation of Protein Fragments: The Molecular Dissection of Dihydrofolate Reductase

Colin V. Gegg, Katherine E. Bowers and C. Robert Matthews

Department of Chemistry and Center for Biomolecular Structure and Function
The Pennsylvania State University, University Park, PA 16802

I. Introduction

The ability to cleave proteins into specific fragments has proven to be a powerful tool for probing complex sequence/structure/function relationships that occur in protein folding, protein-protein interactions and enzyme activity. The reassociation of complementary protein fragments has provided insights into the folding mechanisms of many proteins, including cytochrome *c* (1), the β_2 subunit of tryptophan synthase (2), *trp* aporepressor (3), chymotrypsin inhibitor-2 (4) and yeast phosphoglycerate kinase (5). Protein fragments have also been useful in identifying molecular interactions that are essential for recognition and binding of steroid hormones to human serum albumin (6) and for the binding of immunoglobulin G to staphylococcal protein A (7). Molecular dissection has similarly aided in defining enzymatic regulatory elements of aspartate transcarbamoylase (8).

At present, there are a variety of ways to generate protein fragments, each with inherent advantages and disadvantages. Peptide synthesis is expedient and reliable for making small peptides (<50 residues), but can become costly and complex when producing larger polypeptides. Enzymatic proteolysis of a purified protein can be used to produce larger fragments, but multiple cleavage sites with variable reactivities can make the production of sufficient quantities of the desired protein fragments difficult. Recombinant techniques have been used to produce protein fragments by inserting stop-codons at predetermined locations, however, the yields of marginally-stable fragments are often compromised. Chemical cleavage methods can also provide access to large fragments (9), but are often associated with undesirable side reactions (10).

This communication describes a combination of chemical cleavage and affinity chromatography, which can be used to produce large quantities of protein fragments that vary considerably in size, stability and solubility. The approach utilizes alkaline catalyzed cleavage following cyanylation of cysteine residues by 2-nitro-5-thiocyanobenzoic acid (NTCB)[1] (11) . Cysteine-specific cleavage reagents are particularly attractive because of the relatively unique reactivity of cysteines and the scarcity of free cysteines in many proteins (12).

---

[1]**Abbreviations:** NTCB, 2-nitro-5-thiocyanobenzoic acid; EDTA, ethylenediaminetetraacetic acid; DTNB, dithionitrobenzoic acid; DTE, dithioerythritol; SDS, sodium dodecyl sulfate; PAGE, polyacrylamide gel electrophoresis; IMAC, immobilized metal affinity chromatography; IDA, iminodiacetic acid; AS-DHFR, cysteine free mutant of dihydrofolate reductase; PCR, polymerase chain reaction; HPLC, high-performance liquid chromatography; CD, circular dichroism.

Fragment diversity can be enhanced by the selective placement of the target cysteine residue using PCR-mediated mutagenesis. Immobilized metal affinity chromatography (IMAC), which has proved to be a versatile and specific technique for the purification of poly-histidine tagged fusion proteins (13, 14), can then be used under strongly denaturing conditions (8 M urea) to isolate the resulting fragments.

The application of this approach to dihydrofolate reductase from *E. coli* reveals that several large fragments, which correspond to observed structural domains, adopt significant secondary structure in solution. These fragments apparently contain sufficient information to act as autonomous folding units and may provide clues to the structures of transient folding intermediates.

II. Materials and Methods

A. Reagents

2-nitro-5-thiocyanobenzoic acid (NTCB) was obtained from Sigma (St. Louis, MO) and Chelating Sepharose Fast Flow resin was purchased from Pharmacia (Piscataway, NJ).

B. Mutagenesis

A vector (pTZdhfr20-1) encoding the mutant, cysteine-free (C85A, C152S), form of dihydrofolate reductase (AS DHFR) from *E. coli* (15) was used as a template for subsequent PCR-mediated mutagenesis. Three tagged host vectors were prepared using the two flanking p(His)$_6$ primers and the non-mutagenic flanking primers α and β (Figure 1) in conventional 2-primer PCR mediated mutagenesis (16). To ensure the initiation of protein synthesis, the amino-terminal p(His)$_6$ primer was engineered to encode the homologous tripeptide Met-Ile-Ser preceding the poly-histidine sequence, which was followed by a silent mutation to remove the original *Bcl* 1 restriction site. These three vectors code for cysteine-free DHFR fusion proteins incorporating single poly-histidine affinity tags at either the amino or carboxy termini, or double tags at both the amino- and carboxy- termini. Next, cysteine point mutations were introduced using the 'megaprimer' method of PCR mediated mutagenesis (17, 18), with the three tagged host vectors as templates. Combinations of both the mutagenic primers (N37C, D87C and Q108C) and the flanking primers (α and β) were used to create a series of single and double cysteine mutations in the three tagged host vectors (Figure 1, Table I). The final constructs were prepared using routine cloning and recombinant molecular biological techniques (19). The entire gene for each construct was sequenced to confirm the cysteine mutations, affinity tag configuration and the fidelity of the PCR.

Table I: Summary of mutations and targeted fragments of AS DHFR

Clone	Cysteine Mutation	Tagged Termini	Fragment
5.1d	N37C	-CO$_2$H	1-36
5.2e	D87C	-CO$_2$H	1-86
8.3d	Q108C	-CO$_2$H	1-107
9.1f	N37C	-NH$_2$	37-159
9.2f	D87C	-NH$_2$	87-159
8.4b	Q108C	-NH$_2$	108-159
1.1.2	N37C, D87C	-NH$_2$, -CO$_2$H	37-86
7.1b	N37C, Q108C	-NH$_2$, -CO$_2$H	37-107

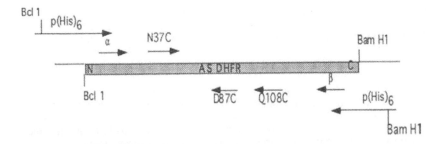

Figure 1:   Primers used in PCR-mediated mutagenesis to introduce terminal poly-histidine affinity tags and cysteine point mutations at positions N37, D87 and Q108. Two p(His)$_6$ extended flanking primers were used to incorporate in-frame affinity tags of six histidines at either, or both, the amino or carboxy termini, by cloning at *Bcl* 1 or *Bam* H1 restriction sites, respectively. The α and β primers are non-mutagenic internal flanking primers used for megaprimer mutagenesis: each contains a unique restriction site to facilitate directional cloning of the mutagenized product into any of the three tagged, host vectors. N37C, D87C and Q108C are mutagenic primers.

C.   Protein Expression

Each clone was expressed in the Ag-1 strain of *E. coli* grown overnight in 2xYT medium (16 g tryptone, 10 g yeast extract and 10 g NaCl in 1 liter at pH 7.0). The cells from 6-10 liters of culture were harvested by centrifugation, incubated for 30 min at room temperature in 200 mL of 20 mM potassium phosphate buffer, pH 7.2 containing 0.5 mg/mL lysozyme. The cell suspension was disrupted by sonication and the debris removed by centrifugation at 17,700 g for 20 min. Streptomycin-sulphate was stirred with the cell extract at a final concentration of 1.9% (w/v) for 30 min at 4 °C and the extract clarified by centrifugation at 48,400 g for 20 min.

D.   Affinity Chromatography

The Chelating Sepharose Fast Flow resin utilizes iminodiacetic acid chelating groups (IDA), which are tridentate. The packed resin, 50 mL in 2.5 x 12 cm column, was charged with NiCl (Ni^{2+}-IDA) and equilibrated in 20 mM potassium phosphate, 0.5 M KCl, 10 mM imidazole, pH 7.2 prior to loading the cell extract. Once the cell extract had been loaded, the Ni^{2+}-IDA column was washed with equilibration buffer and eluted with a 2 liter, linear 10-300 mM imidazole gradient in 20 mM potassium phosphate, 0.5 M KCl, pH 7.2. Between each of the affinity purifications, the Ni^{2+}-IDA column was washed, stripped of Ni^{2+} and recharged as described by the manufacturer.

In preliminary IMAC purifications of full-length, mutant fusion proteins, a significant amount of contaminating protein was found to bind the Ni^{2+}-IDA column when the cell extract was loaded in the absence of imidazole. This contaminant typically eluted at ≈25 mM imidazole and was well resolved from the affinity-tagged protein. The binding of this potential contaminant was minimized by loading the extract in 10 mM imidazole, with no corresponding loss in binding of the affinity-tagged, full-length mutant protein.

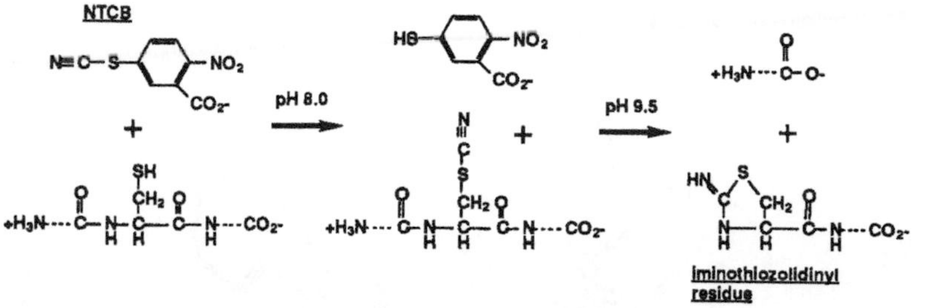

Figure 2: Polypeptide cleavage mechanism at cysteine by 2-nitro-5-thiocyanobenzoic acid (NTCB).

E.   Chemical Cleavage

Chemical cleavage at the engineered cysteines of the mutant DHFR proteins by NTCB was achieved as described by Jacobson, et al. (11), with the following modifications. The affinity-tagged, mutant DHFR proteins isolated with the Ni^{2+}-IDA column were reduced by dialysis in 0.1 M Tris-acetate, pH 8.0 with 5 mM DTE, followed by dialysis in rigorously degassed 0.1 M Tris-acetate, pH 8.0 buffer to remove the DTE. Total thiol concentration was determined with DTNB after dialysis. The reduced proteins were then denatured in 0.1 M Tris-acetate, 6 M urea, pH 8.5 and cyanylated with a 10-15 fold molar excess of NTCB. After incubation for 30 min at room temperature, the base catalyzed cleavage reaction was accelerated by adjusting the pH to 9.5 and incubating the reaction at 37 °C for 16 hrs. The reaction scheme is shown in Figure 2.

Once the cleavage reaction was complete, the yellow colored reaction bi-product, 2-nitro-5-thiobenzoate, and unreacted NTCB were removed by gel filtration on Sephadex G-15 in 0.1 M Tris-acetate, 6 M urea, pH 8.0. The cleared reaction mixture was then loaded on the Ni^{2+}-IDA column equilibrated in 0.1 M Tris-acetate, 6 M urea, pH 8.0 and the untagged fragments were isolated in the column flow-through. In some cases with fragments containing native histidine residues, the fragments bound the column weakly and were eluted at $\approx$25 mM imidazole with a 1 liter, 0-300 mM imidazole gradient. In all cases the untagged fragments were well resolved from the complementary tagged fragment and uncleaved full-length protein. The isolated fragments were concentrated to $\approx$30 µM by diafiltration with 3000 MWCO membranes (Amicon, Beverly, MA).

F.   Fragment Folding

The isolated fragments at an initial concentration of 20 µM in 0.1 M Tris-acetate, 6 M urea, pH 8.0 were refolded by exhaustive dialysis in 10 mM potassium phosphate, 0.2 mM potassium EDTA, pH 7.8 with 1 mM β-mercaptoethanol using 2000 MWCO membranes. The dialysates were centrifuged at 48,400 g for 30 min to remove precipitates and the fragment concentration in the supernatant was determined from the absorbance at 280 nm using measured extinction coefficients (20).

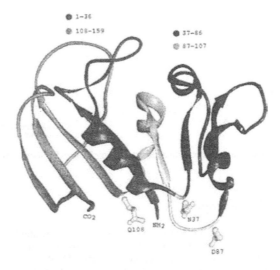

Figure 3: A 3-dimensional image of DHFR illustrating sites for Cys point mutations and several targeted fragments. Structure derived from Brookhaven protein data base file 7dfr.pdb.

G. Physiochemical Characterization

Each step in the isolation of the targeted fragments was monitored by SDS-PAGE using 16% SDS-PAGE gels (21). Fragment purity was estimated to be ≥95% by both SDS-PAGE and reverse-phase HPLC. The isolated fragments were submitted for electrospray mass spectrometric analyses for further characterization. The full-length mutant, fusion proteins and the isolated fragments were each assayed for dihydrofolate reductase activity at both 30 °C and 15 °C by the method of Hillcoat et al. (22).

Circular dichroism spectra of the isolated fragments were collected from 320 -185 nm with an Aviv 62DS circular dichroism spectrometer using 1 or 2 mm cells at 15 °C with a 1 second averaging time and are reported as the average of 8 scans. Spectra were collected with fragment concentrations ranging from 30-160 µg/mL, depending on their solubility, in 10 mM potassium phosphate, 0.2 mM potassium EDTA, pH 7.8 and 1 mM β-mercaptoethanol.

III. Results

A. Fragment Design

The folate-NADP+ ternary complex of DHFR (23) was examined using the MIDAS Plus computer program (24) run on a Silicon Graphics Indigo 2 workstation. From this structure, two discontinuous subdomains (residues 1-37 + 89-159 and 38-88), joined by a flexible hinge, have been previously proposed (23). In the present study, three contiguous sequences (residues 1-36, 37-107 and 108-159) have been selected to represent putative, autonomous folding domains based on the connectivity of their β-strands, intradomain tertiary interactions and the incorporation of ligand binding sites (Figure 3). A fourth subdomain (residues 37-86) was selected because it had been previously defined as the adenine binding domain for NADP+ (23). The junctures of the proposed domains, Asn37, Asp87 and Gln108, all map to surface loops and are relatively

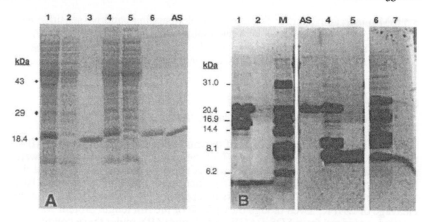

Figure 4: Representative SDS-PAGE analyses of: (**A**) Full-length, mutant fusion protein purification. Lanes 1 and 4 are crude cell extracts, lanes 2 and 5 are IMAC column flow-throughs and lanes 3 and 6 are the full-length, mutant fusion proteins, clones 8.3d and 7.1b, respectively, eluted with imidazole, (**B**) Chemical cleavage reaction and isolated protein fragments. NTCB cleavage reactions (50 μg each) of DHFR 5.1d in lane 1, DHFR 5.2e in lane 4 and DHFR 7.1b in lane 6. Isolated fragments (100 μg each) 1-36 DHFR in lane 2, 1-86 DHFR in lane 5 and 37-107 DHFR in lane 7. AS is full length cysteine-free AS DHFR. M is a set of molecular weight markers.

solvent accessible in the 3-dimensional structure. By introducing cysteine mutations at these sites, as summarized in Table I, it was possible to produce 8 different fragments of DHFR that span the entire protein sequence and include the 4 putative subdomains.

B. Expression and Isolation of the Full-Length, Mutant Fusion Proteins

The expression of each full-length, mutant fusion protein varied by clone and was monitored by SDS-PAGE (Figure 4A). Overall yields of purified protein ranged from 1.6 to 16.2 mg/g cells. As a rule, those fusion proteins carrying an amino-terminal affinity tag were more poorly expressed than the fusion proteins tagged at the carboxy terminus. Cysteine point mutations also perturbed the expression of the mutant fusion proteins to varying degrees, the most notable being the mutation N37C in the amino-terminally tagged vector (DHFR 9.1f) and the double tagged, double mutant N37C/D87C (DHFR 1.1.2). Interestingly, the point mutation N37C had little effect on expression when placed in a carboxy-terminally tagged vector (DHFR 5.1d), suggesting that the more proximal, amino-terminal, poly-histidine affinity tag may interact with cysteine at position 37.

Despite the different point mutations, the singly tagged mutant fusion proteins were consistently eluted at ≈100 mM imidazole, whereas the doubly tagged, double mutant, fusion proteins did not elute until ≈200 mM imidazole. The Ni^{2+} - IDA column (2.5x12 cm) demonstrated a high capacity for histidine-tagged mutant DHFR proteins (≥850 mg), which were specifically eluted ≥95% pure as judged by SDS-PAGE (Figure 4). The small amount of disulphide linked dimer observed (Figure 4A, lane 3, MW ≈40 kDa) is considerably increased in the absence of reducing agent, yet persists despite the addition of

3.5 mM β-mercaptoethanol. Similarly, doublet banding patterns, sensitive to reducing agent, were observed with the double cysteine mutants (Figure 4A, lane 6). These doublets were presumably due to intramolecular disulphide bonds formed between the proximally engineered cysteine residues.

C.   Chemical Cleavage and Isolation of Protein Fragments

After reduction and dialysis of the mutant fusion proteins, the DTNB assay was used to determine that >90% of the mutant cysteine side chains were available as the free thiol for reaction with NTCB. Experiments to optimize the NTCB reagent concentration and reaction incubation times indicated that a 10-15 fold molar excess of NTCB and 16 hrs incubation at 37 °C were optimal for the cleavage of DHFR. Even though only ≈50% of the full length protein was cleaved, as judged from SDS-PAGE gels (Figure 4B), under these conditions, higher NTCB concentrations or pH ≥10 resulted in additional nonspecific fragmentation.

A Sephadex G-15 column was used to remove the unreacted NTCB and reaction bi-product, 2-nitro-5-thiobenzoate (Figure 2) to prevent reduction of the Ni^{2+}-IDA resin by excess thiol and spectral contamination by these aromatic reaction components. The chromatographic separation was easily monitored by ultraviolet absorbance of the protein and direct observation of the yellow reaction components.

The cleaved target fragments have a much lower affinity for the Ni^{2+}-IDA resin and were easily separated from the complementary tagged fragments and the uncleaved fusion protein using the same IMAC column that was used previously to purify the full-length, mutant fusion protein. The untagged target fragments were isolated either in the Ni^{2+}-IDA column flow-through, or if they contained a sufficient number of native histidine residues, the fragments were eluted before the tagged components with an imidazole gradient. The advantage in allowing the untagged fragment to bind the Ni^{2+}-IDA column and eluting with imidazole is that the fragment in the eluted fraction is generally more concentrated than in the flow-through fraction.

D.   Fragment Characterization

The isolated fragments were analyzed by 6 M urea/SDS-PAGE gels and appeared to be ≥95% pure (Figure 4B). There was no indication of contamination by either the complementary fragment or uncleaved fusion protein. Reverse phase HPLC analyses further confirmed that these fragments were ≥95% pure. Occasionally a fragment preparation showed several minor bands in SDS-PAGE gels that were not detected by HPLC. These minor contaminants appeared occasionally, had apparent molecular weights greater than the fragment and represented less than 5% of the total protein. Because these trace contaminants typically appeared after the fragment had been refolded, they may reflect aggregation of the fragments.

Although the SDS-PAGE analyses confirmed the purity of the DHFR fragments, their small size and aberrant mobilities made it difficult to estimate their relative mass. Therefore, mass values of each fragment were determined by electrospray mass spectrometry. The measured values were in good agreement with the values calculated from the amino acid composition of the respective fragments (Table II).

Table II: Electrospray-mass spectroscopy analyses of isolated DHFR fragments

| DHFR Fragments | Mass (daltons) | | |
	Calculated	ES-MS[a] Determined	ES-MS error
1-36	4,043.1	4,042.5	-0.6
1-86	9,451.5	9,445.7	-5.8
1-107	11,752.4	11,756.1	+3.7
37-159	13,942.2	13,949.5	+7.3
87-159	8,532.7	8,535.3	+2.6
108-159	6,218.8	6,218.5	-0.3
37-86	5,441.6	ND	ND
37-107	7,742.4	7,742.5	+0.1
AS DHFR	17,952.2	17,958.5	+6.3

[a] Electrospray-mass spectroscopy.
[b]ND, not determined

Refolding by dialysis was most productive for the carboxy-terminal fragments; however, considerable differences were observed in the solubility of the isolated fragments. The smallest of the amino-terminal fragments were the least soluble, probably a consequence of the extreme hydrophobicity of the first 20 residues in DHFR.

No DHFR-like catalytic activity was detectable in any of the isolated, refolded fragments when assayed at either 30 °C or 15 °C.

After refolding, the isolated fragments were surveyed by CD spectroscopy to determine if they contained regular elements of secondary structure (Figure 5). The largest fragments, 1-107 and 37-159 DHFR, provided the best evidence of secondary structure, with strong minima at ≈218 nm and maxima at ≈195 nm. Fragment 37-159 DHFR had the most nearly wild-type spectrum of those fragments examined. Although the intermediate sized fragments, 37-86 and 37-107 DHFR, appeared largely unfolded based on their minima at ≈200 nm, both also showed shoulders in their spectra at ≈220 nm, suggesting the presence of a modest amount of secondary structure.

IV. Discussion

Immobilized metal affinity chromatography has now become a standard in the arsenal of protein purification techniques (25). Recently, these metal chelating resins have proved particularly useful in purifying recombinant fusion proteins that incorporate poly-histidine affinity tags (13, 14). Advances in PCR-mediated mutagenesis (17, 18) and cloning (16) now permit poly-histidine tagged, mutant fusion proteins to be rapidly engineered and efficiently purified. In this work, these common techniques have been combined with a little-used method for the chemical cleavage of polypeptides at cysteine residues (11) to generate and isolate a large array of protein fragments from DHFR.

The novel combination we describe between PCR-mediated mutagenesis, IMAC purification of poly-histidine tagged fusion proteins and chemical cleavage of polypeptides to produce protein fragments, offers many advantages over conventional approaches. The iterative cycles of mutant re-engineering are accelerated and simplified using PCR to mutate pre-engineered host vectors coding for poly-histidine tags at either, or both, termini of the protein. The IMAC system of purification was selected not only for its specificity for poly-histidine affinity tags, but also because it binds these tags tightly in the presence of strong denaturants, high salt concentrations and detergents. The utility of high affinity binding in strong denaturant is particularly evident when isolating relatively insoluble mutants or

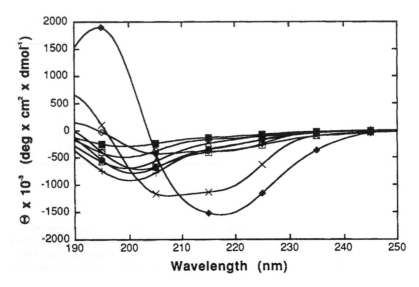

Figure 5: Circular dichroism spectra of refolded DHFR fragments plotted as molar ellipticity. Individual fragment spectra are: 1-36 (O), 1-86 (□), 1-107 (◊), 37-159 (x), 87-159 (+), 108-159 (Δ), 37-86 (●), 37-107 (■) and AS DHFR (◆).

protein fragments. The Chelating Sepharose Fast Flow IMAC resin used in this work was economical, bound poly-histidine tagged protein in high capacity (≥850 mg/50 mL), performed well at fast flow rates (5 mL/min) and seems well suited to large scale, low pressure, affinity purifications. Finally, because this approach to protein fragment production is relatively insensitive to fragment size and solubility, it provides rapid access to potentially large quantities of isolated fragments in a wide variety of lengths.

The cysteine specific chemical cleavage method of Jacobson, et al. (11) was chosen for the fragmentation of DHFR because a well characterized, cysteine-free mutant of DHFR was available and found to be nearly indistinguishable from the wild-type DHFR (15). Because the cleavage reaction is performed in 6 M urea, the IMAC resin is particularly well suited to the separation of the targeted fragments from the affinity tagged fragments and uncleaved proteins. The NTCB mediated chemical cleavage method is limited by relatively moderate yields of ≈50% cleavage and results in the addition of an iminothiazolidinyl residue to the amino terminus of the carboxy-terminal fragment (Figure 2). Cleavage inefficiency may derive from β-elimination of the cyanylated intermediate resulting in an intact polypeptide with an unreactive dihydroalanine residue at the intended cleavage site (9).

The application of this method to the molecular dissection of DHFR demonstrates the facility with which a variety of different fragments may be obtained. The IMAC method consistently produced, in a one-step purification, numerous full-length, mutant fusion proteins that varied significantly in their expression, solubility and enzymatic activity. Similarly, the targeted fragments, some of which later proved marginally soluble, were readily separated from the complementary poly-histidine tagged fragment and uncleaved mutant, fusion protein by IMAC in 6 M urea.

The value of this approach to fragment production is that the protein of interest may be dissected into a diverse group of fragments representing any part of the protein, regardless of size or solubility. In terms of protein folding, the fragments from DHFR exhibit CD spectral properties indicative of varied levels of folding (Figure 5) and provide a starting point in the search for minimal autonomous folding units in this protein. This method for the production and isolation of a broad assortment of protein fragments should be generally applicable to a variety of other proteins.

Acknowledgments

Special thanks to Dr. Jill A. Zitzewitz for her critical review of this manuscript and to Matthew J. Houser for his technical assistance. Mass spectrometry was performed by The Protein Chemistry Laboratory, Department of Molecular Biology and Pharmacology, Washington University School of Medicine (St. Louis, MO). This research was supported by NSF MCB9317273 and NIH GM23303 to CRM and NIH GM16516 to CVG.

References

1.   Fisher, A. and Taniuchi, H. (1992) *Arch. Biochem. Biophys.* **296**, 1-16.
2.   Linkens, H.J., Djavadi-Ohaniance, L. and Goldberg, M.E. (1993)
      FEBS Lett. **320**, 224-228.
3.   Tasayco, M.L. and Carey, J. (1992) *Science* **255**, 594-597.
4.   de Prat Gay, G., Ruiz-Sanz, J. and Fersht, A.R. (1994)
      Biochemistry **33**, 7964-7970.
5.   Pecorari, F., Minard, P., Desmadril, M. and Yon, J., M. (1993)
      Protein Eng. **6**, 313-325.
6.   Fischer, M.J.E., Bos, O., van der Linden, R.F., Wilting, J. and Janssen, L. (1993)
      Biochem. Pharmacol. **45**, 2411-2416.
7.   Kato, K., Gouda, H., Takaha, W., Yoshino, A., Matsunaga, C. and Arata, Y. (1993)
      FEBS Lett. **328**, 49-54.
8.   Powers, V.M., Yang, Y., R., Fogli, M., J. and Schachman, H.K. (1993)
      Protein Sci. **2**, 1001-1012.
9.   Fontana, A. and Gross, E. (1986) in *Practical Protein Chemistry - A Handbook.*,
      (Darbe, A., Ed.), 67-120, John Wiley and Sons, New York.
10.   Means, G.E. and Feeney, R.E. (1971) in *Chemical Modification of Proteins,*
      pg 58-60, Holden-Day, San Francisco.
11.   Jacobson, G.R., Schaffer, M.H., Stark, G.R. and Vanaman, T., C. (1973)
      J. Biol. Chem. **248**, 6583-6591.
12.   Klapper, M.H. (1977) *Biochem. Biophys. Res. Commun.* **78**, 1018-1024.
13.   Hochuli, E., Bannwarth, W., Dobeli, H., Gentz, R. and Stuber, D. (1988)
      Bio/Technology **6**, 1321-1325.
14.   Smith, M.C., Furman, T.C., Ingolia, T.D. and Pidgeon, C. (1988)
      J. Biol. Chem. **263**, 7211-7215.
15.   Iwakura, M., Jones, B.E., Luo, J. and Matthews, C.R. (1995)
      J. Biochem. **117**, 480-488.
16.   Scharf, S.J. (1990) in *PCR protocols: a guide to methods and applications.*,
      (Innis, M.A., *et al.*, Ed.), 84-91, Academic Press, Inc., San Diego.
17.   Sarkar, G. and Sommer, S.S. (1990) *BioTechniques* **8**, 404-407.
18.   Landt, O., Grunert, H.-P. and Hahn, U. (1990) *Gene* **96**, 125-128.
19.   Sambrook, J., Fritsch, E.F. and Maniatis, T. (1989) in *Molecular cloning,*
      2nd Edition, Cold Spring Harbor Laboratory, Cold Spring Harbor.
20.   Gill, S.C. and von Hippel, P.H. (1989) *Anal. Biochem.* **182**, 319-326.
21.   Schägger, H. and von Jagow, G. (1987) *Anal. Biochem.* **166**, 368-379.
22.   Hillcoat, B., Nixon, P. and Blakely, R.L. (1967) *Anal. Biochem.* **21**, 178-189.
23.   Bystroff, C., Oatley, S.J. and Kraut, J. (1990) *Biochemistry* **29**, 3263-3277.
24.   Ferrin, T.E., Huang, C.C., Jarvis, L.E. and Langridge, R. (1988)
      J. Mol. Graphics **6**, 13-27.
25.   Ostrove, S. and Weiss, S. (1990) *Methods Enzymol.* **182**, 371-379.

The Association of Unfolding Intermediates during the Equilibrium Unfolding of Recombinant Murine Interleukin-6

Jacqueline M. Matthews, Larry D. Ward*, Jian-Guo Zhang,
and Richard J. Simpson

Joint Protein Structure Laboratory, Ludwig Institute for Cancer Research and the
Walter and Eliza Hall For Medical Research, Parkville, Victoria 3050, Australia

I. Introduction

Interleukin-6 (IL-6) is a multifunctional cytokine which acts by binding to a specific cell membrane receptor protein (IL-6R) and a signal transducing protein, gp130. IL-6 is a putative member of the long chain, 4-α-helical bundle cytokine/growth factor family (1), of which growth hormone (GH), leukemia inhibitory factor (LIF), granulocyte colony - stimulating factor (G-CSF) and ciliary neutrotrophic factor (CNTF) are examples with known structure (2,3,4,5). IL-6, along with CNTF, oncostatin M (OSM), LIF and IL-11 are defined as a family of pleiotropic cytokines with overlapping biological activities. They share gp130 as a common component for signal transduction which may explain their functional redundancy. The selective inhibition of IL-6 activity, through designed protein or peptide anatogonists of IL-6, may have clinical significance in the treatment of IL-6 associated diseases such as psoriasis and multiple myeloma (6).

As yet, IL-6 has not been successfully crystallised for X-ray crystallographic studies, and the determination of its solution structure by NMR has been impeded by its behaviour at the high protein concentrations required (15 - 20 mg/ml). Initial NMR studies at these concentrations showed a time dependent decrease in signal, consistent with protein aggregation (7), but did not result in precipitation of the protein. This aggregate is readily dispersable by RP-HPLC utilising TFA/acetonitrile.

Our study of the equilibrium unfolding of murine IL-6 (mIL-6) stemmed, at least in part, from our need to determine what caused this aggregation, whether we could prevent it, or engineer less aggregation-prone IL-6 molecules, which would be more conducive to NMR studies (8).

II. Materials and Methods

The mIL-6 used in these studies is an *E. coli* expressed fusion protein, where the first 8 residues originate from β-galactosidase and the polylinker region of

---

* Current address: AMRAD Laboratories, Burnley, Victoria 3121, Australia.

TECHNIQUES IN PROTEIN CHEMISTRY VII

pUC9. The remaining 176 residues correspond to residues Thr12-Thr187 of native IL-6 (9).

The equilibrium unfolding of mIL-6 as a function of denaturant concentration (urea or guanidinium hydrochloride (GdnHCl)) was monitored by both far-UV CD spectroscopy (222 nm) and fluorescence spectroscopy (excitation 295 nm). Lyophilised mIL-6 was dissolved in water and diluted into solutions containing buffer and denaturant. The reversibility of the unfolding transition was assessed by performing the experiments with either native (stock solution in water) or denatured (stock solution in 6 M denaturant) mIL-6 as the starting material. The buffers used were either 10 mM sodium acetate, pH 4.0, or 10 mM Tris-HCl, pH 7.4.

Assuming a two-state unfolding transition, when S is the measured property of the protein at a given denaturant (D) concentration, data can be analysed according to Eqn 1 (10),

$$S = \frac{S_N + S_U.\exp(A)}{1 + \exp(A)} \tag{1}$$

where $A = \{m[D] - \Delta G_U(H_2O)\}/RT = m([D] - [D]_{50\%})/RT$. U represents the unfolded state, N represents the native state, m is the slope of the curve in the transition state, $\Delta G_U(H_2O)$ is the free energy of unfolding in the absence of denaturant, $[D]_{50\%}$ is the denaturant concentration at the midpoint of unfolding, R is the gas constant and T is absolute temperature. Data were fitted using the non-linear regression analysis program KaleidaGraph (version 3.0 Synergy Software; PCS Inc).

III. Results and Discussion

A. *Far-UV CD Monitored Unfolding*

Our initial experiments using far-UV CD spectroscopy were performed at different pH values (7.4 and 4.0), using GdnHCl and urea as denaturants (Fig. 1). Under three of these four conditions, (**I**, pH 7.4, GdnHCl; **II**, pH 7.4, urea; and **III**, pH 4.0, GdnHCl) we observed broad unfolding transitions, which for II and III are clearly biphasic. In the final set of conditions (**IV**, pH 4.0, urea), however, the unfolding transition was sharp and monophasic, which is indicative of a two-state unfolding process. The differential behaviour of I, II and III compared to IV, is obvious from the analysis of these data sets by Eqn 1 (Table I). For denaturing conditions I, II and III, artefactually low values of $\Delta G_U(H_2O)$ and m, which is a measure of the cooperativity of unfolding, were obtained because these conditions do not ensure the assumed two-state unfolding transition of Eqn 1. In comparison, IV has significantly higher values of $\Delta G_U(H_2O) = 6.9$ kcal/mol and $m = 1.9$ kcal/mol/M.

Table I. Analysis of far-UV CD monitored unfolding of mIL-6[a]

pH	Denaturant	$\Delta G_U(H_2O)$ (kcal/mol)	m (kcal/mol/M)	$[D]_{50\%}$ (M)
4.0	Urea	6.9±0.5	1.9±0.1	3.60±0.02
7.4	Urea	3.4±0.6	0.7±0.1	4.8±0.2
4.0	GdnHCl	1.8±0.4	0.7±0.1	2.7±0.1
7.4	GdnHCl	1.8±0.3	1.0±0.2	1.7±0.1

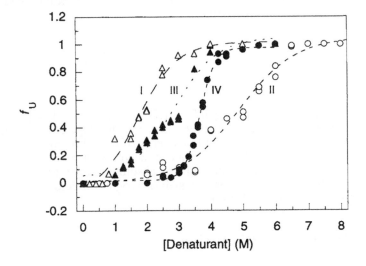

Figure 1: *Denaturant-induced equilibrium unfolding of mIL-6 monitored by far-UV CD at 222 nm.* The fraction of unfolded protein (f_U) is plotted against the denaturant concentration. Curves shown are the data fitted to Eqn 1. **I**, GdnHCl, pH 7.4 (– –Δ– –); **II**, urea, pH 7.4 (- -o- -); **III**, GdnHCl, pH 4.0 (···▲···); **IV**, urea pH 4.0 (—●—). Adapted from Fig. 1 (8).

B.   *Fluorescence   Monitored   Unfolding*

The presence of equilibrium unfolding intermediates can be detected by the non-coincidence of unfolding transitions as monitored by fluorescence and far-UV CD spectroscopy (11). To determine whether the unfolding transition at IV was indeed two-state, and to investigate the nature of any unfolding intermediates in conditions I, II and III, we used fluorescence, with an excitation wavelength of 295 nm to preferentially excite Trp side-chains, to monitor the unfolding of mIL-6.

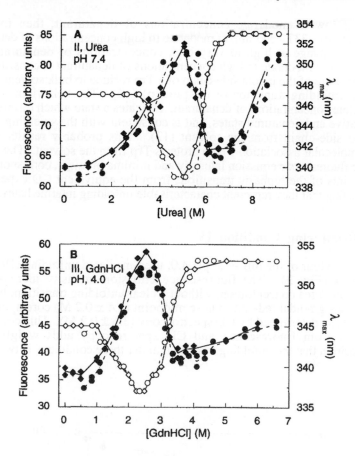

Figure 2: *Denaturant-induced equilibrium unfolding of mIL-6 monitored by fluorescence.* Relative fluorescence emission at 345 nm (closed symbols) and λ_{max} (open symbols) in the region 310-400 nm were plotted against denaturant concentration. Unfolding equilibria (native starting material) are shown as diamonds, refolding equilibria (denatured starting material) are shown as circles. Panel A, condition **II**, urea, pH 7.4; Panel B, condition **III**, GdnHCl, pH 4.0. Adapted from Fig. 2 (8).

1. Denaturing Conditions I, II and III

Under the same conditions where we detected non-two state unfolding by far-UV CD, we also see unusual fluorescence behaviour. This behaviour is similar for conditions I, II and III (Results for II and III are shown in Fig. 2). For example, in the case of condition III (Fig. 2B) the fluorescence emission intensity at 345 nm increases from low to moderate concentrations (0 - 2.5 M) of GdnHCl, decreases from moderate to high concentrations (2.5 - 3.5 M), then increases slowly at high to very high concentrations of GdnHCl (3.5 - 6.5 M). At the same time the λ_{max} shifts to shorter wavelengths (345 - 337 nm) at low to moderate concentrations of denaturant, then to longer wavelengths (337 - 353 nm) at moderate to high concentrations of denaturant, with no change at high to very high concentrations of denaturant. The behaviour at moderate to high concentrations of denaturant is consistent with the exposure of the indole side-chain of a Trp residue to bulk solvent (i.e., the major unfolding transition of the protein) (12). The behaviour at low to moderate concentrations of denaturant, indicates a state which differs to both the native and denatured states, and is consistent with the shielding of a Trp indole side-chain from the solvent (12). This probably results from an intermolecular association which protects Trp from the solvent. The maxima of the fluorescence emission intensity and minima of λ_{max}, correspond to the midpoints of the biphasic transitions from the equivalent CD experiments, suggesting that they represent characterisable unfolding intermediates.

2. Denaturing Condition IV

The behaviour of the mIL-6 at pH 4.0, urea-mediated unfolding (IV), is less complex (Fig. 3). The fluorescence emission intensity increases with increasing urea concentration, without a clear unfolding transition, but λ_{max} shows a sigmoidal red-shift whose midpoint (3.4 ± 0.2 M) corresponds with the $[D]_{50\%}$ determined by CD spectroscopy (3.6 ± 0.2 M). This coincidence of data from fluorescence and CD spectroscopy adds weight to our supposition that the unfolding transition under these conditions is two-state.

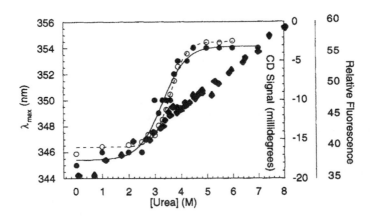

Fig. 3: *Superimposition of far-UV CD monitored and fluorescence monitored urea-induced denaturation of mIL-6 at pH 4.0.* λ_{max} (—●—), raw CD data at 222 nm (- -o- -) and relative fluorescence emission at 345 nm (■), were plotted against denaturant concentration. λ_{max} and raw CD data were fitted to Eqn 1. Adapted from Fig. 2 (8).

C. Effects of Salt on the Unfolding Transition at pH 4.0, Urea

The major difference between the two denaturants used in this study is that GdnHCl is ionic, whereas urea is not. We wanted to determine whether the different behaviour of mIL-6 in these two denaturants at pH 4.0 (III and IV) was due to ionic effects. Previously, we carried out fluorescence experiments at pH 4.0, using urea as the denaturant, in the presence of 0.4 M and 0.8 M salt, where the salt was NaCl or GdnHCl (8). Under these conditions the unusual fluorescence behaviour, described above for conditions I, II and III, was observed. The effects were more marked at 0.8 M compared to 0.4 M salt, and occurred at slightly lower urea concentrations when GdnHCl rather than NaCl was present, which is expected for the addition of extra denaturant.

These studies were extended to include far-UV CD monitored urea-induced unfolding of mIL-6, at pH 4.3, with the addition of 50 mM, 100 mM and 400 mM NaCl (Fig. 4; Table II). At this pH the protein is more resistant to denaturation by urea, but the unfolding transition in the absence of NaCl remains apparently two-state. With the addition of NaCl the unfolding transition appeared to be non two-state, with increasing deviation from two-state behaviour, reflected by decreasing values of m and $\Delta G_U(H_2O)$ (Table II), occurring with increasing concentrations of salt.

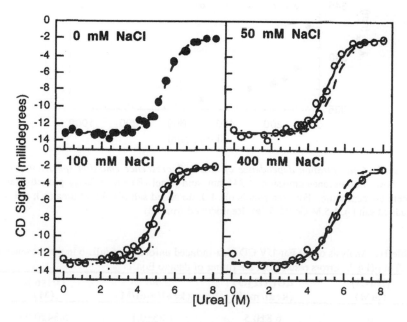

Figure 4: *Effect of NaCl on the urea-induced equilibrium unfolding of mIL-6 at pH 4.3, monitored by far-UV CD at 222 nm.* Amount of added NaCl is indicated. The data and curve fit for mIL-6 in the absence of NaCl (- -●- -, top left panel) are shown in subsequent panels (- - · - -) for comparison.

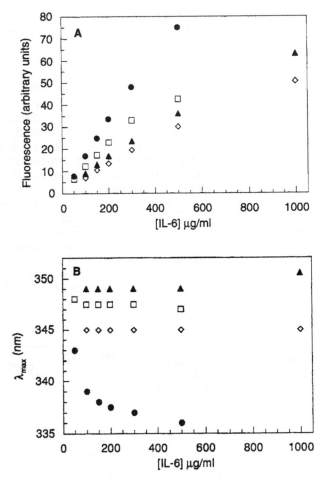

Figure 5: *Concentration dependence of mIL-6 fluorescence emission spectrum.* Panel (A) shows fluorescence emission at 345 nm, and panel (B) shows λ_{max} plotted against protein concentration. For both panels pH 4.0, no added salt (◊), 3.5 M urea (▲); pH 7.4, no added salt ( ), 1,5 M GdnHCl (●). Reproduced from (8).

Table II: Analysis of the Far-UV CD urea-induced unfolding of mIL-6 in the presence of NaCl at pH 4.3. Errors result from the fitting of data to Eqn 1

[NaCl] (mM)	$\Delta G_U(H_2O)$ (kcal/mol)	m (kcal/mol/M)	$[D]_{50\%}$ (M)
0	6.8±0.5	1.25±0.1	5.34±0.05
50	6.1±0.6	1.21±0.1	5.02±0.06
100	5.3±0.4	1.07±0.1	4.97±0.05
400	5.1±0.3	0.92±0.2	5.50±0.15

D. Concentration Dependence of Intermediate Formation

To determine whether intermediate formation was concentration dependent, we monitored the fluorescence emission spectrum of mIL-6 as a function of protein concentration (Fig. 5). The fluorescence emission intensity at each denaturing condition studied showed an essentially linear dependence on mIL-6 concentration over the range 50 to 1000 µg/ml protein (Fig. 5A), however, the concentration dependence of λ_{max} was critically dependent on solution conditions (Fig 5B). Most notably, at pH 7.4 in 1.5 M GdnHCl there is a significant non-linear blue shift from 343 to 336 nm. In addition, under these conditions, the fluorescence emission signal at pH 7.4 is 1.9-fold greater than in the absence of GdnHCl. This is consistent with a concentration dependent association of mIL-6 at intermediate concentrations of GdnHCl at pH 7.4.

IV. Conclusion

A. A Model for Intermediate Formation

Fluorescent properties of the aggregates observed in NMR experiments (protein concentration 15 - 20 mg/ml) are identical to the intermediates seen in these equilibrium experiments (protein concentration 100 µg/ml) (L.D.W., unpublished data), suggesting that the intermediate formation in both cases is the same.

Conformational states intermediate between folded and unfolded states have been observed for several other members of the 4-α-helical-bundle family, including bovine GH (13), porcine GH (14) and G-CSF (15). The properties described here for mIL-6 resemble those of bovine GH for which the following equilibrium unfolding model has been proposed (16).

$$N \leftrightarrow I \leftrightarrow D$$
$$\uparrow\downarrow$$
$$I_{agg}1 \leftrightarrow I_{agg}2 \rightarrow Precipitate$$

Brems, Havel and coworkers propose that at intermediate concentrations of GdnHCl (2-3 M), bovine GH partially unfolds to a state (I), which has a high tendency to aggregate (13,17,18). This unfolding is thought to involve the exposure of the hydrophobic face of helix C to form a "molten globule"-like state (13,19,20). The equilibrium unfolding of mIL-6 probably follows a similar process, although significant precipitation of mIL-6 appears to occur only at elevated temperatures (J.M.M., unpublished).

B. Aggregation of Partially Unfolded Intermediates

We have observed that the intermediate formation of mIL-6 is pH and denaturant dependent (8), that it is increased in the presence of salt (either GdnHCl or NaCl) and that the concentration of intermediates formed is protein concentration dependent. It is probable that mIL-6 is less prone to the aggregation of unfolding intermediates at pH 4 than pH 7.4 because the protein is more charged at acidic pH (estimated molecular charge = +16), than at neutral pH (estimated molecular charge = 0). The presence of salt at pH 4.0, at even low (50 mM) concentrations, appears to mask this charge, allowing aggregation to occur.

References

1. Mott, H.R. & Campbell, I.D. (1995) *Curr. Op. Struct. Biol.* 5, 114-121.
2. deVos A.M., Ultsch, M. & Kossiakoff, A.A. (1992) *Science 255*, 306-312.
3. Robinson, R.C., Grey, L.M., Staunton, D., Vankelecom, H., Vernallis, A.B., Moreau, J.F., Stuart, D.I., Heath, J.K. & Jones E.Y. (1994) *Cell 77*, 1101-1116.
4. Hill, C.P., Osslund, T.D. & Eisenberg, D. (1993) *Proc. Natl. Acad. Sci. U.S.A.* 90, 5167-5171.
5. McDonald, N.Q., Panayotatos, N. & Hendrickson, W.A. (1995) *EMBO J. 14*, 2689-2699.
6. Hirano, T., Akira, S. Taga, T. & Kishimoto, T. (1990) *Immunol. Today 11*, 443-449.
7. Morton, C.J., Simpson, R.J., & Norton, R.S. (1994) *Eur. J. Biochem. 219*, 97-107.
8. Ward, L.D., Matthews, J.M., Zhang, J.G. & Simpson, R.J. (1995) *Biochemistry*, In press.
9. Simpson, R.J., Moritz R.L., Rubira, M.R. & Van Snick, J. (1988) *Eur. J. Biochem. 176*, 187-197.
10. Santoro M.M. & Bolen, D.W. (1988) *Biochemistry 27*, 8063-8068.
11. Brems D.N., Plaisted, S.M., Havel, H.A., Kaufmann, W., Stodola, J.D., Eaton, L.C. & White, R.D. (1985) *Biochemistry 24*, 7662-7668.
12. Lakowicz, J.R. (1983) in *Principles of Fluorescence Spectroscopy*, Plenum Press, New York.
13. Brems, D.N., Plaisted, S.M., Kauffman, E.W. & Havel, H.A. (1986) *Biochemistry 25*, 6539-6543.
14. Bastiras, S. & Wallace, J.C. (1992) *Biochemistry 31*, 9304-9309.
15. Nahri, L.O., Kenney, W.C. & Arakawa, T. (1991) *J. Prot. Chem. 10*, 359-367.
16. Lehrman, S.F., Tuls, J.L., Havel, H.A., Haskell, R.J., Putnam, S.D. & Tanich, C-S.C. (1991) *Biochemistry 30*, 5777-5784.
17. Havel, H.A., Kauffman, E.W., Plaisted, S.M. & Brems, D.N. (1986) *Biochemistry 25*, 6533-6538.
18. Brems, D.N. (1988) *Biochemistry 27*, 4541-4546.
19. Gooley, P.R., Plaisted, S.M., Brems, D.N., & Mackenzie, N.E. (1988) *Biochemistry 27*, 802-809.
20. Brems, D.N., & Havel, H.A. (1989) *Proteins, Struct, Funct. Genet. 5*, 93-95.

Structural Stability of Small Oligomeric Proteins

Craig R. Johnson and Ernesto Freire

Department of Biology and Biocalorimetry Center, The Johns Hopkins University, Baltimore, Maryland 21218

I. Introduction

The thermodynamic characterization of small oligomeric proteins is becoming of increasing interest as efforts to define the forces responsible for protein folding and protein small molecule binding continue to increase. Recently, our laboratory has determined the temperature induced unfolding stabilities of peptide fragments containing the oligomerization domain of the p53 tumor suppressor (p53tet) [Johnson et al., 1995] and the leucine zipper region of the bZIP transcription factor GCN4 [Thompson et al., 1993]. These relatively small proteins form structured oligomers in solution and demonstrate many of the structural properties associated with both protein folding and small peptide-protein interactions. In general, as the size of an oligomeric protein with extensive intersubunit interactions decreases, the contribution to the overall stability attributed to these interactions takes precedence over those of the intrasubunit interactions. The magnitude of these effects can be investigated by studying oligomers like p53tet and the leucine zipper region of GCN4, utilizing methods that determine the structural stability of native and intermediate states that may manifest as the protein is destabilized. Differential scanning calorimetry (DSC) and circular dichroism (CD) are experimental methods that are capable of examining the thermodynamic characteristics of such states.

By investigating a number of structural motifs, particular stabilizing forces and their relation to structural properties can be dissected. Recently, the NMR solution structure of p53tet (64 aa) was determined [Lee et al., 1994] revealing a symmetric 29.8 kilodalton tetramer composed of a dimer of dimers with both β-sheet and α-helical content. This protein contains the entire tetramerization domain plus an additional 34 amino acids that are thought to be mostly unstructured. The GCN4 leucine zipper consists of a coiled coil of 33 amino acids. The N-terminal region of this construct contains a region rich in basic residues that acts as a DNA binding site [O'Shea et al., 1989; O'Shea et al., 1991]. Structural studies of these molecules indicate that a substantial portion of the solvent accessible surface area exposed upon unfolding is found in the interfaces between subunits. In fact, these interfaces are stabilized largely through hydrophobic interactions. In our calorimetric and circular dichroism studies, these oligomers have been found to have highly stable structures displaying cooperative two state thermal unfolding processes. In this paper, we outline the experimental methods used to thermodynamically characterize these types of oligomeric systems and summarize general conclusions regarding the mechanisms of stabilization of these proteins.

II. Methods

A. Differential Scanning Calorimetry (DSC)

Calorimetric experiments were conducted using the newly developed series of solid state high precision differential scanning calorimeters DS-92, DS-93 (Biocalorimetry Center, The Johns Hopkins University, Baltimore, MD) and DASM-4M. These instruments have been carefully constructed to allow for absolute heat capacity determination and are computer interfaced for automated instrument control and data collection. Proteins were scanned at 1 °C min^{-1} with repeated scans revealing approximately 98% reversibility. Samples and reference solutions were properly degassed and carefully loaded into the calorimeter to eliminate bubbling effects. Data analysis of absolute and excess heat capacity measurements were performed using the nonlinear least squares fitting software developed in this laboratory and modified for dissociative unfolding. [Freire & Biltonen, 1978; Montgomery et al., 1993; Thompson et al., 1993; Xie et al., 1994].

B. Circular Dichroism Spectroscopy (CD)

CD experiments were conducted using the computer automated Jasco-710 spectropolarimeter modified to operate at variable temperatures. Wavelength scans were performed in a 5 mm rectangular cell in the appropriate temperature range. Complete spectra were obtained at specific temperatures by taking the average of five consecutive scans collected from 190 to 240 nm at 1nm intervals. For these experiments the instrument was set with a scan rate of 5 nm min^{-1}, a response time of 8 s per point and a bandwidth of 1 nm. The mean residue elipticity is computed by subtracting buffer scans from those of the sample. Temperature was controlled by placing the cell in a water-jacketed cell holder utilizing a Haake F3 circulating water bath. Temperature readings were obtained by using a S/N117.C temperature probe connected to a Micro-therm 1006 thermometer placed in contact with the sample (Hart Scientific).

A series of CD temperature scans (scan rate 1 °C min^{-1}) were also performed using the time scan mode of the Jasco J710 by scanning continuously as the temperature is varied from about 10 to 80 °C. Using a response time of 0.5 s and a bandwidth of 1 nm, elipticity and temperature values were recorded every 20 seconds at a wavelength of 222 nm. The temperature was controlled by a Haake PG 20 temperature programmer interfaced to the Haake F3 circulating water bath. Complete elipticity (θ) vs temperature curves were generated for a series of known protein concentrations. The fraction of unfolded tetramer F_U was calculated from the elipticity by the standard relationship

$$F_U = (\theta - \theta_N)/(\theta_U - \theta_N) \tag{1}$$

θ_N and θ_U represent the elipticity values at temperatures where the fully folded and fully unfolded states exist respectively.

III. Theory

The analysis of two state concentration dependent self-dissociating thermal unfolding can be performed by describing the system in an elementary statistical mechanical framework [Freire, 1989; Thompson et al., 1993; Johnson et al.,

1995]. We outline here the simplest case, that of a folded oligomer of i sequence identical monomeric units (N_i) unfolding to i separate unfolded monomers (U) upon increasing temperature with no intermediate states ($N_i \leftrightarrow iU$). The equilibrium constant is defined as

$$K = \frac{[U]^i}{[N_i]} \tag{2}$$

The total protein monomer concentration (P_T) is

$$P_T = i[N_i] + [U] \tag{3}$$

The fractional populations are given by

$$F_N = \frac{i[N_i]}{P_T} \qquad F_U = \frac{[U]}{P_T} \tag{4, 5}$$

Substitution and rearrangement of these equations results in

$$F_U^i + \frac{KF_U}{iP_T^{(i-1)}} - \frac{K}{iP_T^{(i-1)}} = 0 \tag{6}$$

The roots of this equation give the value of the fraction of molecules in the unfolded state as a function of the equilibrium constant and the total protein concentration. The value of the total protein concentration is known and by definition

$$K = \exp\left(\frac{-\Delta G^\circ}{RT}\right) \tag{7}$$

where ΔG° is the intrinsic free energy of stabilization per mole of i mer defined in the standard way

$$\Delta G^\circ = \Delta H^\circ(T^\circ) + \Delta C_p(T - T^\circ) - T\left[\Delta S^\circ(T^\circ) + \Delta C_p \ln\left(\frac{T}{T^\circ}\right)\right] \tag{8}$$

Here T° is the reference temperature which in our work has been chosen as the temperature at which $\Delta G^\circ = 0$. $\Delta H^\circ(T^\circ)$ and $\Delta S^\circ(T^\circ)$ are the enthalpy and entropy changes evaluated at the temperature T°, respectively, and ΔC_p is the change in heat capacity upon unfolding. It should be noted that for an oligomeric protein ($i \geq 2$) the temperature at which $\Delta G^\circ = 0$ does not coincide with the temperature at which the transition is half completed or the temperature of the peak maximum in the calorimetric scan.

DSC data is obtained in the form of the excess heat capacity ($<\Delta C_p>$) verses temperature [Freire & Biltonen, 1978]. This function can be generated by taking the temperature derivative of the average excess enthalpy function ($<\Delta H>$) given by

$$< \Delta H > = F_U \Delta H^\circ(T) \tag{9}$$

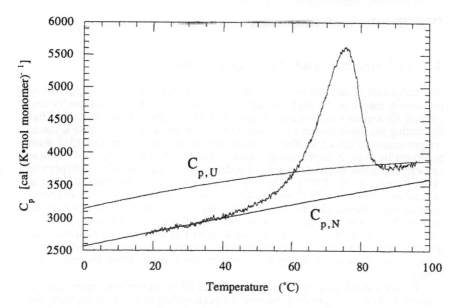

Figure 1: Partial molar heat capacity of p53tet as a function of temperature. The experiment was performed at pH 4.0 in 25 mM sodium acetate with a tetramer concentration of 146 μM. The solid lines represent the heat capacities of the native ($C_{p,N}$) and unfolded ($C_{p,U}$) states respectively (adapted from [Johnson et al., 1995]).

Here $\Delta H^{o}(T)$ is the total unfolding enthalpy at temperature T.

$$< \Delta C_p > = \frac{\partial < \Delta H >}{\partial T} = \Delta H^{o}(T)\frac{\partial F_U}{\partial T} + F_U \Delta C_p \qquad (10)$$

After appropriate substitution the calorimetric data is analyzed by nonlinear least squares optimization. In general, the equation for F_U is solved numerically for i > 2. If the model is correct a complete thermodynamic description of the unfolding event is obtained. If necessary, partly folded intermediates can be introduced in the equations as demonstrated before [Thompson et al., 1993].

Several characteristics distinguish unfolding transitions involving oligomeric proteins from those of monomeric proteins. First, these transitions are concentration dependent. For an oligomer that dissociates upon unfolding, the transition temperature increases upon increasing the protein concentration. It must be realized also, that the definition of the transition temperature must be explicitly given since for oligomeric proteins, the temperature at which the transition is half completed, the temperature of the maximum in the heat capacity function and the temperature at which $\Delta G = 0$ are not the same. For example, for a dimeric protein the maximum in the excess heat capacity function occurs when $F_U = (2 - 2^{1/2})$; and the value of K when the transition is half completed is equal to P_T. The exact concentration dependence of the transition is obtained by solving equation 6 numerically since an analytical solution is not available due to the fact that the two quantities that define K (ΔH and ΔS) are temperature dependent.

IV. Results and Discussion

A. *Calorimetric and CD Experiments*

The temperature induced unfolding of p53tet and GCN4 leucine zipper were previously measured by high sensitivity differential scanning calorimetry and circular dichroism spectroscopy. Figure 1 shows a typical DSC scan of p53tet illustrating the linear increase in absolute heat capacity of the native state at lower temperatures, followed by the transition region and finally the unfolded state nonlinear increase at higher temperatures. The absolute magnitude of the heat capacity is a good index of the level of hydration and structure of the polypeptide chain [Gomez et al., 1995]. The heat capacity of the native state is 2.8 kcal $(K \cdot mol\ monomer)^{-1}$ at 25°C and is slightly higher than what would be expected for a globular protein of similar molecular weight. This observation is consistent with the NMR structural determination indicating that close to 30 amino acids per monomer are unstructured in the native state of this protein. As expected, the absolute heat capacity of the unfolded state is similar to that of an unstructured polypeptide of the same amino acid composition [Freire, 1994; Gomez et al., 1995].

Figure 2 displays a series of DSC scans of p53tet at various concentrations along with their corresponding theoretical fits according to equation 10, using the solution of equation 6 for a tetramer (Table I). In this figure, the excess heat capacity function obtained by subtracting the heat capacity of the native state

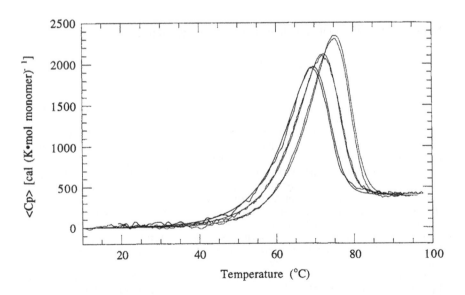

Figure 2: Excess heat capacity of p53tet as a function of protein concentration. Experiments were performed in 25 mM sodium acetate at pH 4.0. Tetramer concentrations are from left to right 70.5 μM, 93.5 μM and 145.8 μM. Theoretical curves generated from the fitted parameters given in Table I are shown as solid lines along with their corresponding experimental data (adapted from [Johnson et al., 1995]).

464   Craig R. Johnson and Ernesto Freire

Table I: Thermodynamic parameters of p53tet unfolding as a function tetramer concentration Solution conditions are 25 mM sodium acetate at pH 4.0.

p53tet Tetramer Conc. [μM]	Tm (°C)	T° (°C) [a]	ΔH (T°) [kcal (mol monomer)$^{-1}$]	ΔS (T°) [cal (K•mol monomer)$^{-1}$]	ΔCp (T°) [cal (K•mol monomer)$^{-1}$]	SSR [b]
70.5	69.5	110.8	46.2	120.4	399	68.9
93.5	71.8	110.0	46.8	122.2	405	20.3
145.8	75.3	110.0	47.1	123.0	387	32.1

[a] $T°$ is the reference temperature at which the intrinsic free energy $\Delta G°$ is equal to zero.
[b] SSR is the sum of the square of the residuals.

clearly shows the transition temperature increase with concentration. It is also apparent that the transition curves are skewed to the low temperature side of the transition which is a characteristic feature for an unfolding transition coupled to dissociation [Freire, 1989]. The data is well represented by a two state transition in which the folded tetramer is converted directly to unfolded monomers ($N_4 \leftrightarrow 4U$). This type of two state behavior is also observed for dimeric GCN4 leucine zipper [Thompson et al., 1993]. Transition curves at p53tet concentrations lower than those shown in Figure 2 were examined by CD as the calorimetric response

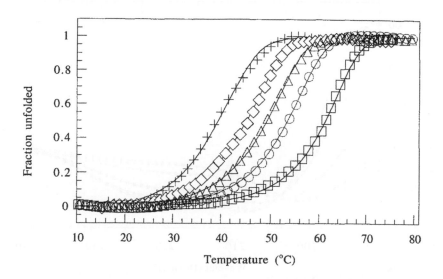

Figure 3: CD data showing the fraction of p53tet unfolded as a function of temperature at pH 4.0. Concentrations are from left to right 0.5 μM (plus signs), 1.25 μM (diamonds), 2.5 μM (triangles), 5 μM (open circles) and 20 μM (squares) tetramer (adapted from [Johnson et al., 1995]).

does not allow accurate measurements at lower concentrations. These measurements were performed, as outlined above, and are displayed in Figure 3. The figure shows the experimentally determined values along with the calculated curves generated from the parameters obtained calorimetrically at higher concentrations under the same solution conditions (Table I).

Figure 4 displays a series of CD wavelength scans of p53tet from 190 to 240 nm at a number of different temperatures at a concentration of 2.5 μM tetramer. The curves are found to be well described by a population weighted linear combination of the pure component conformational spectra for α-helix, β-sheet and random coil. The presence of a well defined isodichroic point is consistent with a system having only two states; in this case, native tetramers and unfolded monomers. At low temperatures the structure is approximately 40% random coil with the remaining structure having both α-helical and β-sheet like character, consistent with the NMR structural determination. As the temperature is increased the proportion of α-helix and β-sheet decreases eventually revealing only random coil.

B. Structural Determinants of Stability

The extent of polar and apolar surfaces that are buried from the solvent are a major determinant of the stability of the native state of proteins since they reflect the extent of non-bonded interactions (van der Waals, hydrogen bonding) within the protein and the loss of interactions with the solvent upon folding. Protein accessible surface area calculations for p53tet and GCN4 leucine zipper were

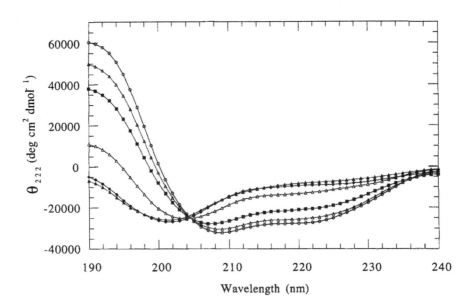

Figure 4: CD spectra of p53tet as a function of temperature. Experimental conditions are 2.5 μM tetramer in 25 mM sodium acetate at pH 4.0. Temperatures are from partially structured to unstructured 21.3 °C, 39.3 °C, 45.0 °C , 51.1 °C, 59.4 °C and 70.4 °C (adapted from [Johnson et al., 1995]).

Table II: Calculated Changes in Accessible Surface Area (per monomer) for p53tet Unfolding

	ΔASA_{Total} $\mathring{A}^2$	ΔASA_{ap} $\mathring{A}^2$	ΔASA_{pol} $\mathring{A}^2$	%Apolar
Subunit Interface Dissociation	1616	1152	464	71%
Unfolding of Isolated Monomer	1266	750	516	59%
Total for Unfolding of Tetramerization Domain	2882	1902	980	66%

Table III: Calculated Changes in Accessible Surface Area (per monomer) for GCN4 Zipper Region Unfolding

	ΔASA_{Total} $\mathring{A}^2$	ΔASA_{ap} $\mathring{A}^2$	ΔASA_{pol} $\mathring{A}^2$	%Apolar
Subunit Interface Dissociation	852	687	165	81%
Unfolding of Isolated Monomer	1798	783	1015	44%
Total for Unfolding of the Dimer	2649	1470	1180	56%

obtained as previously described [Murphy et al., 1992] using the implementation of the Lee and Richards algorithm [Lee & Richards, 1971] in the program ACCESS (Scott R. Presnell, University of California) with a probe radius of 1.4 Å and a slice width of 0.25 Å. These parameters are summarized in Tables II and III. For p53tet approximately 56% of the total buried surface area resides at the interface between subunits compared with 32% found for GCN4 leucine zipper. Most of the surface buried at the interfaces (71% for p53tet and 81% for GCN4 leucine zipper) is apolar indicating that the driving force for oligomerization is hydrophobic. In contrast the surface area buried by isolated monomers is only 59% apolar for p53tet and 44% apolar for GCN4 leucine zipper. Proportionally, p53tet buries a larger fraction of apolar surface (66%) than most globular proteins (~59%) with GCN4 zipper region burying 56%.

According to the calorimetric data, the intrinsic free energy of stabilization for p53tet and GCN4 leucine zipper at 25 °C are 6.5 and 8.5 kcal (mol monomer)$^{-1}$ respectively. Within these totals, the entropic contributions due to solvent reorganization (hydrophobic effect) to the free energy of stabilization are approximately 30 kcal (mol monomer)$^{-1}$ for p53tet and 20 kcal (mol monomer)$^{-1}$ for GCN4 leucine zipper. In both cases, the free energy contributions of the dimer interface are larger than the overall free energy of stabilization indicating that the isolated monomers of these proteins are not thermodynamically stable. This result emphasizes the role of intersubunit interactions in the stabilization of these two proteins. It appears that oligomerization might not be of functional importance only, but play an important role in the stabilization of otherwise unstable structures.

V. Conclusions

Using the techniques of differential scanning calorimetry and circular dichroism, it is possible to thermodynamically characterize the temperature induced unfolding of oligomeric proteins. It has been demonstrated that the thermal unfolding of both p53tet and GCN4 leucine zipper can be described by two state processes in which the folded oligomers unfold cooperatively into unfolded monomers. Under all conditions studied the concentration of intermediate species consisting of folded dimers or monomers for p53tet or folded monomers for GCN4 could not be detected. This description is consistent with the observation that the majority of interfacial contacts in these oligomers are hydrophobic, making isolated dimeric or monomeric structures highly unstable. The hydrophobic effect appears to be a major stabilizing force for both of these oligomers. These results point to a stabilization mechanism whereby intrinsically unstable structures are stabilized by quaternary structural interactions.

Acknowledgment

This work was supported by NIH grant RR-04328.

References

Freire, E. (1989) *Comments Mol. Cell. Biophys.* **6**, 123-140.

Freire, E. (1994) *Meth. Enzym.* **240**, 502-568.

Freire, E. & Biltonen, R. L. (1978) *Biopolymers* **17**, 463-479.

Gomez, J., Hilser, V. J., Xie, D., & Freire, E. (1995) *Proteins* in press.

Johnson, C. R., Morin, P. E., Arrowsmith, C. H., & Freire, E. (1995) *Biochemistry* **34**, 5309-5316.

Lee, B. & Richards, F. M. (1971) *J. Mol. Biol.* **55**, 379-400.

Lee, W. T., Harvey, T. S., Yin, Y., Yau, P., Litchfield, D., & Arrowsmith, C. H. (1994) *Nature Struct. Biol.* **1**, 877-890.

Montgomery, D., Jordan, R., McMacken, R., & Freire, E. (1993) *J. Mol. Biol.* **232**, 680-692.

Murphy, K. P., Bhakuni, V., Xie, D., & Freire, E. (1992) *J. Mol. Biol.* **227**, 293-306.

O'Shea, E. K., Klemm, J. D., & Kim, P. S. (1989) *Science* **243**, 538-542.

O'Shea, E. K., Klemm, J. D., & Kim, P. S. (1991) *Science* **254**, 539-544.

Thompson, K. S., Vinson, C. R., & Freire, E. (1993) *Biochemistry* **32**, 5491-5496.

Xie, D., Fox, R., & Freire, E. (1994) *Protein Science* **3**, 2175-2184.

The Conformational Consequences of Mutations to the H1 Helix of the Prion Protein Explored by Molecular Dynamics Simulations

Steven L. Kazmirski, Darwin O. V. Alonso, Fred E. Cohen, Stanley B. Prusiner, and Valerie Daggett

Department of Medicinal Chemistry
University of Washington
Seattle, Washington 98195-7610
and
Department of Pharmaceutical Chemistry
Department of Medicine
Departments of Cellular and Molecular Pharmacology
Department of Biochemistry and Biophysics
Department of Neurology
University of California, San Francisco
San Francisco, California 94143

I. Introduction

Prion diseases describe a wide range of fatal neurodegenerative disorders. In humans, prion diseases have been subdivided into four classes and are referred to as Creutzfeldt-Jakob disease (CJD), Gerstmann-Sträussler-Scheinker syndrome (GSS), and Fatal Familial Insomnia (FFI) (DeArmond and Prusiner, 1995). These can be sporadic, inherited, or infectious disorders. They differ from other infectious diseases in that the pathogen is a proteinaceous particle (infectious prion) (Prusiner, 1991), and the essential component of prions is the scrapie prion protein (PrPSc). PrPSc is conformationally indistinguishable from the normal cellular prion protein (PrPC) (Hsiau et al., 1992). However, the secondary and tertiary structures differ (Basler et al., 1986; Stahl and Prusiner, 1991; Caughey and Raymond, 1991; Pan et al., 1993; Kocisko et al., 1994). Difficulties with the purification of PrP and its propensity to aggregate have prevented the use of X-ray crystallography and multi-dimensional nuclear magnetic resonance (NMR) spectroscopy for structure determination of PrPSc and PrPC. However, Fourier transform infrared (FTIR) and circular dichroism (CD) spectroscopy studies have provided some information (Pan et al., 1993; Safar et al., 1993). For example, these studies indicate that PrPC is highly helical (42%) with virtually no β-sheet structure (3%). In contrast, PrPSc contains a large amount of β-sheet structure (43%) and less helical structure (30%). These results, together with those of structure prediction studies, suggest that a conversion of α-helices to β-sheets may occur upon formation of PrPSc from PrPC (Gasset et al., 1992; Pan et al., 1993).

Modeling studies predict that PrPC is a four-helix bundle (Huang et al., 1995). Many of the mutations in the prion protein leading to prion diseases lie in or near these four putative helices,

ADVANCES IN PROTEIN CHEMISTRY, Vol. 57

Copyright © 2001 by Academic Press.
All rights reserved.

The Conformational Consequences of Mutations to the H1 Helix of the Prion Protein Explored by Molecular Dynamics Simulations

Steven L. Kazmirski[1], Darwin O. V. Alonso[1, 2], Fred E. Cohen[2, 3, 4, 5]
Stanley B. Prusiner[5, 6], and Valerie Daggett[1*]

[1] Department of Medicinal Chemistry
University of Washington
Seattle, Washington 98195-7610
and
[2] Department of Pharmaceutical Chemistry
[3] Department of Medicine
[4] Department of Cellular and Molecular Pharmacology
[5] Department of Biochemistry and Biophysics
[6] Department of Neurology
University of California, San Francisco
San Francisco, California 94143

I. Introduction

Prion diseases describe a wide range of fatal neurodegenerative disorders. In humans, prion diseases have been subdivided into four classes and are referred to as kuru, Creutzfeldt-Jakob disease (CJD), Gerstmann-Sträussler-Scheinker disease (GSS), and fatal familial insomnia (FFI) (DeArmond and Prusiner, 1995). These diseases can be sporadic, inherited, or infectious disorders. They differ from other infectious diseases in that the pathogen is a proteinaceous particle, termed a prion (Prusiner, 1991), and the essential component of prions is the scrapie prion protein (PrP^{Sc}). PrP^{Sc} is chemically indistinguishable from the normal, cellular prion protein (PrP^{C}) (Stahl et al., 1993); however, their secondary and tertiary structures differ (Basler et al., 1986; Stahl and Prusiner, 1991; Caughey and Raymond, 1991; Pan et al., 1993; Kocisko et al., 1994).

Difficulties with the purification of PrP and its proclivity to aggregate have prevented the use of X-ray crystallography and multi-dimensional nuclear magnetic resonance (NMR) spectroscopy for structure determination of PrP^{C} and PrP^{Sc}. However, Fourier transform infrared (FTIR) and circular dichroism (CD) spectroscopy studies have provided some information (Pan et al., 1993; Safar et al., 1993). For example, these studies indicate that PrP^{C} is highly helical (42%) with virtually no β-sheet structure (3%). In contrast, PrP^{Sc} contains a large amount of β-sheet structure (43%) and less helical structure (30%). These results, together with those of structure prediction studies, suggest that a conversion of α-helices to β-sheets may occur upon formation of PrP^{Sc} from PrP^{C} (Gasset et al., 1992; Pan et al., 1993).

Modeling studies predict that PrP^{C} is a four helix bundle (Figure 1) (Gasset et al., 1992; Huang et al., 1994). Many of the mutations in the prion protein leading to prion diseases lie in or near these four putative helices

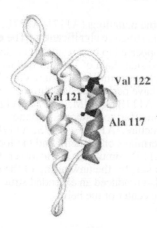

Figure 1. Favored three-dimensional model of the four helix bundle of PrP^C (Huang et al., 1994) containing residues 109-218. The H1 helix is shown in dark gray and specific residues are colored black and labeled. The C_β of Ala 117 is displayed in the middle of H1.

(DeArmond and Prusiner, 1995). For example, the mutation of an alanine to valine at position 117 of H1 (residues 109-122, MKHMAGAA**A**AGAVV) leads to GSS in humans (Doh-ura et al., 1989; Hsiao et al., 1991). The wild type H1 peptide adopts a helical structure under various solvent conditions (Gasset et al., 1992; Zhang et al., 1995; Nguyen et al., 1995). We have previously performed molecular dynamics (MD) simulations of the this peptide at 298 K and have shown that the valine mutation destabilizes the helical state leading to the formation of a more extended structure (Kazmirski et al., 1995) (Figure 2). In

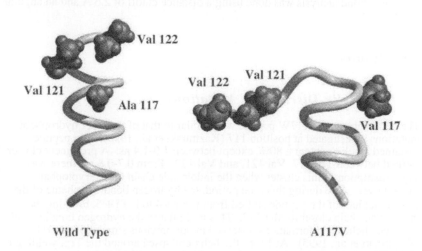

Wild Type   A117V

Figure 2. Final structures of the 2 ns simulation of the wild type and A117V H1 peptides. The wild type peptide maintained a helical conformation, whereas the A117V mutant unfolded at the C-terminus leading to the extended strand. Val 121 and Val 122 are shown in all structures for reference.

contrast, other hydrophobic mutations (A117I, A117L, A117Nle, and A117F) did not perturb the peptide structure significantly. The destabilization caused by the Ala → Val mutation appeared to be a result of the decreased conformational freedom of the β-branched side chain of Val 117 and its inability to form a helix stabilizing hydrophobic cluster with Val 121 and Val 122.

In this study, we have performed MD simulations of nine new mutants of the H1 peptide (A117G, A117P, A117M, A117D, A117E, A117W, A117S, A117N, and A117T) in an effort to better understand how changes at position 117 destabilize the helical structure. A117D, A117S, A117N, and A117T lead to a destabilization of the C-terminus of the helix and the formation of extended structure similar to the A117V simulations. However, these four simulations display different mechanisms for the unwinding of the C-terminus. Furthermore, the A117M simulation also produced an extended structure; however, the disruption occurred in the center of the helix.

II. Methods

All MD simulations were performed using the program ENCAD (Levitt, 1990) and a previously described force field (Levitt, 1983, 1989; Levitt et al., 1995). The starting conformation for all the simulations was an ideal helix built by ENCAD containing the appropriate H1 peptide sequence. Minimization and addition of solvating water molecules was performed as described previously (Kazmirski et al., 1995). Each simulation contained ~1000 water molecules (the numbers differed slightly for each simulation). Periodic boundary conditions were employed throughout each of the 2 ns simulations.

For analysis, the percentage of secondary structure was monitored, via (φ,ψ) angles, throughout the simulation using a previously described method (Daggett and Levitt, 1992; Daggett et al., 1991). The helix content listed in Table I also requires that at least 3 consecutive residues adopt local helical structure. Hydrogen bond analysis was done using a distance cutoff of 2.6 Å and an angular cutoff of 35°.

III. Results

A. Non-Helix-Disrupting Mutations

The behavior of the A117W peptide was similar to that of the other hydrophobic mutations investigated at position 117 (Kazmirski et al., 1995). The peptide sustained a helix content ≥ 80%, except between 1.0-1.4 ns. A prominent cluster formed between Trp 117, Val 121, and Val 122. From 0.7-0.8 ns, there was a slight disruption of this cluster when the indole side chain of the tryptophan flipped over. Also during this time period, the hydrogen bonding scheme of the C-terminal half of the peptide shifted from i → i + 4 to i → i + 5, bringing the indole side chain closer to Met 112. This is similar to the hydrogen bonding and "relaxed" helical conformations observed in our previous simulations of H1 (Kazmirski et al., 1995). At 1.0 ns the helix collapsed around the Trp, yielding a major cluster composed of the residues cited above and Met 109. As a result, a large kink developed centered at Ala 115. After 0.4 ns, however, water disrupted the cluster and the helix reformed (~80%) and was maintained (Figure 3).

Over the 2 ns simulation of the A117G peptide, there were no major structural changes. Although the helix content dropped to 65%, the peptide remained coiled (kinked and compact, not random coil) (Table I). Ala 120 showed a large decline in helix content, averaging ~30% from 0.4-0.6 ns, ~10%

Table I. Summary of H1 Prion Peptide Simulations Performed and the α-Helix Content over the Final 0.2 ns of Each 2 ns Simulation

Peptide	α-Helix Content (%)	Extended Structure Observed
WT[a]	83	
A117V[a]	39	√
A117V-2[a]	41	√
A117W	84	
A117G	65	
A117P	49	
A117E	75	
A117M	53	√
A117D	31	√
A117N	51	√
A117S	67	√
A117S-2	38	
A117T	54	√

[a]Previous results from Kazmirski et al. (1995).

from 1.5-1.9 ns, and ~60% for the rest of the simulation. This adjustment of (φ, ψ) angles resulted in a shift to i → i +5 hydrogen bonding.

The helical content of A117P was comparable to that of A117V (Table I); however, the final structures after 2 ns of MD differ dramatically (Figures 2 and 3). While the A117V simulations developed extended structure that would presumably be problematic in the intact protein, the A117P peptide maintained a coiled structure. Without its normal i → i + 4 hydrogen bonding partner, the carbonyl at Ala 113 bent out into solution and formed a hydrogen bond with a water molecule, distorting the helix at residues 113-114. The resulting structure had a large gap below Pro 117 allowing the entrance of water (Figure 3).

The A117E peptide was stable throughout the simulation (Table I). Val 121 and 122 formed close contacts with the methylene portion of Glu 117. Other

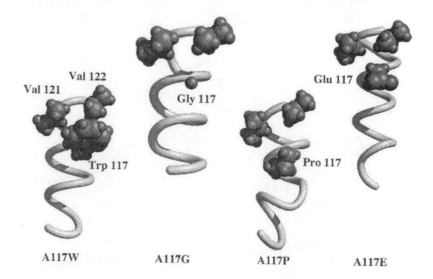

Figure 3. Final structures of mutants that did not form extended structure.

than this minor perturbation, the i → i + 4 α-helical hydrogen bonding scheme was maintained along the chain, except at Val 121, Val 122, and the amide cap which formed i → i - 5 hydrogen bonds (Figure 3).

B. Helix-Disrupting Mutations

Through the first nanosecond of the simulation, the A117M peptide experienced only a minor loss of helical structure centered around Ala 120. The disturbance at Ala 120 allowed for the clustering of the Met 117, Val 121, and Val 122 side chains. After 1.0 ns, the helix content dropped and leveled off at ~40%. During this transition, Ala 115 and Gly 119 left the helical conformation and a larger cluster formed, including Met 109 and Met 112. Shifts in Gly 119 and Ala 120 then led to closer contacts between Val 121 and Met 112 causing the kink to become even more pronounced (Figure 4).

The A117D peptide displayed the most dynamic behavior of all of the mutant peptides, with the helix content plummeting to ~10% during the first 0.5 ns of the simulation. The resulting structure, however, exhibited no extended structure and looked quite helical. All of the residues except for His 111, Ala 115, and Gly 119 had dihedral angles that were out of the helical range, but only barely. From 0.5-1.0 ns, the carbonyl oxygens of Met 112, Ala 113, Ala 116, and Asp 117 pointed out into solution and maintained hydrogen bonds with water molecules solvating the aspartic acid side chain. From 1.0-1.1 ns, the A117D briefly regained most of its helical structure (~75%) when the residues cited above reformed their main chain hydrogen bonds. Over the next 0.8 ns, the helix steadily unraveled and formed an extended strand at the C-terminus. The shell of

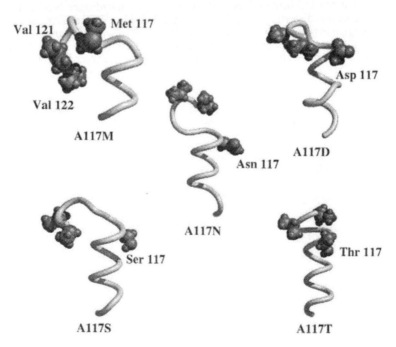

Figure 4. Structures of mutants displayed some degree of unfolding. All of these structures are at 2 ns. The A117M displays a large kink in the middle of the helix while the other four experienced unraveling at the C-terminus.

waters around Asp 117 seemed to crowd the valine side chains at 121 and 122, and from 1.5-1.6 ns, Ala 118 rotated moving the valines away from the aspartic acid. This disruption was propagated to Gly 119 and Ala 120 over the next 0.1, ns leading to the formation of an extended strand (Figure 4).

The A117N helix was stable for the first 1.1 ns of the simulation, then the helix content dropped to ~40% with a kinking of the helix occurring at Ala 111 and Ala 115. These distortions were caused by the tight shell of water molecules around the side chain of Asn 117 that precluded the methionine side chains from forming hydrophobic clusters. The disrupted helix reformed at 1.4 ns only to unfold again at 1.7 ns, when Val 121 and Val 122 moved away from of Asn 117. An extended strand formed at the C-terminus as a result. After the second disruption, the local helical structure was not recovered (Table I, Figure 4).

After 0.3 ns, the helix content of the A117S peptide stabilized at 60-70% (Table I). Within the first 0.1 ns of the simulation, a serine cap interaction formed through a hydrogen bond between the hydroxyl group of Ser 117 and the carbonyl oxygen of Ala 113. The combination of the serine cap interaction and the main chain-main chain hydrogen bond between these same residues resulted in increased rigidity of the helix. During the A117S simulation, the interactions between Ala 113 and Ser 117 were preserved and the $i \rightarrow i + 5$ hydrogen bond shift did not occur. Instead, Gly 119 altered its dihedral angles to adopt a left-handed helical conformation, thus positioning Ala 120 to the C-terminus out into solution as was observed with the A117V peptide (Figure 2 & 4).

Due to the unexpected similarity between the A117S and A117V simulations, another A117S simulation was performed without any *in vacuo* minimization (A117S-2) to produce a slightly different starting structure for molecular dynamics. In this simulation, there was a drastic decline in the helix content (Table I), however no extended structure formed. The hydroxyl group of Ser 117 never formed a hydrogen bond with the carbonyl oxygen of Ala 113 and the $i \rightarrow i + 5$ hydrogen bonding network was established at the C-terminus. During the first 1.3 ns of the simulation, the helix content remained at ~80%. Between 1.3-1.4 ns, a kink developed at Ala 113 that moved Met 112 away from Ser 117 allowing for better solvation of its hydroxyl group and that of the carbonyl group of Met 112. This kink persisted for the remainder of the simulation leading to the low helix content (Table I).

The A117T simulation showed results similar to the A117S simulation, in that some extended structure formed (0.7-1.4 ns) due to a conformational change at Gly 119 (Figure 4). Furthermore, a hydrogen bond was present between the hydroxyl group of Thr 117 and the carbonyl oxygen of Ala 113, which increased the rigidity of the helix and pre-empted the $i \rightarrow i + 5$ hydrogen bonding shift described above. The A117T peptide formed less extended structure than the A117S peptide due to the formation of an $i \rightarrow i + 6$ main chain hydrogen bond between Ala 116 and Val 122, which kept the C-terminus of the peptide closer to the main chain of the peptide. This hydrogen bond was never detected in the A117S simulation. At 1.4 ns the peptide was able to adjust the C-terminus to form $i \rightarrow i + 5$ hydrogen bonds. This was accomplished through distortion of the main chain at Thr 117 and Ala 118. The resulting structure was coiled but only contained 54% helix content (Figure 4).

IV. Discussion

The infectious agent causing prion diseases appears to be the so-called scrapie, or protease resistant form of the prion protein (Prusiner, 1991). There are now numerous pieces of evidence supporting this hypothesis and the assertion that PrP^{Sc} and PrP^{C} are conformational isomers (Basler et al., 1986; Caughey and Raymond, 1991; Pan et al., 1993; Stahl et al., 1993; Kocisko et al., 1994). PrP^{C}

is rich in helical structure (42%) with negligible β-structure (3%), while PrP^{Sc} is high in β-structure (43%) and has a lower helix content (43%). Currently, however, there is little information on the mechanism of the conversion of PrP^{C} to PrP^{Sc}, or the putative α → β conversion. In any case, according to this hypothesis, any changes in the cellular environment or mutations that destabilize PrP^{C} and/or stabilize PrP^{Sc} would shift the equilibrium in favor of the PrP^{Sc} conformer and thereby predispose one to prion diseases.

We are using MD simulations to investigate the α → β transition at the atomic level in fragments of the prion protein. Our initial studies are focusing on the first helix (H1) of the PrP^{C} model for two reasons: (1) it adopts helical or β-structure depending on the experimental conditions (Gasset et al., 1992; Zhang et al., 1995; Nguyen et al., 1995); and (2) it contains a site conferring GSS in humans (A117V) (Doh-ura et al., 1989; Hsiao et al., 1991). Through MD simulations, we can study how certain mutations might destabilize the helical conformation of H1. Although we cannot address the thermodynamic stability of the different helices (too short of time period), we can comment on the sequence dependence of the kinetic stabilization of the peptide and the resulting structures.

In previous simulations, it was found that the A117V mutation was unique in destabilizing the helical structure of the H1 peptide compared to wild type and other hydrophobic mutations (Kazmirski et al., 1995). The destabilization was believed to be caused by decreased conformational freedom of the β-branched side chains of Val 117, Val 121, and Val 122 in the helical state. In the final structure from the simulation, the last turn of the helix converted to an extended strand allowing greater rotation of these side chains. Higher fluctuation in the χ_1 side chain dihedral angles were noted for these residues over the final 0.3 ns of the simulation. Furthermore, we surmise that the unwinding of this portion of the H1 helix in the predicted model of PrP^{C} could have disastrous effects upon protein stability and structure and may represent an early step in the conversion of PrP^{C} to PrP^{Sc} (Kazmirski, et al., 1995).

The simulations presented here are a continuation of our earlier H1 peptide work and we have presented the results of simulations of nine new mutations at position 117. Our findings suggest that the valine mutation is not the only helix destabilizing mutation, but it is relatively unique in its overall behavior. Five of the mutations investigated here were found to be helix destabilizing and generated some form of extended structure in the helix. The most surprising of these was the methionine mutant, which represents the introduction of another hydrophobic group at position 117. In simulations of other hydrophobic mutants at 117, stable helices were observed that contained only transient (~0.2 ns) hydrophobic clusters centered around 117 (Kazmirski et al., 1995). In the A117M simulation, however, a major cluster formed and persisted throughout the last 0.9 ns of the simulation. This behavior was unexpected, as Met is considered to be less hydrophobic than some of the other mutations made at 117 (A117W and A117F).

The other four of the five mutations that lead to extended structure are hydrophilic residues (Asp, Asn, Ser, Thr). The A117D and A117N peptides unfolded due to the involvement of water molecules. The solvation of these two side chains by water crowds the bulky side chains of Val 121 and Val 122, leading to the extended structure. This is similar to the outcome in the A117V simulations, where the rotation of the Val 117 side chain pushes the other valines away. In the A117S and A117T simulations, however, the unfolding of the last turn of the helix occurs by a different mechanism. When extended structure is not formed, the H1 peptide forms i → i + 5 hydrogen bonds at its C-terminus. However in the A117S and A117T simulations, the added hydrogen bond between Ala 113 and the side chain of 117 either does not allow or delays this shift and puts added strain on the peptide. This strain is relieved through formation of an extended strand facilitated by the conformational transition of Gly 119. A second trajectory of the A117S mutant (A117S-2) showed a similar decrease in helical structure but no extended structure formed. The discordance of these results can be explained in that the A117S-2 simulation did not form the side chain-main chain hydrogen bond that leads to the formation of the extended structure at the C-terminus. Also, it must be kept in mind that we are limited by simulation time and the number of simulations that we can perform, so that we are not representing a true ensemble of peptide conformations that may well contain both of these conformers. In any case, serine destabilized the helical nature of the H1 peptide relative to wild type in both simulations.

Of the four remaining simulations, two showed a decline in helix content, yet their structures remained coiled and compact. The structures of the A117G and A117P peptides were perturbed enough to place them outside of our defined helical range but there were only minor structural disruptions that could probably be tolerated in the full protein, PrP^C. The final two mutants are A117W and A117E. The A117W peptide simulation displayed properties typical of the other hydrophobic mutants investigated previously. It formed a stable helix with brief hydrophobic clusters around the indole side chain. Surprisingly, unlike the Asp mutant, the A117E helix was stable. The additional methylene group of the glutamate side chain protected the valine residues from the solvating water surrounding the acid moiety.

In conclusion, A117W and A117E would appear to be well tolerated in the H1 peptide and may even be tolerated in PrP^C, neglecting possible packing difficulties and the introduction of a charge. Furthermore, even though the A117G and A117E mutations lower the helix content of H1, they do not lead to overall changes in the helical structure and may be tolerated in PrP^C. However, the other five mutants would most likely disturb the fold of PrP^C and lead to alternate structural states of the protein either by disrupting the H1 helix or the hydrophobic core (Figure 1). Experimental work is in progress to see if these predictions are borne out.

In the simulations that formed extended structure, Gly 119 is the weak link in the peptide chain. Glycines are highly mobile and are "helix breakers" in many proteins (Chou & Fasman, 1974). Based on our simulation, Gly 119 plays an important role in the destabilization of the H1 helix and may be instrumental in the early steps of the local unfolding and conversion of $\alpha \rightarrow \beta$ structure in PrP^C. Substitution of this glycine by a more constrained residue, such as an alanine, may have a stabilizing effect upon the helical state of H1 and its various mutants and preliminary simulations support this idea (data not presented).

Although five new mutations led to extended structure in the H1 peptide, none of these mutants have been observed in humans. One reason for this is that these mutations may be too drastic to allow the protein to fold. Studies of peptide fragments necessarily neglect the long-range effects of these mutations and how the mutations would be accommodated in the hydrophobic core of the protein. The introduction of a charged or hydrophilic group at 117 could upset folding and lead to rapid degradation. Whereas, the A117V mutation is consistent with the hydrophobic nature of the core but local unfolding of H1 in PrP^C may trigger the conversion to PrP^{Sc}.

Acknowledgments

All molecular graphics images were produced using the MidasPlus software from the Computer Graphics Laboratory, University of California, San Francisco (supported by NIH RR-01081) (Ferrin et al., 1988). This work was supported by the National Institutes of Health (AG-02132 to SBP and FEC) and start-up funds from the Dept. of Medicinal Chemistry to VD. SLK acknowledges a Magnuson Scholar Award funded by the U. S. Dept. of Education.

References

Basler, K., Oesch, B., Scott, M., Westaway, D., Wälchli, M., Groth, D. F., McKinley, M. P., Prusiner, S. B. & Weissman, C. (1986). *Cell* **46**: 417.

Caughey, B. & Raymond, G. J. (1991). *J. Biol. Chem.* **266**, 18217.

Chou, P. Y. & Fasman, G. D. (1974). *Biochem.* **13**, 211.

Daggett, V. & Levitt, M. (1992). *J. Mol. Biol.* **223**, 1121.

Daggett, V., Kollman, P. A. & Kuntz, I. D. (1991). *Biopolymers* **31**, 1115.

DeArmond, S. J. & Prusiner, S. B. (1995). *Am. J. Pathol.* **146**, 785.

Doh-ura, K., Tateishi, J., Sasaki, H., Kitamoto, T. & Sakaki, Y. (1989). *Biochem. Biophys. Res. Commun.* **163**, 974.

Ferrin, T. E., Huang, L. E., Jarvis, L. E. & Langridge, R. (1988). *J. Mol. Graphics* **6**, 13.

Gasset, M., Baldwin, M. A., Lloyd, D. H., Gabriel, J.-M., Holtzman, D. M., Cohen, F., Fletterick, R. & Prusiner, S. B. (1992). *Proc. Natl. Acad. Sci. U. S. A.* **89**, 10940.

Hsiao, K. K., Cass, C., Schellenberg, G. D., Bird, T., Devine-Gage, E., Wisniewski, H. & Prusiner, S. B. (1991). *Neurology* **41**, 681.

Huang, Z., Gabriel, J.-M., Baldwin, M. A., Fletterick, R. J., Prusiner, S. B. & Cohen, F. E. (1994). *Proc. Natl. Acad. Sci. U. S. A.* **91**, 7139.

Kazmirski, S. L., Alonso, D. O. V., Cohen, F. E., Prusiner, S. B. & Daggett, V. (1995). *Chem. & Biol.* **2**, 305.

Kocisko, D. A., Come, J. H., Priola, S. A., Chesebro, B., Raymond, G. J., Lansbury, P. T. & Caughey, B. (1994). *Nature* **370**, 471.

Levitt, M. (1990). ENCAD---Energy Calculations and Dynamics. Molecular Applications Group. Stanford University, Palo Alto, CA.

Levitt, M. (1983). *J. Mol. Biol.* **168**, 595.

Levitt, M. (1989). *Chemica Scripta* **29A**, 197.

Levitt, M., Hirshberg, M., Sharon, R., & Daggett, V. (1995). *Comp. Phys. Commun.* In Press.

Nguyen, J. T., Baldwin, M. A., Livshits, T. L., Jew, S., Zhang, H., Cohen, F. E. & Prusiner, S. B. (1995). *Biochem.* In Press.

Pan, K.-M., Baldwin, M., Nguyen, J., Gasset, M., Serban, A., Groth, D., Huang, Z. Fletterick, R. J., Cohen, F. E. & Prusiner, S. B. (1993). *Proc. Natl. Acad. Sci. U. S. A.* **90**, 10962.

Prusiner, S. B. (1991). *Science* **252**, 1515.

Safar, J., Roller, P. P., Gajdusek, D. C. & Gibbs, C. J. (1993). *J. Biol. Chem.* **268**, 20276.

Stahl, N. & Prusiner, S. B. (1991). *FASEB J.* **5**, 2799.

Stahl, N., Baldwin, M. A., Teplow, D. B., Hood, L., Gibson, B. W., Burlingame, A. L. & Prusiner, S. B. (1993). *Biochem.* **32**, 1991.

Zhang, H., Kaneko, K., Nguyen, J. T., Livshits, T. L., Baldwin, M. A., Cohen, F. E. , James, T. L. & Prusiner, S. B. (1995). *J. Mol. Biol.* **250**, 514.

SECTION IX

Methods and Uses for Synthetic Proteins

Identification of Truncated *E. coli*-Expressed Proteins with a Novel C-Terminal 10Sa RNA Decapeptide Extension

Richard J. Simpson[1], Jian-Guo Zhang[1], Donna S. Dorow[2], Gavin E. Reid[1], Robert L. Moritz[1] and Guo-Fen Tu[1]

[1]Joint Protein Structure Laboratory, Ludwig Institute for Cancer Research and the Walter and Eliza Hall Institute of Medical Research, PO 2008 Royal Melbourne Hospital, Parkville, Victoria, Australia 3050
[2]Peter MacCallum Institute for Cancer Research, Melbourne, Victoria, Australia 3000

1. INTRODUCTION

Recombinant proteins produced in *E. coli* are an important source of material for use as therapeutic drugs or reagents for structure-function relationship studies. Occasionally, during large-scale process development, many of these biosynthetic proteins have been found to contain structural modifications which restrict their usefulness. These include N- and C-terminal truncations (1,2), extensions (3), incomplete removal of N-terminal initiator methionine (4), trisulfide derivatives (5), ε-N-acetyllysine (6), misincorporation of norleucine for methionine (7), and lysine for arginine (8). During the large-scale purification of recombinant murine interleukin-6 (mIL-6)(9) we identified, using a combination of peptide mapping, amino acid compositional analysis, electrospray ionization mass spectrometry (ESI-MS) and automated N- and C-terminal sequencing, a sub-population (5-10%) of mIL-6 molecules that contained a novel C-terminal modification (10).

In addition to being progressively truncated at their C-termini, this sub-population of mIL-6 molecules contains a novel modification in the form of an appended peptide (-Ala-Ala-Asn-Asp-Glu-Asn-Tyr-Ala-Leu-Ala-Ala-COOH) at their C-termini. A search of the available databases revealed that the amino acid sequence of this C-terminal "tag" peptide is identical to that encoded by a small metabolically stable RNA of *E. coli* (10Sa RNA) (11). Disruption of *ssrA*, the gene that encodes 10Sa RNA abrogated the formation of mIL-6-"tag"-peptide chimeras (10). Our discovery that its formation in mIL-6-"tag" chimeras in *E. coli* is not dependent upon the recombinant protein, bacterial strain or plasmid employed has novel structural and functional implications and reveals a hitherto undescribed biochemical process.

II.   EXPERIMENTAL PROCEDURES

Expression and Purification of mIL-6 Analogues T1 and T3: Recombinant mIL-6 and analogues T1 and T3 were expressed in *E. coli* (12), extracted from inclusion bodies with 6M GdnHCl and purified as outlined in Fig. 1.   Following preparative RP-HPLC (Fig.2), fractions containing pure IL-6, T1 and T3 were pooled on the basis of their analytical RP-HPLC profile.

Cyanogen Bromide peptide mapping of T3, T1 and mIL-6:   Proteins (50 µg) were incubated with a 500-fold excess of CNBr (w/w) in 70% aq. HCOOH containing 0.02% Tween 20 under nitrogen for 18 h in the dark at 22°C.

Capillary RP-HPLC/ESI-MS: Mass spectra were recorded on a Finnigan-MAT TSQ-700 (San Jose, CA) triple quadrupole mass spectrometer equipped with an electrospray ion source and a capillary RP-HPLC column (0.2 mm I.D. x 150 mm) slurry-packed, in-house (13,14), with Brownlee RP-300 (300Å-pore size) 7-µm dimethyloctylsilica.

N- and C-Terminal Amino Acid Sequence Analyses: N-terminal amino acid sequence analyses were performed using Applied Biosystems (models 470A and 477A) instruments.   Automated C-terminal protein sequence analyses of T1 and T3 were performed using a Hewlett-Packard (model HP G1009A) sequencing instrument (15).

Bacterial Strains and Growth Conditions: The genotypes of the *E. coli* strains and growth conditions used in this study are described elsewhere (10). The chromosomal 10Sa gene (*ssrA*) was disrupted by insertion of the 0.8-kb chloramphenicol acetyl transferase (*cat*) gene into the 10Sa RNA coding sequence (11). *E. coli* strain NM101 containing a disrupted *ssrA* gene was derived from strain JM101 by P1 transduction (16) using N2211 as a donor strain.

Antibody and Immunoblot Analysis: Polyclonal antiserum was raised against a synthetic "tag" peptide (AANDENYALA), synthesized as described elsewhere (17). For immunoblot analysis, proteins were electroblotted from pre-cast 10-20% Tricine SDS-acrylamide gels (Novex) onto nitrocellulose paper, blocked with 3% BSA and 0.02% Tween-20 in PBS for 1 h and then reacted with the anti-tag peptide polyclonal antiserum (1:15000 dilution) followed by incubation with a 1:20000 dilution of goat antibody to rabbit IgG crosslinked horseradish peroxidase. Immunoblots were developed using the enhanced chemiluminescence procedure (Amersham).

Northern Blot Analysis: Total *E. coli* RNA was isolated using the "hot-phenol" extraction method (18). Northern blot analyses were performed as described elsewhere (10). Prehybridization of the blots were performed at a temperature 5°C below the estimated melting temperature for the individual oligonucleotide. Antisense oligonucleotide probes used for hybridizing to various regions of mIL-6 mRNA (19) were: probe 1: 5'-GAAACCATCTGGCTAGGT-3', nucleotides 970-953; probe 2: 5'-GTGTCCCAACATTCATATTGTCAG-3', nucleotides 798-772; probe 3: 5'-ACTAGGTTTGCCGAGTAGA-3', nucleotides 670-650; probe 4: 5'-ACCTCTTGGTTGAAGATATG-3', nucleotides 503-484; probe 5: 5'-ATATCCAGTTTGGTAGCA-3', nucleotides 345-328; probe 6: 5'-ACAGGTCTGTTGGGAGTGGTATCCTC-3', nucleotides 167-142. The antisense oligonucleotide from *ssrA* ("tag" probe) was 5'-TGCTAAAGCGTAGTTTTCGTC-3' (nucleotides 247-267) (11).

III. RESULTS AND DISCUSSION

E. coli-expressed mIL-6 (20) was purified from inclusion bodies by RP-HPLC and characterized by a combination of peptide mapping, microsequencing and ESI-MS (9). In addition to full-length mIL-6, two other analogues, designated T1 and T3, with identical N-terminal sequences to mIL-6 were observed. Although the amino acid compositions of T1 and T3 (Table 1) were consistent with C-terminally truncated forms of IL-6 (residues 1-164/165 and ~1-120, respectively) the molar ratio of alanine was consistently high. Although apparently homogeneous when chromatographed on a Brownlee C8 column, SDS-PAGE gel analysis showed that T1 and T3 were of lower M_r than the 20-kDa mIL-6, and exhibited mass heterogeneity (9). Capillary RP-HPLC/ESI-MS confirmed that T1 and T3 were complex mixtures, comprising molecules in the mass range 18-20 kDa and 13.5-15 kDa, respectively. Intriguingly, automated

C-terminal amino acid sequence analysis yielded a single sequence, -Leu-Ala-Ala-COOH, which bore no relationship to mIL-6 (20) (Fig.3). Thus, we were confronted with a series of mIL-6 proteins with the same N-terminal and C-terminal sequences, but with a spectrum of molecular masses.

To resolve this paradox, CNBr peptide mapping was performed and the peptide maps of T1 and T3 compared with that of full-length mIL-6 (Fig.4). This study showed that both T1 and T3 lack the parent C-terminal CNBr peptide, CN0 (Lys_{104}-Thr_{187}). Sequence analysis of the C-terminal peptide from T3 (CN3) confirmed the native sequence of mIL-6 from residues 104 to 122 and established a novel "tag" peptide extension

Expression of IL-6 in E. Coli as a β-galactosidase fusion protein using pUC vector
(37g wet cells)
↓ Cell lysis
Isolation of inclusion bodies
↓
Selective urea washing of inclusion bodies
(4M urea, 1% Triton X-100)
↓
Solubilisation with 8M Gdn.HCl
↓
Gel-permeation chromatography in 6M
Gdn. HCl (Fractogel TSK HW-55(S))
↓
Preparative RP-HPLC (Vydac C4)
↓
Pooling of major IL-6-containing fractions
(Based on microbore RP-HPLC (Brownlee RP-300)
analysis of preparative RP-HPLC fractions)
↓
Pure IL-6 (~50-60 mg) Characterization by Bioassay, N-terminal
sequence analysis and mass spectrometry

Fig. 1.   *Flow diagram for the isolation and purification of recombinant mIL-6 from E. coli inclusion bodies.*

Table 1. *Amino acid composition of murine IL-6 and analogues T1 and T3*

Amino Acid	mIL-6		T1		T3	
	mol %					
Asx	23.6	(24)[a]	26.0	(24)	21.3	(20)
Thr	17.6	(18)	13.6	(15)	10.7	(10)
Ser	9.8	(10)	7.6	(8)	6.7	(6)
Glx	24.2	(24)	21.3	(21)	15.6	(15)
Pro	5.7	(6)	6.9	(6)	4.8	(4)
Gly	6.2	(6)	7.0	(6)	6.7	(6)
Ala	5.4	(5)	9.4	(5)	8.6	(4)
Cys	ND	(4)	ND	(4)	ND	(4)
Val	7.1	(8)	7.3	(7)	5.4	(5)
Met	3.3	(4)	4.4	(4)	3.9	(4)
Ile	10.4	(12)	9.7	(11)	6.1	(6)
Leu	24.4	(25)	20.7	(21)	14.8	(14)
Tyr	5.6	(6)	7.0	(6)	6.5	(6)
Phe	3.0	(3)	1.5	(1-2)	0.4	(0)
His	4.0	(4)	4.4	(4)	2.7	(2)
Lys	14.3	(15)	11.2	(13)	8.3	(8)
Trp	ND	(2)	ND	(2)	ND	(1)
Arg	7.6	(8)	6.0	(6)	5.0	(5)

[a] *Values in parenthesis are the calculated values based on the amino acid sequence (20).*

(sequence: AANDENYALAA-COOH) from residues 123 to 133 (Table 2). Edman degradation of the C-terminal CNBr peptides CN1,2 and 4-8, together with ESI-MS, confirmed that they each contained the sequence of mIL-6 (residue 104 onwards) with the "tag" peptide extension at their C-termini (Table 2). In summary, T1 and T3 comprise mIL-6 molecules that have been progressively truncated in a "ladder-like" manner from the C-terminus (in the regions of residues 158-169 and 112-125 respectively) and modified by appendage of a "tag" peptide. There was no evidence of any T1 and T3 deletants that did not contain the "tag" peptide and no "tag" peptide shorter than 11-amino acid residues.

Inspection of the "tag" peptide sequence showed that it is unrelated to either mIL-6 (20) or pUC9 sequences (21), but is identical to the last 10 amino acid residues of a putative polypeptide encoded by a small metabolically stable RNA (10Sa RNA) in *E. coli* (11). The gene encoding 10Sa RNA (*ssrA*) has its own promoter and terminator and an open reading frame encoding a putative 25-residue polypeptide (11). Thus far, attempts to identify the 10Sa RNA polypeptide in *E. coli* extracts have been unsuccessful (22). While the function of 10Sa RNA is not clear, disruption of the *ssrA* gene has been shown to retard cell growth (22).

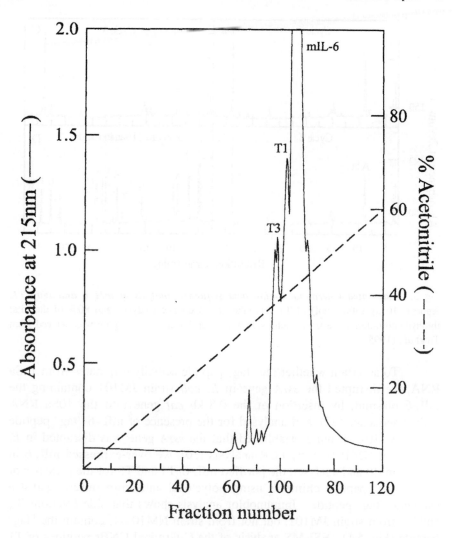

Fig. 2. ***Preparative RP-HPLC of recombinant mIL-6..*** Chromatographic conditions: Vydac column C4 (22.5 mm I.D. x 100 mm), linear 60-min gradient from 0-100%B, where solvent A was 0.1% aq. TFA and solvent B was 60% acetonitrile / 40% water containing 0.08% TFA; flow rate, 20-ml/min. 10-ml fractions were collected and assessed for mIL-6, T1 and T3 purity by analytical RP-HPLC on a Brownlee RP-300 column (4.6mm I.D. x 100mm) developed with TFA/acetonitrile at 1.0 ml/min.

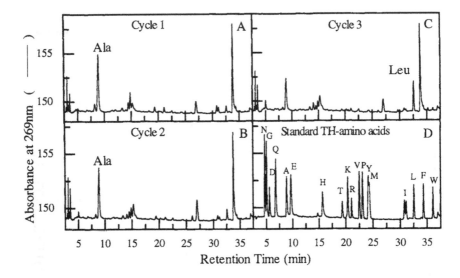

Fig. 3. *Automated C-terminal amino acid sequence analysis of mIL-6 analogue T1.*
Approx. 10 µg (500 pmol) of T1 was taken for sequence analysis, and 80% of the total
thiohydantoin-derivative from each cycle was taken for analysis. Figure was derived from
Tu *et al.*, (1995).

To ascertain whether the "tag" peptide actually originates from 10Sa
RNA, we disrupted the *ssrA* gene in *E. coli* strain JM101 containing the
mIL-6 plasmid, by insertion of the 0.8-kb *cat* gene into the 10Sa RNA
coding sequence (22), and analysed for the presence of mIL-6-"tag" peptide
chimeras (10). Having established that the *ssrA* gene was disrupted in *E.
coli* (strain NM101-1), by PCR analysis (10), we over-expressed mIL-6 in
this strain (9) and analysed the purified mIL-6, T1 and T3 for the presence of
mIL-6-"tag" peptide chimeras using polyclonal antiserum raised against a
synthetic "tag" peptide. Immunoblot analysis shows that side fractions T1
and T3 from strain JM101, but not from strain NM101-1, contain the "tag"
peptide (Fig. 5A). ESI-MS analysis of the C-terminal CNBr peptides of T1
and T3 from NM101-1 confirmed that they were progressively truncated at
the C-terminus, but did not contain the "tag" peptide extension (data not
shown). It is apparent from the data shown in Fig. 5B that the protein-"tag"-
peptide chimera formation is not dependent upon the recombinant protein,
bacterial host strain or plasmid employed. We have also expressed a human
mixed-lineage kinase (MLK) SH3 domain (23) in E. coli and found
expressed "tag" peptide by Western analysis in total cell lysate (data not
shown). Recently, a full-length form of *E. coli*-expressed human
interleukin-2- tag peptide chimera was reported (24).

To explore whether the "tag" is added to mIL-6 during RNA synthesis via a hitherto unknown co-transcriptional process, we attempted, by Northern (RNA) blot analysis of total RNA from *E. coli* strain JM101 carrying the mIL-6 plasmid p9HP1B5B12 (20), to identify mIL-6-"tag" chimeras at the RNA level. Whilst we found no evidence of mIL-6-"tag" chimera transcripts, our data cannot completely exclude the possible existence of a very low level of chimeric RNA or highly unstable chimeric RNA. Earlier attempts, by others (25), to demonstrate direct binding of 10Sa RNA to ribosomes proved unsuccessful, hence it appears unlikely that 10Sa RNA is a mRNA.

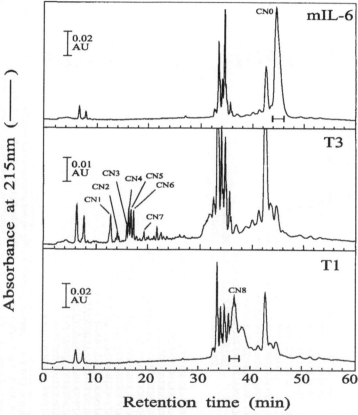

Fig. 4. *CNBr peptide mapping of mIL-6 analogues T1 and T3.* The CNBr digest mixtures (~ 50 µg protein) of T3, T1 and mIL-6 were evaporated to near dryness, diluted with 0.1% aq. TFA and applied to a Brownlee RP-300 column (4.6 mm I.D. x 100 mm) which was developed at 1 ml/min using a linear 60-min gradient from 0-100% B, where solvent A was 0.1% aq. TFA and solvent B was 60% acetonitrile/ 0.09% aq. TFA. Annotated peaks were collected manually and subjected to automated Edman degradation and ESI-MS. Figure was derived from Tu *et al.* (1995).

Table 2. *Summary of sequence and mass analysis data for C-terminally modified peptides from mIL-6*

Peptide	Sequence*	Calculated mass (Da)	Observed mass (Da)	mIL-6 residues‡
Fraction T1				
CN8	KNNLK....SNALLTDKLESQKEWLRTKTIQFIAANDENYALAA	8914.2	8912	101-166
CN8†	KNNLK....SNALLTDKLESQKEWLRTKTIQFAANDENYALAA	8801.1	8800	101-165
CN8†	KNNLK....SNALLTDKLESQKEWLRTKTIQAANDENYALAA	8653.9	8654	101-164
CN8†	KNNLK....SNALLTDKLESQKEWLRTKTKAANDENYALAA	8525.8	8525	101-163
CN8	KNNLK....SNALLTDKLESQKEWLRTKTAANDENYALAA	8412.6	8413	101-162
CN8†	KNNLK....SNALLTDKLESQKEWLRTKAANDENYALAA	8311.5	8313	101-161
CN8	KNNLK....SNALLTDKLESQKEWLAANDENYALAA	7926.0	7926	101-158
CN8	KNNLK....SNALLTDKLESQKEWAANDENYALAA	7812.9	7813	101-157
CN8	KNNLK....SNALLTDKLESQKEAAANDENYALAA	7626.7	7625	101-156
CN8†	KNNLK....SNALLTDKLESQKAANDENYALAA	7497.5	7497	101-155
D9§	DKLESQKEWLRTKTIQFAANDENYALAA			
D8§	DKLESQKEWLRTKTIQAANDENYALAA			
Fraction T3				
CN7	KNNLKDNKKDKARVLQRDTETLAANDENYALAA	3732.1	3732	101-122
CN4/5	KNNLKDNKKDKARVLQRDTETAANDENYALAA	3619.0	3619	101-121
CN4	KNNLKDNKKDKARVLQRDTEAANDENYALAA	3517.9	3517	101-120
CN3	KNNLKDNKKDKARVLQRDTAANDENYALAA	3388.8	3388	101-119
CN3/4	KNNLKDNKKDKARVLQRDAANDENYALAA	3287.6	3287	101-118
CN5	KNNLKDNKKDKARVLQRAANDENYALAA	3172.6	3172	101-117
CN5/6	KNNLKDNKKDKARVLQAANDENYALAA	3016.4	3016	101-116
CN4/5	KNNLKDNKKDKARVLAANDENYALAA	2888.2	2888	101-115
CN2	KNNLKDNKKDKARVAANDENYALAA	2775.1	2774	101-114
CN1	KNNLKDNKKDKARAANDENYALAA	2675.9	2675	101-113
CN1	KNNLKDNKKDKAAANDENYALAA	2519.7	2519	101-112
CN1	KNNLKDNKKDKAANDENYALAA	2448.7	2448	101-111
CN1	KNNLKDNKKDAANDENYALAA	2320.5	2320	101-110
CN1	KNNLKDNKKAANDENYALAA	2205.4	2205	101-109

Legend to Table 2. CNBr peptides CN1-8 were purified by RP-HPLC (Fig. 4) and subjected to automated Edman degradation as well as ESI-MS. Single letter abbreviations for amino acid residues are: A, Ala; D, Asp; E, Glu; F, Phe; G, Gly; H, His; I, Ile; K, Lys; L, Leu, N, Asn; P, Pro; Q, Gln; R, Arg; S, Ser; T, Thr; V, Val; W, Trp; and Y, Tyr. * Sequences determined by Edman degradation are underlined. † underlined sequences were obtained by analysis of Asp-N peptides derived from sub-digestion of CN8 with trypsin as described elsewhere (10). ‡ residue numbers are from Simpson *et al.* (20). § underlined sequences were obtained by analysis of Asp-N peptides from mIL-6 fraction T1 (9); Peptides found in fractions CN3 and CN4; CN5/6 in fractions CN5 and CN6; and CN4/5 were found in fractions CN4 and CN5. "Tag" sequence peptide is designated by bold italic lettering.

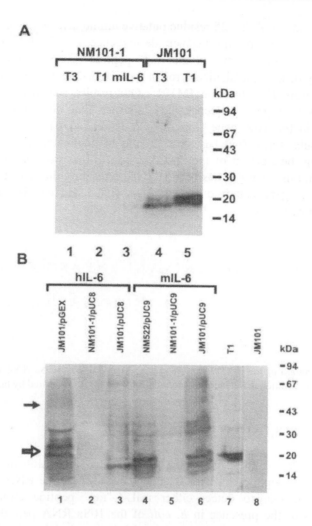

Fig. 5. *Immunoblot analysis of C-terminally truncated forms of murine- and human IL-6 from various strains of E. coli harboring different plasmids with polyclonal antiserum to the "tag" peptide.* (A) RP-HPLC-purified T1 and T3 (~2 μg per lane) were analysed by SDS-PAGE using pre-cast 10-20% Tricine gels (Novex). Lane 1; T3 from NM101-1, lane 2; T1 from NM101-1, lane 3; full-length mIL-6 from NM101-1, lane 4;, T3 from JM101 containing p9HP1B5B12, lane 5; T1 from JM101 containing p9HP1B5B12. (B) Immunoblot of insoluble crude cell lysates (~20 μg protein per lane) from various strains of *E. coli* harboring different plasmids. Lane 1; strain JM101 transformed with *pGEX* containing hIL-6, lane 2; strain NM101-1 transformed with pUC8 containing hIL-6, lane 3; strain JM101 transformed with pUC8 containing hIL 6, lane 4; strain NM522 (cont)

Alignment of the 25 residue putative amino acid sequence of the *ssrA* gene product (11) and the "tag"peptide sequence, reveals the presence of an additional alanine residue in the "tag" peptide (Fig 6). To exclude the possibility of a valine/alanine error in the original sequence, we sequenced the *ssrA* from *E. coli* strain JM101. Our results, which concur with the published *ssrA* gene sequence (11), confirm that no codons have been replaced in the *ssrA* gene from *E. coli* strain JM101. It appears that a Val/Ala substitution (one base change) may be implicated in the mechanism underlying the ligation of the "tag" peptide to the progressively truncated mIL-6. In this regard, it has been suggested recently that the RNA encoded by the *ssrA* gene can fold into a *t*RNA shape and can be charged with alanine *in vitro* (26).

Fig 6. *Alignment of the putative amino acid sequence of the 10 Sa RNA with the "tag" peptide.* The additional alanine residue in the "tag" peptide is indicated by bold lettering.

IV.   CONCLUSION

Taken together, our studies suggest that formation of recombinant protein-"tag" peptide chimeras is dependant on the 10Sa RNA gene. Since we were unable to detect either mIL-6-"tag" peptide chimera mRNA transcripts or the presence in *E. coli* of the 10Sa RNA peptide product or precursor protein, the biochemical mechanism underlying the fusion of the IL-6 and the translated 10Sa RNA peptide must involve a hitherto undescribed biochemical process.

---

transformed with pUC9 containing mIL-6, lane 5; strain NM101 transfected with pUC9 containing mIL-6 . lane 6; strain JM101 transformed with pUC9 containing mIL-6, lane 7; fraction T1 from Fig. 1A; lane 8; parent strain JM101. The open arrow indicates the position of full-length IL-6 in pUC8 and -9 plasmids; the solid arrow indicates the position of the 46-kDa full-length glutathione-S transferase-hIL-6 fusion protein. Molecular weight markers are given to the right in kilodaltons. Figure was derived from Tu *et al.*, (1995).

REFERENCES

1.   Daumy, G.O., Merenda, J.M., McColl, A.S., Andrews, G.C., Franke, A.E., Geoghegan, K.F. & Otterness, I.G. (1989) *Biochem. Biophys. Acta.* **998**, 32-42

2.   Paborsky, L.R., Tate, K.M., Harris, R.J., Yansura, D.G., Band, L., McCray, G., Gorman, C.M., O'Brien, D.P., Chang, J.Y., Swartz, J.R., Fung, V.P., Thoman, J.N. & Vehar, G.A. (1989) *Biochemistry.* **28**, 8072-8077

3.   Danley, D.E., Strick, C.A., James, L.C., Lanzetti, A.J., Otterness, I.G., Grenett, H.E. & Fuller, G.M. (1991) *FEBS Lett.* **283**, 135-139

4.   Ben-Bassat, A., Bauer, K., Chang, S.-Y., Myambo, K., Boosman, A. & Chang, S. (1987) *J. Bacteriol.* **169**, 751-757

5.   Jespersen, A.M., Christensen, T., Klausen, N.K., Nielsen, P.F. & Sørensen, H.H. (1994) *Eur. J. Biochem.* **219**, 365-373

6.   Harbour, G.C., garlick, R.L., Lyse, S.B., Crow, F.W., Robins, R.H., & Hoogerheide, J.G., (1992) Techniques in Protein Chemistry III, 487-495.

7.   Bogosian, G., Violand, B.N., Dorward-King, E.J., Workman, W.E., Jung, P.E. & Kane J.F. (1989) *J. Biol. Chem.* **264**, 531-539

8.   Seetharam, R., Heeren, R.A., Wong, E.Y., Braford, S.R., Klein, B.K., Aykent, S., Kotts, C.E., Mathis, K.J., Bishop, B.F., Jennings, M.J., Smith, C.E. & Siegel, N.R. (1988) *Biochem. Biophys. Res. Commun.* **155**, 518-523.

9.   Zhang, J.-G., Moritz, R.L., Reid, G.E., Ward, L.D. & Simpson, R.J. (1992) *Eur. J. Biochem.* **207**, 903-913

10.   Tu G-F, Reid, G.E., Moritz, R.L., & Simpson, R.J. (1995) *J. Biol. Chem.* **270**, 9322-9326.9.

11.   Chauhan, A.K. & Apirion, D. (1989) *Mol. Microbiol.* **3**, 1481-1485

12.   Simpson, R.J., Moritz, R.L., Rubira, M.R. & Van Snick, J. (1988) *Eur. J. Biochem* **176**, 187-197

13.   Moritz, R.L. & Simpson, R.J. (1992) *J. Microcol.* **4**, 485-489

14.   Moritz, R.L., Reid, G.E. & Simpson, R.J. (1994) *Methods. A Companion to Methods in Enzymol.* **6**,213-226.

15.   Miller, C.G., Bailey, J.M., Tso, J., Early, S., and Hawke, D.H. (1995) Methods in Protein Structure Analysis, Atassi, M.Z. and Appella, E. (Eds.), Plenum Press, New York, In Press

16.   Miller, J.H. (1992) A Short Course in Bacterial Genetics: A Laboratory Manual and Handbook for *Escherichia Coli* and Related Bacteria, Cold Spring Harbor Laboratory, Cold Spring Harbor, N.Y.

17.   Reid, G.E. & Simpson, R.J. (1992) *Anal. Biochem.* 200, 301-309

18.   Brumlik, M.J. & Storey, D.G. (1992) *Mol. Microbiol.* **6**, 337-344

19.   Van-Snick, J., Cayphas, S., Szikora, J.-P., Renauld, J.-C., Van-Roost, E., Boon, T. & Simpson, R.J. (1988) *Eur. J. Immunol.* **18**, 193-197

20.   Simpson, R.J., Moritz, R.L., Van Roost, E. & Van Snick, J. (1988) *Biochem. Biophys. Res. Commun.* **157**, 364-372

21.   Vieira, J. & Messing, J. (1982) *Gene* **19**, 259-268

22.   Oh, B.-K. & Apirion, D. (1991) *Mol. Gen. Genet.* **229**, 52-56

23.   Dorow, D.S. (1994) *J. Prot. Chem.* **13**, 458-460.

24.   Ahmad, Z., Ciolek, D., Pan, Y-C. E., Michel, H. & Khan, F. R. (1994) J. Prot. Chem. 13, 591-598.

25.   Ray, B.K. & Apirion, D. (1979) *Molec. Gen. Genet.* **174**, 25-32

26.   Komine, Y., Kitabatake, M., Yokogawa, T., Nishikawa, K., and Inokuchi, H. (1994) *Proc. Natl. Acad. Sci. U.S.A.* **91**, 9223-9227

Simplifying the fragment condensation semisynthesis of protein analogs

Carmichael J.A. Wallace, Anthony C. Woods
and J. Guy Guillemette[a]

Department of Biochemistry, Dalhousie University, Halifax, Canada, B3H 4H7 and
[a]Department of Microbiology, University of Sherbrooke, Quebec J1H 5N4

I   Introduction

Generating protein analogs by either total synthesis or semisynthesis has been complicated in the past by difficulties inherent in the fragment condensation steps. These difficulties are primarily lack of specificity, requiring extensive protection of other reactive groups, and inefficiency of coupling methods, which had evolved from classical organic chemical methods of forming peptide bonds, often under harsh conditions. They appear to have been overcome quite recently with the evolution of several chemoselective and conformationally-directed methodologies, reviewed in ref. (1). In the chemoselective approach unique mutually reactive groups, unrelated to the -COOH and -NH2 components of the peptide bond, are introduced at termini to be joined, forming a linkage that is not peptidic. This may be stable or stabilizable, giving a proteomimetic, or may represent the first step in a 'capture-activation' strategy, where the linkage brings together the conventional terminal groups, and a facile rearrangement forms a peptide bond.

Conformationally directed methods have been developed and exploited by us in the synthesis of very many cytochrome c mutants (2). This phenomenon, which we call Autocatalytic Fragment Religation (AFR), depends on the common property of large protein fragments to mutually reassociate to give the native backbone fold. Thus, the termini are brought into proximity, in analogy with the 'capture-activation' strategy, but in this case by non-covalent forces only. If the termini are activated, then complex formation both ensures specificity and can catalyse that reaction. The aminolysis of the C-terminal homoserine lactone of CNBr fragments is normally slow and uncompetitive with hydrolysis, but when a complex forms between fragments 1-65 and 66-104 of horse cytochrome c, the missing peptide bond spontaneously reforms, in 60-80% yield over 24h, in neutral phosphate buffer, giving a fully functional holoprotein. The manifest advantages of AFR compared to traditional peptide condensation chemistry, including the avoidance of conditions that denature native protein structures, are so great that we would wish to extend the strategy to other sites in cytochrome c and other proteins.

We have examined other systems where limited CNBr cleavage yields a two-fragment complex. In some cases yields of religated protein are high, in others negligable, and it was not clear why (3). Another difficulty is that a methionine residue may not be conveniently located for the development of a semisynthetic strategy. We have addressed these problems by developing other AFR tactics based on serine protease fragments (4). But the seductive simplicity of the CNBr fragment condensation has led us to attempt the synergistic use of site-directed mutagenesis to tailor the protein for semisynthesis, to shuffle methionine residues in the sequence to optimize cleavage and religation. We have demonstrated the feasibility of this idea with the non-productive complex obtained from yeast cytochrome c: in this case the breakpoint is 64-65, just displaced by one residue. By moving the breakpoint in the yeast protein to 65-66, we obtained comparable religation yields to the horse system (5,6). We suggested the difference was due to local secondary structural factors: thus our present goal is to try and establish a set of principles to govern the choice of secondary and tertiary structural locales for the introduction of novel cleavage and religation sites.

II   Methods

We introduced novel methionine residues into a background yeast sequence in which Cys^{102} was replaced by Thr, and Met^{64} by Leu. The gene encoding this protein was incorporated into the pING4 plasmid and mutated, inserted and expressed in yeast by established methods. These techniques, and the methods for extraction and purification of the protein are fully described in a previous volume of this series (5). Many of the methods for physical and biological characterization of the mutant proteins are referenced in that paper. ATP affinity chromatography takes advantage of a specific binding site on the cytochrome c surface in the groove between the 60s helix and the 80s strand (6). ATP affinity is dependent on interaction with a number of residues in this region and is thus sensitive to disruption of surface conformation here, although there is also a simple ion-exchange component in the binding to the column. Relative affinity is guaged by elution time when the protein bound to a column of agarose-hexane ATP is developed with a gradient of phosphate buffer, pH 7 [20-200mM] over 100 minutes (7). In addition to the whole-mitochondrion succinate oxidase assay previously employed (5), in which transfer between reductase and cytochrome c is the limiting step (6), we also assayed the mutants with isolated cytochrome c oxidase; preparation and assay methods are fully described in (6).

CNBr fragmentation of mutant cytochromes was generally performed on 6mg (0.5μMole) batches in 0.5ml 90% formic acid, with a 3-fold molar excess of CNBr. This proved inadequate in one case (L68M) and a 6-fold excess was later employed. Cleavage mixtures were fractionated on a column of sephadex G50 (1200 x 26mm) equilibrated in 7% HCOOH, and fragments that coeluted further resolved by cation-exchange chromatography on a sulfopropyl matrix in a gradient of 20-200mM potassium phosphate buffer, pH 7, in 7M urea. Religation reactions were performed in the standard manner (5).

III   Results and Discussion

In order to be a useful site for autocatalytic religation, the local structures must be suitable, but it is also important that the methionine substitution, or the homoserine that replaces it in the ligated product, does not compromise the

Figure 1: A model of the tertiary structure of yeast cytochrome c emphasizing the sites of mutations to methionine residues. The peptide backbone of the whole molecule is shown together with the side-chains of the residues that have been mutated. Note the variety of secondary structural motifs into which these residues fall. Model generated on an Iris Indigo workstation using Biosym Insight/Discover software.

Table 1: Physicochemical properties of the mutant cytochromes

Cytochrome	Class of mutated residue	HPLC r.t.[1] Fe2+ (mins)	Fe3+	ATP affinity[1] t. elution (mins)	E'$_M$[2] (mV)	pK$_{695nm}$[3]
Native Yeast [4]		17.6	18.6	47	279	8.5
P25M [5]	V [6]	17.8	18.8	46	288	8.5
V28M	V	17.7	18.7	54	268	8.1
L35M	C	17.6	18.5	54	276	8.0
K55M	V	**17.3**	**18.2**	**41**	284	**7.8**
S65M	V	17.6	18.6	49	270	8.5
L68M	I	17.5	18.4	52	268	**7.6**
I75M	C	17.7	18.5	52	**247**	8.1

1 Significant shifts (bold) indicate changes in surface charge or charge distribution (conformation)
2 Indicator of internal structural integrity at pH 7
3 Indicator of structural resistance of heme crevice to chaotropic conditions
4 Includes C102T mutation to prevent dimerisation
5 All mutants also incorporate C102T and M64L changes
6 I = invariant, V = variable, C = conservative substitutions only

structure or function of the resulting protein. Sites of mutation were selected with this constraint in mind, as well as the need to explore a variety of structural contexts for suitability. The positions chosen are shown in Figure 1. We selected for test sites in helices, turns and Ω-loops; fully or partly buried, or surface exposed. The mutant molecules were modelled using Biosym's Insight/Discover programs; in no case did energy minimization procedures propose a structure that differed substantially from that of a native protein [J.C. Parrish and C.J.A. Wallace, unpublished data]. These inferences were supported by detailed physical characterization of the protein [table 1]. None of the mutants showed UV/Vis spectra that differed from the native form. Shifts in absorbance maxima would indicate change in the heme environment or ligation pattern. The K55M mutant has one less positive charge which is reflected in shorter retention times upon ion-exchange or ATP-affinity chromatography, but none of the others has properties that would suggest an altered surface charge distribution, or ATP binding site topology, that can signal surface conformation change. Redox potential is an indicator of the internal structural stability. The only significant deviation is shown by the I75M mutant. The 695nm spectral band is due to charge-transfer between Fe^{3+} and the ligating Met^{80} sulfur. The pK for the ligand-exchange reaction, in which a lysine residue inserts, indicates the stability of the heme crevice structure.

Table 2 sets out the results of biological assays of the mutant gene products, which were all expressed at levels greater than 3 mg of purified protein per litre of culture. Four of the mutants showed some diminution of relative biological activity that could usually be related to a specific change in the physicochemical properties described above. Nonetheless, all mutants were sufficiently similar to the parent protein to be useful bases for semisythetic routes, and it is of interest that the absolutely conserved Leu^{68} can be changed to Met with limited loss of functionality, yet change at the very variable K55 [it is M in plant species] caused the largest drop in electron transfer efficiency.

In addition to the introduced methionine residue, all cytochromes must contain Met^{80}, which is essential to function as the 6th heme iron ligand. Thus, cleavage can occur at two sites giving, at limiting CNBr concentration, five cleavage products. Generally cleavage occurs with equal facility at any site, but it was noted that Met^{68} seemed less susceptible than Met^{80}. Mixtures of cleavage products were separated by gel exclusion chromatography, a typical fractionation is shown in figure 2.

Table 2: Relative biological activities of the mutant cytochromes

Cytochrome	Class of mutated residue	Bioassay Succinate oxidase	Bioassay Cytochrome oxidase
Native			
Yeast		100%	100%
P25M	V	100%	n.d.
V28M	V	63%	85%
L35M	C	100%	86%
K55M	V	55%	46%
S65M	V	100%	n.d.
L68M	I	73%	52%
I75M	C	48%	92%

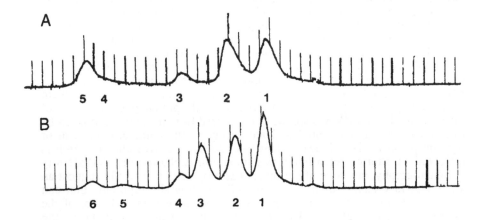

Figure 2: A comparision of the cleavage profiles of two mutant cytochromes. A is P25M, in which cleavage can occur at either residue 25 or 80. Peak 1 is uncleaved cytochrome; 2 is (-5 - 80H) [yeast cytochrome c has an N-terminal 5-residue extension] coeluted with (26-103); 3 is (26-80); 4, (81-104); and 5, (-5 - 25H). B is K55M, in which the shift of cleavage point from 25 to 55 yields a very different pattern. Peak 1 is uncleaved protein, and 2 is (-5 - 80H) alone. Peak 3 is (-5 - 55H); 4, 56-103; 5, 56-80; and 6, 81-103.

Table 3: Relationship between location of the site and religation efficiency

Cytochrome	Location of Cleavage site	CNBr Cleavage Sensitivity	Coupling yield [1]	Corrected yield [2]	Bioactivity of product
Native yeast [M^{64}]	Hydrophobic face of helix	Normal	0%	0%	n.a.
P25M	Surface-exposed extremity of 20s loop	Normal	25%	40%	91%
V28M	Surface-exposed γ - turn	Normal	20%	40%	45%
L35M	Buried residue i of ß-turn	Normal	<10%	<20%	17%
K55M	Outer face of short α - helix	Normal	12%	60%	36%
S65M	Hydrophilic face of amphipathic helix	Normal	50%	50%	100%
L68M	Interface of amphipathic helix	Low	15%	15%	60%
I75M	Partly exposed residue i of ß-turn	Normal	50%	50%	58%

1   Yield relative to heme fragment content of reaction mix
2   Yield relative to minority fragment content of reaction mix

Complexes of purified fragments were prepared in phosphate buffer, pH 7, at sub-millimolar peptide concentrations. These were reduced with sodium dithionite and kept anaerobic at room temperature for 24 h prior to gel-exclusion chromatography. Yields of ligated products were calculated and are shown in Table 3. All of the complexes derived from the mutants showed some religation. The small amount of material obtained from the L35M complex was inadequate for further purification, but its low biological activity suggests that it may be substantially contaminated and the yield shown is an overestimate. The other low yielding mutant, L68M, shares with it a partially buried location for the peptide bond being formed. All other cases gave efficient ligation. We had suggested as a result of our earlier study (5) that the reason for the failure of the native protein complex to catalyse the reaction was that the two termini would be required to penetrate the hydrophobic interior of the protein to make contact. This supposition is strongly supported by the pattern of reactivity displayed by the present set of mutants.

Conclusions

(1) Site-directed mutagenesis can be used with facility to create mutants with shuffled methionine residues. The degree of phylogenetic variability at a potential religation site is not a strict guide to the permissibility of a mutation there. Despite the wide variety of secondary structural locations chosen, mutations at the selected sites caused no major disruptions of protein structure or function.

(2) Differences in susceptibility to CNBr cleavage of some sites is apparent. Autocatalytic religation can occur at many types of site, giving functional products. However, efficiency is generally much greater when the peptide bond to be formed is surface-exposed than when it is buried. This survey supports the proposal made earlier (5) to explain the difference in religation efficiency between positions 64 and 65.

(3) We have developed the ability to predict and create sites for facile peptide condensation by the Autocatalytic Fragment Religation method. This permits the engineering by semisynthesis of yeast cytochrome c and, potentially, many other proteins.

Acknowledgments

We thank Angela Brigley for technical assistance, NSERC of Canada for financial support and the Dalhousie Medical Research Foundation for a scholarship to Anthony C. Woods.

References

1.   Wallace C.J.A. (1995) *Curr. Opin. Biotechnol.* 6, 403-410.
2.   Wallace, C.J.A. (1993) *FASEB. J.* 7, 505-515.
3.   Wallace, C.J.A. (1991) in Peptides 1990 [Giralt, E. and Andreu, D., Eds.] *ESCOM*, Leiden, 260-263.
4.   Proudfoot, A.E.I., Rose, K. and Wallace, C.J.A. (1989) *J. Biol. Chem.* 264, 8764-8770.
5.   Wallace, C.J.A., Guillemette, J.G., Smith, M. and Hibaya, Y. (1992) in Techniques in Protein Chemistry III (R.H. Angeletti, Ed.) *Academic Press*, San Diego, 209-217.
6.   Craig, D.B. and Wallace, C.J.A. (1995) *Biochemistry* 34, 2686-2693.
7.   Craig, D.B. and Wallace, C.J.A. (1991) *Biochem. J.* 279, 781-786.

Using semisynthesis to insert heavy-atom labels in functional proteins

Carmichael J.A. Wallace and Ian Clark-Lewis[a]

Departments of Biochemistry, Dalhousie University, Halifax, Nova Scotia, B3H 4H7 and
[a]University of British Columbia, Vancouver, B.C. V6T 1W5, Canada

I   Introduction

In protein semisynthesis fragments of the natural protein are used as preformed intermediates in the synthesis of a mutant structure (1). A variety of means can be used to effect the mutation, but the most common is the replacement of one fragment by a totally synthetic substitute in the fragment condensations used to make the holoprotein analog. Thus, semisynthesis can be used in ways that the genetic approach to specific mutants can not (or cannot easily) be used: to introduce d-amino acids, labels for spectroscopy at specific sites, or other non-coded side chain structures. In the present example we have used the method to incorporate residues including heavy atoms - here Selenium and Bromine - into cytochrome c for experiments using X-ray standing-waves to visualize protein orientation. While heavy-atom derivatives are useful in conventional X-ray crystallographic determinations of protein three-dimensional structure, this new technique is absolutely dependent on the presence of an atom that is X-ray fluorescent at the frequencies generated in synchrotron radiation.

X-rays reflected at a shallow angle from a silver mirror set up standing waves. The heights of the field intensity maxima can be tuned by varying the angle (figure 1). If a monolayer of X-ray fluorescent heavy atom is present above the mirror surface, as when a regular array of labelled macromolecules forms on it, the fluorescence will be at a maximum when a field intensity maximum coincides with the monolayer (2). The fluorescence is detected and the monolayer's distance from the mirror can be calculated. With two heavy atoms, two distances can be measured and a precise orientation determined. Since artificial membranes can be layered on the mirror surface, the method is ideally suited to studies of the interaction of membranes and associated proteins or peptide hormones.

II   Methods

A variety of heavy atom labels are suitable. We chose selenium and bromine, which have good X-ray fluorescence characteristics and are readily available incorporated in amino acids. Peptides corresponding to residues 66-104 of horse cytochrome c were prepared by solid-phase peptide synthesis using methods described in detail elsewhere (3), purified by preparative HPLC, and checked by analytical HPLC, amino acid analysis and mass spectrometry. The

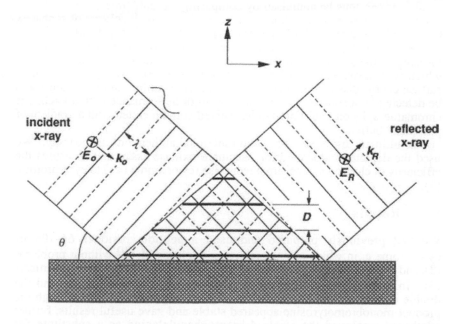

Figure 1. The X-ray standing wave field formed by the interference between the incident and specular-reflected plane waves above an X-ray mirror surface. Note, to satisfy the total external reflection condition, the incident angle 0 is normally less than 1°. In the figure, E_O and k_O are the electric field amplitude and wave vector of the incident X-ray beam while E_R and k_R are those for the reflected wave; D represents the standing wave period. The position of the antinodes are marked by the horizontal bars.

novel analogs of this peptide that we made were bromotyrosine 97 (66-104); selenomethionine 80, bromotyrosine 97 (66-104); selenomethionine 68, bromophenylalanine 74 (66-104); selenomethionine 68, bromophenylalanine 97 (66-104); and bromophenylalanine 74, selenomethionine 80 (66-104). These substitutions replaced leucine at position 68 [naturally invariant but methionine is acceptable (4)], tyrosine at 74 [found as phenylalanine in some yeasts], methionine at 80 [invariant but selenomethionine can replace it with partial functionality] and tyrosine at 97 [also phenylalanine in a few species].

Protein analogs were prepared by condensation of the synthetic peptides with fragment (1-65H) derived from the native protein by cyanogen bromide cleavage. Equimolar mixtures (0.4mM) of the two fragments were dissolved in 50mM potassium phosphate buffer, pH 7.0. The non-covalent complex formed is reduced with sodium dithionite and sealed in a glass-walled syringe. The mixture is kept anaerobic, in the dark and at room temperature for 24 h before resolution of the religated product from unreacted fragments by gel-exclusion chromatography. The leading fraction was purified by cation-exchange chromatography and checked by HPLC.

Among the functional tests of protein integrity employed were UV-visible spectroscopy, including titration of the 695nm charge-transfer band. Loss of this band with rising pH signals a ligand exchange reaction (lysine replaces methionine) that is normally resisted by the stability of the heme crevice. The strength of the non-covalent forces stabilizing internal structure is normally so great that O_2 is slow to penetrate the protein matrix protecting the heme. This parameter can thus be appraised by comparing the half-time of autoxidation when the reduced protein is buffer-exchanged on a small column of sephadex G25 into O_2-saturated 50mM potassium phosphate buffer, pH 7.0. Collapse of the characteristic 550nm α-band is follow spectrophometrically. Another measure of the stability of internal structure is oxidation-reduction potential, which was measured by the method of mixtures (3). Any perturbation of surface charge distribution that would result from conformational change will be detected by a shift in retention time of high performance cation-exchange chromatography on a sulfopropyl-derivatized column eluted with a gradient of phosphate buffer.

The ultimate test of structural and functional authenticity is bioassay. We used the depleted-mitochondrion succinate oxidase assay that measures the efficiency of electron transfer from reductase (cytochrome bc_1) to cytochrome c (3).

III   Results

We had previously prepared and used a peptide, residues 66-104 of cytochrome c, in which selenomethionine replaces Met^{80} without problems (3), and demonstrated the usefulness of the protein analog in XSW experiments (5). In making the 3,5-dibromotyrosine^{97} analog of this peptide, and the double mutant, we noted bromine loss by mass spectrometry. Although the product monobromotyrosine appeared stable and gave useful results, further syntheses employed the stable 4-bromophenylalanine as a substitute for tyrosine at positions 74 and 97. Selenomethionine oxidation in SeM^{80} mutants was not significant, but a slight tendency to air oxidation of SeM^{68} was noted. Since this residue is not crucial, it was not considered a serious problem, as the selenoxide form can be resolved from unmodified peptide by HPLC, and no evidence of the selenoxide form of the protein was observed upon high-performance cation-exchange chromatography. In general, religations involving these proteins occurred with yields of >50% with 1:1 component ratios, though the yield of SeM^{68}, BrF^{74} cytochrome c was only 25%.

Studies of the analogs to validate their suitability as models for the natural cytochrome are tabulated below.

Table 1: Properties of the cytochrome analogs

Cytochrome	Heme Spectra λmax Fe^{3+}/Fe^{2+}	pK 695nm	T_{12} autoxidation	E'_m	HPIEC r.t. red ox	Bioassay
Native Horse	529,408/549,520,415	9.25	>100h	266mV	13.0 13.9(a) 13.6 14.6(b)	100%
SeM80[1]	Red shifted	9.20 (691nm)	n.d.	212mV	13.0 13.9(a)	44%
BrY97 SeM80,	minor blue shift	8.85	9h	260mV	13.0 14.0(a)	95±5%
BrY97 SeM68,	minor red shift	9.00	<1h	202mV	13.0 14.0(a)	34±5%
BrF74 SeM68,	minor blue shift	7.80	8h	254mV	13.6 14.6(b)	90±10%
BrF97 BrF74,	minor blue shift	8.75	10h	268mV	13.7 14.7(b)	97±3%
SeM80	minor red shift	8.75	1.25h	205mV	13.6 14.6(b)	30±1%

1   Data from Ref 3

Some mutants are less resistant to denaturing conditions, but all indications are that they adopt the native fold; this is reflected in near-full biological activity of all but those containing SeMet[80]. In this case replacement of sulphur by selenium as the 6th heme iron ligand changes redox potential, and consequently driving force for electron transfer, but leaves the protein otherwise unaffected (3). Thus, the choice of labels is compatible with the strategy and semisynthesis provides a simple means to precise and non-invasive labelling.

Data on the double mutants is currently being collected at the Brookhaven synchrotron and analysed at Ohio State University [M. Caffrey and J. Wang, unpublished results].

Acknowledgments

Thanks are due to Angela Brigley for technical assistance and NSERC and NIH for financial support.

References

1.   Wallace, C.J.A. *FASEB J.* 1993 **7**, 505-515.
2.   Wang, J., Bedzyk, M.J., Penner, T. and Caffrey, M. *Nature* 1991, **354**, 377-380.
3.   Wallace, C.J.A. and Clark-Lewis, I. *J. Biol. Chem.* 1992, **267**, 3852-3861.
4.   Wallace, C.J.A., Woods, A.C. and Guillemette, J.G. (1995) This volume, pp 000-000.
5.   Wang, J., Wallace, C.J.A., Clark-Lewis, I. and Caffrey, M. *J. Mol. Biol.* 1994, **237**, 1-4.

Optimized Methods for Chemical Synthesis of Bovine Pancreatic Trypsin Inhibitor (BPTI) Analogues

George Barany,* Christopher M. Gross,*,† Marc Ferrer,*
Elisar Barbar,† Hong Pan,† and Clare Woodward†

Departments of *Chemistry and †Biochemistry,
University of Minnesota, *Minneapolis, MN 55455 and †St. Paul, MN 55108

I. Introduction

Solid-phase peptide synthesis has advanced to the challenge of preparing small proteins. In 1992, we published an efficient procedure for the stepwise synthesis of the 58-residue protein bovine pancreatic trypsin inhibitor (BPTI) and of two analogues which replace paired half-cystines by paired α-amino-n-butyric acid (Abu) isosteres [1, 2]. These experiments used chemistry with the base-labile N^{α}-9-fluorenylmethyloxycarbonyl (Fmoc) group and compatible side-chain protecting groups, as well as an appropriate anchoring handle, then-optimized methods for activation/coupling and for final cleavage/deprotection, and novel polyethylene glycol-polystyrene (PEG-PS) supports with advantageous physicochemical and mechanical properties. The linear precursor proteins were assembled by automated synthesis and purified to homogeneity, and reproducible methods were devised to achieve proper folding and formation of the disulfide bridges (three in native sequence, one or none in analogues). Overall isolated yields in these initial studies were 2 to 5%, and purities were in the 92 to 99% range, depending on the analogue. Subsequently, we have made further refinements in various aspects of the protocols, and are now able to generate, on a semi-routine basis, synthetic BPTI analogues in 3 to 6% overall yields and > 98% purity (Figure 1).

In a separate phase of our research program, proteins accessed by total synthesis (Table 1) are characterized by one and two-dimensional nuclear magnetic resonance (NMR), circular dichroism (CD), fluorescence enhancement upon ANS binding, native gel electrophoresis, and biological assay studies, as is being described elsewhere [1-6]. We have shown that BPTI analogues with fewer than the full complement of disulfide bridges serve as instructive models for partially folded intermediates in protein folding. Partially folded species are currently of great interest, as they are believed to have structures very similar to transient intermediates formed during the first 10 to 20 ms of the protein folding process. The BPTI analogue termed [14-38]$_{Abu}$, which contains a single disulfide that connects the flexible loops via Cys^{14} and Cys^{38}, is a highly ordered β-sheet molten globule under appropriate conditions [3-6], whereas the analogue with all disulfides eliminated, [R]$_{Abu}$, is a prototype collapsed "molten coil" lacking stable secondary structure but containing extensive non-random tertiary contacts [4, 7].

The targets that we describe generally cannot be produced by molecular biology techniques, as they are degraded rapidly by cellular proteases. A further advantage of synthesis is the capability to introduce, at selected positions, unnatural amino acid residues as well as isotopically labelled residues, thereby extending the scope of structure/function questions that can be addressed, and in some cases facilitating the spectroscopic studies. Thus, ^{15}N residues incorporated at judiciously chosen positions in [14-38]$_{Abu}$ made possible complete assignments of NMR signals due to both folded and unfolded conformations [5, 6]. In addition, we have devised an alternative approach to access labelled BPTI analogues: parent proteins are reduced, followed by alkylation with ^{13}C-iodomethane [5, 7]. This approach replaces Cys residues with S-methylcysteine (Smc). The resultant protein, [R]$_{Smc}$, has proven to be functionally quite similar to [R]$_{Abu}$ [7]. The present communication summarizes our recent progress with the synthesis and modification chemistries, including aspects related to labelling of proteins, and provides accompanying analytical data.

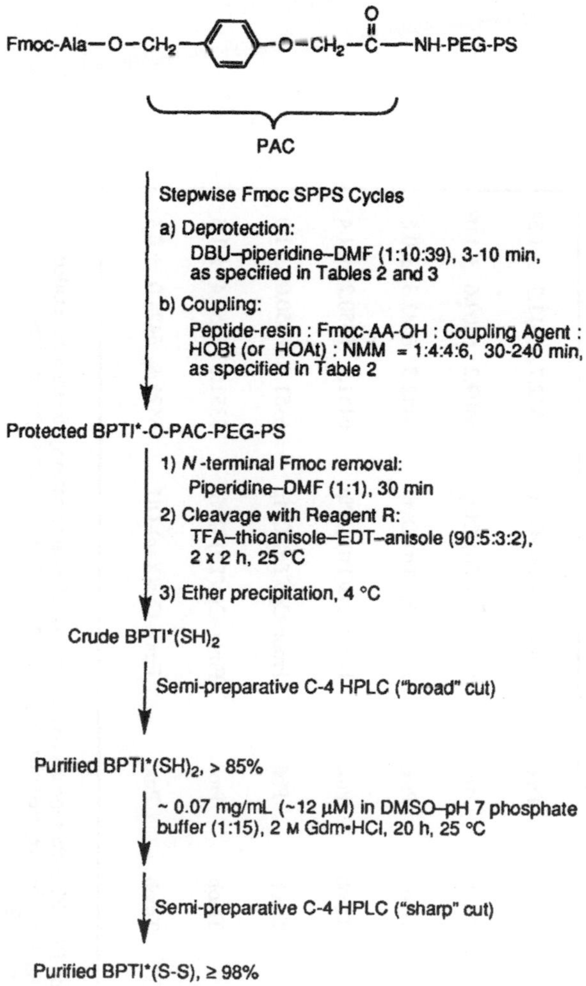

Fmoc-Ala—O—CH$_2$—⟨ ⟩—O—CH$_2$—C—NH-PEG-PS
(with O double bond on C)

PAC

Stepwise Fmoc SPPS Cycles

a) Deprotection:
   DBU–piperidine–DMF (1:10:39), 3-10 min,
   as specified in Tables 2 and 3

b) Coupling:
   Peptide-resin : Fmoc-AA-OH : Coupling Agent :
   HOBt (or HOAt) : NMM = 1:4:4:6, 30-240 min,
   as specified in Table 2

Protected BPTI*-O-PAC-PEG-PS

1) *N*-terminal Fmoc removal:
   Piperidine–DMF (1:1), 30 min

2) Cleavage with Reagent R:
   TFA–thioanisole–EDT–anisole (90:5:3:2),
   2 x 2 h, 25 °C

3) Ether precipitation, 4 °C

Crude BPTI*(SH)$_2$

Semi-preparative C-4 HPLC ("broad" cut)

Purified BPTI*(SH)$_2$, > 85%

~ 0.07 mg/mL (~12 µM) in DMSO–pH 7 phosphate
buffer (1:15), 2 M Gdm•HCl, 20 h, 25 °C

Semi-preparative C-4 HPLC ("sharp" cut)

Purified BPTI*(S-S), ≥ 98%

Figure 1. A general scheme for solid-phase synthesis of protein analogues, BPTI*, with one disulfide bridge. Further details in Tables 1-3, and text.

Table 1. Synthetic BPTI Protein Analogues[a]

BPTI Species	Coupling Chemistry	Asn/Gln Protection	Arg Protection	Cys Derivative	Mass Calcd.	Mass Found
Naturally Occurring[b]	BOP	Tmob	Pmc	Fmoc-Cys(Trt)-OPfp	6512.7	6511.7 ± 1.3[b]
[R]Abu[a,b]	BOP	Tmob	Pmc	Fmoc-Cys(Trt)-OPfp	6409.2	6409.6 ± 1.0[b]
[14-38]Abu[a]	HATU	Tmob	Pmc	Fmoc-Cys(Trt)-OPfp	6442.3	6442.3 ± 0.6
[15N]8-[14-38]Abu[c]	BOP	Tmob	Pmc	Fmoc-Cys(Trt)-OPfp	6451.6	6450.2 ± 1.4
[14-38]Abu Tyr23Ala	BOP	Tmob	Pbf	Fmoc-Cys(Trt)-OPfp	6351.6	6350.0 ± 1.0
[14-38]Abu Tyr21Ala	BOP	Tmob	Pmc	Fmoc-Cys(Trt)-OPfp	6351.6	6350.4 ± 1.3
[30-51]Abu Met52Nle	HBTU	Tmob	Pmc	Fmoc-Cys(Xan)-OH	6425.6	6424.0 ± 0.4

a   All proteins were assembled on PAC-PEG-PS supports on a continuous-flow automated synthesizer as detailed herein. [R]Abu and [14-38]Abu were also prepared manually [2], and [14-38]Abu was also prepared by batchwise automated procedures [1, 2].

b   As reported in ref. 1.

c   [15N] labels at Leu6, Gly12, Ala16, Leu29, Gly37, Ala40, Ala48, and Gly56, as outlined in ref. 6.

II. Materials and Methods

A. *General*

The essential materials and methods have been reported already [1, 2, 4, 7]; hence descriptions that follow cover only modifications, improvements, or additional details. Naturally occurring BPTI is from Novo BioLabs (Bagsvaerd, Denmark), and is used without further purification by dialyzing to the appropriate final buffer. Dialysis is now carried out using Spectra/Por® CE (Cellulose Ester) membrane tubing (Houston, TX), with a nominal weight cutoff of 500; concentrations are achieved in Amicon Model 8050 or 8400 ultrafiltration stirred cells (Beverly, MA) with aid of Amicon YM1 membrane, cutoff 1000. Suitably protected Fmoc-amino acid derivatives are primarily from the Biosearch Division of PerSeptive Biosystems (Framingham, MA; formerly Millipore in Bedford, MA), Advanced Chemtech (Louisville, KY), Star Biochemicals (Torrance, CA), Chem-Impex (Wood Dale, IL), Richelieu Biotechnologies (St. Hyacinthe, Quebec, Canada), or Nova Biochem (Nottingham, United Kingdom). In addition, Fmoc-Asn/Gln(Tmob)-OH are from Bachem Feinchemikalien AG (Bubendorf, Switzerland), Fmoc-Arg(Pbf)-OH is from Sygena (Liestal, Switzerland), and Fmoc-Cys(Xan)-OH is prepared in our laboratory by our recently described method [8]. The ^{15}N-labelled Fmoc-amino acids are from Isotec, Inc. (Miamisburg, OH). The solvent for peptide synthesis, *N,N*-dimethylformamide (DMF) is a Dupont product, 99%, distributed by Aldrich (Milwaukee, WI) and obtained in 18 L drums; prior to use, it is checked for the absence of amines by a bromphenol blue test [9]. Piperidine, *N*-methylmorpholine (NMM) and 1-hydroxybenzotriazole (HOBt) are "Sequencing Grade" from Fisher (Pittsburgh, PA), 1,8-diazabicyclo[5.4.0]undec-7-ene (DBU) is from Aldrich, benzotriazole-1-yloxytris(dimethylamino)-phosphonium hexafluorophosphate (BOP reagent) is from AminoTech (Nepean, Ontario, Canada) or Chem-Impex, and *N*-[(1*H*-benzotriazol-1-yl)(dimethylamino)methylene]-*N*-methylmethanaminium hexafluorophosphate *N*-oxide (HBTU), *N*-[(dimethylamino)-1*H*-1,2,3,triazolo[4,5-*b*]pyridin-1-ylmethylene]-*N*-methylmethanaminium hexafluorophosphate *N*-oxide (HATU) and 3-hydroxy-3*H*-1,2,3-triazolo[4,5-*b*]pyridine (HOAt) are from Biosearch-PerSeptive.

Analytical high performance liquid chromatography (HPLC) is performed using a Vydac analytical C-4 reversed-phase column (214TP10415, 10 μm particle size; 0.46 x 15 cm) on a Beckman system configured with two Model 125 pumps and a Model 115 detector, controlled from an IBM computer with Beckman System Gold software. Protein samples are chromatographed at 1.2 mL/min using 0.1% aqueous TFA–CH$_3$CN (9:1 to 1:1 over 40 min), detection at 220 nm. Semi-preparative HPLC is performed on a Waters Deltaprep system, using either a Vydac semi-preparative C-4 reversed-phase column (214TP1010; 10 μm particle size; 1.0 x 25 cm), or preferably a PrePak® cartridge (25 x 100 mm) containing Delta-Pak™ C-4 (15 μm, 300 Å) packing material compressed with a Waters 25 x 10 Radial Compression Module set to 600 psi. All [14-38] species are chromatographed at 5 mL/min in 0.1% aqueous TFA–CH$_3$CN (26:74 to 40:60 over 45 min for reduced; 25:75 to 34:66 over 35 min once oxidized). Reduced [30-51]$_{Abu}$ behaves somewhat differently (27:73 to 37:63 over 60 min is ideal), as does [R]$_{Smc}$ (27:73 to 32:68 over 60 min). Detection is usually at 280 nm for the reduced protein and at 220 nm for oxidized protein, reflecting the amount of material available at each step (Scheme 1). For concentration/purification of those analogues with a single disulfide generated by DMSO oxidation [1, 10], the DMSO-containing reaction mixture is sometimes applied directly onto the reversed-phase column through the pump head at a high flow rate (10-15 mL/min). As described previously [1, 2], individual chromatography fractions are evaluated by capillary zone electrophoresis (CZE) to identify pure material for further pooling and lyophilization. The purities of the chemically synthesized as well as chemically modified proteins are determined by analytical HPLC and CZE. Masses of all purified proteins are verified by ion electrospray mass spectrometry (ESMS), and compared to calculated values (details in Table 1 and text).

B. *Total Chemical Protein Synthesis*

The PEG-PS supports, previously made in our own laboratories [1, 11-13], are now commercially available from the Biosearch Division of PerSeptive Biosystems, complete with a norleucine "internal reference" amino acid (IRAA), and the C-terminal residue Fmoc-Ala loaded

(0.22 mmol/g) as a p-alkoxybenzyl (PAC) ester manufactured by our recommended preformed handle procedure. PEG-PS supports are ideal for continuous-flow peptide syntheses, which are carried out on a PerSeptive (formerly MilliGen) Model 9050 instrument. The support should be shrunk in methanol (rather than swollen in DMF) to facilitate packing into the synthesis column. Chain assemblies to introduce successive Fmoc-amino acids (Table 2) are carried out typically on a scale starting with 0.75 g of Fmoc-Ala-O-PAC-PEG-(Nle)-PS packed into a standard 1 x 10 cm synthesis column, and provide typically 1.7 g of protected-BPTI*-O-PAC-PEG-PS (uncorrected for samples withdrawn throughout the chain assembly process for analytical purposes). Whereas previously Fmoc removals were accomplished with piperidine–DMF (1:4), we now routinely add 2% DBU to the deprotection reagent (Table 3). In addition to the previously described BOP/HOBt/NMM protocols [1, 2, 14, 15], we are also using HBTU/HOBt/NMM or HATU/HOAt/NMM [16-20]. The molar excess of activated Fmoc-amino acid is 4-fold, and the effective concentration circulating through the column is ~ 0.1 M. Irrespective of the coupling chemistry used (Table 1), we estimate on the basis of weight gain and amino acid ratios (with respect to Nle IRAA) that approximately 75% of the initial chains are still growing at the end of the syntheses. Throughout, back pressures are observed to remain below 100 psi, well below the acceptable level of 200 psi or less. [During earlier syntheses with more resin in the same standard column, it was necessary once or twice to unscrew the top of the column and then reclose it.] At certain stages, the flow rate is customized (details in Table 3), but generally for deprotections, washings, and couplings, it is scaled to 8.3 mL/min. This value, which differs from manufacturer recommendations by a factor of 2.2, is achieved by programming higher than actual amounts of resin, and correspondingly lower than actual loading levels.

Upon completion of chain assembly, the dried protected peptide-resin is cleaved portion-wise (~ 200 to 300 mg per cleavage) with freshly prepared Reagent R: trifluoroacetic acid–thioanisole–1,2-ethanedithiol–anisole (90:5:3:2), as described previously [1, 2]. Release of the chains from the support proceeds in high yields (~ 88 to 95%), as judged by amino acid ratios (with respect to Nle IRAA) of the recovered resin. Other than for the protein with all six Cys residues, we no longer add β-mercaptoethanol at the ether precipitation step, nor do we carry out a reduction step prior to semi-preparative C-4 reversed-phase HPLC. For those analogues with one disulfide, the HPLC-purified linear precursor in the reduced form (~ 3 to 5 mg, ~ 0.4 to 0.6 µmol obtained; "broad" cut taken which shows a major peak of ~ 85% homogeneity upon analytical

*   *   *   *   *   *   *   *   *   *

Notes to Table 2 (Table 2 appears on the next page)

General

The synthesis of [30-51]$_{Abu}$ Met52Nle was carried out starting with 0.75 g of Fmoc-Ala-O-PAC-PEG-PS. Couplings were mediated by HBTU/HOBt/NMM (Figure 1), with the activated protected Fmoc-amino acid solution circulating through the column at 8.3 mL/min for the indicated time. See text for further information. Deprotections were according to code I (initial), D (diketopiperazine reduction), S (standard), or E (extended), as described further in Table 3. Other analogues were synthesized by closely related protocols (Table 1 and text).

Specific Notes

A   The first amino acid Ala was already on the support as a PAC ester, through coupling of a preformed handle derivative.

B   Initial Fmoc deprotection difficult.

C   Potential diketopiperazine formation minimized by use of short deprotection at a high flow-rate, followed by quick coupling.

D   Drop in Fmoc-dibenzofulvene adduct peak height, as explained further in text.

E   Double-coupling of indicated Fmoc-amino acid [i.e., no intervening deprotection before second coupling].

F   Consecutive Ile-Ile.

G   Arg[1] is notoriously difficult to introduce quantitatively. Despite three couplings, final Fmoc-BPTI*-PEG-PS resin remained ninhydrin-positive.

Table 2. Protocol for Continuous-Flow Fmoc Solid-Phase Synthesis of [30-51]Abu Met52Nle

Cycle No.	Wild Type Residue	Derivative	Coupling time (min)	Deprotection Protocol	Notes
0	A58	Fmoc-Ala-O-PEG-PS	N/A	N/A	A
1	G57	Fmoc-Gly-OH	60	1	B
2	G56	Fmoc-Gly-OH	30	D	C
3	C55	Fmoc-Abu-OH	60	S	
4	T54	Fmoc-Thr(tBu)-OH	60	S	
5	R53	Fmoc-Arg(Pmc)-OH	90	S	
6	M52	Fmoc-Nle-OH	60	S	
7	C51	Fmoc-Cys(Xan)-OH	60	S	
8	D50	Fmoc-Asp(OBu)-OH	60	S	
9	E49	Fmoc-Glu(OBu)-OH	60	S	
10	A48	Fmoc-Ala-OH	90	S	
11	S47	Fmoc-Ser(tBu)-OH	60	S	
12	K46	Fmoc-Lys(Boc)-OH	75	E	D
13	F45	Fmoc-Phe-OH	60	S	
14	N44	Fmoc-Asn(Tmob)-OH	75	S	
15	N43	Fmoc-Asn(Tmob)-OH	60	E	
16	R42	Fmoc-Arg(Pmc)-OH	90	E	
17	K41	Fmoc-Lys(Boc)-OH	60	S	
18	A40	Fmoc-Ala-OH	60	S	
19	R39	Fmoc-Arg(Pmc)-OH	90	S	
20	C38	Fmoc-Abu-OH	90	S	
21	G37	Fmoc-Gly-OH	60	S	
22	G36	Fmoc-Gly-OH	75	S	
23	Y35	Fmoc-Tyr(tBu)-OH	75	S	D
24	V34	Fmoc-Val-OH	90	E	
25	F33	Fmoc-Phe-OH	90	E	
26	T32	Fmoc-Thr(tBu)-OH	90	E	
27	Q31	Fmoc-Gln(Tmob)-OH	72	E	
28	C30	Fmoc-Cys(Xan)-OH	72	E	
29	L29	Fmoc-Leu-OH	72	E	
30	G28	Fmoc-Gly-OH	72	E	
31	A27	Fmoc-Ala-OH	75	E	
32	K26	Fmoc-Lys(Boc)-OH	90	E	
33	A25	Fmoc-Ala-OH	72	E	
34	N24	Fmoc-Asn(Tmob)-OH	90	E	
35	Y23	Fmoc-Tyr(tBu)-OH	90	E	
36	F22	Fmoc-Phe-OH	90	E	
37	Y21	Fmoc-Tyr(tBu)-OH	90	E	
38	R20	Fmoc-Arg(Pmc)-OH	120	E	
39	I19	Fmoc-Ile-OH	60	E	
40		Fmoc-Ile-OH	30	N/A	E
41	I18	Fmoc-Ile-OH	60	E	F
42		Fmoc-Ile-OH	30	N/A	E
43	R17	Fmoc-Arg(Pmc)-OH	240	E	
44	A16	Fmoc-Ala-OH	90	E	
45	K15	Fmoc-Lys(Boc)-OH	90	E	
46	C14	Fmoc-Abu-OH	90	N/A	E
47	P13	Fmoc-Pro-OH	90	E	
48	G12	Fmoc-Gly-OH	90	E	
49	T11	Fmoc-Thr(tBu)-OH	90	E	
50		Fmoc-Thr(tBu)-OH	72	N/A	
51	Y10	Fmoc-Tyr(tBu)-OH	60	E	
52	P9	Fmoc-Pro-OH	72	E	
53		Fmoc-Pro-OH	72	N/A	
54	P8	Fmoc-Pro-OH	72	E	
55		Fmoc-Pro-OH	72	N/A	
56	E7	Fmoc-Glu(OBu)-OH	60	E	
57		Fmoc-Glu(OBu)-OH	78	N/A	
58	L6	Fmoc-Leu-OH	72	E	
59		Fmoc-Leu-OH	40	N/A	
60	C5	Fmoc-Abu-OH	120	E	
61	F4	Fmoc-Phe-OH	120	E	
62	D3	Fmoc-Asp(OBu)-OH	120	E	
63	P2	Fmoc-Pro-OH	78	E	
64		Fmoc-Pro-OH	72	N/A	
65	R1	Fmoc-Arg(Pmc)-OH	240	E	
66		Fmoc-Arg(Pmc)-OH	60	N/A	
67		Fmoc-Arg(Pmc)-OH	60	N/A	G

Table 3. Deprotection Protocols[a]

	Code	First Wash (min)	Flow Rate (mL/min)	Second Wash (min)	Flow Rate (mL/min)
Initial	I	6.0	30.0	2.3	30.0
DKP Reduction	D	1.5	24.9	2.3	24.9
Standard	S	1.0	8.3	5.0	3.0
Extended	E	3.0	10.0	7.0	3.0

[a] The deprotection reagent was DBU–piperidine–DMF (1:10:39).

HPLC) is dissolved in 3 mL of 0.01 N aqueous HCl, and diluted further with ~ 40 mL of a mixture of dimethyl sulfoxide (DMSO)–2 M guanidinium hydrochloride (Gdm·HCl)–0.1 M sodium phosphate buffer, pH 7.0 (1:15). Oxidation is carried out for 20 h at 25 °C. The addition of Gdm·HCl is a recent innovation [the conditions published in ref. 1 involve no Gdm·HCl, double the DMSO concentration, a pH of 6, and a shorter reaction time]. The presence of Gdm·HCl is believed to aid disulfide formation by creating a partially denaturing environment. It is also noteworthy that under our oxidation conditions, the amount of precipitation is negligible. The oxidation solution is then concentrated to ~ 15 mL and applied in 3 batches for HPLC purification, or alternatively loaded directly onto the column prior to purification (see "General"). The collected fractions with the proper purity (≥ 98%) are lyophilized to provide, typically, white powders (2.5 to 3.5 mg) in 3 to 5 % overall yields.

C. Chemical Modification of BPTI

The conversion of Cys residues in BPTI to Smc residues is carried out as follows: Naturally occurring BPTI (15-20 mg) in 5 mL of 6 M Gdm·HCl, 1 mM EDTA, 0.1 M Tris·HCl, pH 8.7, is reduced with 10 mM dithiothreitol (DTT) for 8 h at 25 °C. The resultant reduced BPTI is desalted on Sephadex G-10 equilibrated with H_2O/HCl at pH 3, lyophilized, and taken up in 5 mL of 1 mM EDTA, 0.1 M Tris·HCl, pH 8.7 containing ^{13}C-iodomethane (~ 3 to 5 μL, ~ 25-fold excess), and the alkylation is conducted for 30-45 min at 25 °C. Desalting as before, followed by semi-preparative C-4 reversed-phase HPLC, provides the desired [R]$_{Smc}$ protein product with six ^{13}C-methyl groups in ~ 50% yield, ESMS calcd.: 6608.7; mass found 6609.9 ± 2.0. A variant of [R]$_{Smc}$ with the ^{13}C-label only at positions 14 and 38 is prepared by omitting the Gdm·HCl denaturant from the first step of the overall procedure. [It is known, see refs. 21-23, that the 14-38 disulfide is susceptible to selective reduction since the 5-55 and 30-51 disulfide bridges are buried; the transformation is usually carried out with $NaBH_4$ rather than with DTT as in the present work.] The [13]C-[Smc[14], Smc[38]]-BPTI is purified by reversed-phase HPLC, and then reduced in the presence of Gdm·HCl, followed by alkylation with ^{12}C-iodomethane exactly as described above. The final overall yield of purified protein is typically 30 to 40%.

III. Results and Discussion

Since this research program was started, seven proteins from the BPTI family have been chemically synthesized (Tables 1 and 2), some several times due to material needs of the biophysical studies. Much was learned from manual syntheses [2], but in current work, an automated synthesizer is used for carrying out the repetitive chain assembly steps. Products from the earlier manual efforts included substantial levels (up to 35%) of the by-products in which Met[52] was oxidized to the corresponding sulfoxide. These eluted as shoulders on the front edges of the main BPTI analogue peaks, and were isolated by careful HPLC (ESMS showed masses 16 more than expected for the desired products) [2]. A separate reduction step with ~ 0.7 M N-methylmercaptoacetamide (MMA), in 10 mM sodium phosphate, 2 M Gdm·HCl, pH 7.0, for at least two days at 37 °C, under N_2 was inserted prior to the first purification step, in order to regenerate the Met to its thioether form [2, 24]. Automated syntheses are conducted under a more readily controlled inert atmosphere, and over a more rapid time frame (6 days vs. 3 weeks); hence we do not observe sulfoxide by-products in our more recent work. Nevertheless, the problem with methionine may be side-stepped entirely by assembling BPTI analogues that contain the norleucine (Nle) isostere instead of Met (e.g., Table 1, bottom line; Table 2), as has been done fruitfully in a number of other synthetic structure/function studies.

As improved chemistries have been developed from our laboratories or reported in the literature, we have evaluated their applications to the BPTI system. For example, considerable progress in the formulation and production of PEG-PS supports [11-13] has confirmed the importance of these materials for reproducible continuous-flow solid-phase syntheses of small proteins in the BPTI size range. On the other hand, initial purities and final yields of homogeneous BPTI analogues have not been affected to any significant extent by modifications in the reagents and/or protocols used for coupling (Table 1, Figure 1). Thus, the BOP/HOBt/NMM protocol reported originally is equally effective as protocols involving the uronium salts HBTU or HATU. Note that our conditions simply substitute the uronium salt for BOP, use the appropriate corresponding triazole (HOBt or HOAt), and retain NMM as the base; in this sense, they differ from literature protocols with these reagents [16-20].

We observe that as the BPTI chain grows, Fmoc removal becomes increasingly difficult, and our optimized protocol (Tables 2 and 3) takes this into consideration. The problem is reflected by broadening of the Fmoc adduct peak observed during continuous-flow monitoring, and confirmed by weaker Kaiser ninhydrin tests from separate manual syntheses [2]. Generally, for the first thirty cycles, the peaks are sharp (half-width ≤ 0.5 min); peaks from later synthesis cycles show some broadening and tailing but still return to baseline well within the total time allotted for Fmoc removal (i.e., half-width ~ 1 min; baseline achieved in ≤ 5 min) (see Figure 2). Therefore, longer reaction times and/or higher % of piperidine [up to piperidine–DMF (1:1)] are used when needed to remove the Fmoc group unambiguously at "difficult" points. Our current practice for continuous-flow deprotection is to add 2% DBU to the standard piperidine–DMF (1:4) mixture, applied in two pulses for a total of 3 to 10 min depending on where in the sequence the step is applied. As known from the work of Wade *et al.* [25], DBU is a much more effective base for abstraction of the labile proton of Fmoc, and the β-elimination which follows is hence more rapid. Piperidine is still needed to scavenge the resultant dibenzofulvene co-product, as noted by Fields *et al.* [26], among others (these authors find that 2% piperidine is sufficient). The known tendency of DBU to promote aspartimide formation [27, 28] is not taken as a serious detriment with regard to its specific application to assembly of the BPTI sequence (Table 1). We find that our DBU–piperidine deprotection conditions lead to sharp peak shapes *throughout* the syntheses.

Our accumulated experiences suggest that the key to success in BPTI chain assembly is to address properly the *consistent* "difficult" sequences. The points which require extra attention, i.e., residues Gly[57] and Gly[56], Ser[47] to Phe[45], Tyr[35], and Tyr[10], have been identified by real-time monitoring, as well as Edman sequential degradation preview analysis of completely assembled BPTI analogues [1, 2]. At the very start of the synthesis, procedures are modified (see Tables 2 and 3), due to sluggish deprotection of the starting Fmoc-Ala-O-PAC-PEG-(Nle)-PS, and the risk of diketopiperazine formation subsequent to deprotection of the second residue (Gly[57]). During the cycles for incorporation of Lys[46], Tyr[35], and Tyr[10], respectively, the Fmoc deprotection peak *heights* dropped by almost 50%, ~ 50%, and ~ 25% with respect to the previous cycles. Since heights are not necessarily proportional to area, these monitoring data are just a qualitative gauge of the progress of the synthesis. The "difficult" points are not affected by alterations in Fmoc removal conditions (Table 3) or coupling reagents (earlier paragraph).

Other aspects of the design and execution of successful syntheses in the BPTI family include the choice of side-chain protecting groups, and the corresponding cleavage conditions. For our original protocols, we used acid-labile groups as follows: *tert*-butyl (*t*Bu) ethers and esters for Ser, Thr, Tyr, Asp, and Glu; *tert*-butyloxycarbonyl (Boc) for Lys; 2,2,5,7,8-penta-methylchroman-6-sulfonyl (Pmc) for Arg; 2,4,6-trimethoxybenzyl (Tmob) for Asn and Gln, and trityl (Trt) for Cys. The 2,2,4,6,7-pentamethyldihydrobenzofuran-5-sulfonyl (Pbf) protecting group, reported recently [29, 30] as an alternative for Arg protection, appears to provide equivalent overall results (both structures shown in top portion of Figure 3). More crucial are the choices for Asn, Gln, and Cys, since three families of protecting groups have been reported for these residues [8, 27] (see bottom portion of Figure 3). Fields and Fields [31, 32] attempted to synthesize BPTI

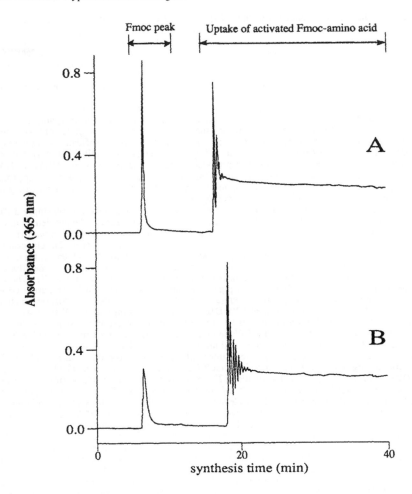

Figure 2. Post-reactor UV traces at 365 nm from continuous-flow synthesis of naturally occurring BPTI. Deprotection with piperidine–DMF (1:4). Panel A shows a sharp Fmoc peak from deprotection of Met[52], and the pattern for uptake of activated Fmoc-amino acid during incorporation of Cys[51]. Panel B shows the broader Fmoc peak from the deprotection of Leu[6] at later stages, and the pattern for uptake of Cys[5].

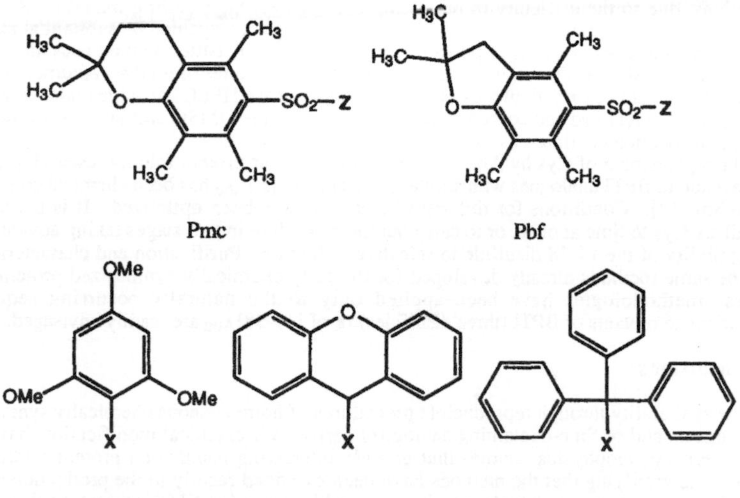

Figure 3. Structures of guanidino, ω-amide, and sulfhydryl protecting groups. *Top:* Arginine protecting groups, pentamethylchroman-6-sulfonyl (Pmc) and pentamethyldihydrobenzofuran-5-sulfonyl (Pbf). **Z** = NHC(=NH)NH- of Arg. *Bottom:* Asparagine, Glutamine or Cysteine protecting groups, trimethoxybenzyl (Tmob), xanthenyl (Xan), and trityl (Trt). **X** = (C=O)NH- of Asn, Gln or S- of Cys.

using Trt for all three amino acids Asn, Gln, Cys; poor couplings documented at Asn[43]-Asn[44] and Cys[30]-Gln[31] presumably reflect the steric hindrance of consecutive Trt-protected residues. The same effort led to the isolation of a triple deletion protein, des[Tyr[10], Gln[31], Asn[43 or 44]]-BPTI. When we use Tmob for Asn and Gln, and Trt for Cys, such difficulties are not observed. However, use of Trt for Cys protection is not ideal for another reason. The insolubility of Fmoc-Cys(Trt)-OH makes it necessary to use the corresponding pentafluorophenyl ester for incorporation of this residue. The Fmoc-Cys(Xan)-OH derivative developed recently in our laboratory [8] can be used directly in the standard HBTU/HOBt/NMM coupling protocols. Irrespective of what combination of acidolysable protecting groups are used, the optimal cleavage cocktail appears to be Reagent R (Figure 1). As judged by HPLC analysis directly afer cleavage, the trace from Reagent R [15] shows a higher level of homogeneity than when either Reagent K [33] or B [11, 34] are used. Reagent R was originally optimized for removal of Pmc from Arg, and has the added virtue of providing a strong reducing environment that allows Cys residues to be obtained in the free sulfhydryl form.

    Once chain assembly is complete and the cleavage has been carried out, further purification occurs by C-4 reversed-phase HPLC. We currently use relatively flat gradients (0.1 to 0.3% CH₃CN per min), which allow for higher loadings onto the column, and better resolution of close peaks. Changes in experimental details, and generally increased experience, has resulted in improved efficiencies and overall higher isolated yields [e.g., we are now able to recover correct product by recycling impure fractions]. An initial purification of the reduced form of the protein is essential (in the case of [R]_{Abu}, this is the only step). The homogeneity requirement for material to be pooled after such an initial step is not as stringent as for later stages (Figure 1); hence, broader cuts can be tolerated to ultimately increase yields. Oxidation of analogues that require a single disulfide is best achieved by Tam's DMSO method [10], in part because of protein insolubilites at the optimal pH values required for other methods. Nevertheless, use of DMSO introduces new complications, due to the difficulty in removing this viscous, high-boiling material. As noted earlier [1, 2], attempts to use dialysis for DMSO removal were accompanied by substantial material losses (perhaps, the DMSO damages pores in the membrane). A better method uses an Amicon stirred cell with a YM1 membrane, which gives material in recoveries of ~ 80%. Alternatively, the oxidation mixture may be loaded directly onto the reversed-phase HPLC cartridge under conditions where the protein is retained and concentrated at the top while the DMSO and salts are washed out; subsequent purification continues as usual.

The replacement of Cys by Abu has been useful for our protein folding research [1, 2, 4]. Another avenue to BPTI analogues with similar properties to [R]$_{Abu}$ has been chemical conversion of Cys to Smc [7]. Conditions for this transformation have been optimized. It is possible to convert all six Cys to Smc at once, or to carry out the procedure in two stages taking advantage of the susceptibility of the 14-38 disulfide to selective reduction. Purification and characterization follows the same methods already developed for the fully chemically synthesized proteins. To date, these methodologies have been applied only to the naturally occurring sequence; generalizations to mutants of BPTI (three disulfides) or of [14-38]$_{Abu}$ are readily envisaged.

Conclusions

The availability through reproducible procedures of homogeneous chemically synthesized BPTI analogues, and of Smc-containing analogues derived *via* chemical modification, has made possible a series of biophysical studies that provide interesting insights on protein folding and dynamics. It is gratifying that the methods have been extended readily to the production of ^{15}N and/or ^{13}C-labelled materials which are of considerable value for NMR spectroscopic studies. Further improvements in the chemistries are an ongoing goal, particularly since the better the "crude" product directly after chain assembly and cleavage, the higher yields and final purities that can be expected from the subsequent fractionation processes. In particular, studies currently underway seek more convenient methods for disulfide formation in BPTI analogues, including exploration of resin-bound oxidation procedures. Finally, we wish to re-emphasize the central role of high resolution chromatographic methods, and the importance of using a variety of complementary techniques for the careful characterization of synthetic intermediates and final products [1, 2].

Acknowledgments

We thank NIH grants GM 26242 (C.W.), GM 43552 (G.B.), and GM 51628 (G.B. and C.W.) for support of this work. C.G. is a graduate student on the training grant "Chemical Basis of Molecular Biology," GM 07323, for which we are grateful. Marc Ferrer held a Fulbright Fellowship through the Ministry of Education of Spain, and is currently at the Department of Biochemistry, Harvard University, Cambridge, MA 02138.

References

1. Ferrer, M., Woodward, C., and Barany, G.. *Int. J. Peptide Protein Res.* **40**, 194-207 (1992), and references cited therein.
2. Ferrer, M. "Chemical synthesis and structural characterization of native sequence and partially folded analogs of bovine pancreatic trypsin inhibitor (BPTI)," Ph.D. Thesis, University of Minnesota, June 1994, *Dissertation Abstr.* **55**, 4356-B (1994).
3. Barbar, E., Barany, G., and Woodward, C. Poster at "Gibbs Conference on Biothermodynamics," October 1-4, 1994. Carbondale, Illinois.
4. Ferrer, M., Barany, G., and Woodward, C. *Nature Structural Biology* **2**, 211-217 (1995).
5. Barbar, E., Pan, H., Gross, C.M., Woodward, C., and Barany, G. *In* "Proceedings of the Fourteenth American Peptide Symposium" (P.T.P. Kaumaya and R.S. Hodges, eds.), in press (1995).
6. Barbar, E., Barany, G., and Woodward, C. *Biochemistry* **34**, 11423-11434 (1995).
7. Pan, H., Barbar, E., Barany, G., and Woodward, C. *Biochemistry* **34**, in press (1995).

8. Solé, N.A., Han, Y., Vágner, J., Gross, C.M., Tejbrant, J., and Barany, G. *In* "Proceedings of the Fourteenth American Peptide Symposium" (P.T.P. Kaumaya and R.S. Hodges, eds.), in press (1995).
9. Krchňák, V., Vágner, J., and Lebl, M. *Coll. Czech. Chem. Commun.* **53**, 2542-2548 (1988).
10. Tam, J.P., Wu, C.-R., Liu, W., and Zhang, J.-W. *J. Am. Chem. Soc.* **113**, 6657-6662 (1991).
11. Barany, G., Solé, N.A., Van Abel, R.J., Albericio, F., and Selsted M.E. *In* "Innovation and Perspectives in Solid Phase Synthesis: Peptides, Polypeptides and Oligonucleotides 1992" (R. Epton, ed.), Intercept Limited, Andover, England, 1992, pp. 29-38.
12. Barany, G., Albericio, F., Solé, N.A., Griffin, G.W., Kates, S.A., and Hudson, D. *In* "Peptides 1992: Proceedings of the Twenty-Second European Peptide Symposium" (C.H. Schneider and A.N. Eberle, eds.), ESCOM Science Publishers, Leiden, The Netherlands, 1993, pp. 267-268.
13. Zalipsky, S., Chang, J.L., Albericio, F., and Barany G. *Reactive Polymers* **22**, 243-258 (1994), and references cited therein.
14. Hudson, D. *J. Org. Chem.* **53**, 617-624 (1988).
15. Albericio, F., Kneib-Cordonier, N., Biancalana, S., Gera, L., Masada, R.I., Hudson, D., and Barany G. *J. Org. Chem.* **55**, 3730-3743 (1990), and references cited therein.
16. Dourtoglou, V., Gross, B., Lambropoulou, V., and Zioudrou C. *Synthesis*, pp. 572-574 (1984).
17. Knorr, R., Trzeciak, A., Bannwarth, W., and Gillessen, D. *In* "Peptides 1990" (E. Giralt and D. Andreu, eds), ESCOM, Leiden, The Netherlands, 1991, pp. 62-64.
18. Fields, C.G., Lloyd, D.H., Macdonald, R.L., Otteson, K.M., and Noble, R.L. *Peptide Res.* **4**, 95-101 (1991).
19. Carpino, L.A. *J. Am. Chem. Soc.* **115**, 4397-4398 (1993).
20. Carpino, L.A., El-Faham, A., Minor, C.A., and Albericio, F. *J. Chem. Soc., Chem. Commun.*, pp. 201-203 (1994).
21. Kress, L.F., and Laskowski Sr., M.S. *J. Biol. Chem.* **242**, 4925-4929 (1967).
22. Schwartz, H., Hinz, H.-J., Mehlich, A., Tschesche, H., and Wenzel, H.R. *Biochemistry* **26**, 3544-3551 (1987).
23. Goldenberg, D.P., Frieden, R.W., Haack, J.A., and Morrison, T.B. *Nature* **328**, 127-132 (1987), and references cited therein.
24. Houghten, R.A., and Li, C.H. *Anal. Biochem.* **98**, 36-46 (1979).
25. Wade, J.D., Bedford, J., Sheppard, R.C., and Tregear, G.W. *Peptide Res.* **4**, 194-199 (1991).
26. Fields, C.G., Mickelson, D.J., Drake, S.L., McCarthy, J.B., and Fields, G.B. *J. Biol. Chem.* **268**, 14153-14160 (1993).
27. Fields, G.B., Tian, Z., and Barany, G. *In* "Synthetic Peptides: A User's Guide" (G.A. Grant, ed.), W.H. Freeman & Co., New York, 1992, pp. 77-183.
28. Lauer, J.L., Fields, C.G., and Fields, G.B. *Lett. Peptide Sci.* **1**, 197-205 (1995)
29. Carpino, L.A., Shroff, H., Triolo, S.A., Mansour, E.-S.M.E., Wenschuh, H., and Albericio, F. *Tetrahedron Lett.* **34**, 7829-7832 (1993).
30. Fields, C.G. and Fields, G.B. *Tetrahedron Lett.* **34**, 6661-6664 (1993).
31. Fields, C.G. and Fields, G.B. *In* "Innovation and Perspectives in Solid Phase Synthesis: Peptides, Polypeptides and Oligonucleotides 1992" (R. Epton, ed.), Intercept Limited, Andover, England, 1992, pp. 153-162.
32. Fields, C.G., VanDrisse, V.L., and Fields, G.B. *Peptide Res.* **6**, 39-46 (1993).
33. King, D.S., Fields, C.G., and Fields, G.B. *Int. J. Peptide Protein Res.* **36**, 255-266 (1990).
34. Van Abel, R.J., Tang, Y.-Q., Rao, V.S.V., Dobbs, C.H., Tran, D., Barany, G., and Selsted, M. E. *Int. J. Peptide Protein Res.* **45**, 401-409 (1995).

On the Use of Novel Coupling Reagents for Solid-Phase Peptide Synthesis

Steven A. Kates,[1] Elke Diekmann,[1] Ayman El-Faham,[2] Lee W. Herman,[1] Dumitru Ionescu,[2] Brian F. McGuinness,[1] Salvatore A. Triolo,[1] Fernando Albericio,[3] and Louis A. Carpino[2]

[1]PerSeptive Biosystems Biosearch Products, 500 Old Connecticut Path, Framingham, MA 01701
[2]Department of Chemistry, University of Massachusetts, Amherst, MA 01003 USA, and [3]Department of Organic Chemistry, University of Barcelona, 08028-Barcelona, Spain.

I. Introduction

Common coupling reagents used for the construction of peptide bonds include carbodiimides[1] [N,N-dicyclohexylcarbodiimide (DCC) and N,N-diisopropylcarbodiimide (DIPCDI) for Boc and Fmoc strategies, respectively], active esters such as pentafluorophenyl derivatives (OPfp)[2] and uronium and phosphonium salts based on N-hydroxybenzotriazole (HOBt)[3]. The carbodiimide and active ester techniques are improved by the incorporation of N-hydroxybenzotriazole (HOBt)[4] which, as is also the case for onium salts, leads to the intermediate formation of an OBt ester[5]. Recently, two new coupling reagents, Fmoc amino acid fluorides[6] and urethane-protected N-carboxyanhydrides (UNCA's)[7] have been shown to be efficient reagents for rapid peptide coupling under both solution and solid-phase conditions[8].

In 1993, an additive more efficient than HOBt, 1-hydroxy-7-azabenzotriazole (HOAt), which had first been described twenty years earlier, was recommended for use in peptide synthesis[9]. HOAt incorporates into the HOBt structure a pyridine nitrogen atom, strategically placed so as to enhance coupling rates and maintenance of chiral integrity. Uronium and phosphonium salts based on HOAt {N-[(dimethylamino)-1H-1,2,3-triazolo[4,5-b]pyridin-1-ylmethylene]-N-methylmethanaminium hexafluorophosphate N-oxide (HATU)[10], 7-azabenzotriazol-1-yloxytris(pyrrolidino)phosphonium hexafluorophosphate (PyAOP) and 1-(1-pyrrolidinyl-1H-1,2,3-triazolo[4,5-b]pyridin-1-yl methylene) pyrrolidinium hexafluorophosphate N-oxide (HAPyU), O-(7-azabenzotriazol-1-yl)-1,3-dimethyl-1,3-dimethyleneuronium hexafluorophosphate (HAMDU), O-(7-azabenzotriazol-1-yl)-1,3-dimethyl-1,3-trimethyleneuronium hexafluorophosphate (HAMTU), O-(7-azabenzotriazol-1-yl)-1,1,3,3-bis(pentamethylene)uronium hexafluorophosphate (HAPipU), 7-azabenzotriazol-1-yloxytris(dimethylamino)phosphonium hexafluorophosphate (AOP)} were synthesized and their reactivity compared with that of the analogous HOBt-derived reagents such as benzotriazol-1-yloxytris(dimethylamino)phosphonium hexafluorophosphate (BOP), benzotriazol-1-yloxytris(pyrrolidino)phosphonium hexafluorophosphate (PyBOP) and O-(benzotriazol-1-yl)-1,1,3,3-tetramethyluronium hexafluorophosphate (HBTU) (Figure 1). Most recently, a second type of onium salt derived from tetramethylurea, namely tetramethyl fluoroformamidinium hexafluorophosphate (TFFH), has been synthesized and shown to be an excellent coupling reagent, especially in the case of hindered substrates[11]. TFFH acts by prior in situ conversion of the Fmoc amino acids to the corresponding acid fluorides.

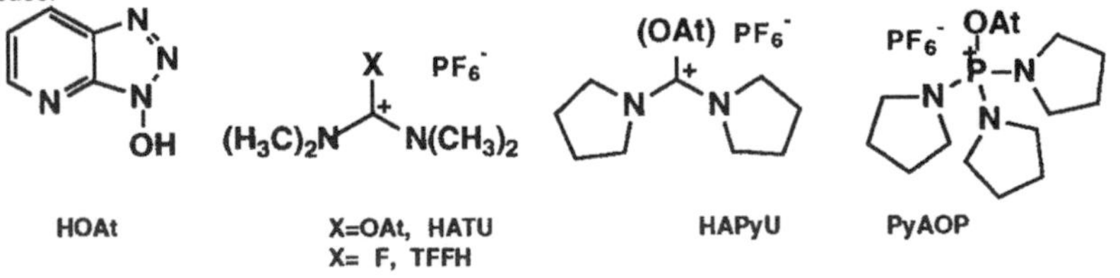

HOAt

X=OAt, HATU
X= F, TFFH

HAPyU

PyAOP

Figure 1. Structures of HOAt-based coupling reagents.

515

II. Materials and Methods

Fmoc-amino acids, resins, activators, and all peptide synthesis reagents were obtained from PerSeptive Biosystems Biosearch Products Division. The *t*Bu group was used for protection of the side-chains of Asp, Glu, Thr, and Tyr, Boc for Lys and Trp, Trt for Asn and Gln and the recently developed Pbf[12] for Arg. All solvents were HPLC grade or of equivalent purity and used without further purification. For amino acid analysis, peptides and peptide resins were hydrolyzed with gaseous HCl, the analyses being carried out using a Waters Femtotag chemical kit[13]. Mass spectra were recorded on a prototype matrix assisted laser desorption time-of-flight instrument. HPLC was carried out on a Waters apparatus with two model 600 solvent delivery systems, a Wisp model 712 automatic injector, a model 490 programmable UV wavelength detector or a model 994 programmable photodiode array detector and an 860 Networking Computer for control of system operation and collection of data. Separations were carried out on Waters Delta Pak C_{18} 100 Å analytical columns (3.9 x 150 mm, 5 µm) using a gradient system (solution A: 0.1% TFA in H_2O; solution B: 0.1% TFA in CH_3CN).

Peptide Synthesis: Continuous-flow solid-phase syntheses were carried out automatically using a PerSeptive Biosystems 9050 Plus PepSynthesizer. The flow rate of the unit pump was set at 5.0 mL/min with the following synthetic protocol: Fmoc group deblocking with 20% piperidine in DMF (7 min), DMF washing (12 min), amino acid coupling (30 min), and DMF washing (8 min). Syntheses were carried out either on Fmoc-PAL-PEG-PS (0.18 mmol/g)[14] or PAC-PEG-PS. For the coupling step, four equivalents of Fmoc-amino acid and activator were dissolved in DMF in a 0.6 M solution of DIEA to give a final concentration of 0.3 M.

Cleavage Conditions: Peptide-resin samples were treated with TFA–H_2O (9:1) for 2 h at 25 °C. The filtrates were collected and the resin was further washed with TFA. Cold ether was added to the combined extracts and the solution was cooled to -70 °C. After removing the supernatant liquid, the resulting precipitate was washed several times with cold ether, dissolved in acetic acid and lyophilized.

III. Results and Discussion

Comparison of Coupling Techniques. Early studies[9] demonstrated that HOAt as an additive for carbodiimides [*N*-ethyl-*N'*-(3-dimethylamino)propylcarbodiimide (EDC) and DCC] or built into stand-alone coupling reagents [HATU, HAPyU] caused significant reductions in both coupling times and loss of configuration for solution syntheses. For example, in the coupling of Bz-Val-OH with H-Val-OMe, HOAt-based reagents reduced racemization to 1/3-1/2 the level found for analogous HOBt reactions. Advantages were also shown for segment coupling processes and following such encouraging results the applicability of these new reagents to solid-phase syntheses was examined.

In order to demonstrate the effectiveness of HOAt-based reagents and compare performance with HOBt-analogs, assembly of the common decapeptide model ACP (65-74), H-VQAAIDYING-NH2 derived from the acyl carrier protein sequence, was examined[15,16]. This sequence is known to exhibit difficult couplings at Ile[72], Ile[69], Val[75], and Asn[73] and the various deletion peptides which result can be used to evaluate the success of the synthesis. Peptide elongation was carried out with deliberately reduced coupling times and excesses of reagents in order to magnify differences among the various coupling techniques. After cleavage, reversed-phase high performance liquid chromatography (HPLC) analysis was used to assess the purity of the crude products. Data are collected in Table I.

For carbodiimide (entries 1-3) and pentafluorophenyl ester (entries 4-6) couplings (4 equiv. of protected amino acid/activator for 3 min), the use of HOAt as an additive greatly enhances synthetic efficiency. Under similar conditions, uronium salt couplings (entries 7-8), give slightly better results. To further exaggerate the differences among these reagents, all uronium and phosphonium-induced couplings were carried out with only 1.5 equiv. of protected amino acid/activator and only 1.5 min of coupling time. Syntheses carried out under such forcing conditions demonstrate clearly the superiority of the azabenzotriazole-based uronium and phosphonium reagents relative to the corresponding HOBt-based analogs (entries 9 vs. 10, 20-22 vs. 23-25). Furthermore, the addition of HOXt in HXTU or PyXOP couplings did not enhance the coupling efficiency except for Fmoc-Asn(Trt)-OH. A difficulty was noted for the incorporation of trityl-protected asparagine since relatively large amounts of des-Asn-ACP were formed. Better results were obtained if one equivalent of HOAt was added during coupling for this step only (entries 10 vs. 13, 15; 23 vs. 24, 25; 26 vs. 27, 28). An alternate method of handling the Asn(Trt) problem is to switch from trityl to dimethylcyclopropylmethyl (Dmcp) protection for the amide nitrogen[17]. Excellent incorporation of Asn was obtained with TFFH when Fmoc-Asn(Dmcp)-OH was used for introduction of the Asn unit.

Table I. Percent distribution of products, including various deletion peptides, according to HPLC analysis of the assembly of ACP(65-74) *via* different coupling methods

	Coupling Method	equiv	Time (min)	ACP	-2Ile	-Ile^{72}	-Ile^{69}	-Val	-Asn
1	DIPCDI	4	3	14	4	7	22	2	32
2	DIPCDI-HOBt	4	3	31	13	15	18	3	1
3	DIPCDI-HOAt	4	3	65	2	7	9	1	2
4	Pfp	4	3	0	-	-	-	-	-
5	Pfp-HOBt	4	3	53	1	1	33	5	1
6	Pfp-HOAt	4	3	82	-	-	15	1	1
7	HBTU	4	3	73	-	1	2	3	5
8	HATU	4	3	84	-	-	-	3	8
9	HBTU	1.5	1.5	18	16	11	19	2	7
10	HATU	1.5	1.5	53	3	6	12	3	16
11	PyBrOP	1.5	1.5	0	-	-	-	-	-
12	HBTU-HOBt	1.5	1.5	17	18	12	18	3	3
13	HATU-HOAt	1.5	1.5	50	5	9	12	2	8
14	HBTU-HOAt	1.5	1.5	42	8	11	14	3	5
15	HATU-HOBt	1.5	1.5	29	12	13	18	3	2
16	AOP	1.5	1.5	49	3	6	13	1	18
17	HAPipU	1.5	1.5	53	1	7	14	3	13
18	HAMDU	1.5	1.5	53	1	6	13	3	13
19	HAMTU	1.5	1.5	41	1	7	12	7	16
20	PyBOP	1.5	1.5	10	22	13	13	3	<1%
21	PyBOP-HOAt	1.5	1.5	19	13	12	12	3	2
22	PyBOP-HOBt	1.5	1.5	11	21	14	14	3	<1%
23	PyAOP	1.5	1.5	46	3	5	5	2	18
24	PyAOP-HOAt	1.5	1.5	60	3	6	7	4	2
25	PyAOP-HOBt	1.5	1.5	45	9	12	12	2	2
26	HAPyU	1.5	1.5	44	2	5	3	1	13
27	HAPyU-HOAt	1.5	1.5	74	0	2	3	2	5
28	HAPyU-HOBt	1.5	1.5	52	4	8	8	10	1

To determine if these reagents were moisture sensitive, Boc-Ile-OH•1/2H$_2$O was coupled to the amino terminus of ACP(65-74) using 1.5 equiv. of reagent and activator and 3.0 equiv. of base for 1.5 min (Table II). The results demonstrate that aza derivatives incorporate amino acids more efficiently and water has no influence on the coupling process.

Table II. Coupling of Boc-Ile-OH•1/2H$_2$O onto H-VQAAIDYING-PAL-PEG-PS

Coupling Method	H-IVQAAIDYING-NH$_2$
HATU	71%
HBTU	50%
PyAOP	73%
PyBOP	50%

HATU-mediated couplings are compatible with sequences containing multiple Arg and Trp residues. In the case of Arg, the well known formation of δ-lactam from most activated derivatives of Fmoc-Arg(Pbf)-OH is slow enough not to interfere during normal coupling times. The sequence, H-Ile-Leu-Pro-Trp-Lys-Trp-Pro-Trp-Trp-Pro-Trp-Arg-Arg-OH[18] was assembled on PAC-PEG-PS *via* 4 equiv. of activator and HATU and 8 equiv. of base with 30-min coupling times. Cleavage of the peptide from the support and removal of the side-chain protecting groups with TFA–anisole–β-mercaptoethanol (95:3:2) for 2 h at 25 °C gave the desired peptide in 44% yield and excellent purity (Figure 2). In view of the well known instability of the tosyl histidine residue toward HOBt during attempts to incorporate His *via* Boc-His(Tos)-OH, it was expected that the same problem would arise with HOAt-based coupling reagents. This was confirmed.

Hindered amino acids. In a demanding example H-Tyr-Aib-Deg-Phe-Leu-NH$_2$ was assembled with 4 equiv. amino acid/activator, 8 equiv. base, 2 h coupling for Tyr, Aib, and Deg and 30 min for Phe and Leu. For HATU the result was a 1:1 mixture of the desired peptide and the des-Aib derivative (Figure 3). With HBTU activation, none of the pentapeptide was formed. Multiple extended-time couplings (3 x 18 h) with HATU were required in order to obtain the desired 5-mer in good yield. While clearly not a routine synthesis, this example illustrates that under drastic conditions some very difficult sequences can be assembled *via* HATU. Manual solid-phase syntheses gave a 20:1 ratio of penta- to tetrapeptide for the case of H-Tyr-Aib-Aib-

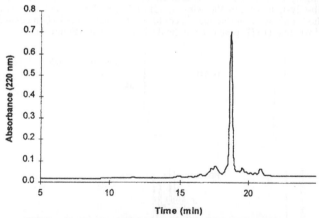

HPLC: C$_{18}$, linear gradient over 30 min of CH$_3$CN/0.1% TFA and
        H$_2$O/0.1% TFA from 1:9 to 3:2, flow rate: 1.0 mL/min.

Figure 2. HPLC chromatogram of crude H-Ile-Leu-Pro-Trp-Lys-Trp-Pro-Trp-Trp-Pro-Trp-Arg-Arg-OH directly after cleavage reagent.

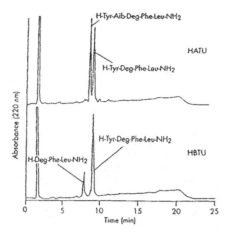

Figure 3. HPLC chromatogram of crude peptide mixtures containing H-Tyr-Aib-Deg-Phe-Leu-NH₂ directly after extraction with CHCl₃ and H₂O-HOAc (7:3) of the cleavage reagent [TFA-H₂O (9:1)].

Phe-Leu-NH₂. With TFFH couplings, the same peptide was obtained giving from 10:1 to 3:1 mixtures of the desired pentapeptide and the des-Aib tetrapeptide, respectively, depending on the conditions. The highly challenging 20-mer, alamethicin amide, Ac-Aib-Pro-Aib-Ala-Aib-Ala-Gln-Aib-Val-Aib-Gly-Leu-Aib-Pro-Val-Aib-Aib-Glu-Gln-Phe-NH₂, previously assembled by solid-phase methods only *via* acid fluorides[19], could be obtained *via* HATU-mediated coupling in excellent yield whereas HBTU gave none of the desired sequence (Figure 4).

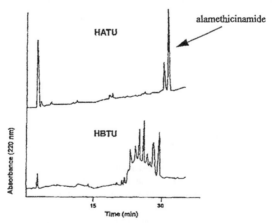

HPLC: C₁₈, linear gradient over 30 min of CH₃CN/0.1% TFA and H₂O/0.1% TFA from 3:7 to 7:3, flow rate: 1.0 mL/min.

Figure 4. HPLC chromatogram of crude alamethicinamide directly after cleavage reagent.

N-Substituted amino acids represent another class of highly hindered amino acids. An appropriate model bearing multiple units of such amino acids is the hexapeptide segment of cyclosporin, H-DAla-MeLeu-MeLeu-MeVal-Phe-Val-OH, which was assembled on a hyperacid-labile resin (HAL-PEG-PS) using 2 x 2 h couplings for the final three amino acids[20]. The desired peptide was obtained in yields of 85% and 8% for HATU and HBTU, respectively (Figure 5). These results suggest that backbone protected amino acids (2-hydroxy-4-methoxybenzyl [Hmb][21], 2,4,6-trimethoxybenzyl [Tmob]) which are of interest in peptide synthesis for their ability to inhibit aggregation, might also be coupled readily *via* HOAt-based methodology.

Stability. In order to determine the compatibility of HOAt-based coupling reagents with automated peptide synthesizers, their stability in solution and in the solid state was examined *via* HPLC and ^{1}H NMR analysis. In solid form, HOAt, HATU, PyAOP, HAPyU and TFFH are stable as solids at 25 °C. Solutions of HOAt and HATU (0.3 and 0.5 M in DMF, respectively) stored under nitrogen were stable at 25 °C for 3 weeks. Solutions of HAPyU and PyAOP (0.5 M in DMF) exposed to the atmosphere were stable for 5 d and 2 d, respectively. Although the instability of PyAOP is similar to that of PyBOP, the former is the more reactive coupling species.

Since there is no ultraviolet chromophore associated with TFFH, an assay was developed to determine its stability *via* HPLC analysis. Treatment of TFFH with benzylamine in H_2O converts the salt to the known tetramethylbenzyl guanidine which can be detected at 220 nm. Solutions of TFFH in DMF (0.6 M) are stable for 24 d. On the other hand, a DMF solution containing TFFH (0.3 M) and DIEA (0.6 M) is stable for 30 min.

Reagent and Base Concentration. For the PerSeptive Biosystems Biosearch 9050Plus peptide synthesizer, DMF solutions of HOAt (0.3 M), HATU, PyAOP, HAPyU (0.5 M) and TFFH (0.6 M) are recommended. To determine the proper base concentration, the dual syringe

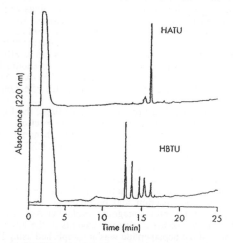

HPLC: C$_{18}$, linear gradient over 20 min of CH$_3$CN/0.1% TFA and H$_2$O/0.1% TFA from 0:1 to 1:0, flow rate 1.0 mL/min

Figure 5. HPLC chromatograms of crude peptide mixtures containing H-DAla-MeLeu-MeLeu-MeVal-Phe-Val-OH directly after evaporation of the cleavage reagent. The peak at 16.1 min corresponds to the title peak. Peaks at 12.9, 13.7, 14.6, and 15.3 min correspond to H-MeVal-Phe-Val-OH, H-DAla-MeVal-Phe-Val-OH, H-MeLeu-MeVal-Phe-Val-OH, H-DAla-MeLeu-MeVal-Phe-Val-OH, respectively.

Table III. HATU-mediated couplings (0.5 M) leading to ACP (65-74) in the presence of 1.5 equiv. of amino acid and 1.5 min couping time

Concentration of Base	ACP Purity	Concentration of Base	ACP Purity
0.5 M DIEA	72%	0.4 M DIEA–0.4 M collidine	48%
1.0 M DIEA	72%	0.45 M DIEA–0.45 M collidine	61%
1.5 M DIEA	66%	0.5 M DIEA–0.5M collidine	74%

protocol was used to synthesize ACP (65-74) with previously described conditions that incorporate reduced excesses of reagents and coupling times. For HATU-mediated couplings, DIEA in DMF concentrations of 0.5 and 1.0 M gave similar results whereas increasing the concentration to 1.5 M lowered the peptide purity (Table III). The use of collidine in place of DIEA for uronium-based segment couplings in solution reduced the amount of racemization[22], but this technique is not applicable to stepwise solid-phase synthesis since activation with this base is too slow. A collidine-DIEA mixture (1:1) in DMF (1.0 M) is an effective combination for rapid activation since only a single equivalent of the strong base (DIEA) is required. Phosphonium salts are more sensitive to base concentration and a 1.0 M solution of DIEA in DMF is preferred. For HAPyU-mediated couplings, 0.5-1.0 M solutions can be used.

To determine the proper base concentration for TFFH couplings, the model H-Tyr-Aib-Aib-Phe-Leu-NH$_2$ was prepared using the single syringe protocol under conditions described above. The ratio of desired pentapeptide to des-Aib tetrapeptide was determined from HPLC analysis. Using 1 and 2 equiv. of DIEA as base, the desired 5-mer was obtained in yields of 71% and 73%, respectively. These results demonstrate the compatibility of TFFH for the coupling of hindered amino acids *via* automated instrumentation.

Activation. The time required to activate Fmoc-amino acids was examined with HOAt-based uronium and phosphonium reagents. In both cases, hindered amino acids such as Val and Deg were converted to the corresponding Fmoc-XX-OAt ester within 2 min in higher yields than the corresponding HOBt analogs. In carbodiimide couplings, the reaction proceeded at a slower rate corroborating earlier results that this coupling method is not as effective as that involving uronium and phosphonium salts. Extended preactivation times should be avoided in order to minimize possible side reactions. In the construction of H-Tyr-Aib-Aib-Phe-Leu-NH$_2$ using TFFH-couplings, varied preactivation times (2-13 min) had little effect on the purity of the crude pentapeptide.

Guanidine formation. In slow uronium salt couplings, the formation of a guanidine residue at the amino terminus is a potential detrimental side reaction since the amino group is blocked from further chain extension[23]. To examine this phenomenon, Fmoc-Deg-OH was coupled to H-Phe-OFm in the presence of 2 equiv of DIEA and HXTU (X=A, B). After 75 min, the desired dipeptide was formed in yields of 94% and 85% (after 2 min, 60% and 35%) from HATU and HBTU, respectively with < 0.5% formation of the guanidine derivative in either case. In the linear assembly of peptides, this side reaction is typically not encountered. This is not the case in resin- or solution-based cyclization reactions and therefore phosphonium salts, such as PyAOP, which do not give this side reaction, may be advantageous.

Cyclizations. Tachykinin peptide antagonist *cyclo*(Tyr-Asp-Arg-DTrp-DTrp-Val-DTrp) was chosen as a model to study azabenzotriazole reagents in the key resin-bound cyclization process[24]. Linear assembly of the heptapeptide was accomplished using HATU–DIEA mediated coupling with side-chain anchoring *via* Fmoc-Asp(O-PAC-PEG-PS)-OAl. Allyl removal with Pd(PPh$_3$)$_4$ in CHCl$_3$–HOAc–NMM (20:1:0.5) for 2 h at 25 °C followed by Fmoc deprotection with piperidine–DMF (1:4) exposed the carboxy and amino termini, respectively. For this model peptide, azabenzotriazole-based reagents performed slightly better than the benzotriazole analogs with PyAOP giving the crude cyclo-peptide in highest yield (74%) and purity (61%). For solution-based cyclizations, a recent study showed HAPyU to be the most effective reagent for the synthesis of all L-cyclohexapeptides devoid of glycine or proline, amino acid residues which promote cyclization and thereby reduce the amount of C-terminal epimerization[25].

Fragment couplings. Segment condensations onto resins are somewhat more demanding than analogous reactions carried out in solution in terms of configurational control. Pre-activation times, identity of the base and choice of solvent are all critical. For the coupling of Fmoc-Phe-Ser(tBu)-OH onto H-Pro-PAL-PEG-PS and a number of related models, HATU/collidine/DMF-CH_2Cl_2 proved to be most effective, epimerization being at least 1/3 that of comparable HBTU couplings[26]. In a related example, HAPyU activation of Fmoc-Leu-OH in the presence of collidine for 2 min followed by addition to H-Pro-PAL-PEG-PS and subsequent peptide cleavage gave optimal results (0.03% DL-isomer).

Incorporation of first amino acid onto a hydroxyl resin. Incorporation of the first amino acid onto a hydroxyl resin as opposed to an amino resin is more difficult since the hydroxyl group is not as effective a nucleophile as an amino group. Typical conditions for loading a first amino acid onto a hydroxyl resin include the tedious preparation of preformed handles or the use of DIPCDI in the presence of the strong base, 4-(dimethylamino)pyridine (DMAP)[27]. The latter method suffers from the difficulty of obtaining high yields with minimal racemization. Fortunately, PyAOP is an excellent reagent for such first amino acid incorporation (Table IV). Yields are high and racemization is low especially for His (Trt) which leads to 80% incorporation with ~3% racemization as opposed to other methods that give yields < 60% with 20-40% racemization[28].

Table IV. Incorporation of the first amino acid *via* different coupling methods

Reagent	Yield	
	Cys(Trt)	His(Trt)
PyAOP	100%	80%
PyBOP	82%	58%
HATU	-	51%
DIPCDI/DMAP	57%	-

IV. Conclusion

HOAt, HATU, PyAOP, HAPyU and TFFH are efficient coupling reagents for solution- and solid-phase peptide synthesis. These derivatives enhance reactivity, reduce racemization and are suited for peptides containing either natural or hindered amino acids. These reagents are compatible with both manual and automated batch and continuous-flow techniques.

References

1.   Sheehan, J.C., Hess, G.P. J. Am. Chem. Soc. 77 (1955) 1067.
2.   (a) Kovács, K., Penke, B. *In* (1973) Peptides 1972 (Hanson, H., Jakubke, Eds.) North Holland Publ., Amsterdam, pp. 187-188. (b) Atherton, E., Sheppard, R.C. J. Chem. Soc., Chem. Commun. (1985) 165.
3.   (a) Dourtoglou, V., Ziegler, J.C., Gross, B. Tetrahedron Lett. (1978) 1269. (b) Dourtoglou, V., Gross, B., Lambropoulou, V., Zioudrou, C. Synthesis (1984) 572. (c) Knorr, R., Trzeciak, A., Bannwarth, W., Gillessen, D. Tetrahedron Lett. 30 (1989) 1927. (d) Dormoy, J.R., Castro, B. Tetrahedron Lett. (1979) 3321. (e) Castro, B., Dormoy, J.R., Evin, G., Selve, C. Tetrahedron Lett. (1975) 1219.
4.   König, W., Geiger, R. Chem. Ber. 103 (1970) 788.
5.   Hudson, D. J. Org. Chem. 53 (1988) 617.
6.   Carpino, L.A., Sadat-Aalee, D., Chao, H.G., DeSelms, R.H. J. Am. Chem. Soc. 112 (1990) 9651.
7.   (a) Fuller, W.D., Cohen, M.P., Shabankareh, M. Blair, R.K. J. Am. Chem. Soc. 112 (1990) 7414. (b) Xue, C., Naider, F. J. Org. Chem. 58 (1993) 350.

8.   (a) Wenschuh, H., Beyermann, M., Krause, E., Brudel, M., Winter, R., Schümann, M., Carpino, L.A.,
      Bienert, M. J. Org. Chem. 59 (1994) 3275. (b) Fehrentz, J.A., Genu-Dellac, C., Amblard, H., Winternitz,
      F., Loffet, A., Martinez, G. J. Peptide Sci. 1 (1995) 124.
9.   Carpino, L.A. J. Am. Chem. Soc. 115 (1993) 4397.
10.   As noted earlier[9], abbreviations for these new reagents will maintain the style previously recommended.
      For example, HBTU becomes HATU. The systematic names given here for HATU and HAPyU differ
      from those previously given in the literature. Recent X-ray structure determinations for these compounds
      as well as HBTU have clarified the crystal structures as guanidinium N-oxides rather than uronium salts.
      See, Abdelmoty, I., Albericio, F., Carpino, L. A., Foxman, B. M., Kates, S. A. Lett. Pept. Sci. 1 (1994) 57.
11.   Carpino, L.A., El-Faham, A. J. Am. Chem. Soc. 117 (1995) 5401.
12.   (a) Carpino, L.A., Shroff, H., Triolo, S.A., Mansour, E.M.E., Wenschuh, H., Albericio, F. Tetrahedron Lett.
      34 (1993) 7829. (b) Fields, C.G., Fields, G.B. Tetrahedron Lett. 34 (1993) 6661.
13.   Bidlingmeyer, B. A., Cohen, S. A., Tarvin, T. L. J. Chromatogr. 336 (1984) 93.
14.   (a) Albericio, F., Kneib-Cordonier, N., Biancalana, S., Gera, L., Masada, I., Hudson, D., Barany, G. J. Org.
      Chem. 55 (1990) 3730. (b) Barany, G., Albericio, F., Solé, N. A., Griffin, G. W., Kates, S. A., Hudson, D.
      (1993) In Peptides 1992: Proceedings of the Twenty-second European Peptide Symposium (Schneider,
      C.H., Eberle, A.N., Eds.) Escom, Leiden, pp. 267-268.
15.   Carpino, L.A., El-Faham, A., Minor, C., Albericio, F. J. Chem. Soc., Chem. Commun. (1994) 201.
16.   Kates, S.A., Triolo, S.A., Diekmann, E., Carpino, L.A., El-Faham, A., Ionescu, D., Albericio, F. In
      Peptides: Chemistry, Structure and Biology, Proceedings of the Fourteenth American Peptide Symposium,
      Escom, Leiden, 1996, in press.
17.   Carpino, L.A., Shroff, H.N., Chao, H.-G., Mansour, E.M.E., Albericio, F. (1995) In Peptides 1994:
      Proceedings of the Twenty-Third European Peptide Symposium (Maia, H.L.S., Ed.) Escom, Leiden, pp.
      155-156.
18.   Selsted, M.E., Novotny, M.J., Morris, W.L., Tang, Y.Q., Smith, W., Cullor, J.S. J. Biol. Chem. 267 (1992)
      4292.
19.   Wenschuh, H., Beyermann, M., Haber, H., Seydel, J.K., Krause, E., Bienert, M., Carpino, L.A., El-Faham,
      E., Albericio, F. J. Org. Chem. 60 (1995) 405.
20.   In a related study, see: Angell, Y.M., García-Echeverría, C., Rich, D.H. Tetrahedron. Lett. 35 (1994) 5981.
21.   Hyde, C., Johnson, T., Owen, D., Quibell, M., Sheppard, R.C. Int. J. Peptide Prot. Res. 43 (1994) 431.
22.   Carpino, L.A., El-Faham, A. J. Org. Chem. 59 (1994) 6955.
23.   Gausepohl, H., Pieles, U., Frank, R.W. (1992) In Peptides: Chemistry and Biology (Proceedings of the
      Twelvth American Peptide Symposium (Smith, J.A., Rivier, J.E., Eds.) Escom, Leiden, pp. 523-524.
24.   Kates, S.A., Daniels, S.B., Albericio, F. Anal. Biochem. 212 (1993) 303.
25.   Ehrlich, A., Rothemund, S., Brudel, M., Beyermann, M., Carpino, L.A., Bienert, M. Tetrahedron Lett. 34
      (1993) 4781.
26.   Carpino, L.A., El-Faham, A., Albericio, F. Tetrahedron Lett. 35 (1994) 2279.
27.   (a) Atherton, E., Benoiton, N.L., Brown, E., Sheppard, R.C., Williams, B.J. J. Chem. Soc., Chem.
      Commun. (1981) 336.
28.   Mergler, M., Nyfeler, R., Gosteli, J., Tanner, R. Tetrahedron Lett. 30 (1989) 6745.

Index

ISBN 0-12-473555-X

90018